T0142384

# Wireless Networks

**Series Editor**

Xuemin Sherman Shen, University of Waterloo, Waterloo, ON, Canada

The purpose of Springer's Wireless Networks book series is to establish the state of the art and set the course for future research and development in wireless communication networks. The scope of this series includes not only all aspects of wireless networks (including cellular networks, WiFi, sensor networks, and vehicular networks), but related areas such as cloud computing and big data. The series serves as a central source of references for wireless networks research and development. It aims to publish thorough and cohesive overviews on specific topics in wireless networks, as well as works that are larger in scope than survey articles and that contain more detailed background information. The series also provides coverage of advanced and timely topics worthy of monographs, contributed volumes, textbooks and handbooks.

** Indexing: Wireless Networks is indexed in EBSCO databases and DPLB **

Jun Du • Chunxiao Jiang

# Cooperation and Integration in 6G Heterogeneous Networks

## Resource Allocation and Networking

 Springer

Jun Du 🆔
Department of Electronic Engineering
Tsinghua University
Beijing, China

Chunxiao Jiang 🆔
Information Science and Technology
Tsinghua University
Beijing, China

ISSN 2366-1186                    ISSN 2366-1445    (electronic)
Wireless Networks
ISBN 978-981-19-7650-6         ISBN 978-981-19-7648-3    (eBook)
https://doi.org/10.1007/978-981-19-7648-3

This Springer imprint is published by the registered company Springer Nature Singapore Pte Ltd.
The registered company address is: 152 Beach Road, #21-01/04 Gateway East, Singapore 189721,
Singapore

# Foreword

To provide ubiquitous and various services, 6G networks tend to be more comprehensive and multidimensional by integrating current terrestrial networks with space-/air-based information networks and marine information networks; then, heterogeneous network resources, as well as different types of users and data, will be also integrated. Driven by the exponentially growing demands of multimedia data traffic and computation-heavy applications, 6G heterogenous networks are expected to achieve a high quality of service (QoS) with ultra-reliability and low latency. In response, resource allocation has been considered an important factor that can improve 6G performance directly by configuring heterogeneous communication, computing, and caching resources effectively and efficiently.

In this book, we deliver a range of technical issues in cooperative resource allocation and information sharing for the future 6G heterogenous networks, from the terrestrial ultra-dense networks and space-based networks to the integrated satellite-terrestrial networks, as well as introducing the effects of cooperative behavior among mobile users on increasing capacity, trustworthiness, and privacy. For the cooperative transmission in heterogeneous networks, we commence with the traffic offloading problems in terrestrial ultra-dense networks, and the cognitive and cooperative mechanisms in space-based networks. Moreover, for integrated satellite-terrestrial networks, we present a pair of dynamic and adaptive resource allocation strategies of traffic offloading, cooperative beamforming, and traffic prediction-based cooperative transmission. Later, we discuss the cooperative computation and caching resource allocation in heterogeneous networks, with the highlight of providing our current studies on the game theory, auction theory, and deep reinforcement learning-based approaches. Meanwhile, we introduce the cooperative resource and information sharing among users, in which capacity-oriented, trustworthiness-oriented, and privacy-oriented cooperative mechanisms are investigated. Finally, the conclusion is drawn.

**Object of This Book** This book is aimed at graduate students, researchers, and engineers who work/study electronic engineering, wireless communications, and information science, especially in the area of the next-generation wireless networks. This book can educate systems engineers to carve out the critical role that analytical and experimental engineering play in the research and development of 6G networks. The prerequisite knowledge of the readers is probability and wireless communications.

This book adopts the total-point-total writing format.

Part I: Introduction

Part I introduces the heterogeneous architecture of 6G networks, and many challenges and mathematic tools for resource allocation in 6G, so that readers have a preliminary understanding of knowledge in this field.

Part II: Cooperative Transmission in Heterogeneous Networks

Part II provides an introduction to the cooperative resource allocation mechanisms in heterogeneous cellular networks and heterogeneous space-based networks to enhance the transmission capability.

Part III: Cooperative Transmission in Integrated Satellite-Terrestrial Networks

Part III includes a description of cooperative transmission mechanism design in integrated satellite-terrestrial networks from aspects of transmission capability enhancement, secure transmission with interference control, and adaptivity to traffic properties.

Part IV: Cooperative Computation and Caching in Heterogeneous Networks

Part IV contains a description of QoS-aware computational resource allocation, QoS-aware caching resource allocation, priority-aware computational resource allocation, and energy-aware computational resource allocation in 6G heterogeneous networks.

Part V: Cooperative Resource and Information Sharing Among Users

Part V provides study cases to show how to solve the key challenges in resource and information sharing among users for data transaction, trustworthiness evaluation, and privacy protection in mobile networks.

Part VI: Conclusion

Part VI provides a brief summary of the book.

**Acknowledgment**  Dr. Jun Du and Dr. Chunxiao Jiang would like to thank those who helped to get this book published. Moreover, we would like to acknowledge the supports of the National Natural Science Foundation of China and the Young Elite Scientist Sponsorship Program by CAST.

Beijing, China                                                                   Jun Du
November 2022                                                         Chunxiao Jiang

# Contents

**Part I Introduction**

**1 Introduction of 6G Heterogeneous Networks** .......................... 3
  1.1 Heterogeneous Architecture of 6G Networks ...................... 3
  1.2 Challenges of Heterogeneous Resource Allocation ............... 5
      1.2.1 Heterogeneous Resource Modeling and
            Performance Evaluation ................................... 5
      1.2.2 Task Adaptation and Resource Efficiency ................. 6
      1.2.3 Interference Control and Secure Communications ........ 6
  1.3 Mathematic Tools for Resource Allocation ...................... 7
      1.3.1 Information Economics Theory ........................... 7
      1.3.2 Machine Learning and Artificial Intelligence ............ 8
  References ....................................................... 9

**Part II Cooperative Transmission in Heterogeneous Networks**

**2 Introduction of Cooperative Transmission in Heterogeneous
  Networks** ...................................................... 15

**3 Traffic Offloading in Heterogeneous Networks** ................... 17
  3.1 Introduction .................................................. 17
  3.2 Architecture of SDWN ......................................... 19
  3.3 Contract Formulation for Traffic Offloading .................. 20
      3.3.1 Transmission Model Formulation ......................... 21
      3.3.2 Economic Models Formulation ........................... 22
  3.4 Contract Design for Traffic Offloading ....................... 24
      3.4.1 Contract Design with Information Asymmetry ........... 25
      3.4.2 Contract Design Without Information Asymmetry ....... 27
      3.4.3 Contract Design by Linear Pricing ...................... 28
  3.5 Conditions for Contract Feasibility ........................... 29

3.6     Simulation Results .................................................    35
3.7     Conclusion ..........................................................    42
References ...................................................................    42

**4   Cooperative Resource Allocation in Heterogeneous
Space-Based Networks** ......................................................    45
4.1     Introduction .........................................................    45
4.2     Related Works .......................................................    47
4.3     System Model .......................................................    48
        4.3.1     ON/OFF Model ...........................................    51
        4.3.2     Physical Channel Model .................................    53
4.4     Cooperative Resource Allocation Protocol .........................    55
        4.4.1     GEO Relay ...............................................    56
        4.4.2     LEO Relay ...............................................    57
4.5     Stability Analysis ...................................................    58
        4.5.1     GEO Relay ...............................................    58
        4.5.2     LEO Relay ...............................................    64
        4.5.3     Multiple Users Case .....................................    71
4.6     Simulation Results ..................................................    72
4.7     Conclusion ..........................................................    77
4.8     Proof of Lemma 4.1 ................................................    78
4.9     Proof of Lemma 4.2 ................................................    81
References ...................................................................    83

**Part III   Cooperative Transmission in Integrated
Satellite-Terrestrial Networks**

**5   Introduction of Cooperative Transmission in Integrated
Satellite-Terrestrial Networks** .............................................    89

**6   Traffic Offloading in Satellite-Terrestrial Networks** ...................    91
6.1     Introduction .........................................................    91
6.2     Related Works .......................................................    93
6.3     Architecture of SDN ...............................................    95
        6.3.1     Service Plane .............................................    95
        6.3.2     Control Plane .............................................    96
        6.3.3     Management Plane .......................................    96
6.4     System Model of Traffic Offloading in H-STN ....................    97
        6.4.1     Fully-Loaded Transmission ..............................    98
        6.4.2     Satellite's Transmission Rate Through Each Channel ....    98
        6.4.3     BSs' Cooperative and Competitive Modes ................    99
6.5     Second-Price Auction Based Traffic Offloading
        Mechanism Design .................................................   100
        6.5.1     Second-Price Auction ...................................   100
        6.5.2     Auction Operation .......................................   100
        6.5.3     Outcomes of Auction-Based Traffic Offloading ..........   101

6.6     Satellite's Equilibrium Bidding Strategies .......................... 103
        6.6.1    Bidding Strategy for $R_{\text{thr}} \in (\mu_{\text{min}}, \mu_{\text{max}}]$ ................ 104
        6.6.2    Bidding Strategy for $R_{\text{thr}} \in \left(\mu_{\text{max}}, \left(1 + \frac{1-\beta}{N}\right)\mu_{\text{max}}\right)$ ... 105
        6.6.3    Bidding Strategy for $R_{\text{thr}} \in \left[\left(1 + \frac{1-\beta}{N}\right)\mu_{\text{max}}, +\infty\right)$ .... 106
        6.6.4    Bidding Strategy for $R_{\text{thr}} \in [0, \mu_{\text{min}}]$ ...................... 107
6.7     Expected Utility Analysis for MNO ................................ 108
        6.7.1    Utility Analysis for $R_{\text{thr}} \in (\mu_{\text{min}}, \mu_{\text{max}}]$ ................ 108
        6.7.2    Utility Analysis for $R_{\text{thr}} \in \left(\mu_{\text{max}}, \left(1 + \frac{1-\beta}{N}\right)\mu_{\text{max}}\right)$ .... 110
        6.7.3    Utility Analysis for $R_{\text{thr}} \in \left[\left(1 + \frac{1-\beta}{N}\right)\mu_{\text{max}}, +\infty\right)$ ..... 111
        6.7.4    Utility Analysis for $R_{\text{thr}} \in [0, \mu_{\text{min}}]$ ...................... 111
6.8     Simulation Results ............................................... 111
        6.8.1    Beam Group's Strategy of the Satellite ................... 112
        6.8.2    Expected Utility of the MNO ........................... 112
6.9     Conclusion ....................................................... 118
6.10    Proof of Lemma 6.1 .............................................. 118
6.11    Proof of Theorem 6.1 ............................................ 119
        6.11.1   $\mu_n \in [R_{\text{thr}}, \mu_{\text{max}}]$ ........................................... 119
        6.11.2   $\mu_n \in (\tilde{\mu}_a(R_{\text{thr}}), R_{\text{thr}})$ .................................. 121
        6.11.3   $\mu_n = \tilde{\mu}_a(R_{\text{thr}})$ ..................................... 123
        6.11.4   $\mu_n \in [\mu_{\text{min}}, \tilde{\mu}_a(R_{\text{thr}}))$ ................................ 123
6.12    Proof of Theorem 6.3 ............................................ 124
References ............................................................... 125

7   **Cooperative Beamforming for Secure Satellite-Terrestrial**
    **Transmission** ........................................................ 129
    7.1    Introduction ..................................................... 129
    7.2    Related Works ................................................... 131
           7.2.1    Satellite Terrestrial Networks ........................... 131
           7.2.2    Physical Layer Security ................................. 132
    7.3    System Model ................................................... 133
           7.3.1    Channel Model .......................................... 135
           7.3.2    Received Signal Model ................................... 137
           7.3.3    Signal-to-Interference Plus Noise Ratio .................. 138
           7.3.4    Achievable Secrecy Rate ................................. 139
    7.4    Secure Transmission Beamforming Schemes for Satellite
           Terrestrial Networks ............................................ 139
           7.4.1    Non-Cooperative Beamforming for Secure
                    Transmission ............................................ 140
           7.4.2    Cooperative Secure Beamforming for Secure
                    Transmission ............................................ 140
    7.5    Solutions of the Optimization Problems .......................... 141
           7.5.1    Feasible Solution of the Optimization Problems .......... 142
           7.5.2    Path-Pursuit Iteration Based Approach ................... 142

7.5.3    Feasibility of Path-Pursuit Iteration Based Solution ...... 148
7.5.4    Complexity Analysis ..................................... 150
7.6    Simulation Experiments and Analysis ........................... 150
7.7    Conclusion.................................................... 158
7.8    Proof of Theorem 7.1 ......................................... 159
7.9    Proof of Theorem 7.2 ......................................... 161
References............................................................ 162

8    **Traffic Prediction Based Transmission in Satellite-Terrestrial**
     **Networks**........................................................ 165
8.1    Introduction.................................................. 165
8.2    Related Works ................................................ 167
8.3    System Model ................................................. 168
       8.3.1    The Traffic Model ..................................... 169
       8.3.2    Physical Channel Model ................................ 169
       8.3.3    The Cloud-Based Predictive Service Model .............. 170
       8.3.4    The Queueing Model .................................... 172
8.4    Wavelet Based Backpropagation Prediction for Traffic ........... 173
       8.4.1    Multi-Level Wavelet Decomposition ..................... 174
       8.4.2    Backpropagation Neural Network Prediction ............. 175
       8.4.3    Wavelet Based Backpropagation Prediction............... 177
8.5    Resource Allocation Based on the Predictive Backpressure ....... 179
       8.5.1    Dynamic Evolution of Queues ........................... 179
       8.5.2    Prediction Based Backpressure ......................... 180
8.6    Simulation Results and Analysis ............................... 182
       8.6.1    Video Traffic Model ................................... 183
       8.6.2    Performance of Wavelet Based Backpropagation
                Prediction ............................................ 183
       8.6.3    Performance of Predictive Backpressure ................ 186
8.7    Conclusion.................................................... 188
References............................................................ 189

**Part IV    Cooperative Computation and Caching in Heterogeneous**
            **Networks**

9    **Introduction of Cooperative Computation and Caching**............... 195

10   **QoS-Aware Computational Resource Allocation** ...................... 199
10.1    Introduction................................................. 200
10.2    Related Works ............................................... 202
10.3    SDN Architecture Design for Edge/Cloud Computing Systems ... 204
        10.3.1    Infrastructure Plane................................. 205
        10.3.2    Control Plane ....................................... 205
        10.3.3    Management Plane .................................... 206
10.4    System Model and Hierarchical Game Framework................. 206
        10.4.1    System Model......................................... 207

        10.4.2   Hierarchical Game Framework ........................... 209
   10.5  Evolutionary Game for Service Selection of User Devices ........ 210
        10.5.1   Evolutionary Game Based Service Selection ............. 211
        10.5.2   Existence and Uniqueness of Equilibrium................. 212
        10.5.3   Analysis of Evolutionary Stable State (ESS).............. 214
   10.6  Stackelberg Differential Game Based Dynamic
        Computational Power Pricing and Allocation ..................... 215
        10.6.1   Formulation of Stackelberg Differential Game ........... 215
        10.6.2   Open-Loop Stackelberg Equilibrium Solutions ........... 218
   10.7  Simulation Results .................................................... 224
        10.7.1   Evolution of Population Distribution ..................... 225
        10.7.2   Dynamic Pricing and Allocation of Computing
                 Resource ...................................................... 229
        10.7.3   Influence of Delay in Replicator Dynamics .............. 231
   10.8  Conclusion............................................................ 231
   References................................................................... 232

11  QoS-Aware Caching Resource Allocation .............................. 237
   11.1  Introduction........................................................... 237
   11.2  Related Works ........................................................ 239
   11.3  System Model ........................................................ 241
        11.3.1   Network Model ............................................. 241
        11.3.2   Video Popularity ........................................... 243
        11.3.3   VSP Preference ............................................ 244
   11.4  Caching Problem Formulation and Profit Analysis ................ 244
        11.4.1   Caching Procedure.......................................... 245
        11.4.2   Benefit Analysis ........................................... 246
   11.5  Double Auction Mechanism Design for Small-Cell Based
        Caching System ...................................................... 248
        11.5.1   Social Welfare Maximization Problem .................... 248
        11.5.2   Iterative Double Auction Mechanism Design ............. 250
   11.6  Implementation of I-DA Mechanism............................... 254
        11.6.1   I-DA Mechanism Based Algorithm........................ 254
        11.6.2   Convergence of I-DA Algorithm........................... 255
        11.6.3   Economic Properties of I-DA Mechanism ................ 258
   11.7  Evaluation Results.................................................... 260
   11.8  Conclusion............................................................ 267
   References................................................................... 268

12  Priority-Aware Computational Resource Allocation ................... 271
   12.1  Introduction........................................................... 271
   12.2  Related Work ......................................................... 274
        12.2.1   Computation Offloading Optimization In VEC .......... 274
        12.2.2   Computation Offloading Optimization in VFC .......... 274
        12.2.3   DRL-Based Computation Offloading
                 Optimization in VFC ....................................... 275

12.3   System Model ........................................................... 276
       12.3.1   System Architecture ........................................ 276
       12.3.2   Mobility Model ............................................. 277
       12.3.3   Communication Model ...................................... 278
       12.3.4   Computation Model ......................................... 279
       12.3.5   Task Model ................................................ 279
       12.3.6   Service Availability ........................................ 281
       12.3.7   Pricing Model .............................................. 283
12.4   Formulation of Optimization Problem for Task Offloading ........ 284
12.5   SAC Based DRL Algorithm for Task Offloading ................... 285
       12.5.1   State Space ................................................ 286
       12.5.2   Action Space ............................................... 286
       12.5.3   Reward Function ........................................... 287
       12.5.4   Policy and Value Function ................................. 287
       12.5.5   Policy Evaluation .......................................... 288
       12.5.6   Policy Improvement ........................................ 289
       12.5.7   Algorithm Design Based on SAC ........................... 290
       12.5.8   Complexity Analysis ....................................... 293
12.6   Performance Evaluation ............................................... 293
       12.6.1   Simulation Setup .......................................... 294
       12.6.2   Average Utility ............................................ 295
       12.6.3   Completion Ratio .......................................... 298
       12.6.4   Average Delay ............................................. 298
12.7   Conclusion ............................................................ 301
References ..................................................................... 303

13   Energy-Aware Computational Resource Allocation ................... 307
13.1   Introduction ........................................................... 307
13.2   Related Works ......................................................... 310
13.3   System Model ......................................................... 311
       13.3.1   Task Model ................................................ 312
       13.3.2   Local Computing .......................................... 313
       13.3.3   Offloading Computing ...................................... 314
       13.3.4   Energy Harvesting ......................................... 315
13.4   Hybrid Decision Based DRL For Dynamic Computation
       Offloading ............................................................ 316
       13.4.1   MDP Modeling ............................................. 317
       13.4.2   Hybrid Decision Based DRL Method ...................... 318
13.5   Multi-Device Hybrid Decision Based DRL for Dynamic
       Computation Offloading .............................................. 321
13.6   Performance Evaluations ............................................. 324
13.7   Simulation Results .................................................... 331
       13.7.1   General Setups ............................................. 331
       13.7.2   Performance of Convergence and Generalizability ....... 331

     13.7.3  Performance Evaluation of Hybrid-AC with
              Different System Parameters ................................. 334
     13.7.4  Performance Evaluation of MD-Hybrid-AC with
              Different System Parameters .............................. 338
  13.8   Conclusion ......................................................... 342
  References ................................................................. 343

**Part V   Cooperative Resource and Information Sharing Among
Users**

**14  Introduction of Cooperative Resource and Information
Sharing** ..................................................................... 349

**15  Cooperative Data Transaction in Mobile Networks** .................... 351
  15.1   Introduction ........................................................ 351
     15.1.1  Motivation .................................................. 352
     15.1.2  Contribution ................................................ 354
  15.2   Related Work ....................................................... 355
  15.3   Data Allocation of Single Data Provider ......................... 356
     15.3.1  Basic Auction Mechanism ................................. 356
     15.3.2  Data Allocation for Single-Auctioneer Transaction ...... 360
  15.4   Networked Auction Model for Data Transaction
     with Multiple Auctioneers ......................................... 363
     15.4.1  Networked Auction Model ................................. 364
     15.4.2  Mobility Model .............................................. 366
     15.4.3  Expected Income of Networked Systems ................. 367
     15.4.4  Data Allocation for Networked Data Transaction ......... 370
  15.5   Operation of Data Allocation for Data Transaction Systems ....... 375
     15.5.1  Approximate Solution of Optimization Problems ........ 375
     15.5.2  Data Allocation for Data Transaction ..................... 376
  15.6   Performance Evaluation ............................................ 377
     15.6.1  Data Transaction Systems with Single Auctioneer ....... 378
     15.6.2  Data Transaction Systems with Multi-Auctioneer ........ 382
  15.7   Conclusion ......................................................... 385
  References ................................................................. 385

**16  Cooperative Trustworthiness Evaluation and Trustworthy
Service Rating** ........................................................... 389
  16.1   Introduction ........................................................ 389
  16.2   Related Works ...................................................... 391
  16.3   Mathematical Model for Service Rating Based on User
     Report Fusion ...................................................... 392
     16.3.1  System Model ............................................... 393
     16.3.2  Service Rating Based on User Report Fusion ............. 394
  16.4   Peer Prediction for User Trustworthiness ......................... 395
     16.4.1  Private-Prior Peer Prediction Mechanism ................. 395

        16.4.2   Incentive Compatibility ................................... 401
  16.5  User Trustworthiness and Unreliability Based Service Rating..... 404
        16.5.1   Unreliability of User Report ............................. 404
        16.5.2   Peer Prediction Based Service Rating .................... 406
  16.6  Performance Evaluation............................................ 407
        16.6.1   Simulation Settings ...................................... 407
        16.6.2   Accumulative Trustworthiness and Unreliability ......... 408
        16.6.3   Influence of $\varepsilon$, Scoring Rules and User Structure ......... 409
  16.7  Conclusions...................................................... 414
  References............................................................. 414

17  **Cooperative Privacy Protection Among Mobile User** ................. 417
  17.1  Introduction..................................................... 418
  17.2  Related Works ................................................... 419
  17.3  Community Structure Based Evolutionary Game Formulation .... 421
        17.3.1   Basic Concept of Evolutionary Game ..................... 421
        17.3.2   Community Structured Evolutionary Game
                 Formulation .............................................. 422
  17.4  Privacy Protection Among Users Belonging to
        $K$ Communities .................................................. 427
        17.4.1   Evolution of Security Behavior on Communities ......... 427
        17.4.2   Finding the Critical Ratio ............................... 430
  17.5  Privacy Protection Among Users with $L$-Triggering Game........ 441
        17.5.1   $L$-Triggering Game....................................... 442
        17.5.2   Analysis of Cost Performance ........................... 444
  17.6  Performance Evaluation........................................... 445
  17.7  Conclusions...................................................... 451
  References............................................................. 453

**Part VI   Conclusion**

18  **Conclusion** ................................................................ 459

# About the Authors

**Assist. Prof. Jun Du** received her B.S. in information and communication engineering from Beijing Institute of Technology, in 2009, and her M.S. and Ph.D. in information and communication engineering from Tsinghua University, Beijing, in 2014 and 2018, respectively. From Oct. 2016 to Sept. 2017, Dr. Du was a sponsored researcher, and she visited Imperial College London. Currently she is an assistant professor in the Department of Electrical Engineering, Tsinghua University. Her research interests are mainly in communication, networking, resource allocation, and system security problems of heterogeneous networks and space-based information networks. She has authored/co-authored 70+ technical papers in renowned international journals and conferences, including 30+ renowned IEEE journal papers. Dr. Du is the recipient of the Best Student Paper Award from IEEE Global SIP in 2015, the Best Paper Award from IEEE ICC 2019, the Best Paper Award from IWCMC in 2020, and the WuWenJun Young Elite Scientist Award from CAAI in 2020.

**Assoc. Prof. Chunxiao Jiang** is an associate professor in the School of Information Science and Technology, Tsinghua University. He received his B.S. degree in information engineering from Beihang University, Beijing, in 2008 and his Ph.D. degree in electronic engineering from Tsinghua University, Beijing, in 2013, both with the highest honors. His research interests include application of game theory, optimization, and statistical theories to communication, networking, and resource allocation problems, in particular space networks and heterogeneous networks. Dr. Jiang has served as an Editor of *IEEE Internet of Things Journal*, *IEEE Network*, *IEEE Communications Letters*, and a Guest Editor of *IEEE Communications Magazine*, *IEEE Transactions on Network Science and Engineering*, and *IEEE Transactions on Cognitive Communications and Networking*. He has also served as a member of the technical program committee as well as the Symposium Chair for a number of international conferences, including IEEE CNS 2020 Publication Chair, IEEE WCSP 2019 Symposium Chair, IEEE ICC 2018 Symposium Co-Chair, IWCMC 2020/19/18 Symposium Chair, WiMob 2018 Publicity Chair, ICCC 2018 Workshop Co-Chair, and ICC 2017 Workshop Co-Chair. Dr. Jiang is the recipient of the Best Paper Award from IEEE GLOBECOM in 2013, the Best Student Paper Award from IEEE GlobalSIP in 2015, IEEE Communications Society Young Author Best Paper Award in 2017, the Best Paper Award IWCMC in 2017, IEEE ComSoc TC Best Journal Paper Award of the IEEE ComSoc TC on Green Communications & Computing 2018, IEEE ComSoc TC Best Journal Paper Award of the IEEE ComSoc TC on Communications Systems Integration and Modeling 2018, the Best Paper Award from ICC 2019, and IEEE VTS Early Career Award 2020. He received the Chinese National Second Prize in Technical Inventions Award in 2018 and Natural Science Foundation of China Excellent Young Scientists Fund Award in 2019. www.jiangchunxiao.net.

# Part I
# Introduction

# Chapter 1
# Introduction of 6G Heterogeneous Networks

**Keywords** 6G · Heterogeneous Networks · Architecture · Resource Allocation · Artificial Intelligence (AI) · Game Theory

## 1.1 Heterogeneous Architecture of 6G Networks

Recently, the fifth-generation (5G) of wireless network is developed to support enhanced mobile broadband (eMBB), massive machine-type communications (mMTC) and ultra-reliable and low-latency communications (uRLLC) [1], according to the report of International Telecommunication Union (ITU). Benefitting from such high performance, 5G has opened new doors of opportunity towards emerging applications, e.g., augmented reality (AR), virtual reality (VR), tactile reality, mixed reality, etc. However, the new media, such as holographic communications, will request much higher transmission speeds up to Tera bits per second (Tbit/s) than AR and VR. Then 5G is far from enough to support the faster, more reliable and larger scale communication requirements of these services. In response, the investigation of future generation wireless networks (6G) has been triggered, which promises more powerful capacities in terms of ultra-broadband, supper massive access, ultra-reliability and low-latency than 5G does, as listed in Table 1.1 [1].

To provide ubiquitous and various services, 6G networks tend to be more comprehensive and multi-dimensional by integrating current terrestrial networks with space/air-based information networks and marine information networks, and then heterogeneous network resources, as well as different types of users and data, will be also integrated, as shown in Fig. 1.1. According to such architecture, 6G networks are conceived to be cell-free, which means that users will move from one network to another seamlessly and automatically to pursue the most suitable and qualified communications, without manual managements and configurations [2]. On the contrary, current 5G networking technologies still mainly focus on a macro-cell and small-cell based heterogeneous architecture, which will be broken by the cell-free operation of 6G, and their performance will be deteriorated when applied to 6G with brand new architectures. In addition, how to manage and control 6G networks to realize the promising capacities of ultra-broadband, ultra-

J. Du, C. Jiang, *Cooperation and Integration in 6G Heterogeneous Networks*, Wireless Networks, https://doi.org/10.1007/978-981-19-7648-3_1

**Table 1.1** Comparison of key performance indexes between 4G, 5G and 6G

| Performance | 4G | 5G | 6G |
|---|---|---|---|
| Peak data rate | 1 Gbit/s | 20 Gbit/s | $\geq 1$ Tbit/s |
| User experienced data rate | 10 Mbit/s | 100 Mbit/s | 1 Gbit/s |
| Spectrum efficiency | $1\times$ | $3\times$ | $15–30\times$ |
| Mobility | 350 km/h | 500 km/h | $\geq 1000$ km/h |
| Latency | 10 ms | 1 ms | $\leq 100\mu$s |
| Connection density (devices/km$^2$) | $10^5$ | $10^6$ | $10^7$ |
| Network energy efficiency | $1\times$ | $100\times$ | $100–10,000\times$ |
| Area traffic capacity | 0.1 Mbit/s/m$^2$ | 10 Mbit/s/m$^2$ | $\geq 1$ Gbit/s/m$^2$ |

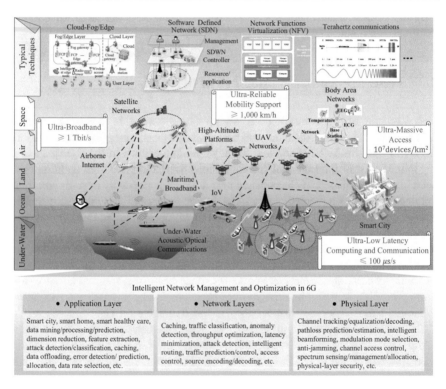

**Fig. 1.1** Illustration of heterogeneous resource applications in 6G to support ultra-broadband, ultra-massive access and ultra-reliability/ low-latency

massive access, ultra-reliability and low-latency also meet great challenges brought by increasing ultra-dense, heterogeneous and dynamic characteristics. Specifically, different kinds of satellite internet consisting of a large number of satellites were proposed and implemented in recent years. For instance, the project of SpaceX, known as Starlink, initially planned to build a constellation of 12,000 satellites in the low Earth orbit, which has been expanded to 42,000 recently.

In addition, mobile network operators are accelerating the dense deployment of small-cell base stations, in order to reduce the service latency by avoiding backhaul transmissions. Moreover, future large-scale Internet of Things (IoT) systems in 6G prospect will also bring challenges of spectrum management and massive or super access control. Furthermore, the integration of high-dynamic satellites, Unmanned Aerial Vehicle (UAVs) and Internet of Vehicles (IoV) will result in more frequent handovers, more uncertain user requirements and more unpredictable wireless communication environments than any previous generation of networks, which makes it difficult to guarantee the ultra-reliability and low-latency of services.

Therefore, 6G networks are developing into more multi-dimensional, heterogenous, large-scale and high-dynamic systems. All these characteristics make it urgent to explore new techniques that is adaptive, flexible and intelligent to bring a revolutionary leap of communications with ultra-broadband, ultra-massive access supporting, ultra-reliability and low-latency. In addition, enormous amounts of widely heterogeneous data generated from 6G networks will request advanced mathematical tools to extract meaningful information from these data, and then take decisions including resource management and access control pertaining to the proper functioning of 6G, which are hardly achieved by traditional network optimization techniques. In recent years, artificial intelligence (AI) is emerging as a fundamental paradigm to orchestrate the communication and information systems from bottom to top. For the foreseeable future, AI-enabled networks will open up new opportunities for the smart and intelligent 6G networking.

As illustrated in Fig. 1.1, network economics based game theory, contact, auction theory, as well as AI and machine learning (ML) techniques are expected to help 6G networks make more optimized and adaptive data-driven decisions, alleviate communication challenges, and meet requirements from emerging services. In this book, we will focus on the scope of applying these advanced technologies to networking and resource allocation optimization, aiming to bring about significant innovation of communications on ultra-broadband, ultra-massive access, ultra-reliability and low-latency.

## 1.2  Challenges of Heterogeneous Resource Allocation

### 1.2.1  Heterogeneous Resource Modeling and Performance Evaluation

Heterogeneous resource modeling and performance evaluation are particularly important for the 6G network to reveal how the transmission environment, spectrum interference, resource conflict and transmission delay impact its service's capacity. As shown in Fig. 1.1, the 6G heterogenous network tends to be an integration of traditional cellular network with satellites, high-altitude platforms, UAVs, IoV, and maritime and underwater networks. These infrastructures and devices present obvious heterogeneity characteristics referring to motion characteristics, stability, transmission capacity, temporal and spatial coverage of services, as well as their characteristics of wireless propagation environment. For instance, the wireless propagation environment over the sea is quite different from that of terrestrial communications, due to its special reflection from ocean surfaces, evaporation duct property and atmospheric absorption loss [3]. In addition, the dynamic propagation environment and large transmission delay will further lead to limited and insufficient channel information acquisition [4, 5]. On the other hand, there is quite a difference between these platforms in the communication protocols and standards, especially when they are expected to provide comprehensive services of communications, detections, navigation and data processing. Such heterogeneous characteristics reflected both in resource and services bring great challenges to generic and standard resource modeling and performance evaluation for 6G networks.

## 1.2.2 Task Adaptation and Resource Efficiency

Driven by the exponentially growing demands of multimedia data traffic and computation-heavy applications, 6G networks are expected to achieve a high quality of service (QoS) with efficient transmission, ultra-reliability and low latency. In response, task-adaptive resource allocation has been considered as an important factor which can improve the 6G performance directly by configuring heterogenous resources effectively and efficiently. In 6G, the allocated resource can be divided into communication resource, which includes channels and bandwidth, and computing resource, such as memory and processing power. In recent years, various traffic offloading, caching and cloud/fog/edge computing mechanisms are designed to allocate these communication, storage and computing resources in heterogenous networks, respectively, which become promising solutions to handle the increasing data and computational requirements with low-latency and on-demand services [6, 7].

## 1.2.3 Interference Control and Secure Communications

6G heterogeneous networks integrate billions of digital sensors, smart nodes, people, services and other physical objects which are capable of realizing a seamless information connection, interaction and exchanging with each other [8]. Concerning more and more human-related data processed in current 6G applications, as well as the tremendous business value behind the privacy-sensitive and patent-sensitive data, security issues, such as privacy preserving, secure data storage, exchanging and analysis, are becoming significant challenges for these applications and designing resource allocation mechanisms. In 2013, the exposure of NSA's PRISM program triggered off vigorous debate on privacy. In addition, InBloom, a company worked at student data storage, closed down in 2014 due to the public's worries of the privacy security of students' data, although there was no indication that InBloom used students' data illegally. Therefore, it is rather necessary for 6G networks to protect the private information contained in data against being exposed, meanwhile to fulfill the functional requirements and ensure the service quality of 6G applications. To achieve this goal, introducing security measures to processes of data transfer and analysis among service providers and data owners plays an important role, since that these processes pose potential threats to the privacy [9–12]. Moreover, mechanisms with incentive capability should be designed to encourage data owners to provide their data honestly [13, 14]. Considering these aforementioned problems, this book will introduce some state-of-the-art technologies of secure resource and information sharing mechanisms in 6G networks.

## 1.3   Mathematic Tools for Resource Allocation

### 1.3.1   Information Economics Theory

As a typical information economics theory, game theory has offered a set of mathematical tools to establish and analyze the complex interactions among interdependent and rational players, and then predict their choices of strategies [15–18]. Therefore, game theory has been considered as an effective and suitable tool to study the resource allocation and cooperative networking problems for 6G heterogeneous networks. In addition, game theoretic models can be developed to obtain solutions for spectrum management, channel assignment and power control among heterogeneous devices.

Contract theory, as a powerful microeconomics framework of modeling multilateral labor/employment relations, is proposed to essentially deal with the information asymmetry in the market, regarding the service capability of "employees" which cannot be observed by "employers" before they are employed [19, 20]. According to contract theory, an incentive mechanism that encourages every employee to consciously choose the contract designed for its service capability will be realized by offering a set of contracts that includes a required performance and a corresponding reward. Due to such property, there is great potential to utilize the concepts from contract theory to ensure cooperations and incentive mechanisms in 6G heterogeneous networks, especially for cellular traffic offloading, spectrum sharing, edge computing and caching [21–24]. To apply contract theory, a set of contracts for the employee need to be designed delicately, the object of which is maximizing the employer's payoff or utility. In most studies, the problem is usually formulated as maximizing an objective function that represents the employer's utility, subject to the following two constraints: (1) the incentive compatibility constraint, which refers to that the employee's expected payoff is maximized when signing the contract; (2) the individual rationality constraint, which requires that the employee's payoff under this contract is larger than or equal to its reservation payoff when not participating.

Auction theory, as a branch of game theory, can be widely used in trading if the price of a commodity and service is undetermined, and has also become an important tool in network economics to model resource supply and demand, especially for networks with heterogeneous transmission resource [25, 26]. In addition, auctions can also be used in a network to automatically carry out objectives which may satisfy either the sellers or the buyers [27], and in resource allocation for admission control based on resource availability [28–31]. In recent years, any classic auction models, such as Vickrey-Clarke-Groves (VCG) auction, share auction, double auction, and combinatorial auction, have been designed and introduced to resource allocation and cooperative networking in heterogeneous networks. In the auction-based resource management, efficient mechanism design for a given auction, equilibrium and optimal bidding strategies, revenue comparison, etc., are typical issues to meet the resource requirements and achieve efficient resource utilization at the same time.

## 1.3.2   Machine Learning and Artificial Intelligence

In recent years, AI is emerging as a fundamental paradigm to orchestrate the communication and information systems from bottom to top. For the foreseeable future, AI-enabled networks will open up new opportunities for the smart and intelligent 6G networking. As a major branch of AI, ML can establish an intelligent system which operates in the complicated environment [32]. Recently, ML has mainly developed into many branches such as classical ML, including supervised learning and unsupervised learning, deep learning (DL) and reinforcement learning (RL). DL aims to understand the representations of data, and can be modeled in supervised, unsupervised, and reinforcement learning. Different AI/ML techniques, such as RL, DRL, Double DRL (DDRL), etc., have been introduced to cooperative networking and resource allocation techniques, in order to deal with the sophisticated optimization of decision making resulting from the multi-dimension, randomly uncertainty and dynamics of 6G. By applying AI/ML tools, the valuable information can be extracted through training observed data, and then different functions for prediction, optimization, and decision making in traffic offloading, caching and cloud/fog/edge computing can be learned to support ultra-reliable and low-latency services [33, 34].

However, most current RL or DRL based resource allocation approaches were modeled in a discrete action space, which restricts the optimization of offloading decisions in a limited action space [35]. Such model assumption is unreasonable in practice, where the action space of offloading decision is often continuous-discrete hybrid. To be specific, in a task offloading enabled 6G network, the strategies of which node should be selected to implement traffic/computation offloading or caching constitute a discrete action space. On the other hand, the possible resource volume should be provided by the selected node for offloading is a continuous value usually. Such resource allocation problem with continuous-discrete hybrid decision spaces tends to be extremely complex, especially when time-varying tasks, energy harvesting and security issues are also considered. To provide low-latency computing services, this book will introduce some preliminary work focusing on the hybrid decision of computation offloading in 6G networks based on DRL. To be specific, the sever selection problem is modeled in a discrete action space, and meanwhile the decision spaces of offloading the ratio and local computation capacity are continuous.

# References

1. Z. Zhang, Y. Xiao, Z. Ma, M. Xiao, Z. Ding, X. Lei, G. K. Karagiannidis, and P. Fan, "6G wireless networks: Vision, requirements, architecture, and key technologies," *IEEE Veh. Technol. Mag.*, vol. 14, no. 3, pp. 28–41, Sept. 2019.
2. X. Zhu and C. Jiang, "Integrated satellite-terrestrial networks toward 6G: Architectures, applications, and challenges," *IEEE Internet Things J.*, vol. 9, no. 1, pp. 437–461, Jan. 2022.
3. F. B. Teixeira, T. Oliveira, M. Lopes, C. Leocádio, P. Salazar, J. Ruela, R. Campos, and M. Ricardo, "Enabling broadband internet access offshore using tethered balloons: The bluecom+ experience," in *OCEANS 2017 - Aberdeen*. Aberdeen, UK, 19–22 Jun. 2017.
4. X. Zhu, C. Jiang, L. Kuang, and Z. Zhao, "Cooperative multilayer edge caching in integrated satellite-terrestrial networks," *IEEE Trans. Wireless Commun.*, vol. 21, no. 5, pp. 2924–2937, May 2022.
5. J. Du, J. Song, Y. Ren, and J. Wang, "Convergence of broadband and broadcast/multicast in maritime information networks," *Tsinghua Sci. Techno.*, vol. 26, no. 5, pp. 592–607, Oct. 2021.
6. Y. Zhang, C. Jiang, N. H. Tran, S. Bu, F. R. Yu, and Z. Han, "Insurance plan for service assurance in cloud computing market with incomplete information," *J. Commun. Inform. Networks*, vol. 7, no. 1, pp. 11–22, Mar. 2022.
7. L. Xu, Z. Yang, H. Wu, Y. Zhang, Y. Wang, Li, Wang, and Z. Han, "Socially Driven Joint Optimization of Communication, Caching, and Computing Resources in Vehicular Networks," *IEEE Transa. Wireless Commun.*, vol. 21, no. 1, pp. 461–476, Jan. 2022.
8. J. Du, C. Jiang, H. Zhang, X. Wang, Y. Ren, and M. Debbah, "Secure satellite-terrestrial transmission over incumbent terrestrial networks via cooperative beamforming," *IEEE J. Sel. Areas Commun.*, vol. 36, no. 7, pp. 1367–1382, Jul. 2018.
9. L. Xiao, X. Wan, X. Lu, Y. Zhang, and D. Wu, "IoT security techniques based on machine learning: How do iot devices use AI to enhance security?," *IEEE Signal Proc. Mag.*, vol. 35, no. 5, pp. 41–49, Sept. 2018.
10. J. Du, C. Jiang, K.-C. Chen, Y. Ren, and H. V. Poor, "Community-structured evolutionary game for privacy protection in social networks," *IEEE Trans. Inf. Forens. Security*, vol. 13, no. 3, pp. 574–589, Mar. 2018.
11. J. Du, C. Jiang, S. Yu, K.-C. Chen, and Y. Ren, "Privacy protection: A community-structured evolutionary game approach," in *IEEE Global Conf. Signal Inform. Process. (GlobalSIP'16)*. Washington, DC, USA, 07–09 Dec. 2016.
12. A. Ferdowsi and W. Saad, "Deep learning for signal authentication and security in massive internet-of-things systems," *IEEE Trans. Commun.*, vol. 67, no. 2, pp. 1371–1387, 2018.
13. J. Du, E. Gelenbe, C. Jiang, H. Zhang, Y. Ren, and H. V. Poor, "Peer prediction-based trustworthiness evaluation and trustworthy service rating in social networks," *IEEE Trans. Inf. Forens. Security*, vol. 14, no. 6, pp. 1582–1594, Nov. 2018.

14. J. Du, C. Jiang, J. Wang, S. Yu, and Y. Ren, "Trustable service rating in social networks: A peer prediction method," in *IEEE Global Conf. Signal Inform. Process. (GlobalSIP'16)*. Washington, DC, USA, 07–09 Dec. 2016.

15. G. Zhang, K. Yang, and H.-H. Chen, "Resource allocation for wireless cooperative networks: a unified cooperative bargaining game theoretic framework," *IEEE Wireless Commun.*, vol. 19, no. 2, pp. 38–43, Apr. 2012.

16. J. Du, C. Jiang, A. Benslimane, S. Guo, and Y. Ren, "SDN-based resource allocation in edge and cloud computing systems: An evolutionary stackelberg differential game approach," *IEEE/ACM Trans. Networking*, vol. 30, no. 4, pp. 1613–1628, Aug. 2022.

17. L. Song, D. Niyato, Z. Han, and E. Hossain, "Game-theoretic resource allocation methods for device-to-device communication," *IEEE Wireless Commun.*, vol. 21, no. 3, pp. 136–144, Jun. 2014.

18. J. Du, E. Gelenbe, C. Jiang, Z. Han, and Y. Ren, "Auction-based data transaction in mobile networks: Data allocation design and performance analysis," *IEEE Trans. Mobile Computing*, vol. 19, no. 5, pp. 1040–1055, May 2020.

19. Y. Zhang, M. Pan, L. Song, Z. Dawy, and Z. Han, "A Survey of Contract Theory-Based Incentive Mechanism Design in Wireless Networks," *IEEE Commun. Mag.*, vol. 24, no. 3, pp. 80–85, Jun. 2017.

20. J. Du, C. Jiang, Z. Han, H. Zhang, S. Mumtaz, and Y. Ren, "Contract mechanism and performance analysis for data transaction in mobile social networks," *IEEE Trans. Network Sci. Eng.*, vol. 6, no. 2, pp. 103–115, Apr. - Jun. 2019.

21. D. H. N. Nguyen, Y. Zhang, and Z. Han, "Contract-based spectrum allocation for wireless virtualized networks," *IEEE Trans. Wireless Commun.*, vol. 17, no. 11, pp. 7222–7235, Nov. 2018.

22. X. Zhang, S. Sarkar, A. Bhuyan, S. K. Kasera, and M. Ji, "A non-cooperative game-based distributed beam scheduling framework for 5G millimeter-wave cellular networks," *IEEE Trans. Wireless Commun.*, vol. 21, no. 1, pp. 489–504, Jan. 2022.

23. H. Chergui, L. Blanco, and C. Verikoukis, "Statistical federated learning for beyond 5g sla-constrained ran slicing," *IEEE Trans. Wireless Commun.*, vol. 21, no. 3, pp. 2066–2076, Mar. 2022.

24. J. Du, E. Gelenbe, C. Jiang, H. Zhang, and Y. Ren, "Contract design for traffic offloading and resource allocation in heterogeneous ultra-dense networks," *IEEE J. Sel. Areas Commun.*, vol. 35, no. 11, pp. 2457–2467, Nov. 2017.

25. C. Jiang, Y. Chen, Q. Wang, and K. R. Liu, "Data-driven auction mechanism design in IaaS cloud computing," *IEEE Trans. Services Computing*, vol. PP, no. 99, pp. 1–14, 2015.

26. J. Du, C. Jiang, E. Gelenbe, H. Zhang, and T. Q. S. Quek, "Double auction mechanism design for video caching in heterogeneous ultra-dense networks," *IEEE Trans. Wireless Commun.*, vol. 18, no. 3, pp. 1669–1683, Mar. 2019.

27. A. Di Ferdinando, R. Lent, and E. Gelenbe, "A framework for autonomic networked auctions," in *Proc. of the 2007 Workshop on Innovative Service Technol.* ICST (Institute for Computer Sciences, Social-Informatics and Telecommunications Engineering), 2007, p. 3.

28. G. Sakellari, T. Leung, and E. Gelenbe, "Auction-based admission control for self-aware networks," in *Computer and Inform. Sci. II*. Springer London, 2012, pp. 223–230.

29. J. Du, C. Jiang, H. Zhang, Y. Ren, and M. Guizani, "Auction design and analysis for SDN-based traffic offloading in hybrid satellite-terrestrial networks," *IEEE J. Sel. Areas Commun.*, vol. 36, no. 10, pp. 2202–2217, Oct. 2018.

30. J. Du, C. Jiang, H. Zhang, Y. Ren, and T. Q. Quek, "Double auction based resource allocation for secure video caching in heterogeneous networks," in *IEEE Int. Wireless Commun. Mobile Comput. Conf. (IWCMC'19)*.

31. J. Du, C. Jiang, H. Zhang, Y. Ren, and V. C. Leung, "Second-price auction based cognitive traffic offloading in heterogeneous networks," in *IEEE Int. Wireless Commun. Mobile Comput. Conf. (IWCMC'19)*.

32. J. Du, C. Jiang, J. Wang, Y. Ren, and M. Debbah, "Machine learning for 6G wireless networks: Carry-forward-enhanced bandwidth, massive access, and ultrareliable/low latency," *IEEE Veh. Technol. Mag.*, vol. 15, no. 4, pp. 123–134, Dec. 2020.
33. M. Min, L. Xiao, Y. Chen, P. Cheng, D. Wu, and W. Zhuang, "Learning-based computation offloading for IoT devices with energy harvesting," *IEEE Trans. Veh. Technol.*, vol. 68, no. 2, pp. 1930–1941, Feb. 2019.
34. Z. Li, C. Jiang, and L. Kuang, "Double auction mechanism for resource allocation in satellite MEC," *IEEE Trans. Cognitive Commun. Networking*, vol. 7, no. 4, pp. 1112–1125, Dec. 2021.
35. Y. He, F. R. Yu, N. Zhao, V. C. Leung, and H. Yin, "Software-defined networks with mobile edge computing and caching for smart cities: A big data deep reinforcement learning approach," *IEEE Commun. Mag.*, vol. 55, no. 12, pp. 31–37, Dec. 2017.

# Part II
# Cooperative Transmission in Heterogeneous Networks

# Chapter 2
# Introduction of Cooperative Transmission in Heterogeneous Networks

**Keywords** 6G · Heterogeneous Networks · Cooperative Transmission

Both current cellular mobile networks and space-based information networks (SBINs) presents heterogeneous characteristics. For these heterogeneous networks, cooperative resource allocation and management are effective manners to enhance the transmission capability. Aiming at the problem of cooperation-based transmission capability enhancement, we introduce two cooperative transmission mechanisms in this chapter.

Firstly, traffic offloading for terrestrial heterogeneous networks. We use the contract-based model to investigate the traffic offloading and resource allocation mechanisms. For such cooperative transmission mechanism, incentive compatibility is an important property to deal with the information asymmetry situation, where the properties of resource providers are not available for the resource requesters. In addition, we propose a valid and effective definition of resource type for the contract-based traffic offloading. The closed-form expression of resource type defined can reflect the offloading service capacity of heterogeneous resource provided. Meanwhile, this definition can ensure the monotonicity and incentive compatibility of the designed contracts. Secondly, cognitive multiple access via conative communications. We investigate the efficient networking and resource allocation mechanisms to support the cooperative transmissions among satellites. In space-based networks, the data relay satellites can assist low-earth-orbit (LEO) satellites in relaying data to other satellites or the ground station and improve the real-time system throughput. However, the transmission resource of the relay satellite is limited, so a key issue in cooperative communication through a relay is the resource allocation for the full use of limited resource. To take full advantage of transmission resource of the cooperative relays, we propose two multiple access and resource allocation strategies, in which relays can receive and transmit simultaneously according to channel characteristics of space-based systems for both the geosynchronous orbit (GEO) and LEO relay systems. Moreover, to reveal the impact

J. Du, C. Jiang, *Cooperation and Integration in 6G Heterogeneous Networks*, Wireless Networks, https://doi.org/10.1007/978-981-19-7648-3_2

of cooperation on important multi-access performance metrics, we further derive the mathematical maximum stable throughput region and the delay performance based on queueing theoretic formulation, which provide the appropriate guidances for the design of the system.

# Chapter 3
# Traffic Offloading in Heterogeneous Networks

**Abstract** In heterogeneous ultra-dense networks (HetUDNs), the software-defined wireless network (SDWN) separates resource management from geo-distributed resources belonging to different service providers. A centralized SDWN controller can manage the entire network globally. In this work, we focus on mobile traffic offloading and resource allocation in SDWN-based HetUDNs, constituted of different macro base stations (MBSs) and small-cell base stations (SBSs). We explore a scenario where SBSs' capacities are available, but their offloading performance is unknown to the SDWN controller: this is the *information asymmetric* case. To address this asymmetry, incentivized traffic offloading contracts are designed to encourage each SBS to select the contract that achieves its own maximum utility. The characteristics of large numbers of SBSs in HetUDNs are aggregated in an analytical model, allowing us to select the *SBS types* that provide the off-loading, based on different contracts which offer rationality and incentive compatibility to different SBS types. This leads to a closed-form expression for selecting the SBS types involved, and we prove the monotonicity and incentive compatibility of the resulting contracts. The effectiveness and efficiency of the proposed contract-based traffic offloading mechanism, and its overall system performance, are validated using simulations.

**Keywords** Traffic Offloading · Software Defined Wireless Networks (SDWNs) · Contract Theory · Heterogeneous Ultra-dense Networks (HetUDNs) · Resource Sharing

## 3.1 Introduction

Fifth Generation (5G) cellular networks were first proposed to meet the increasing mobile data traffic, which will expand one thousand times from 2010 to 2020 [1–3]. To meet this increasing data challenge, ultra-densification, i.e., overlaying macro base stations (MBSs) with a large number of small-cell base stations (SBSs) such as pico base stations (BSs), femto BSs and WiFi hotspots, etc., which constitute heterogeneous networks (HetNets), is one of the "big three" 5G technologies [4]. With assistance of these SBSs, *mobile traffic offloading* technology provides a solution to address the enormous expansion of mobile data, by moving traffic load from cellular networks to alternative wireless networks consisting of densely distributed SBSs. To operate such heterogeneous ultra-dense networks (HetUDNs), an effective and efficient network architecture and resource management mechanisms are indispensable. In recent years, cloud-based Software-defined Wireless Networks (SDWNs) are proposed to control and manage HetUDNs in a central manner efficiently. SDWNs can potentially revolutionize network design and resource management, and enable the applications to manipulate various services by separating the control plane from the data plane [5–8]. In SDWNs, mobile traffic offloading can be enabled by the SDWN at the edge [9], which can exploit knowledge of the data requests and the network resource status of MBSs and SBSs. With a centralized controller, resources in HetUDNs can be managed efficiently to meet data requests from mobile users, and optimize system performance including data rate, load balancing and energy consumption [10–12].

Recently, mobile traffic offloading in HetNets received significant attention for its effectiveness on rescuing the heavy traffic load in cellular networks by switching and exchanging traffic, and using access control and compatibility protocols [13]. Focusing on energy consumption optimization [14, 15], security guarantees [16] and performance analysis [17], much work has paid attention to mechanism design for mobile traffic offloading. As in [5], resource management in SDWNs is a form of competitive market, where resource requesters and providers compete and cooperate to maximize their own utilities. For traffic offloading in SDWN-based HetUDNs, competition and cooperation among resource providing and utilizing entities can be modeled and analyzed through economics theory [18]. Game theory is used to model the supply and demand relationship of resources for traffic offloading in HetNets, and many different game theory based offloading approaches have been applied, including Nash bargaining game [19, 20], coalition game [21], Stackelberg game [22], etc. In HetNets with densely distributed BSs, the computational complexity of such approaches grows exponentially, so that mean-field games can provide low-complexity tractable partial differential equation based solutions for traffic offloading [23–25].

Auction theory is another important tool in network economics to model the heterogeneous resource supply and demand [26]. Analytical work in this area includes the analysis of price and income in single [27] and networked multiple auctions [28], and for sealed bids can be found in [29]. In [30–32], the authors focus on the effect of an auction on the success of bidders who wish to make optimal choices. Auction theory can be also introduced in mobile traffic offloading for efficient cooperative transmission in HetNets. In SDWN-based HetNets, the central controller can act as an auction broker, and traffic offloading can be operated efficiently with appropriate auction mechanisms, such as "double" [33] and reverse auctions [34].

In SDWN-based HetUDNs, despite the existence of a central controller, SBSs are selfish and may hide or fabricate the status of their resource. However, the SBSs' resource availability can be recognized by the central controller, although their offloading performance may be unknown. To offer an incentive compatibility, contract theory can be introduced to the traffic offloading mechanism designs. Contract theory, as a powerful microeconomics framework, is proposed to essentially deal with the information asymmetry in the market, regarding the service capability of "employees" which cannot be observed by "employers" before they are employed [35]. According to contract theory, an incentive mechanism that encourages every employee to consciously choose the contract designed for its service capability will be realized. This approach has already been employed in resource allocation problems for device-to-device communications [36], heterogeneous Long-term Evlution-Advanced (LTE-A) networks [37] and heterogeneous cloud-based radio access networks [38]. Classic contract theory is based on the definition of different *employee types*, which is only considered as an abstract index without any specific definition in the aforementioned studies. Especially in HetUDNs with a large number of SBSs, this broad definition of "SBS type" can cause difficulties when contract models are applied to the real network environment. Thus in this work, we pay special attention to the SBS types and investigate how they can affect the performance of a contract-based traffic offloading mechanism.

The remainder of this part is organized as follows. Section 3.2 describes the SDWN framework for resource sharing in HetUDNs. The contract formulation and three contract-based traffic offloading mechanisms are designed in Sects. 3.4 and 3.5, respectively. Conditions for contract feasibility are analyzed and derived in Sect. 3.3. Simulations are presented in Sect. 3.6, and Sect. 3.7 summarizes the conclusions of this part.

## 3.2 Architecture of SDWN

SDWN is an emerging network framework which separates the control plane from the data plane. The architecture of the SDWN-based resource sharing system of HetUDNs is shown as Fig. 3.1. In the resource and application level, the network provides data services with distributed MBSs and SBSs. These heterogeneous BSs are operated by the same or different operators (service providers) and deployed with a high density, which means that their coverage areas are overlapped seriously. The MBSs' mobile user equipments (MUEs) and small cell user equipments (SUEs) are randomly distributed in the coverage of the BSs. Through traffic offloading, the throughput and other performance of the system can be improved.

As shown in Fig. 3.1, the SDWN separates resource management from the geo-distributed resource cloud, which forms a virtual network topology in the control plane. In the control plane, the centralized *SDWN Controller* discovers the traffic demands of MUEs, available transmission resource and the channel status in of the HetUDN through the *Access Network Discovery and Selection Function (ANDSF)*. The ANDSF fulfils this mission above by requiring to the MBS and SBSs, the current LTE/cellular network operators of which are more than willing to share the status information above to maximize their service capability and resource utilization. After receiving the supply and demand status of network resource, the SDWN controller designs a bundle of contracts for different types of SBSs, and

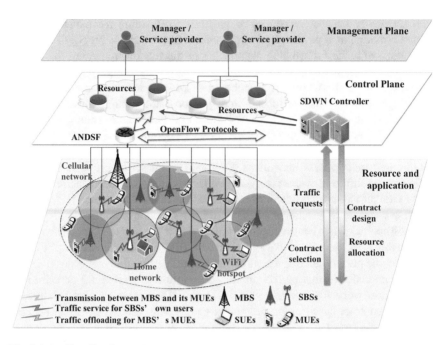

**Fig. 3.1** Traffic offloading and resource allocation for SDWN-based HetUDNs

then broadcasts the contracts to the MBS and all SBSs through the ANDSF. Every SBS distributed within the coverage of the MBS selects one contract to maximize its own payoff, and reports its selection to the SDWN controller that it will provide the certain amount of traffic offloading and get the certain payments from the MBS specified by the selected contract. According to the SBS's contract selection, Then the SDWN controller allocates this SBS's bandwidth resource for MUEs covered by this SBS, and requests the MBS to pay the SBS for its offloading service. During the process above, the ANDSF performs as a medium for the information interaction between the HetUDN and the SDWN Controller, and an executive of resource allocation. The ANDSF can interact with the SDWN controller for traffic offloading and resource allocation by standardized interfaces such as OpenFlow-enable switches [9, 18], which need some corresponding modifications for the requested and released information mentioned above. Moreover, in this work, we focus on the bandwidth resource allocation of the heterogeneous SBSs in the system.

## 3.3  Contract Formulation for Traffic Offloading

Consider an SDWN-based HetUDN with one MBS and a number of SBSs randomly distributed in the coverage of the MBS. These SBSs are not owned by the MBS operator, which means that the MBS cannot obtain the local information, such as transmission capacity, load status, operation and offloading cost, etc., of these SBSs. This model is flexible to be applied into a system with multiple MBSs, in which all SBSs can be associated to their respective MBS according to a certain association strategy, and then each subsystem consisting of an MBS and its associating SBSs can be analyzed by the system model of this work. We denote with $\mathcal{N} = \{1, 2, \cdots, N\}$ as the set of SBSs. Consider a set of $\mathcal{M} = \{1, 2, \cdots, M\}$ MUEs who are randomly distributed within the coverage of the MBS. In addition, let $\mathcal{M}_n$ be the set of MUEs in the coverage of SBS $n \in \mathcal{N}$, then we have $\bigcap_{n \in \mathcal{N}} \mathcal{M}_n = \mathcal{M}$. Let $\mathcal{N}_i$ denote the set of SBSs who can cover MUE $i \in \mathcal{M}$, which means that SBS $n \in \mathcal{N}_i$ can provide the traffic offloading service for MUE $i$.

Assume that the time is slotted. During the duration of a time slot, the location of MUEs, the offloading decision of SBSs and resource allocation are considered to be fixed. Denote $\mathbf{s}_n = \{s_{ni}\}_{i \in \mathcal{M}_n}$ as the scheduling vector, where $s_{ni} = 1$ indicates that the traffic of MUE $i$ is allocated to be offloaded by SBS $n$, and $s_{ni} = 0$, otherwise. Assume that each MUE can be associated with at most one SBS, i.e., $\forall i \in \mathcal{M}$, $\sum_{n \in \mathcal{N}} s_{ni} \leq 1$. The case of $\sum_{n \in \mathcal{N}} s_{ni} = 0$ indicates that the traffic requested by MUE $i$ is not offloaded by any SBS and is delivered by the MBS directly. Let $s_{0i} = 1$ denote that MUE $i$ is served by the MBS without any SBS offloading for it.

### 3.3.1 Transmission Model Formulation

The transmission data rate can be used to evaluate the performance of the HetUDN, and is related to the signal-to-interference-plus-noise ratio (SINR). In this work, we model the channel between MUEs and BSs as a Rayleigh fading channel. Then $\forall n \in \mathcal{N}$ and $i \in \mathcal{M}$, the SINR from SBS $n$ and the MBS to MUE $i$ is defined as

$$\gamma_{ni} = \frac{p_n |h_{ni}|^2}{\sum_{n' \in \mathcal{N}_i} \kappa_{n'i} p_{n'} |h_{n'i}|^2 + \kappa'_{0i} p_0 |h_{0i}|^2 + \sigma^2}, \tag{3.1a}$$

$$\gamma_{0i} = \frac{p_0 |h_{0i}|^2}{\sum_{n \in \mathcal{N}} \kappa'_{ni} p_n |h_{ni}|^2 + \sigma^2}, \tag{3.1b}$$

respectively. In (3.1a) and (3.1b), $\sigma^2$ is the constant addictive noise power, while $p_n$ and $p_{n'}$ are the transmission power consumption of SBS $n$ and $n'$, respectively. $\kappa_{n'i} \in [0, 1]$ is the interference parameter among SBSs, and $\kappa'_{0i}, \kappa'_{ni} \in [0, 1]$ are the interference parameters between the MBS and SBSs. Considering different licensed spectrum applied for direct transmission by the MBS and traffic offloading by SBSs, the interference between the MBS and SBSs can be ignored, i.e., $\kappa'_{0i} = \kappa'_{ni} = 0$. Then considering the channel allocation, the achievable service rate for MUE $i$ can be presented by

$$r_i = \omega_0 s_{0i} \log(1 + \gamma_{0i}) + \sum_{n \in \mathcal{N}} \omega_n s_{ni} \log(1 + \gamma_{ni}), \tag{3.2}$$

where $\omega_n$ and $\omega_0$ are the bandwidths of spectrum used by SBS $n$ and the MBS. Take LTE-A for instance, the bandwidth for one resource block is $\omega = 180\,\text{kHz}$. Considering all SBSs in the network utilize the common spectrum for traffic offloading, then let $\omega_n = \omega$, $\forall n \in \mathcal{N}$. Let $y_n$ denote the traffic offloading accepted by or allocated to SBS $n$, i.e., $y_n = \sum_{i \in \mathcal{M}_n} r_{ni} s_{ni}$.

### 3.3.2  Economic Models Formulation

The offloading quality provided by heterogeneous SBSs is different. On the other hand, the benefit for the MBS from different SBSs is different as well. For instance, the MBS tends to get much more benefit from SBSs those located closed to the edge of MBS's coverage. Therefore, it is better for the SDWN controller to design diverse contracts for the heterogeneous BSs, to improve the performance of the HetUDN.

**Utility of MBS**  Let $T_n(y_n)$ be the payment for SBS $n$ when it helps to offload the amount of $y_n$ traffic. Assume $T_n(0) = 0$, and in addition, $T_n(y_n)$ is a strictly increasing function of $y_n$, $\forall n \in \mathcal{N}$. Then we define the utility of MBS as

$$U(\mathbf{s}, \mathbf{y}, \mathbf{T}(y)) = \delta \sum_{i \in \mathcal{M}} r_i - \sum_{n \in \mathcal{N}} T_n(y_n), \qquad (3.3)$$

where $\mathbf{s} = \{s_{0i}, s_{ni}\}_{(N+1) \times M}$ denotes the association matrix, $\mathbf{y} = \{y_n\}_{n \in \mathcal{N}}$ is the traffic offloading vector, $\mathbf{T}(y) = \{T_n(y)\}_{n \in \mathcal{N}}$ denotes the vector of payment bundles for different types of SBSs, and $\delta$ is the MBS's unit monetary gain through the traffic rate.

**Utility of SBSs**  Let $x_n$ ($n \in \mathcal{N}$) denote the SBS $n$'s own traffic demands. Assume that traffic requests arrival for different SBSs are independent and identically distributed (i.i.d.), and follows a probability distribution function $f_n(x)$. In this work, we consider the traffic requests are sequences of Poisson arrivals with arrival rate $\lambda_n$, $\forall n \in \mathcal{N}$.

We define $\psi_n$, transmission efficiency of SBS $n$, as the average amount of data traffic (bits) can be delivered by one unit of bandwidth resource per time unit, which is given by

$$\psi_n = \frac{\sum_{i \in \mathcal{M}_n} r_{ni}}{\omega_n \left\| \{r_{ni}\}_{i \in \mathcal{M}_n} \right\|_0}, \qquad (3.4)$$

where $r_{ni} = \omega_n \log(1 + \gamma_{ni})$ denotes the achievable data rate of MUE $i$ receiving from SBS $n$, and $\|\cdot\|_0$ calculates the number of non-zero elements. Then the average bandwidth resource consumption for SBS $n$ on delivering one unit traffic is $1/\psi_n$.

Let $\Omega_n$ denote the resource capacity of SBS $n$. Then we have $y_n \leq \Omega_n \psi_n$, $\forall n \in \mathcal{N}$. In addition, denote $w_n > 0$ as SBS $n$'s average revenue achieved from one unit of its bandwidth resource consumption caused by its own traffic demands. Let $c_n$ $(0 < c_n < w_n)$ represent SBS $n$'s average cost on one unit of bandwidth utilization. Then the expected revenue of SBS $n$ resulting from serving its own traffic demands is given by

$$P_n(\Omega_n) = (w_n - c_n) E\left(\frac{x_n}{\psi_n}\right)$$

$$= (w_n - c_n)\left[\int_0^{\Omega_n \psi_n} \frac{x}{\psi_n} f_n(x)\, dx + \int_{\Omega_n \psi_n}^{\infty} \Omega_n f_n(x)\, dx\right] \tag{3.5}$$

$$= a_n\left(1 - e^{-\frac{\Omega_n \psi_n}{\lambda_n}}\right),$$

where $a_n = \lambda_n (w_n - c_n)/\psi_n$, and $f_n(x) = \lambda_n^{-1} e^{-\lambda_n^{-1} x}$.

Furthermore, given feasible amount of traffic offloading by SBS $n$, the expected revenue from the rest bandwidth resource for serving this SBS's own traffic demands can be obtained as

$$P_n\left(\Omega_n - \frac{y_n}{\psi_n}\right) = a_n\left(1 - e^{-\frac{\Omega_n \psi_n - y_n}{\lambda_n}}\right). \tag{3.6}$$

Then the utility of SBS $n$ from traffic offloading is given by

$$V_n = P_n\left(\Omega_n - \frac{y_n}{\psi_n}\right) + T_n(y_n) - c_n y_n/\psi_n, \quad \forall n \in \mathcal{N}. \tag{3.7}$$

In addition, we define the net utility of SBS $n$ as the SBS utility improvement when offloading traffic for the MBS:

$$V'_n = V_n - P_n(\Omega_n), \quad \forall n \in \mathcal{N}. \tag{3.8}$$

In (3.7), we assume that the total revenue of SBS $n$:

$$\upsilon_n(y_n) = P_n\left(\Omega_n - \frac{y_n}{\psi_n}\right) + T_n(y_n) \tag{3.9}$$

is a strictly increasing concave function of $y_n$, i.e., $\upsilon'(y_n) > 0$, and $\upsilon''(y_n) < 0$. This setting is reasonable, due to that as the amount of offloading traffic increasing, payment $T(y)$ from the MBS increases slowly, and meanwhile the income brought to the SBSs grows slowly, which also results from less service for SBSs' own traffic requests. This property of revenue function will be further analyzed in Sect. 3.5.

**Social Welfare**  The social welfare of HetUDN is defined as the aggregate utility of the MBS and SBSs, denoted by

$$
\begin{aligned}
W &= \sum_{n \in \mathcal{N}} U_n + \sum_{n \in \mathcal{N}} V_n \\
&= \sum_{n \in \mathcal{N}} \left[ \delta \sum_{i \in \mathcal{M}_n} r_i - T_n\left(y_n\right) \right] \\
&\quad + \sum_{n \in \mathcal{N}} \left[ P_n\left(\Omega_n - \frac{y_n}{\psi_n}\right) + T_n\left(y_n\right) - \frac{c_n y_n}{\psi_n} \right] \\
&= \underbrace{\delta \sum_{i \in \mathcal{M}} r_i}_{\substack{\text{MBS: Profit from} \\ \text{MUEs' throughput}}} + \underbrace{\sum_{n \in \mathcal{N}} P_n\left(\Omega_n - \frac{y_n}{\psi_n}\right)}_{\substack{\text{SBS: Profit from serving} \\ \text{its own traffic demands}}} - \underbrace{\sum_{n \in \mathcal{N}} \frac{c_n y_n}{\psi_n}}_{\substack{\text{SBS: Cost} \\ \text{of offloading}}}.
\end{aligned}
\tag{3.10}
$$

## 3.4  Contract Design for Traffic Offloading

According to contract theory, a reasonable definition of SBS's type is very important to realize the contract-based traffic offloading. So first of all, we propose a new definition of the *SBS type* for traffic offloading in the HetUDN as Definition 3.1, based on the models established previously.

**Definition 3.1  SBS Type:**  In the HetUDN with multiple SBSs, the definition of SBS $n$'s type, which is determined by SBS's transmission efficiency $\psi_n$, resource capability $\Omega_n$, average revenue achieved from per unit of its bandwidth resource consumption caused by SUEs' traffic demands $w_n$, average cost on one unit of bandwidth utilization $c_n$ and the arrival rate of SUEs' traffic requests $\lambda_n = \lambda$, is given by

$$
\theta_n = \frac{\psi_n}{c_n + (w_n - c_n)\, e^{-\frac{\Omega_n \psi_n}{\lambda}}}.
\tag{3.11}
$$

*Remark 3.1*  Notice that the definition of the SBS type is reasonable since that (3.11) gives an index which can reflect the SBS's capability of providing traffic offloading service for the MBS. To be specific, Definition 3.1 indicates that a larger value of SBS type $\theta_n$, which means a smaller $c_n/\psi_n$ (SBS $n$'s cost by one unit of traffic transmission), smaller $(w_n - c_n)/\psi_n$ (SBS $n$'s net benefits from one unit of traffic transmission for SUEs), larger $\Omega_n \psi_n$ (SBS $n$'s maximum resource can be provided for transmission) and lower $\lambda$ (SBS $n$'s traffic load from its own users), indicates a stronger capability of providing the traffic offloading service for the MBS.

According to Definition 3.1, each of the $N$ SBSs in the HetUDN belongs to one of the $N$ types. In the SDWN-based HetUDN, the SDWN controller needs to design a bundle of contracts $\{\mathbf{T}(\mathbf{y}), \mathbf{y}\}$ for these $N$ types of SBSs. Consequently, based on the definitions above, the traffic offloading contract for SBS $n$ with type $\theta_n$ can be expressed by $\{T_n(y_n), y_n\}$. Next, we will introduce necessary principles that ensure a contract to be valid and feasible.

### 3.4.1 Contract Design with Information Asymmetry

#### 3.4.1.1 Individual Rationality (IR)

No matter whether the MBS and the SDWN controller can identify the types of SBSs, the designed traffic offloading contract must ensure that every SBS has an incentive to provide the traffic offloading service for MUEs. Therefore, the following *Individual Rationality (IR)* constraint must be satisfied when designing the contracts.

**Definition 3.2 Individual Rationality (IR):** Any type of SBSs in the HetUDN will only select the traffic offloading contract that can guarantee that the utility received is not less than its utility can be received when it does not provide the traffic offloading service, i.e., $\forall n = 1, 2, \cdots, N$,

$$V_n = P_n \left( \Omega_n - \frac{y_n}{\psi_n} \right) + T_n(y_n) - \frac{c_n y_n}{\psi_n} > P_n(\Omega_n). \qquad (3.12)$$

#### 3.4.1.2 Incentive Compatibility (IC)

Under the situation with information asymmetry, SBSs tends to request high payment and provide the traffic offloading service as little as possible, according to (3.7). To ensure that every SBS will select the right contract designed for its type specially, the designed bundle of contracts must make sure that the maximum utility can be achieved if and only if the SBS selects the contract for its type specially. This principle of contract designing is called *Incentive Compatibility (IC)*, which is defined as Definition 3.3.

**Definition 3.3 Incentive Compatibility (IC):** Any type of SBSs in the HetUDN will obtain the maximum utility if and only if it selects the contract for its own type specially. In other words, selecting the traffic offloading contract designed for its type will bring to this SBS more utility than any other contracts in the contract bundle, i.e., $\forall n, m = 1, 2, \cdots, N$,

$$
\begin{aligned}
V_n (T_n, y_n) &= P_n \left( \Omega_n - \frac{y_n}{\psi_n} \right) + T_n (y_n) - \frac{c_n y_n}{\psi_n} \\
&\geq V_n (T_m, y_m) = P_n \left( \Omega_n - \frac{y_m}{\psi_n} \right) + T_m (y_m) - \frac{c_n y_m}{\psi_n}.
\end{aligned}
\tag{3.13}
$$

Due to the case of information asymmetry, the types of SBSs cannot be accessed by the ANDSF. However, the knowledge of the probability $\pi_n$, with which an SBS might belong to type $\theta_n$, is available for the SDWN controller, and $\sum_{n \in \mathcal{N}} \pi_n = 1$. Therefore, with the IR and IC constraints, the SDWN controller will formulate the bundle of traffic offloading contracts which will maximize the MBS's utility. Then the contract-based traffic offloading optimization problem in the scenario with information asymmetry is formulated as

$$
\max \ U^* (\mathbf{s}, \mathbf{y}, \mathbf{T}(y)) = \sum_{n \in \mathcal{N}} \pi_n \left[ \delta \sum_{i \in \mathcal{M}_n} r_i - T_n (y_n) \right],
\tag{3.14a}
$$

$$
\text{s.t.} \ s_{0i} + \sum_{n \in \mathcal{N}} s_{ni} = 1, \quad \forall i = 1, 2, \cdots, M,
\tag{3.14b}
$$

$$
y_n = \sum_{i \in \mathcal{M}_n} r_{ni} s_{ni} \geq 0, \quad \forall n = 1, 2, \cdots, N,
\tag{3.14c}
$$

$$
\Omega_n - \frac{y_n}{\psi_n}, \quad \forall n = 1, 2, \cdots, N,
\tag{3.14d}
$$

$$
T_n (y_n) - \frac{c_n y_n}{\psi_n} - a_n e^{-\frac{\Omega_n \psi_n}{\lambda_n}} \left( e^{\frac{y_n}{\lambda_n}} - 1 \right) \geq 0,
$$
$$
\forall n = 1, 2, \cdots, N, \quad \text{(IR)}
\tag{3.14e}
$$

$$
P_n \left( \Omega_n - \frac{y_n}{\psi_n} \right) + T_n (y_n) - \frac{c_n y_n}{\psi_n}
$$
$$
\geq P_n \left( \Omega_n - \frac{y_m}{\psi_n} \right) + T_m (y_m) - \frac{c_n y_m}{\psi_n},
$$
$$
\forall n, m = 1, 2, \cdots, N, \quad \text{(IC)}
\tag{3.14f}
$$

$$
y_n \geq 0, \quad \forall n = 1, 2, \cdots, N.
\tag{3.14g}
$$

The feasibility conditions of the traffic offloading contacts formulated in (3.14) will be analyzed and derived in Sect. 3.5.

### 3.4.2  Contract Design Without Information Asymmetry

Without information asymmetry, the IC constraint is unnecessary because any SBS cannot be disguised as other types of SBSs. Then the optimization problem of traffic offloading processed by the SDWN controller can be formulated as (3.14), with the IC constraint being removed.

We provide the optimal traffic offloading solution for the HetUDN as Lemma 3.1.

**Lemma 3.1** *Without information asymmetry, the optimal traffic offloading contract for type $\theta_n$ ($\forall n = 1, 2, \cdots, N$), which is defined by (3.11), is given by*

$$y_n^{upper} = \lambda_n \left[ \ln \left( \delta \psi_n - c_n \right) - \ln \left( w_n - c_n \right) \right] + \Omega_n \psi_n, \tag{3.15a}$$

$$T_n^{upper} (y_n) = \frac{c_n y_n^{upper}}{\psi_n} + a_n e^{-\frac{\Omega_n \psi_n}{\lambda_n}} \left( e^{\frac{y_n^{upper}}{\lambda_n}} - 1 \right). \tag{3.15b}$$

**Proof** The objective function (3.14a) can be rewritten as

$$U^M = \sum_{n \in \mathcal{N}} \pi_n \left[ \delta \sum_{i \in \mathcal{M}_n, s_{0i}=1} r_i + \delta y_n - T_n (y_n) \right]. \tag{3.16}$$

Consider that the SBSs are selfish and with the IR constraint same as (3.14e) in the main text, we have

$$T_n (y_n) = \lambda_n \varphi_n \left( e^{\frac{y_n}{\lambda_n}} - 1 \right) + \frac{c_n}{y_n} \psi_n. \tag{3.17}$$

Replace $T_n (y_n)$ in (3.16) with (3.17), and take the first derivative:

$$\frac{\partial U^M}{\partial y_n} = \pi_n \left( \delta - \varphi_n e^{\frac{y_n}{\lambda_n}} - \frac{c_n}{y_n} \psi_n \right). \tag{3.18}$$

Then we get the optimal amount of traffic offloaded by SBS $n$ under the non-information asymmetry situation as

$$y_n^{upper} = \lambda_n \left[ \ln \left( \delta - \frac{c_n}{\psi_n} \right) - \ln \varphi_n \right]. \tag{3.19}$$

The same result can also be obtained through the Lagrange function approach.
This completes the proof of Lemma 3.1.

*Remark 3.2* Notice that the value of social welfare in (3.10) is equal to the MBS's utility. In addition, all SBSs receive zero net utility due to the selfish property of the MBS, who tries to extract as much profit from SBSs' offloading as possible when satisfying the IR constraint shown as (3.14e). Solutions given in Lemma 3.1 provide the first best contract solution for the traffic offloading problem, since both the social welfare and MBS utility are maximized and achieve the Pareto efficiency.

### 3.4.3   Contract Design by Linear Pricing

Linear pricing based contracts are designed for the scenario with information asymmetry. To operate a linear pricing based contract, without the IC constraint and the discrimination pricing principle, the SDWN controller designs a optimal payment $\beta^*$ to optimize the MBS utility without the IC constraint, and then requests the MBS to pay $\beta^*$ for every SBS equally for one unit of offloaded traffic. In other words, the SBS requesting more offloading traffic will get more payment linearly. To maximize the SBS utility, every SBS tends to request an appropriate amount of offloading traffic $y_n^{lower}$. We provide the optimal traffic offloading contract selected by the SBS and the optimal unit-price $\beta^*$ in Lemma 3.2.

**Lemma 3.2** *With information asymmetry, the optimal traffic offloading contract for type $\theta_n$ ($\forall n = 1, 2, \cdots, N$) under the linear pricing rule is given by*

$$y_n^{lower} = \lambda_n \left[ \ln \left( \beta^* \psi_n - c_n \right) - \ln \left( w_n - c_n \right) \right] + \Omega_n \psi_n, \tag{3.20a}$$

$$T_n^{lower} = \beta^* y_n^{lower}, \tag{3.20b}$$

*where $\beta^*$, designed by the SDWN controller to maximize the MBS utility, is the solution of the following equation:*

$$(\delta - \beta) \left( \beta - \frac{c_n}{\psi_n} \right)^{-1} = \ln \left( \frac{\psi_n \beta - c_n}{w_n - c_n} \right) + \frac{\Omega_n \psi_n}{\lambda_n}. \tag{3.21}$$

***Proof*** Let $\beta$ denote the payment of one unit of offloaded traffic for every SBS $n = 1, 2, \ldots, N$. Under the linear pricing contract situation, the utility of SBS is

$$V_n = a_n \left( 1 - e^{-\frac{\Omega_n \psi_n - y_n}{\lambda_n}} \right) + \beta y_n - \frac{c_n}{y_n} \psi_n. \tag{3.22}$$

Take the first derivative of $V_n$ and we get

$$\frac{\partial V_n}{\partial y_n} = -\varphi_n e^{\frac{y_n}{\lambda_n}} + \beta - \frac{c_n}{\psi_n}. \tag{3.23}$$

To maximize the SBS utility under the linear pricing contract, the optimal offloading requested by SBS $n$ can be obtained as

$$y_n^{lower} = \lambda_n \left[ \ln \left( \beta - \frac{c_n}{\psi_n} \right) - \ln \varphi_n \right]. \tag{3.24}$$

Replace $y_n$ in (3.16) with (3.24), and the MBS utility is given by

$$U(\beta) = \sum_{n \in \mathcal{N}} \pi_n \left[ \delta \sum_{i \in \mathcal{M}_n, s_{0i}=1} r_i + (\delta - \beta) y_n \right]. \tag{3.25}$$

Take the first derivative with respect to $\beta$, and then we can get the optimal $\beta^*$ as the solution of the following equation:

$$(\delta - \beta) \left( \beta - \frac{c_n}{\psi_n} \right)^{-1} = \ln \left( \beta - \frac{c_n}{\psi_n} \right) - \ln(\varphi_n). \tag{3.26}$$

This completes the proof of Lemma 3.2.

*Remark 3.3* The contract-based traffic offloading designed above is feasible and can be realized under the SDWN framework. The required status information in Definitions 3.1, 3.2 and 3.3 is obtained through the ANDSF, and contracts satisfying IR and IC are designed by the SDWN controller. However, enough computing capacity of the SDWN controller and corresponding modifications of the interface and interaction protocols are still necessary to realize the contract-based traffic offloading.

## 3.5 Conditions for Contract Feasibility

First, we propose the following Lemma 3.3 which provides the condition that ensures the increasing concave property of revenue function $v_n(y_n)$ defined as (3.9).

**Lemma 3.3** *In a traffic offloading system with a set $\mathcal{N}$ of SBSs indicated by $n = 1, 2, \cdots, N$. The arrival rate of SBS's own traffic requests is $\lambda_n = \lambda, \forall n \in \mathcal{N}$. Define*

$$\varphi_n = \frac{w_n - c_n}{\psi_n} e^{-\frac{\Omega_n \psi_n}{\lambda}}. \tag{3.27}$$

*With a bundle of traffic offloading contracts satisfying IR an IC conditions, the traffic offloading allocated to SBS n is $y_n$, $\forall n \in \mathcal{N}$. Given $y_n \leq y_m$, $(n, m \in \mathcal{N})$, if $\varphi_n \leq \varphi_m$, then the revenue function shown in (3.9) is a strictly increasing concave function of the amount of traffic offloading allocated.*

**Proof** Take the first derivative of $v_n$ in (3.9) and we get

$$\frac{\partial v_n(y_n)}{\partial y_n} = \frac{w_n - c_n}{\psi_n} e^{-\frac{\Omega_n \psi_n - y_n}{\lambda}} > 0. \tag{3.28}$$

Therefore, revenue function $v_n(y_n)$ is a strictly increasing function of $y_n$, $\forall n \in \mathcal{N}$.

Given $y_n \leq y_m$, and according to IC conditions, the revenue margin between SBS $n$ and SBS $m$ can be calculated by

$$v_m(y_m) - v_n(y_n)$$

$$= a_m\left(1 - e^{-\frac{\Omega_m \psi_m - y_m}{\lambda}}\right) + T_m(y_m) - \frac{c_m y_m}{\psi_m} + \frac{c_m y_m}{\psi_m} - \left[a_n\left(1 - e^{-\frac{\Omega_n \psi_n - y_n}{\lambda}}\right) + T_n(y_n)\right]$$

$$\geq a_m\left(1 - e^{-\frac{\Omega_m \psi_m - y_n}{\lambda}}\right) + T_n(y_n) - \frac{c_m y_n}{\psi_m} + \frac{c_m y_m}{\psi_m} - \left[a_n\left(1 - e^{-\frac{\Omega_n \psi_n - y_n}{\lambda}}\right) + T_n(y_n)\right]$$

$$= a_m\left(1 - e^{-\frac{\Omega_m \psi_m - y_n}{\lambda}}\right) - a_n\left(1 - e^{-\frac{\Omega_n \psi_n - y_n}{\lambda}}\right) + \frac{c_m}{\psi_m}(y_m - y_n)$$

$$\geq a_m\left(1 - e^{-\frac{\Omega_m \psi_m - y_n}{\lambda}}\right) - a_n\left(1 - e^{-\frac{\Omega_n \psi_n - y_n}{\lambda}}\right) \triangleq F_1(y_n).$$

Take the first derivative of $F_1(y_n)$, and then we get

$$\frac{\partial F_1(y_n)}{\partial y_n} = \left(\frac{w_n - c_n}{\psi_n} e^{-\frac{\Omega_n \psi_n}{\lambda}} - \frac{w_m - c_m}{\psi_m} e^{-\frac{\Omega_m \psi_m}{\lambda}}\right) e^{\frac{y_n}{\lambda}}.$$

When $\varphi_n \leq \varphi_m$, then we have $\partial F_1(y_n)/\partial y_n \leq 0$, which reflects that with $y_n$ and $y_m$ increasing, the revenue margin between $y_n$ and $y_m$ tends to be smaller. Consequently, the revenue function shown in (3.9) is a strictly increasing concave function of the amount of traffic offloading provided by SBSs. This completes the proof of Lemma 3.3.

A feasible traffic offloading contract for the information-asymmetry situation must ensure that without the knowledge of SBS types, all SBSs can receive maximum net utility only if they select the right contracts designed for their types. Based on Lemma 3.3, the following Theorem 3.1 proposes the monotonic property of SBS's offload amount, payment, and net utility. Theorem 3.1 demonstrates the feasibility of the proposed contract based traffic offloading and resource allocation method in Sect. 3.4.1 for the HetUDN with different types of SBSs.

**Theorem 3.1** *Monotonicity: In an SDWN-based HetUDN with $N$ heterogeneous SBSs, the type of each SBS $\theta_n$ ($n \in \mathcal{N}$) is defined by Definition 3.1. Without the information of SBS types, the SDWN controller designs a bundle of traffic offloading contracts $\{T(y), y\}$ for these $N$ types of SBSs and the MBS, according to the optimization problem formulated as (3.14). Consider that the arrival rates of traffic requests from SUEs are equal for every SBS, i.e., $\lambda_n = \lambda$, $\forall n$. Then for each contract $\{T_n(y_n), y_n\}$, the amount of traffic offload $y$ allocated to each SBS and payment $T(y)$ obtained by (3.14) have the monotonicity. Specifically, if and only if $\theta_1 < \theta_2 < \cdots < \theta_N$,*

$$y_1 < y_2 < \cdots < y_N, \tag{3.29a}$$

$$T_1(y_1) < T_2(y_2) < \cdots < T_N(y_N), \tag{3.29b}$$

$$V'_1 < V'_2 < \cdots < V'_N. \tag{3.29c}$$

**Proof** We first prove that $y_1 < y_2 < \cdots < y_N$ if and only if $\theta_1 < \theta_2 < \cdots < \theta_N$. According to the IC constraints in (3.14f), we have $\forall n, m = 1, 2, \cdots, N$,

$$
a_n \left( 1 - e^{-\frac{\Omega_n \psi_n - y_n}{\lambda}} \right) + T_n \left( y_n \right) - \frac{c_n y_n}{\psi_n}
$$
$$
\geq a_n \left( 1 - e^{-\frac{\Omega_n \psi_n - y_m}{\lambda}} \right) + T_m \left( y_m \right) - \frac{c_n y_m}{\psi_n}, \tag{3.30a}
$$

$$
a_m \left( 1 - e^{-\frac{\Omega_m \psi_m - y_m}{\lambda}} \right) + T_m \left( y_m \right) - \frac{c_m y_m}{\psi_m}
$$
$$
\geq a_m \left( 1 - e^{-\frac{\Omega_m \psi_m - y_n}{\lambda}} \right) + T_n \left( y_n \right) - \frac{c_m y_n}{\psi_m}. \tag{3.30b}
$$

**Necessity** Consider that $0 \leq y_n \leq y_m$ ($\forall n, m \in \mathcal{N}, n \neq m$). For the concave property of the revenue function, the condition of $\varphi_n \leq \varphi_m$ is satisfied according to Lemma 3.3, and $\varphi_n = \varphi_m$ if and only if $y_n = y_m$. Then add the two inequalities above in (3.30) together and then we get the following inequality

$$
0 \leq \left( a_m e^{-\frac{\Omega_m \psi_m}{\lambda}} - a_n e^{-\frac{\Omega_n \psi_n}{\lambda}} \right) \left( e^{\frac{y_m}{\lambda}} - e^{\frac{y_n}{\lambda}} \right)
$$
$$
= \lambda \left( \varphi_m - \varphi_n \right) \left( e^{\frac{y_m}{\lambda}} - e^{\frac{y_n}{\lambda}} \right) \leq \left( \frac{c_n}{\psi_n} - \frac{c_m}{\psi_m} \right) \left( y_m - y_n \right). \tag{3.31}
$$

For $0 \leq y_n \leq y_m$, the following inequality is always satisfied:

$$
e^{\frac{y_m}{\lambda}} - e^{\frac{y_n}{\lambda}} \geq \frac{y_m - y_n}{\lambda} \geq 0. \tag{3.32}
$$

According to (3.32) and (3.31) can be transformed to

$$
\frac{1}{\lambda} \left( a_m e^{-\frac{\Omega_m \psi_m}{\lambda}} - a_n e^{-\frac{\Omega_n \psi_n}{\lambda}} \right) \leq \frac{c_n}{\psi_n} - \frac{c_m}{\psi_m}, \tag{3.33}
$$

which can be further derived as

$$
\frac{a_m}{\lambda} e^{-\frac{\Omega_m \psi_m}{\lambda}} + \frac{c_m}{\psi_m} \leq \frac{a_n}{\lambda} e^{-\frac{\Omega_n \psi_n}{\lambda}} + \frac{c_n}{\psi_n}
$$
$$
\Rightarrow \frac{w_m - c_m}{\psi_m} e^{-\frac{\Omega_m \psi_m}{\lambda}} + \frac{c_m}{\psi_m} \leq \frac{w_n - c_n}{\psi_n} e^{-\frac{\Omega_n \psi_n}{\lambda}} + \frac{c_n}{\psi_n}, \tag{3.34}
$$

which is equal to

$$
\frac{\psi_n}{c_n + (w_n - c_n) e^{-\frac{\Omega_n \psi_n}{\lambda}}} \leq \frac{\psi_m}{c_m + (w_m - c_m) e^{-\frac{\Omega_m \psi_m}{\lambda}}}. \tag{3.35}
$$

According to the definition of $\theta$, we can get $\theta_n \leq \theta_m$, and $\theta_n = \theta_m$ if and only if $y_n = y_m$.

**Sufficiency**   Consider Definition 3.1, $0 < \theta_n \leq \theta_m$ is equal to

$$\frac{c_n}{\psi_n} + \frac{w_n - c_n}{\psi_n} e^{-\frac{\Omega_n \psi_n}{\lambda}} \geq \frac{c_m}{\psi_m} + \frac{w_m - c_m}{\psi_m} e^{-\frac{\Omega_m \psi_m}{\lambda}}, \tag{3.36}$$

which can be written as

$$\left( \frac{w_m - c_m}{\psi_m} e^{-\frac{\Omega_m \psi_m}{\lambda}} - \frac{w_n - c_n}{\psi_n} e^{-\frac{\Omega_n \psi_n}{\lambda}} \right) + \left( \frac{c_m}{\psi_m} - \frac{c_n}{\psi_n} \right) \leq 0. \tag{3.37}$$

Hypothesise $y_n > y_m > 0$, then $\varphi_n > \varphi_m$, and inequality

$$\lambda \left( e^{\frac{y_n}{\lambda}} - e^{\frac{y_m}{\lambda}} \right) > y_n - y_m > 0 \tag{3.38}$$

is always satisfied. Then (3.37) can be further derived as

$$0 \leq \left( \frac{w_n - c_n}{\psi_n} e^{-\frac{\Omega_n \psi_n}{\lambda}} - \frac{w_m - c_m}{\psi_m} e^{-\frac{\Omega_m \psi_m}{\lambda}} \right) (y_n - y_m)$$
$$+ \left( \frac{c_n}{\psi_n} - \frac{c_m}{\psi_m} \right) (y_n - y_m) \tag{3.39}$$

$$< \lambda \left( \frac{w_n - c_n}{\psi_n} e^{-\frac{\Omega_n \psi_n}{\lambda}} - \frac{w_m - c_m}{\psi_m} e^{-\frac{\Omega_m \psi_m}{\lambda}} \right) \left( e^{\frac{y_n}{\lambda}} - e^{\frac{y_m}{\lambda}} \right)$$
$$+ \left( \frac{c_n}{\psi_n} - \frac{c_m}{\psi_m} \right) (y_n - y_m). \tag{3.40}$$

However, according to IC constraints and adding (3.30a) and (3.30b) together, then we have

$$\left( a_m e^{-\frac{\Omega_m \psi_m}{\lambda}} - a_n e^{-\frac{\Omega_n \psi_n}{\lambda}} \right) \left( e^{\frac{y_m}{\lambda}} - e^{\frac{y_n}{\lambda}} \right) \leq \left( \frac{c_n}{\psi_n} - \frac{c_m}{\psi_m} \right) (y_m - y_n). \tag{3.41}$$

Considering $y_n > y_m > 0$, (3.41) can be transformed as

$$\left( a_n e^{-\frac{\Omega_n \psi_n}{\lambda}} - a_m e^{-\frac{\Omega_m \psi_m}{\lambda}} \right) \left( e^{\frac{y_n}{\lambda}} - e^{\frac{y_m}{\lambda}} \right) + \left( \frac{c_n}{\psi_n} - \frac{c_m}{\psi_m} \right) (y_n - y_m) \leq 0.$$

As $a_n = \lambda (w_n - c_n)/\psi_n$, we can get

$$\lambda \left( \frac{w_n - c_n}{\psi_n} e^{-\frac{\Omega_n \psi_n}{\lambda}} - \frac{w_m - c_m}{\psi_m} e^{-\frac{\Omega_m \psi_m}{\lambda}} \right) \left( e^{\frac{y_n}{\lambda}} - e^{\frac{y_m}{\lambda}} \right)$$
$$+ \left( \frac{c_n}{\psi_n} - \frac{c_m}{\psi_m} \right) (y_n - y_m) \leq 0, \tag{3.42}$$

which is a contradiction with (3.39). Therefore, the hypothesis $y_n > y_m$ is invalid, which means that $y_n \leq y_m$ if $\theta_n \leq \theta_m$.

Then we have demonstrated the proposition that $y_n < y_m$ if and only if $\theta_n < \theta_m$, and $y_n = y_m$ if and only if $\theta_n = \theta_m$. Next, we will prove that $T_1 (y_1) < T_2 (y_2) < \cdots < T_N (y_N)$ if and only if $y_1 < y_2 < \cdots < y_N$.

**Sufficiency** $\forall n, m = 1, 2, \cdots, N, n \neq m$, we have (3.30b) according to IC constraints, which can be transformed to

$$T_n (y_n) - T_m (y_m) \leq a_m e^{-\frac{\Omega_m \psi_m}{\lambda}} \left( e^{\frac{y_n}{\lambda}} - e^{\frac{y_m}{\lambda}} \right) + \frac{c_m}{\psi_m} (y_n - y_m),$$

and then we get $T_n (y_n) \leq T_m (y_m)$ if $y_n \leq y_m$.

**Necessity** Inequality (3.30a) obtained by the IC constraints can be transformed to

$$a_n e^{-\frac{\Omega_n \psi_n}{\lambda}} \left( e^{\frac{y_n}{\lambda}} - e^{\frac{y_m}{\lambda}} \right) + \frac{c_n}{\psi_n} (y_n - y_m) \leq T_n (y_n) - T_m (y_m).$$

Given $T_n (y_n) \leq T_m (y_m)$, the left part of the inequality above can be written by

$$a_n e^{-\frac{\Omega_n \psi_n}{\lambda}} \left( e^{\frac{y_n}{\lambda}} - e^{\frac{y_m}{\lambda}} \right) + \frac{c_n}{\psi_n} (y_n - y_m) \leq 0, \tag{3.43}$$

which can be satisfied only by $0 \leq y_n \leq y_m$.

Then we have demonstrated the proposition that $T_n (y_n) < T_m (y_m)$ if and only if $y_n < y_m$, and $T_n (y_n) = T_m (y_m)$ if and only if $y_n = y_m$. Due to the transferability of the necessary and sufficient conditions, $T_n (y_n) < T_m (y_m)$ if and only if $\theta_n < \theta_m$, and $T_n (y_n) = T_m (y_m)$ if and only if $\theta_n = \theta_m$.

Last, we will prove the monotonicity of SBS's net utility. According to (3.7), (3.8) and the IC constraints in Definition 3.3, the net utility difference between SBS $n$ and SBS $m$ ($\forall n, m = 1, 2, \cdots, N, n \neq m$) can be calculated as

$$V'_m - V'_n = V_m - P_m(\Omega_m) - (V_n - P_n(\Omega_n))$$

$$= a_m \left(1 - e^{-\frac{\Omega_m \psi_m - y_m}{\lambda}}\right) + T_m(y_m) - \frac{c_m y_m}{\psi_m} - P_m(\Omega_m)$$

$$- \left[a_n \left(1 - e^{-\frac{\Omega_n \psi_n - y_n}{\lambda}}\right) + T_n(y_n) - \frac{c_n y_n}{\psi_n}\right] + P_n(\Omega_n) \tag{3.44}$$

$$\geq a_m \left(1 - e^{-\frac{\Omega_m \psi_m - y_n}{\lambda}}\right) + T_n(y_n) - \frac{c_m y_n}{\psi_m} - P_m(\Omega_m)$$

$$- \left[a_n \left(1 - e^{-\frac{\Omega_n \psi_n - y_n}{\lambda}}\right) + T_n(y_n) - \frac{c_n y_n}{\psi_n}\right] + P_n(\Omega_n)$$

$$= a_m \left(1 - e^{-\frac{\Omega_m \psi_m - y_n}{\lambda}}\right) - a_n \left(1 - e^{-\frac{\Omega_n \psi_n - y_n}{\lambda}}\right)$$

$$+ \left(\frac{c_n}{\psi_n} - \frac{c_m}{\psi_m}\right) y_n - P_m(\Omega_m) + P_n(\Omega_n) \triangleq F_2(y_n) \tag{3.45}$$

Consider that $\theta_n < \theta_m$, then $y_n < y_m$ and $\varphi_n < \varphi_m$ according to (3.29a) in Theorem 3.1 proved previously and Lemma 3.3, respectively. Let $y_n = 0$ in $F_2(y_n)$, and according to the results of $P_m(\Omega_m)$ and $P_m(\Omega_m)$ calculated by (3.5), we have

$$F_2(0) = a_m \left(1 - e^{-\frac{\Omega_m \psi_m}{\lambda}}\right) - a_n \left(1 - e^{-\frac{\Omega_n \psi_n}{\lambda}}\right) - P_m(\Omega_m) + P_n(\Omega_n)$$
$$= 0. \tag{3.46}$$

According to the expression of $\theta_n$ defined in Definition 3.1 and considering that $y_n > 0$, the first derivative of $F_2(y_n)$ with respect to $y_n$ can be written as

$$\frac{\partial F_2(y_n)}{\partial y_n} = \left(\frac{c_n}{\psi_n} - \frac{c_m}{\psi_m}\right) + (\varphi_n - \varphi_m) e^{\frac{y_n}{\lambda}}$$

$$\geq \left(\frac{c_n}{\psi_n} - \frac{c_m}{\psi_m}\right) + (\varphi_n - \varphi_m) \tag{3.47}$$

$$= \left(\frac{c_n}{\psi_n} + \varphi_n\right) - \left(\frac{c_m}{\psi_m} + \varphi_m\right) = \frac{1}{\theta_n} - \frac{1}{\theta_m} > 0.$$

The necessity of (3.29c) can be proved by applying the reduction to absurdity. Since the proving idea is similar to (3.44)–(3.47), we omit the proof of necessity for (3.29c). Therefore, we have $V'_n < V'_m$, if and only if $\theta_n < \theta_m$, and $V'_n = V'_m$, if and only if $\theta_n = \theta_m$, $\forall n, m \in \mathcal{N}, n \neq m$. This completes the proof of Theorem 3.1.

*Remark 3.4*

*(1)* ***Valid of SBS type definition:*** Theorem 3.1 demonstrates that SBS type $\theta_n$ proposed and defined in Definition 3.1 is reasonable, since it can effectively reflect the influence of heterogeneous SBSs' performance and capacity on the contract designed by a competitive market based economics theory.

*(2)* ***Fairness and monotonicity:*** Theorem 3.1 demonstrate that, for both of the service requester and service providers, i.e., the MBS and SBSs, respectively, the proposed contract-based traffic offloading and resource allocation mechanism as (3.14) guarantees the fairness and incentive property of the transmission resource market, in the scenario of information asymmetry and that service providers are heterogeneous. On the one hand, monotonicity of (3.29a) and (3.29b) implies that for the SBSs with higher $\theta$, they are more suitable for offloading traffic, and their best choice to achieve highest payoff is offloading larger amount of traffic. Meanwhile, they will receive more payment. This contract principle can ensure the fairness among the heterogeneous SBSs. On the other hand, monotonicity also provides an incentive for SBSs. Specifically, if a high type of SBS selects the contract designed for low types of SBSs, even though a small amount of traffic offloading will be requested by the SDWN controller, the corresponding low payment will deteriorate this high-type SBS's payment.

*(3)* ***Incentive Compatibility*** Monotonicity of (3.29c) also implies that the incentive for SBSs is compatible, which means that SBSs with high capability will receive more net utility than low ones. For those SBSs whose types cannot be aware of the MBS and SDWN controller, the designed contract is self-revealing for SBSs, since that each type of SBS will receive the maximum net utility, which reflects the net revenue by offloading, if and only if it selects the right traffic offloading contract designed exactly for its type.

## 3.6 Simulation Results

In this part, we will use MATLAB 2016b to evaluate the proposed contract-based traffic offloading and resource allocation. First of all, we introduce the scenario setup of the simulations. In the following simulations, we assume a typical 4G/5G macrocell with a transmission radius of 500 m. The HetUDN consists of one MBS, $N = 100$ heterogeneous SBSs with $N = 100$ different types, and $M = 250$ MUEs. Both SBSs and MUEs are randomly distributed within the macrocell. The distribution of network elements in the simulation is shown as Fig. 3.2. In addition, we set $c_n = 0.6$, $w_n = 1$ and $\delta = 1$. The other main parameters of the HetUDN are shown in Table 3.1.

To demonstrate monotonicity and incentive compatibility of the contract, the indexes of SBSs are sorted according to their values of type obtained by Definition 3.1. By applying the three different contracts designed in Sect. 3.4, we obtain

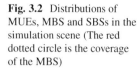

**Fig. 3.2** Distributions of MUEs, MBS and SBSs in the simulation scene (The red dotted circle is the coverage of the MBS)

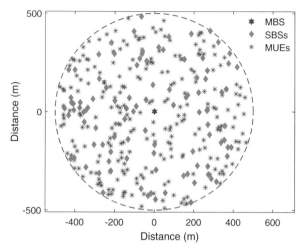

**Table 3.1** Simulation parameters

| System parameters | Value setting |
|---|---|
| Transmission power of MBS | 46 dBm |
| Transmission power of SBSs | $\sim U$ [15, 35] dBm |
| Path loss of MBS | $28.3 + 22.0\log_{10}l$, $l$ (km) |
| Path loss of SBSs | $30.5 + 36.7\log_{10}l$, $l$ (km) |
| MBS / SBS bandwidth | 20 MHz |
| MBS / SBS operating frequency | 2.6 GHz/2.4 GHz |
| SBSs' own traffic requests arrival rate | $\lambda = 10$ Mbps |
| Power spectral density of thermal noise | $-174$ dBm/Hz |

the amount of traffic offloading requested by SBSs and payments required to the MBS, which are shown as Fig. 3.3a and b, respectively. In Fig. 3.3, results illustrate that both the amount of traffic offloading and payment increase with the value of SBS type increasing, for the three different contracts, which reflects the fairness of the contracts. In addition, among the three contracts, the no information asymmetry contract requires the highest amount of traffic offloading and the highest payment for SBSs, followed by the incentive contract proposed in Sect. 3.4.1. The lowest traffic offloading and payment are requested by the linear pricing contract.

Moreover, the incentive compatibility of the contract designed in Sect. 3.4.1 is verified by results shown in Fig. 3.3c. Figure 3.3c presents the net utility received by selecting $N = 100$ different contracts in the contract bundle for four sample SBSs $n = 1$, 10, 50, 90. The pentagram marks in Fig. 3.3c are the maximum net utility received for the four SBSs, and the corresponding horizontal axes points are the indexes of SBS types that contracts are designed for. Results indicate that for each type of SBS, the maximum net utility can be achieved only by selecting the right contract designed for this type.

By applying three different contracts, the system performance of HetUDN is shown as Fig. 3.4, which presents that the MBS utility, SBS net utility and social

**Fig. 3.3** The contract monotonicity and incentive compatibility versus different SBS types. (**a**) Traffic offloading $y_n$ versus SBS types. (**b**) Payments $T_n (y_n)$ versus SBS types. (**c**) Net utility received by different contracts

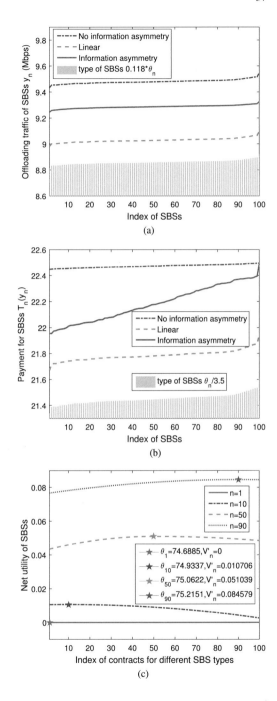

**Fig. 3.4** System
performance of different
types of SBSs when applying
different traffic offloading and
resource allocation
mechanisms. (**a**) Utility of
MBS $U_n$. (**b**) Net utility of
SBSs $V'_n$. (**c**) Social welfare
$W_n$

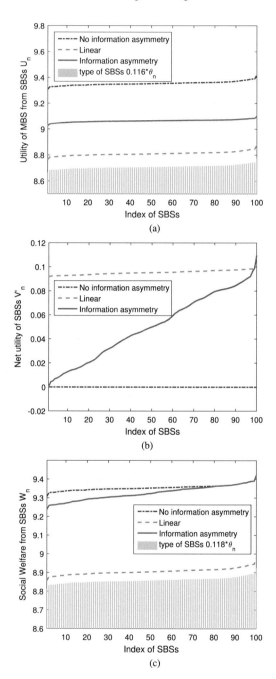

welfare increase monotonically with the value of SBS type growing. Results in Fig. 3.4a show that the contract for the scenario without information asymmetry brings the maximum utility for the MBS. Under the case that the SBS types are unavailable, the designed IC-based contract can only bring an approximate optimal utility for the MBS, which is upper bounded by the no information asymmetry situation. With information asymmetry, the linear pricing based contract does not treat differently to all types of SBSs. Therefore, without the knowledge of SBS type, the linear pricing performs worst on the MBS utility.

Since the MBS is selfish, when it is aware of the type of every BSS, the designed contract only need to satisfy the IR constraints when maximizing the MBS utility. Then every SBS can only get the utility equal to that of providing no offloading service, which means that the net utility is zero for every SBS, as shown in Fig. 3.4b. By applying the contract with IC constraints, only the SBS with the lowest type value will receive zero-net utility, and SBSs with lower $\theta$ will receive less net utility than that obtained by linear pricing contract. However, for those SBSs with higher $\theta$, they can receive more net utility than that obtained by linear pricing contract, which demonstrates the incentive compatibility of the IC based contract. The social welfare shown in Fig. 3.4c presents a similar result as Fig. 3.4a. In addition, with the IC based contract for information asymmetry, the SBS with the highest $\theta$ brings the same social welfare as no information asymmetry.

Next, we study that how the contract and system performance change with the changing density of SBSs. Let the number of SBSs in the macrocell system varies from 20 to 100, and other parameters are set as before. The average traffic offloaded and the average payments for each SBS versus the nember of SBSs (types) are shown in Fig. 3.5. The differences between the effects on the amount of traffic offloading by three traffic offloading contracts shown in Fig. 3.5a are similar to that of Fig. 3.3a. In addition, the average amount of traffic offloading for every SBS decreases when the number of SBSs grows, which results that if the amount of total traffic request are fixed, distributing more SBSs will lighten the load of every BS. Figure 3.5b indicates that the average payment for each SBS does not change by applying the contract for the no information asymmetry case and the linear pricing contract, no matter how many SBSs in the HetUDN. By applying the IC based contract for the information asymmetry case, the average payment obtained by per SBS decreases when the number of SBS increases. These results reflect the high effectiveness and efficiency of the contract designed in Sect. 3.4.1. Specifically, when there are more SBSs in the system, which means that more candidates can provide the traffic offloading service and the competitiveness among these SBSs tends to be weak, the average payment provided by the MBS for each SBS will be less than that in a more competitive market.

Due the same reason explained above, the average MBS utility, average SBS net utility and average social welfare by one SBS will all decrease with increasing number of SBS, by applying the three contracts, except that the SBS net utility obtained under the no information asymmetry case is always zero, as shown in Fig. 3.6. In addition, the social welfare shown in Fig. 3.6c also implies the

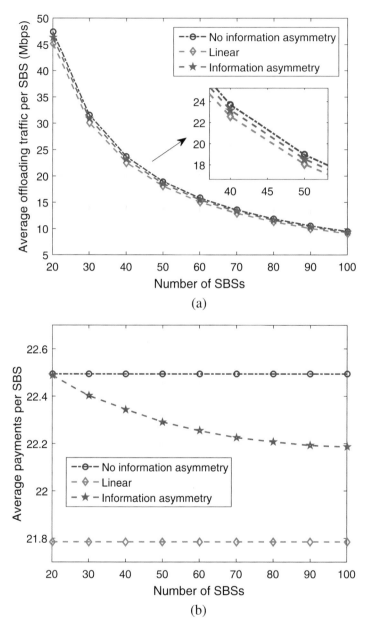

**Fig. 3.5** The contract performance versus the number of different SBS types. (**a**) Average traffic offloading versus the number of SBS types. (**b**) Average payments versus the number of SBS types

**Fig. 3.6** System
performance versus the
number of SBSs when
applying different traffic
offloading and resource
allocation mechanisms. (**a**)
Average MBS utility from
one SBS. (**b**) Average net
utility of SBSs. (**c**) Average
social welfare from one SBS

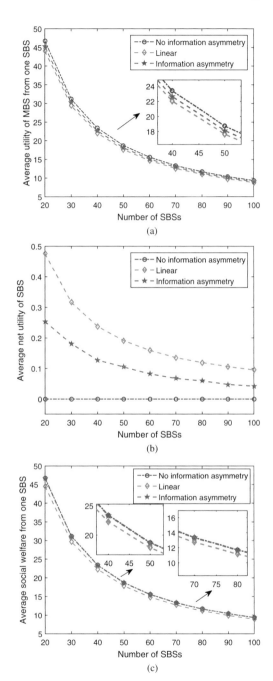

approximate optimization property of the IC based contract for the information asymmetry case.

## 3.7  Conclusion

In this work, we have proposed a contract-based traffic offloading and resource allocation mechanism for the SDWN-cased HetUDN. In the scenario with information asymmetry, the designed IC based traffic offloading contract has the incentive property to encourage every SBS to select the right contract designed personally to it, which specifies the amount of traffic offloading and the payment from the MBS. In addition, the SBS utility, MBS utility and social welfare can achieve an approximate optimization, comparing the situation without information asymmetry, and better than that achieved by linear pricing contract with information asymmetry. Furthermore, the definition of SBS type proposed in this work provides a valid index to measure the offloading performance of heterogeneous SBSs. Meanwhile, the SBS type definition also guarantee the monotonicity and incentive compatibility of contracts. In addition, the defined closed-form expression of SBS type makes this definition enforceable to be applied in HetNets with densely distributed BSs for resource management.

## References

1. X. Costa-Perez, A. Garcia-Saavedra, X. Li, T. Deiss, A. de la Oliva, A. di Giglio, P. Iovanna, and A. Moored, "5G-crosshaul: An SDN/NFV integrated fronthaul/backhaul transport network architecture," *IEEE Wireless Commun.*, vol. 24, no. 1, pp. 38–45, Feb. 2017.
2. C. Jiang, H. Zhang, Y. Ren, Z. Han, K.-C. Chen, and L. Hanzo, "Machine learning paradigms for next-generation wireless networks," *IEEE Wireless Commun.*, vol. 24, no. 2, pp. 98–105, Apr. 2017.
3. C. Jiang, B. Wang, Y. Han, Z.-H. Wu, and K. R. Liu, "Spatial focusing inspired 5g spectrum sharing," in *IEEE Int. Conf. on Acoustics, Speech and Signal Process. (ICASSP)*. New Orleans, LA, USA, 5–9 Mar. 2017, pp. 3609–3613.
4. J. G. Andrews, S. Buzzi, W. Choi, S. V. Hanly, A. Lozano, A. C. Soong, and J. C. Zhang, "What will 5g be?" *IEEE J. Sel. Areas Commun.*, vol. 32, no. 6, pp. 1065–1082, Jun. 2014.
5. J. Ding, R. Yu, Y. Zhang, S. Gjessing, and D. H. Tsang, "Service provider competition and cooperation in cloud-based software defined wireless networks," *IEEE Commun. Mag.*, vol. 53, no. 11, pp. 134–140, Nov. 2015.
6. S. Kang and W. Yoon, "SDN-based resource allocation for heterogeneous LTE and wlan multi-radio networks," *J. Supercomputing*, vol. 72, no. 4, pp. 1342–1362, Feb. 2016.
7. J. Du, C. Jiang, A. Benslimane, S. Guo, and Y. Ren, "SDN-based resource allocation in edge and cloud computing systems: An evolutionary stackelberg differential game approach," *IEEE/ACM Trans. Networking*, vol. 30, no. 4, pp. 1613–1628, Aug. 2022.
8. J. Du, C. Jiang, H. Zhang, Y. Ren, and M. Guizani, "Auction design and analysis for SDN-based traffic offloading in hybrid satellite-terrestrial networks," *IEEE J. Sel. Areas Commun.*, vol. 36, no. 10, pp. 2202–2217, Oct. 2018.

9. Y. Zhang, L. Liu, Y. Gu, D. Niyato, M. Pan, and Z. Han, "Offloading in software defined network at edge with information asymmetry: A contract theoretical approach," *J. Signal Processing Syst.*, vol. 83, no. 2, pp. 241–253, May 2016.
10. H. Zhang, S. Huang, C. Jiang, K. Long, V. C. Leung, and H. V. Poor, "Energy efficient user association and power allocation in millimeter-wave-based ultra dense networks with energy harvesting base stations," *IEEE J. Sel. Areas Commun.*, vol. 35, no. 9, pp. 1936–1947, Sept. 2017.
11. J. Du, C. Jiang, J. Wang, Y. Ren, S. Yu, and Z. Han, "Resource allocation in space multiaccess systems," *IEEE Trans. Aerosp. Electron. Syst.*, vol. 53, no. 2, pp. 598–618, Apr. 2017.
12. J. Du, C. Jiang, J. Wang, S. Yu, and Y. Ren, "Stability analysis and resource allocation for space-based multi-access systems." in *IEEE Global Commun. Conf. (GLOBECOM'15)*. San Diego, CA, USA, 6-10 Dec. 2015.
13. E. Baccarelli, N. Cordeschi, A. Mei, M. Panella, M. Shojafar, and J. Stefa, "Energy-efficient dynamic traffic offloading and reconfiguration of networked data centers for big data stream mobile computing: review, challenges, and a case study," *IEEE Network*, vol. 30, no. 2, pp. 54–61, Mar. 2016.
14. Y. Mao, J. Zhang, and K. B. Letaief, "Dynamic computation offloading for mobile-edge computing with energy harvesting devices," *IEEE J. Sel. Areas Commun.*, vol. 34, no. 12, pp. 3590–3605, Dec. 2016.
15. S. Zhang, N. Zhang, S. Zhou, J. Gong, Z. Niu, and X. Shen, "Energy-aware traffic offloading for green heterogeneous networks," *IEEE J. Sel. Areas Commun.*, vol. 34, no. 5, pp. 1116–1129, May 2016.
16. Y. Wu, K. Guo, J. Huang, and X. S. Shen, "Secrecy-based energy-efficient data offloading via dual connectivity over unlicensed spectrums," *IEEE J. Sel. Areas Commun.*, vol. 34, no. 12, pp. 3252–3270, Dec. 2016.
17. F. Mehmeti and T. Spyropoulos, "Performance analysis of mobile data offloading in heterogeneous networks," *IEEE Trans. Mobile Computing*, vol. 16, no. 2, pp. 482–496, Feb. 2017.
18. N. C. Nguyen, P. Wang, D. Niyato, Y. Wen, and Z. Han, "Resource management in cloud networking using economic analysis and pricing models: A survey," *IEEE Commun. Surveys & Tutorials*, vol. PP, no. 99, pp. 1–49, Jan. 2017.
19. L. Gao, G. Iosifidis, J. Huang, L. Tassiulas, and D. Li, "Bargaining-based mobile data offloading," *IEEE J. Sel. Areas Commun.*, vol. 32, no. 6, pp. 1114–1125, Jun. 2014.
20. M. I. Kamel and K. M. Elsayed, "ABSF offsetting and optimal resource partitioning for eICIC in LTE-Advanced: Proposal and analysis using a nash bargaining approach," in *IEEE Int. Conf. on Commun. (ICC 2013)*. Budapest, Hungary, 9–13 Jun. 2013.
21. B. C. Chung and D.-H. Cho, "Mobile data offloading with almost blank subframe in LTE-LAA and Wi-Fi coexisting networks based on coalition game," *IEEE Commun. Lett.*, vol. 21, no. 3, pp. 608–611, Nov. 2016.
22. T. M. Ho, N. H. Tran, L. B. Le, W. Saad, S. A. Kazmi, and C. S. Hong, "Coordinated resource partitioning and data offloading in wireless heterogeneous networks," *IEEE Commun. Lett.*, vol. 20, no. 5, pp. 974–977, May 2016.
23. A. F. Hanif, H. Tembine, M. Assaad, and D. Zeghlache, "Mean-field games for resource sharing in cloud-based networks," *IEEE/ACM Trans. Networking*, vol. 24, no. 1, pp. 624–637, Feb. 2016.
24. S. Samarakoon, M. Bennis, W. Saad, M. Debbah, and M. Latva-Aho, "Ultra dense small cell networks: Turning density into energy efficiency," *IEEE J. Sel. Areas Commun.*, vol. 34, no. 5, pp. 1267–1280, May 2016.
25. X. Xu, C. Yuan, J. Li, and X. Tao, "Energy-efficient active offloading with collaboration communication and power allocations for heterogeneous ultradense networks," *Trans. Emerging Telecommun. Technol.*, vol. PP, no. 99, pp. 1–9, Feb. 2017.
26. C. Jiang, Y. Chen, Q. Wang, and K. R. Liu, "Data-driven auction mechanism design in IaaS cloud computing," *IEEE Trans. Services Computing*, vol. PP, no. 99, pp. 1–14, 2015.
27. E. Gelenbe, "Analysis of automated auctions," in *Computer and Inform. Sci.–ISCIS*. Springer Berlin Heidelberg, 2006, pp. 1–12.

28. ——, "Analysis of single and networked auctions," *ACM Trans. Internet Technol.*, vol. 9, no. 2, p. 8, 2009.
29. E. Gelenbe and L. Györfi, "Performance of auctions and sealed bids," in *Computer Performance Eng.* Springer Berlin Heidelberg, 2009, pp. 30–43.
30. E. Gelenbe and K. Velan, "An approximate model for bidders in sequential automated auctions," in *Agent and Multi-Agent Systems: Technol. and Applicat.* Springer Berlin Heidelberg, 2009, pp. 70–79.
31. ——, "Modelling bidders in sequential automated auctions," *Proc. of 8th Int. Conf. on Autonomous Agents and Multiagent Syst. (AAMAS 2009), Budapest, Hungary*, 2009.
32. K. Velan and E. Gelenbe, "Analysing bidder performance in randomised and fixed-deadline automated auctions," in *Agent and Multi-Agent Syst.: Technol. and Applications.* Springer Berlin Heidelberg, 2010, pp. 42–51.
33. G. Iosifidis, L. Gao, J. Huang, and L. Tassiulas, "A double-auction mechanism for mobile data-offloading markets," *IEEE/ACM Trans. Networking*, vol. 23, no. 5, pp. 1634–1647, Oct. 2015.
34. D. Zhang, Z. Chang, and T. Hamalainen, "Reverse combinatorial auction based resource allocation in heterogeneous software defined network with infrastructure sharing," in *IEEE 83rd Veh. Technol. Conf. (VTC Spring 2016),*. Nanjing, China, 15–18 May 2016.
35. J. Du, E. Gelenbe, C. Jiang, H. Zhang, and Y. Ren, "Contract design for traffic offloading and resource allocation in heterogeneous ultra-dense networks," *IEEE J. Sel. Areas Commun.*, vol. 35, no. 11, pp. 2457–2467, Nov. 2017.
36. Y. Zhang, L. Song, W. Saad, Z. Dawy, and Z. Han, "Contract-based incentive mechanisms for device-to-device communications in cellular networks," *IEEE J. Sel. Areas Commun.*, vol. 33, no. 10, pp. 2144–2155, Oct. 2015.
37. A. Asheralieva and Y. Miyanaga, "Optimal contract design for joint user association and inter-cell interference mitigation in heterogeneous lte-a networks with asymmetric information," *IEEE Trans. on Veh. Technol.*, vol. PP, no. 99, pp. 1–15, Oct. 2016.
38. M. Peng, X. Xie, Q. Hu, J. Zhang, and H. V. Poor, "Contract-based interference coordination in heterogeneous cloud radio access networks," *IEEE J. Sel. Areas Commun.*, vol. 33, no. 6, pp. 1140–1153, Jun. 2015.

# Chapter 4
# Cooperative Resource Allocation in Heterogeneous Space-Based Networks

**Abstract** Currently, most Landsat satellites are deployed in the low earth orbit (LEO) to obtain high resolution data of the Earth surface and atmosphere. However, the return channels of LEO satellites are unstable and discontinuous intrinsically resulting from the high orbital velocity, long revisit interval and limited ranges of ground-based radar receivers. Space-based information networks (SBIN), in which data can be delivered by the cooperative transmission of relay satellites, can greatly expand the spatial transport connection ranges of LEO satellites. While different types of these relay satellites deployed in orbits of different altitudes represent distinctive performances when they are participating in forwarding. In this chapter, we consider the cooperative mechanism of relay satellites deployed in the geosynchronous orbit (GEO) and LEO according to their different transport performances and orbital characteristics. To take full advantage of the transmission resource of different kinds of cooperative relays, we propose a multiple access and bandwidth resource allocation strategy for GEO relay, in which the relay can receive and transmit simultaneously according to channel characteristics of space-based systems. Moreover, a time slot allocation strategy which is based on the slotted time division multiple access is introduced for the system with LEO relays. Based on the queueing theoretic formulation, the stability of the proposed systems and protocols is analyzed and the maximum stable throughput region is derived as well, which provide the guidance for the design of the system optimal control. Simulation results exhibit multiple factors that affect the stable throughput and verify the theoretical analysis.

**Keywords** Space-based Information Networks · Cooperative Communication · Resource Allocation · Queueing Theory · Stability Analysis

## 4.1 Introduction

Space exploration has developed for nearly 60 years since October 4th 1957, when the first artificial satellite launched. So far there have been more than 1100 active satellites all around the world, which are playing an important role of the

© The Author(s), under exclusive license to Springer Nature Singapore Pte Ltd. 2023  45
J. Du, C. Jiang, *Cooperation and Integration in 6G Heterogeneous Networks*,
Wireless Networks, https://doi.org/10.1007/978-981-19-7648-3_4

earth observation, positioning and navigation, broadcast communication, military premising scout, etc. Currently, most Landsat satellites are deployed in the low earth orbit (LEO), which is around the Earth with an altitude between 300 and 10,000 km. This low orbit altitude can obtain much higher resolution data of the Earth surface and atmosphere. However, the return channels of LEO satellites are unstable and discontinuous intrinsically, which results from extremely high orbital velocity, long revisit interval and the limited ranges of ground-based radar receivers.

Recently, the space-based information network (SBIN) is proposed to increase the transmission and detection capabilities of single satellite [1, 2]. In SBIN, the real time space data acquisition and transmission can be realized through the cooperative detection and transmission among the different satellites, satellite systems and other space facilities. These different types of satellites with varying functionalities perform diverse tasks, or perform the same task but with different capabilities. Therefore, this diversity of functions and activities leads to the SBIN becoming a kind of complex networks, which cause difficulties to the cooperation mechanism design and the detection and transmission resource allocation for the SBIN [3, 4]. In this work, we focus on the analysis and transmission resource allocation for the data relay satellite (DRS), and the diversity of functions and activities among the SBIN is considered, modeled and analyzed in our work.

In the SBIN with the DRS, the problems of the connection between LEO satellites and ground station mentioned above can be solved potentially. Specifically, over an inter-satellite link (ISL), data from LEO satellites is transmitted first to a relay satellite that can establish the connection with the ground station, and then via the satellite-ground station link forwarded from the relay to the ground stations [5–7]. So far, many countries have deployed DRS for stable and quick data return. Such as NASA's Tracking and Data Relay Satellite (TDRS) [8, 9], DRSs are deployed in the geosynchronous orbit (GEO), which can expand the spatial transport connection range of LEO satellites and provide a stable transmission to the ground stations.

Most of current DRSs are deployed in the GEO to greatly improve the space range that supports data return for LEO satellites and provide stable satellite-to-ground downlink, so the DRS always refers specifically to the GEO relay satellite. However, in case of that there are no available DRS or the DRS performing high priority tasks and leaving no extra transmission resource, the LEO satellite carrying urgent data needs to connect with the ground by finding other space transmission resource when it does not in the capture range of the ground radar [10]. In this case, those satellites passing over the ground station and with no data to send back or lower priority tasks can contribute their transmission resource and perform as relays. In this article, this kind of satellites is abbreviated as the LEO relay. In contrast, DRSs specifically refer to the GEO relays in this article.

Main contributions of this part are summarized as follows.

- We establish two cooperative transmission systems for different types of relay satellites. In each model, we formulate the classical queueing models to describe the arrivals and service process in the systems. Then we propose an ON/OFF based probability model, which can describe the link connection status for the LEO satellite, relay satellite and ground station,

- We propose two cooperative multi-access resource allocation strategies based on employing different orbit types of relay satellites in the space-based multi-access systems, to improve resource utilization and satisfy the constraints of different kinds of relays. According to the protocols, the idle bandwidth resource of GEO relay satellite and the time slot resource of LEO relay satellite can be allocated efficiently in the multi-access systems.

- Based on the two resource allocation strategies, we discuss and provide the stable throughput region of the two proposed cooperative protocols for the case with two source nodes in the system. To deal with the interaction between the two queues of the source nodes, a system decomposition method is introduced in our work for the analysis of the interacting queues. According to the dominant system analysis, we obtain the maximum stable throughput regions for both proposed multi-access systems, which can play an important role for the design of the system optimal control.

The remainder of this part is organized as follows. In Sect. 4.2, we review the related works. In Sect. 4.3 the system models is described. Two cooperative multi-access resource allocation protocols are proposed in Sect. 4.4. Then we analyze the stability characteristics in Sect. 4.5. Simulations are shown in Sect. 4.6, and conclusions are drawn in Sect. 4.7.

## 4.2   Related Works

The transmission resource of the both kinds of relay satellites above is limited in different aspects, so a key issue in cooperative communication through a relay is the resource allocation for full utilization of the limited resource. Researches on the resource allocation have been addressed in many kinds of wireless networks. A bandwidth allocation strategy of scalable video multicast in a WiMAX relay system was proposed in [11] to maximize the network throughput and the number of satisfied users. An optimal resource allocation scheme to maximize the achievable rate and enhance the resource allocation efficiency was proposed through jointing exploitation of multi-user diversity and multi-hop diversity in [12]. In [13], researchers summarized many recent channel-aware resource allocation techniques proposed for downlink multicast services in OFDMA systems. Most of the studies above on resource allocation concentrate on the homogeneous wireless network. Yet, data transmission in SBINs with DRSs depends on heterogeneous transport medium, such as ISLs and SGLs. Moreover, the SBIN is a kind of opportunistic networks, of which the above two kinds of links disrupted much more frequently. Hence, the current models for resource allocation are no longer applicable for SBINs. Concerning this issue, this part establishes an ON/OFF model to describe the connection status of ISLs and satellite-ground links.

Meanwhile, the DRSs providing the relay service play an important role for the cooperative communication in the SBIN. So the design of cooperation strategy is an essential part of the operation and performance optimization of the SBIN with DRSs. Currently, many works have started to studied the cooperative strategies for the satellite networks. To improve the energy efficiency, a real-time adaptive cooperative transmission strategy for dynamic selection between the direct and cooperative links based on the channel conditions was designed and evaluated in [14, 15]. In [16, 17], some cooperative relaying strategies which rely on the exploitation of the Delay Diversity technique and the Maximal Ratio Combining (MRC) receive diversity algorithm are proposed for a DVB-SH compliant hybrid satellite/terrestrial network. Fractionated Satellite Networks discussed in [18] allow the satellites to cooperate by exchanging resources to improve the network capability significantly. In [19], the coordination between two coexisting transmitters and a cognitive beamhopping technique were introduced into the cooperative and cognitive satellite systems, respectively. Moreover, for the cooperative communication of the SBIN, many inter-satellite routings were proposed to improve transmission efficiency [20–22], reliability [23], security [24] and the quality of service [25]. However, most of the current researches did not consider or analyze that the uplink and downlink of the DRS use different types of wireless channels. Then DRSs can receive and forward data simultaneously and current cooperation strategies are no longer applicable. In this part, the powerful store-and-forward capability and massive bandwidth resource of DRSs (GEO relay) are considered. On the other hand, the LEO relay subjects to the constraints of bandwidth and connect time with the ground station, and its transmission capacity is not as powerful as GEO relay's. Hence, it is necessary to design different resource allocation protocols to improve resource utilization and satisfy the constraints of different kinds of relays.

## 4.3   System Model

We consider two types of resource allocation mechanisms for space-based information networks in different scenarios. In particular, transmission resource allocation and cooperation protocols are designed for the SBIN with the GEO relay satellite and the LEO relay satellite, respectively. Motivations of the above two aspects are that the proposed space resource allocation methods can be applied to the main network deployment modes: when the transmission resource of GEO relay satellites is available, the protocol proposed for the GEO relay scene can be applied, otherwise, the LEO relay protocol can be the alternate. As shown in Fig. 4.1a, the relay satellite operating in the geosynchronous orbit can establish a consistent and reliable connection with the ground station. Generally, the GEO relay satellite has more powerful storage-and-forwarding capabilities and higher transmission bandwidth. As a result, in the SBIN with a GEO relay, we assume that the relay can forward all the data received from different source nodes simultaneously. Moreover, uplinks and downlinks through the DRS go through different types of channels (ISLs and satellite-ground station links), so relays can receive and forward data at the same time slot. In addition, the number of the accessible satellites for the relay is limited, which results from the limited number of satellites and their orbit constraints. On

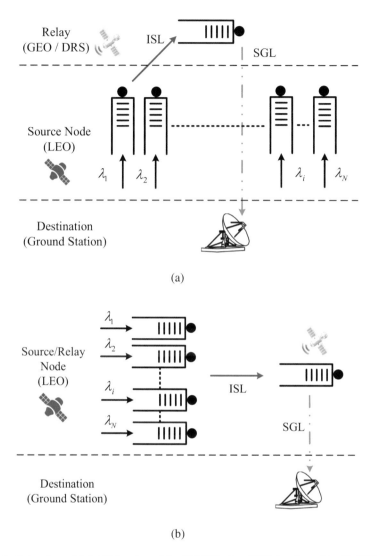

**Fig. 4.1** Network model. (**a**) Network model with GEO relay. (**b**) Network model with LEO relay

the other hand, due to its transmission and access capabilities, the relay satellite also restricts the number of satellites allowed to access. Consequently, we consider that simultaneous transmission can be always successful. These assumptions above are feasible because of the powerful transmission capability of current DRSs. So when there are more than one source satellites have transmission missions through the GEO relay, the bandwidth resource of the relay needs to be allocated appropriately to these sources to improve the transmission efficiency and resource utilization of the relay. In another situation, there is no available GEO data relay satellite, such as when the relay is executing the high priority task and has no extra bandwidth resource for other satellites. Then if the LEO satellite has the urgent data transmission demand when it does not enter the reception range of ground stations, the transmission has to depend on the cooperation with those LEO satellites that meet the transmission condition and have no or low priority transmission missions. This cooperation model is shown as Fig. 4.1b. Different from the GEO relay satellite, the LEO satellite performing as the relay is not deployed for the data relay mission, so the transmission bandwidth is quite limited and the bandwidth allocation mechanism to the GEO relay do not fit for the LEO relay any more. Therefore, we introduce a transmission time allocation mechanism based on the slotted time division multiple access (TDMA) system for the cooperative transmission with a LEO relay satellite.

The two scenes mentioned above are modeled as follows. As shown in Fig. 4.1, the network consists of a finite number $N < \infty$ of source nodes, a relay node $r$ and a destination node $d$. The N source nodes indicate LEO satellites and are numbered by $1, 2, \cdots, N$. The relay node $r$ indicates the data relay satellite, which can be deployed in GEO or LEO as Fig. 4.1a and b. The destination node $d$ has different reception channels allowed to receive LEO and GEO satellites simultaneously. When there are more than one source nodes sending data to the relay node, the transmission resource can be allocated through different appropriate strategies in these two scenes.

Then we describe the queueing model for the multiple access-based transmission resource allocation system. Each of the $N$ source nodes has a finite buffer. The channel is slotted. The arrival process to each source node is independent and identically distributed (i.i.d.) at different time slots, and $\lambda_i$ $(i = 1, 2, \cdots, N)$ denotes the arrival rates at the $i$th queue. Arrival processes to different nodes are independent with each other. Let $\Omega = [\omega_1, \omega_2, \cdots, \omega_N]$ denote the resource sharing vector, where $\omega_i$ $(i = 1, 2, \cdots, N)$ represents the proportion of the relay's total bandwidth resource allocated to the $i$th source node, or it can represent the probility that the $i$th source node is allocated the total time slot. The set of all feasible resource sharing vectors is expressed as:

$$W = \left\{ \Omega = [\omega_1, \omega_2, \cdots, \omega_N] \in R_+^N \;\middle|\; \sum_{i=1}^{N} \omega_i \leq 1 \right\}. \tag{4.1}$$

The stability region of the queue is an important and fundamental performance for the communication network. In this part, we consider the stability as the finiteness of the queue size. Let the vector $Q(t) = [Q_i(t)]$ $(i = 1, 2, \cdots, N)$ be queue sizes of the source nodes at time slot $t$. We use the definition of stability in [26] and [27] as follows. Each queue $i$ $(i = 1, 2, \cdots, N)$ of the network is stable, if

$$\lim_{t \to \infty} \Pr\{Q_i(t) < x\} = F(x) \text{ and } \lim_{x \to \infty} F(x) = 1. \tag{4.2}$$

For a weaker condition:

$$\lim_{x\to\infty}\lim_{t\to\infty}\inf \Pr\{Q_i(t) < x\} = 1, \tag{4.3}$$

then the process is called substable [26].

As for the stability, Loynes provided the theory to judge the stability condition [28]. The theory indicates that when the arrival and service processes are strictly stationary in a queueing system, and the average arrival rate is less than the average service rate, the queue is stable. If the average arrival rate is greater than the average service rate, the queue is unstable.

### 4.3.1   ON/OFF Model

Next, we introduce the ON/OFF probability model to describe the link connection status for the LEO satellite, relay satellite and ground station. Specifically, when the connection can be established, we call the link status is ON, otherwise is OFF. In this part, we assume that the link connection status is a stochastic process as in a long time horizon. In addition, to be general, for a particular source node (LEO satellite), the connection status with the relay and the destination is i.i.d. with an ON/OFF probability, respectively, from one slot to another. Define a binary function to describe the link connection status as (4.4).

$$L_{jk} = \begin{cases} 1, & \text{link between } j \text{ and } k \text{ is ON}, \\ 0, & \text{link between } j \text{ and } k \text{ is OFF}. \end{cases} \tag{4.4}$$

In this part, we focus on the SBIN with a GEO relay satellite, as shown in Fig. 4.2.

#### 4.3.1.1   ISL Connection Status

Consider that source nodes can transmits data to the relay as long as they are visible to each other. Therefore, the blue part of the LEO satellite orbit determined by the tangent from the relay to the Earth denotes the linkable range, and we call the link status is ON, as shown in Fig. 4.2a. Due to the Earth rotation and satellite orbit modes such as polar orbits and Walker Delta Pattern constellation orbits, the movement of a LEO Landsat satellite forms a sphere within a radius of the satellite's orbit radius. Therefore on this sphere, the linkable range is the spherical cap formed from the blue arc in Fig. 4.2a.

According to the analysis above, we set the area ratio of the spherical cap to the entire sphere as the probability that the connection status between LEO satellite $i$ and the relay $r$ is ON:

$$p_{1(i)} \triangleq \Pr\{L_{ir} = 1\} = 0.5\left(1 + \cos\alpha_{1(i)}\right), \tag{4.5}$$

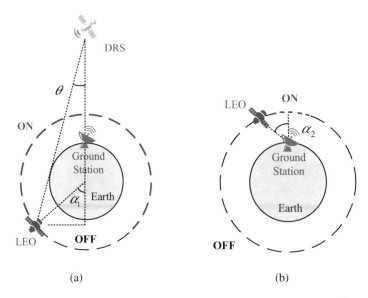

(a)                                        (b)

**Fig. 4.2** ON/OFF Model. (**a**) The blue arc (ON) of the LEO satellite orbit denotes the linkable range between the relay and the LEO satellite limited by the shade of the Earth. (**b**) The blue arc (ON) scopes the range that the LEO satellite can directly connect with the ground station, which is determined by the maximum capture range of the ground-based radar $\alpha_2$

where $\alpha_{1(i)}$ can be calculated through

$$\sin \alpha_{1(i)} = \left( \sqrt{R_i^2 - R_E^2} + \sqrt{R_r^2 - R_E^2} \right) R_E / R_i R_r . \tag{4.6}$$

In (4.6), $R_E$, $R_i$ and $R_r$ are the radius of the Earth, the LEO satellite orbit and the geosynchronous orbit, respectively. In the later sections, let $p_{1(i)}$ denote the probability of the connection between the $i$th LEO satellite and the relay.

### 4.3.1.2   Satellite-Ground Station Link Connection Status

For the data transmission directly from the LEO satellite to the ground station (destination), we assume that the transmission opportunity is determined by the maximum capture range of the ground-based radar. This range can be described with the maximum elevation angle, as $\alpha_2$ ($0 < \alpha_2 \leq \frac{\pi}{2}$) shown in Fig. 4.2b. The blue arc in Fig. 4.2b scopes the range that the LEO satellite can directly connect with the ground station. Similar to the analysis in the previous part, let the area ratio of the spherical cap to the entire sphere denote the probability that the connection status between LEO satellite $i$ and the destination $d$ is ON:

$$p_{2(i)} \triangleq \Pr\{L_{id} = 1\}$$

$$= 0.5\left(1 - \rho_i \sin^2\alpha_2 - \cos\alpha_2\sqrt{1 - (\rho_i \sin\alpha_2)^2}\right), \tag{4.7}$$

where $\rho_i = R_i/R_E$. We use $p_{2(i)}$ to denote the probability of the connection between the $i$th LEO satellite and the ground station.

Then we get the corresponding conditional probabilities which are defined as

$$p_{3(i)} \triangleq \Pr\{L_{id} = 1 \,|\, L_{ir} = 1\}$$

$$= \frac{\Pr\{L_{id} = 1 \cap L_{ir} = 1\}}{\Pr\{L_{ir} = 1\}} \tag{4.8}$$

$$= \frac{\Pr\{L_{id} = 1\}}{\Pr\{L_{ir} = 1\}} = \frac{p_{2(i)}}{p_{1(i)}},$$

$$\Pr\{L_{id} = 0 \,|\, L_{ir} = 1\} = 1 - p_{3(i)}. \tag{4.9}$$

We use $p_{3(i)}$ to denote the probability of the connection between the $i$th LEO satellite and the relay node, under the condition that this satellite can connect with the ground station.

The situation that with a LEO relay is similar, and the only difference is that the DRS in Fig. 4.2 is deployed in the LEO, with $R_r$ representing the radius of the LEO relay satellite, rather than the GEO.

## 4.3.2  Physical Channel Model

In the transmission of the SBIN, the line of sight (LOS) signal is much stronger than the others, which is different from ground networks. Therefore, the wireless channel for the LEO satellites, relay satellite and ground stations is considered as a Rician fading channel model with additive Gaussian noise. The signal received at the relay or the destination at the time slot $t$ is modeled as

$$y_{ij}^t = \sqrt{Gd_{ij}^{-\gamma}}\, h_{ij}^t x_i^t + n_{ij}^t, \tag{4.10}$$

where $i$ is the source node or the relay node when it transmits data to the destination node, and $j$ is the destination or the relay when it receives data from source nodes, $x_i$ is the data transmitted from $i$, $G$ is the transmitting power, $d_{ij}$ is the distance between $i$ and $j$, $\gamma$ denotes the path loss exponent, and $n_{ij}^t$ is i.i.d. additive Gaussian noise between $i$ and $j$ at the time slot $t$ with zero-mean and variance $N_0$ [29, 30]. In (4.10), $h_{ij} = X_1 + jX_2$ is the channel fading coefficient modeled as a circularly symmetric complex Gaussian random variable, in which $X_1 \sim \mathcal{N}\left(\mu_1, \frac{\sigma^2}{2}\right)$ and $X_2 \sim \mathcal{N}\left(\mu_2, \frac{\sigma^2}{2}\right)$ are modeled as Gaussian random variable. Then the distribution of $\left|h_{ij}\right|$ is given by the Rician probability density function (PDF)

$$f_{\left|h_{ij}\right|}(h) = \frac{2h}{\sigma^2} \exp\left\{\frac{-\left(h^2 + s^2\right)}{\sigma^2}\right\} I_0\left(\frac{2sh}{\sigma^2}\right), \tag{4.11}$$

where $s^2 = \mu_1^2 + \mu_2^2$ is the power due to the Line of Sight (LOS) signal, and $I_0(\cdot)$ is the 0th order modified Bessel function of the first kind [31, 32]. Then $\text{SNR}_{ij}$, the Signal-to-Noise Ratio (SNR) between two nodes $i$ and $j$, can be specified as

$$\text{SNR}_{ij} = \frac{\left|h_{ij}\right|^2 d_{ij}^{-\gamma} G}{N_0}, \tag{4.12}$$

where $\left|h_{ij}\right|^2$ follows the non-central chi-squares ($\mathcal{X}^2$) distribution with the PDF as

$$f_{\left|h_{ij}\right|^2}(h) = \frac{K+1}{\Omega} \exp\left\{-K - \frac{(K+1)h}{\Omega}\right\} I_0\left(2\sqrt{\frac{K(K+1)h}{\Omega}}\right). \tag{4.13}$$

In (4.13), $\Omega = s^2 + \sigma^2$ is the total power of the LOS and scattering signal, $K = \frac{s^2}{\sigma^2}$ is the ratio between the power in the direct path and the power in the other scattered paths [32, 33]. Next, we introduce the outage event and outage probability to characterize the success and failure of the packet transmission and reception. The condition of outage is defined as that the SNR is less than the given SNR threshold $\beta$ [29, 34, 35]. Then outage event can be expressed as (4.14).

$$\left\{h_{ij} : \text{SNR}_{ij} < \beta\right\} = \left\{h_{ij} : \left|h_{ij}\right|^2 < \frac{\beta N_0 d_{ij}^{\gamma}}{G}\right\}. \tag{4.14}$$

Then the success probability of the packet between $i$ and $j$ at SNR threshold $\beta$ is

$$f_{ij} \triangleq \Pr\left\{C_{ij}\right\} = \Pr\left\{\left|h_{ij}\right|^2 \geq \frac{\beta N_0 d_{ij}^{\gamma}}{G}\right\}$$

$$= \int_{\frac{\beta N_0 d_{ij}^{\gamma}}{G}}^{+\infty} \frac{K+1}{\Omega} \exp\left\{-K - \frac{(K+1)h}{\Omega}\right\} I_0\left(2\sqrt{\frac{K(K+1)h}{\Omega}}\right) dh, \tag{4.15}$$

where $C_{ij}$ denotes the success transmission between nodes $i$ and $j$.

## 4.4 Cooperative Resource Allocation Protocol

In a LEO satellite detection system without relays, satellites cannot establish stable and continuous data transmission as mentioned in previous sections. With the cooperative transmission through the relay satellite deployed in the geosynchronous or low orbit, LEO satellites can distribute the data back to the Earth even if the direct transmission link between this satellite and ground stations does not exist, especially when with the GEO relay that can connect to the Earth all weather and all time. On the other hand, for the relay satellite, the transmission resource (bandwidth or transmission opportunity) is limited and the opportunity that the data relay satellite can be accessed by LEO satellites ($L_{ir} = 1$) is also limited according to (4.5), particularly for the LEO relay. Therefore, LEO satellites tend to utilize the transmission resource as fully as possible. Without the cognition mechanism, the transmission resource of the relay is allocated for accessible LEO satellites based on some allocation strategies, such as according to the number of accessing LEO satellites. This means that when a source node accesses the relay satellite but does not have a packet to transmit, the transmission resource allocated to this node is wasted.

To solve the problems above, we propose two cooperative multi-access resource allocation strategies based on employing different orbit types of relay satellites in the space-based network, respectively. We assume that the relay satellite can sense the communication channel to detect the idle bandwidth resource or empty time slot. In addition, the errors and delay in packet ACK feedback is neglected in our work.[1] We use the similar transmission failure handling approach in [29], in which source nodes and the relay can confirm whether the packet is transmitted to destination correctly, and if not, both of the source nodes and the relay will store this packet in their queues and resend it at the next time slot. Then we propose different cooperative resource allocation protocols for the cooperative transmission in space-based networks, in which the GEO relay satellite and LEO landsat satellite are performing as the relay, respectively. In the following protocols, the relative variables defined previously are summarized as Table 4.1.

---

[1] In this work, the errors and delay in packet ACK feedback is neglected. However, when these errors and delay happen, the transmission of the relevant packets are considered to be unsuccessful. Then according to the two protocols designed above, these packets considered unsuccessfully transmitted will be resent. This situation will bring packet errors or repetitions to the receivers at the relays and ground stations, but has no effect on the operation of the system. In addition, the stability analysis in the next section is still valid, by considering equally that the success probability $f_{ij}$ decreases resulting from the errors and delay of ACK feedback.

**Table 4.1** Variable definitions in cooperative resource allocation protocols

| Variables | Definitions |
|---|---|
| $N$ | Number of total satellites |
| $M$ | Number of satellites considered in the resource allocation |
| $L_{id}$ | Link between satellite $i$ and the destination (ground station) |
| $L_{ir}$ | Link between satellite $i$ and the relay satellite |
| $L_{rd}$ | Link between relay satellite and the destination (ground station) |
| $\Omega$ | Resource sharing vector |
| $\omega_i$ | resource allocated to source node $i$ |

**Table 4.2** Bandwidth resource allocation (BA) protocol for the GEO relay system

Protocol 1: BA protocol for the GEO relay system

- Each source node transmits the packet at the head of its queue with entire satellite-to-ground transmission resource if it satisfies $L_{id} = 1$. Meanwhile, it does not transmit data to the relay, although it satisfies $L_{ir} = 1$.
- When there are multiple source satellites $i = 1, 2, \cdots, M$ satisfying $L_{ir} = 1 \cap L_{id} = 0$, the resource of the relay will be allocated with $\Omega = [\omega_1, \omega_2, \cdots, \omega_M]$. The relay detects idle transmission resource resulting from two situations:

  – At the time slot when some of these nodes change the link status from $L_{id} = 0$ to $L_{id} = 1$, resource allocated to them is reallocated to the rest whose $L_{id} = 0$.
  – At the time slot when some of accessed source nodes have no packet to send, resource allocated to them is also reallocated to the rest nodes.

- When there is only one source satellite $i$ satisfies $L_{ir} = 1 \cap L_{id} = 0$, resource $\omega_0$ ($\omega_i < \omega_0 \le 1$) is allocated to satellite $i$.

### 4.4.1  GEO Relay

As discussed above, the DRSs deployed in GEO always have much powerful transmission capability and bandwidth resource. Then we assume that the DRS can receive and forward data received from source satellites simultaneously through ISLs and SGLs, respectively. Moreover, we consider that if the LEO satellites can directly connect to the Earth ($L_{id} = 1$), they can use entire satellite-to-ground transmission resource to download data. This assumption is reasonable for current earth resource satellites, as they belong to different departments and have their own ground stations and satellite-to-ground transport channels. So the resource allocated is of relay satellites in this part. We consider that among total $N$ satellites, there are $M$ satellites can connect to the relay but not to the Earth, so we need to allocate resource for these $M$ satellites. $\Omega = [\omega_1, \omega_2, \cdots, \omega_M]$ denotes the resource sharing vector, where $\omega_i$ ($i = 1, 2, \cdots, M$) represents the proportion of the relay's total bandwidth resource to source node $i$. The proposed cooperative resource allocation protocol is described in Table 4.2.

## 4.4.2   LEO Relay

In the space-based network without the GEO relay, when LEO satellites have the urgent transmission tasks but their links with the stations are not available, other LEO satellites that arrive into the capture range of the ground stations can provide temporary data relay services, if they have no or low priority data transmission of their own to the ground station. However, compared with the DRSs, these LEO satellites temporarily performed as relays do not have enough transmission resource such as bandwidth. Moreover, they cannot establish stable and continuous connection with the ground. Therefore, we introduce the time resource allocated protocol based on a slotted TDMA framework. In contrast with the resource allocation for the GEO relay, this protocol rules that each time slot is assigned to only one source satellite. Under the circumstances, we consider that among total $N$ satellites, there are $M$ satellites satisfying $L_{ir} = 1$. Then $\Omega = [\omega_1, \omega_2, \cdots, \omega_M]$ denotes the time sharing vector, where $\omega_i$ $(i = 1, 2, \cdots, M)$ represents the probability that the $i$th source node is allocated the whole time slot. The proposed cooperative resource allocation protocol is described in Table 4.3.

**Table 4.3**  Timeslot Resource Allocation (TA) protocol for the LEO relay system

| Protocol 2: TA protocol for the LEO relay system |
| --- |
| • Each source node transmits the packet at the head of its queue with entire satellite-to-ground transmission resource if it satisfies $L_{id} = 1$. Under this circumstances, the source does not transmit data to the relay, although it satisfies $L_{ir} = 1$.<br>• When there are more than one source satellite $i = 1, 2, \cdots, M$ all satisfing $L_{ir} = 1$, the transmission time of the relay will be allocated as $\Omega = [\omega_1, \omega_2, \cdots, \omega_M]$.<br><br>  – When the ground stations receives a packet correctly, it sends an ACK packet which can be received by both the relay and the corresponding source node, and this packet at the head of the relay and the source is deleted. If the ground stations does not receives a packet correctly but the relay does, then the relay and the corresponding source node store this packet at the end and the head of their queues, respectively.<br>  – At the time slot when some of these nodes change the link status from $L_{id} = 0$ to $L_{id} = 1$, which means that they turn to connect with ground stations, the relay can sense this situation and uses the time slot allocated to these sources to send the packet at the head of its queue.<br>  – At the time slot when some of accessed source nodes have no packet to send, the relay can sense this situation and uses the time slot allocated to these sources to send the packet at the head of its queue.<br><br>• If the packet from the source node is not successfully delivered by the relay in the following $N-1$ time slot, then the corresponding source $i$ node delivers the packet to the ground station if it satisfies $L_{id} = 1$, otherwise the packet will be dropped from the queue of the source. The relay drops the packet from its queue. |

## 4.5  Stability Analysis

In this section, we mainly discuss and provide the stable throughput region of the two proposed cooperative protocols for the case with two source nodes in the system.

### 4.5.1  GEO Relay

In our work, we consider the relay buffer is sufficiently large and can forward entire received data immediately according to DRS's powerful storage and transmission capacity. Based on this assumption, we consider that the queue of the relay is always stable. Next, we will focus on the analysis of the source satellites' queues.

In the network, there are two LEO satellites (source nodes) with the data arrival process $\lambda_1$ and $\lambda_2$, a relay satellite (DRS) deployed in the GEO and a group of ground stations (destinations) which can receive data transmitted from both source nodes and the relay. Let $L_{ir}$ $(i = 1, 2)$ denote the link status of source node $i$ and the relay, and there exist following three access cases for the relay according to the ON/OFF model of our system. According to the bandwidth allocation protocol, Lemma 4.1 states the stability region of the cooperative satellite system with a GEO relay satellite when $\omega_0$ and the resource sharing vector $[\omega_1, \omega_2]$ are fixed.

**Lemma 4.1** *In different link status, given $\omega_0$ and the resource sharing vector $[\omega_1, \omega_2]$, the stability region $\mathscr{R}(S)$ of two LEO satellites and a GEO relay satellite system with the bandwidth allocation protocol protocol is given by*

*1. $L_{1r} = L_{2r} = 0$:*

$$\mathscr{R}(S) = \{[\lambda_1, \lambda_2] = [0, 0]\};  \tag{4.16}$$

*2. $L_{ir} = 1 \cap L_{jr} = 0, \{i, j \in \{1, 2\} : i \neq j\}$:*

$$\mathscr{R}(S) = \left\{[\lambda_i, \lambda_j] \in R_+^2 \,\big|\, \lambda_i < \omega_0 \left(1 - p_{3(i)}\right) f_{ir} + p_{3(i)} f_{id}, \lambda_j = 0\right\};  \tag{4.17}$$

*3. $L_{1r} = L_{2r} = 1$:*

$$\mathscr{R}(S) = \mathscr{R}(S_1) \cup \mathscr{R}(S_2),  \tag{4.18}$$

*where*

$$\mathcal{R}(S_1) = \Big\{ [\lambda_1, \lambda_2] \in R_+^2 \,\Big|\, \lambda_2 < \omega_0 \left(1 - p_{3(2)}\right) f_{2r} + p_{3(2)} f_{2d}$$

$$- \frac{(\omega_0 - \omega_2)\left(1 - p_{3(1)}\right)\left(1 - p_{3(2)}\right) f_{2r}}{\left(1 - p_{3(1)}\right)\left[\omega_0 p_{3(2)} f_{1r} + \omega_1 \left(1 - p_{3(2)}\right) f_{1r}\right] + p_{3(1)} f_{1d}} \lambda_1,$$

$$\text{for } \lambda_1 < \left(1 - p_{3(1)}\right)\left[\omega_0 p_{3(2)} f_{1r} + \omega_1 \left(1 - p_{3(2)}\right) f_{1r}\right] + p_{3(1)} f_{1d}\Big\},$$

$$\mathcal{R}(S_2) = \Big\{ [\lambda_1, \lambda_2] \in R_+^2 \,\Big|\, \lambda_1 < \omega_0 \left(1 - p_{3(1)}\right) f_{1r} + p_{3(1)} f_{1d}$$

$$- \frac{(\omega_0 - \omega_1)\left(1 - p_{3(1)}\right)\left(1 - p_{3(2)}\right) f_{1r}}{\left(1 - p_{3(2)}\right)\left[\omega_0 p_{3(1)} f_{2r} + \omega_2 \left(1 - p_{3(1)}\right) f_{2r}\right] + p_{3(2)} f_{2d}} \lambda_2,$$

$$\text{for } \lambda_2 < \left(1 - p_{3(2)}\right)\left[\omega_0 p_{3(1)} f_{2r} + \omega_2 \left(1 - p_{3(1)}\right) f_{2r}\right] + p_{3(2)} f_{2d}\Big\}.$$

In Lemma 4.1, $S_1$ and $S_2$ are obtained through system decomposition.

***Proof of Lemma 4.1***  See Sect. 4.8.

The region $\mathcal{R}(S_1) \cup \mathcal{R}(S_2)$, as shown in Fig. 4.3, is the stability region of the original system when the fixed sharing vector $[\omega_1, \omega_2]$ is given. Taking the entire feasible resource sharing vectors $W$ in (3.10), the stability of the system with the proposed cooperative resource allocation protocol can be formulated as follows:

$$\mathcal{R}(S) = \bigcup_{\Omega \in W} \{\mathcal{R}_\Omega(S_1) \cup \mathcal{R}_\Omega(S_2)\}, \tag{4.19}$$

where $\mathcal{R}_\Omega(S_i)$ $(i = 1, 2)$ is the stability region of dominant system $S_i$ with a given sharing vector $\Omega$.

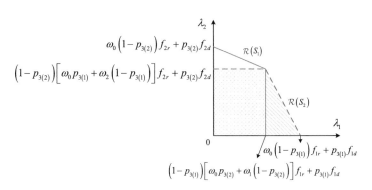

**Fig. 4.3**  Stability region of two-user and geo-relay system for a fixed $[\omega_1, \omega_2]$ given by $\mathcal{R}(S_1) \cup \mathcal{R}(S_2)$

Next, we analyze the complete characterization of the stability region of the two-user system when $L_{1r} = L_{2r} = 1$. The stabile throughput of the system for a given $[\omega_1, \omega_2]$ by $\mathscr{R}(S_1)$ and $\mathscr{R}(S_2)$ through (4.79), (4.82), (4.83) and (4.84). The stability region is given by Theorem 4.1.

**Theorem 4.1** *With the bandwidth allocation protocol, the stability region of the whole system with two-use and an GEO relay is given by*

$$\mathscr{R}(S) = \left\{ [\lambda_1, \lambda_2] \in R_+^2 | \lambda_2 < \Phi_1(\lambda_1) \right\}, \tag{4.20}$$

*where $\Phi_1(\lambda_1)$ is determined by (4.21).*

$$\Phi_1(\lambda_1) = \begin{cases} \left(1 - p_{3(2)}\right) f_{2r} + p_{3(2)} f_{2d}, & \lambda_1 < \lambda_m, \\ -\frac{f_{2r}}{f_{1r}} \lambda_1 + \psi, & \lambda_m \le \lambda_1 \le \lambda_M. \end{cases} \tag{4.21}$$

*In (4.21), $\lambda_m = \left(1 - p_{3(1)}\right) p_{3(2)} f_{1r} + p_{3(1)} f_{1d}$, and $\lambda_M = \left(1 - p_{3(1)}\right) f_{1r} + p_{3(1)} f_{1d}$.*

**Proof** To find the whole stability region for entire feasible resource sharing vector set $W$, a constrained optimization problem to find the maximum feasible $\lambda_2$ corresponding to each feasible $\lambda_1$ can be introduced [29].

Firstly, for the system $\mathscr{R}(S_1)$, the maximum of the arrival rate $\lambda_2$ can be formulated as the following optimization problem according to (4.79) and (4.82) when $\lambda_1$ is fixed (let $\omega_0 = 1$).

$$\max \lambda_2 = \left(1 - p_{3(2)}\right) f_{2r} + p_{3(2)} f_{2d}$$
$$- \frac{(1 - \omega_2)\left(1 - p_{3(1)}\right)\left(1 - p_{3(2)}\right) f_{2r} \lambda_1}{\omega_1 \left(1 - p_{3(1)}\right)\left(1 - p_{3(2)}\right) f_{1r} + \left(1 - p_{3(1)}\right) p_{3(2)} f_{1r} + p_{3(1)} f_{1d}}, \tag{4.22}$$

$$\text{s.t. } \omega_1 + \omega_2 \le 1, \tag{4.23a}$$

$$\lambda_1 < \left(1 - p_{3(1)}\right) \left[ p_{3(2)} + \omega_1 \left(1 - p_{3(1)}\right) \right] f_{1r} + p_{3(1)} f_{1d}. \tag{4.23b}$$

Substituting $\omega_1 + \omega_2 = 1$ in the objective in (4.22), then we get

$$\max \lambda_2 = \left(1 - p_{3(2)}\right) f_{2r} + p_{3(2)} f_{2d}$$
$$- \frac{\omega_1 \left(1 - p_{3(1)}\right)\left(1 - p_{3(2)}\right) f_{2r} \lambda_1}{\omega_1 \left(1 - p_{3(1)}\right)\left(1 - p_{3(2)}\right) f_{1r} + \left(1 - p_{3(1)}\right) p_{3(2)} f_{1r} + p_{3(1)} f_{1d}}, \tag{4.24}$$

Take the first derivative of (4.24) with respect to $\omega_1$, we get

$$\frac{\partial \lambda_2}{\partial \omega_1} = -\frac{\left(1 - p_{3(1)}\right)\left(1 - p_{3(2)}\right) f_{2r} \left[ \left(1 - p_{3(1)}\right) p_{3(2)} f_{1r} + p_{3(1)} f_{1d} \right]}{\Phi^2},$$

where $\Phi = \omega_1 \left(1 - p_{3(1)}\right) \left(1 - p_{3(2)}\right) f_{1r} + \left(1 - p_{3(1)}\right) p_{3(2)} f_{1r} + p_{3(1)} f_{1d}$. Notice that $\partial \lambda_2 / \partial \omega_1 < 0$, so $\lambda_2$ is a monotony decrease function of $\omega_1$. According to (4.23b), we get

$$\omega_1 > \frac{\lambda_1 - p_{3(1)} f_{1d} - \left(1 - p_{3(1)}\right) p_{3(2)} f_{1r}}{\left(1 - p_{3(1)}\right) \left(1 - p_{3(2)}\right) f_{1r}}. \tag{4.25}$$

So we get the optimal value of $\omega$ in different cases:

(1)  $0 < \frac{\lambda_1 - p_{3(1)} f_{1d} - (1 - p_{3(1)}) p_{3(2)} f_{1r}}{(1 - p_{3(1)})(1 - p_{3(2)}) f_{1r}} \leq 1$
   In this case, we get

$$p_{3(1)} f_{1d} + \left(1 - p_{3(1)}\right) p_{3(2)} f_{1r} < \lambda_1 \leq p_{3(1)} f_{1d} + \left(1 - p_{3(1)}\right) f_{1r},$$

then

$$\omega_1^* = \frac{\lambda_1 - p_{3(1)} f_{1d} - \left(1 - p_{3(1)}\right) p_{3(2)} f_{1r}}{\left(1 - p_{3(1)}\right) \left(1 - p_{3(2)}\right) f_{1r}}. \tag{4.26}$$

(2)  $\frac{\lambda_1 - p_{3(1)} f_{1d} - (1 - p_{3(1)}) p_{3(2)} f_{1r}}{(1 - p_{3(1)})(1 - p_{3(2)}) f_{1r}} \leq 0$
   In this case, we get

$$\lambda_1 \leq p_{3(1)} f_{1d} + \left(1 - p_{3(1)}\right) p_{3(2)} f_{1r},$$

then

$$\omega_1^* = 0. \tag{4.27}$$

Hence the optimal solution for the optimization problem in (4.22)–(4.23b) is given by (4.28).

$$\omega_1^* = \begin{cases} 0, & \lambda_1 < \lambda_m, \\ \frac{\lambda_1 - p_{3(1)} f_{1d} - (1 - p_{3(1)}) p_{3(2)} f_{1r}}{(1 - p_{3(1)})(1 - p_{3(2)}) f_{1r}}, & \lambda_m < \lambda_1 \leq \lambda_M, \end{cases} \tag{4.28}$$

where $\lambda_m == p_{3(1)} f_{1d} + \left(1 - p_{3(1)}\right) p_{3(2)} f_{1r}$, and $\lambda_M = p_{3(1)} f_{1d} + \left(1 - p_{3(1)}\right) f_{1r}$.

Next, we solve the other stability region of the system $S_2$. Substituting $\omega_1 + \omega_2 = 1$ and $\omega_0 = 1$ in (4.83), the maximum $\lambda_1$ can be written as

$$\max \lambda_1 = \left(1 - p_{3(1)}\right) f_{1r} + p_{3(1)} f_{1d}. \tag{4.29}$$

Then for a fixed $\lambda_1$, we solve the optimal $\lambda_2$ in $S_2$. According to (4.83), $\lambda_2$ can be written as

$$\lambda_2 < \frac{\left(1 - p_{3(2)}\right) f_{2r} + p_{3(2)} f_{2d} - \omega_1 \left(1 - p_{3(1)}\right) \left(1 - p_{3(2)}\right) f_{2r}}{\left(1 - p_{3(1)}\right) \left(1 - p_{3(2)}\right) f_{1r} - \omega_1 \left(1 - p_{3(1)}\right) \left(1 - p_{3(2)}\right) f_{1r}} \quad (4.30)$$
$$\cdot \left[\left(1 - p_{3(1)}\right) f_{1r} + p_{3(1)} f_{1d} - \lambda_1\right].$$

And take the first derivative of (4.30) with respect to $\omega_1$, we can calculate to get

$$\frac{\partial \lambda_2}{\partial \omega_1} = \frac{\left(1 - p_{3(1)}\right) \left(1 - p_{3(2)}\right) f_{1r} \left[p_{3(1)} \left(1 - p_{3(2)}\right) f_{2r} + p_{3(2)} f_{2d}\right]}{\left[\left(1 - p_{3(1)}\right) \left(1 - p_{3(2)}\right) f_{1r} \left(1 - \omega_1\right)\right]^2}.$$

Notice that $\partial \lambda_2 / \partial \omega_1 > 0$, so $\lambda_2$ is a monotonous increasing function of $\omega_1$. According to (4.84) and (4.30), we get

$$\omega_1 \leq \frac{\lambda_1 - p_{3(1)} f_{1d} - \left(1 - p_{3(1)}\right) p_{3(2)} f_{1r}}{\left(1 - p_{3(1)}\right) \left(1 - p_{3(2)}\right) f_{1r}}. \quad (4.31)$$

Then we discuss and calculate the optimal value of $\omega$ in different cases:

(1) $0 < \frac{\lambda_1 - p_{3(1)} f_{1d} - (1 - p_{3(1)}) p_{3(2)} f_{1r}}{(1 - p_{3(1)})(1 - p_{3(2)}) f_{1r}} \leq 1$
  In this case,

$$p_{3(1)} f_{1d} + \left(1 - p_{3(1)}\right) p_{3(2)} f_{1r} < \lambda_1 \leq \left(1 - p_{3(1)}\right) f_{1r} + p_{3(1)} f_{1d},$$

  then

$$\omega_1^* = \frac{\lambda_1 - p_{3(1)} f_{1d} - \left(1 - p_{3(1)}\right) p_{3(2)} f_{1r}}{\left(1 - p_{3(1)}\right) \left(1 - p_{3(2)}\right) f_{1r}}. \quad (4.32)$$

(2) $\frac{\lambda_1 - p_{3(1)} f_{1d} - (1 - p_{3(1)}) p_{3(2)} f_{1r}}{(1 - p_{3(1)})(1 - p_{3(2)}) f_{1r}} \leq 0$
  In this case,

$$\lambda_1 \leq p_{3(1)} f_{1d} + \left(1 - p_{3(1)}\right) p_{3(2)} f_{1r},$$

  then

$$\omega_1^* = 0. \quad (4.33)$$

So we can get the same optimal solution of $\omega_1^*$ as (4.28).

**Fig. 4.4** The envelope for the stability region of two-user and geo-relay system

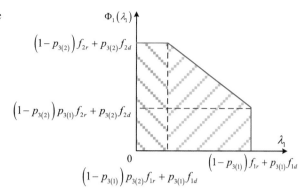

We summarize equation (4.28), (4.24) and (4.30) to describe the envelopes of the system $S_1$ and $S_2$. To these two systems, we get the same branch given by (4.34),

$$\Phi\left(\lambda_1\right) = \begin{cases} \left(1 - p_{3(2)}\right) f_{2r} + p_{3(2)} f_{2d}, & \lambda_1 < \lambda_m, \\ -\frac{f_{2r}}{f_{1r}} \lambda_1 + \psi, & \lambda_m \le \lambda_1 \le \lambda_M, \end{cases} \tag{4.34}$$

where $\psi = f_{2r} + \frac{f_{2r}}{f_{1r}} f_{1d} p_{3(1)} + f_{2d} p_{3(2)} - f_{2r} p_{3(1)} p_{3(2)}$, $\lambda_m = \left(1 - p_{3(1)}\right) p_{3(2)} f_{1r} + p_{3(1)} f_{1d}$, and $\lambda_M = \left(1 - p_{3(1)}\right) f_{1r} + p_{3(1)} f_{1d}$.

This proves the Theorem 4.1.

The envelope for the stability region of the proposed system is shown as Fig. 4.4.

The analysis above is based on the assumption that the buffer of GEO relay is sufficiently large and can forward entire received data immediately, which can ensure that the queue of the relay is always stable. This assumption is reasonable for current DRSs deployed in the GEO, such as TDRSS [36, 37] and Tianlian [38, 39], which are high-performance data relay satellites equipped with powerful store-and-forward facilities. When the storage and transmission capacity of relay is limited, the data overflow and loss may happen for the relay. Then the source satellites need to stop sending data to the relay and wait for new storage space and transmission resource. Consequently, the stability analysis of the system will be different, and the cooperation protocol needs to be designed for the situation of data overflow and loss, which will be studied in our future work. On the other hand, the protocol designed for the systems with LEO relay can be also applied into the GEO relay systems, when the storage and transmission capacity of the GEO relay is not powerful enough to keep the queue of the relay stable.

### 4.5.2  LEO Relay

In the network, $\lambda_1$ and $\lambda_2$ indicate the data arrival process to the queues of two LEO satellites (source nodes). Another LEO satellite performs as the relay node which can receive the packet from only one satellite at a single time slot. In addition, we assume that the relay satellite does not receive packet from the source or send to the ground at the same time slot due to the limited transmission capability of the LEO landsat satellite. According to above assumptions, the classified analysis of $L_{ir}$ ($i = 1, 2$) in the GEO relay system can be simplified. That is the last two situations: $L_{ir} = 1 \cap L_{jr} = 0$, $\{i, j \in \{1, 2\} : i \neq j\}$ and $L_{1r} = 1 \cap L_{2r} = 1$ can be discussed as one situation that there is at least one satellite can establish connection with the relay. In this condition, the time slot allocated to the satellite $i$ is used for the relay to send the packet at its queue head received from the satellite $j$ ($i \neq j$), when the satellite $i$ satisfies $L_{ir} = 0$. According to the time slot allocation protocol, the Lemma 4.2 states the stability region of the cooperative satellite system with an LEO relay satellite when $\omega_0$ and the resource sharing vector $[\omega_1, \omega_2]$ are fixed.

**Lemma 4.2** *In different link status, given $\omega_0$ and the resource sharing vector $[\omega_1, \omega_2]$, the stability region $\mathscr{R}(S)$ of two LEO satellites and a LEO relay satellite system with the time slot allocation protocol protocol is given by*

1. $L_{1r} = L_{2r} = 0$:

$$\mathscr{R}(S) = \{[\lambda_1, \lambda_2] = [0, 0]\}; \tag{4.35}$$

2. $L_{1r} = 1 \cup L_{2r} = 1$:

$$\mathscr{R}(S) = \mathscr{R}(S_1) \cup \mathscr{R}(S_2), \tag{4.36}$$

*where*

$$
\mathscr{R}(S_1) = \left\{ [\lambda_1, \lambda_2] \in R_+^2 \middle| \lambda_2 < \omega_2 p_{3(2)} f_{2d} \right.
$$
$$
+ \omega_1 \omega_2 \left( 1 - \frac{\lambda_1}{\omega_1 p_{3(1)} f_{1d}} \right) \left( 1 - p_{3(2)} f_{2d} \right) f_{rd}, \tag{4.37}
$$
$$
\left. \text{for } \lambda_2 < \omega_2 p_{3(2)} f_{2d} \right\},
$$

$$
\mathscr{R}(S_1) = \left\{ [\lambda_1, \lambda_2] \in R_+^2 \middle| \lambda_1 < \omega_1 p_{3(1)} f_{1d} \right.
$$
$$
+ \omega_1 \omega_2 \left( 1 - \frac{\lambda_2}{\omega_2 p_{3(2)} f_{2d}} \right) \left( 1 - p_{3(1)} f_{1d} \right) f_{rd}, \tag{4.38}
$$
$$
\left. \text{for } \lambda_1 < \omega_1 p_{3(1)} f_{1d} \right\}.
$$

**Fig. 4.5** Stability region of the leo-relay system for a fixed $[\omega_1, \omega_2]$ given by $\mathscr{R}(S_1) \cup \mathscr{R}(S_2)$

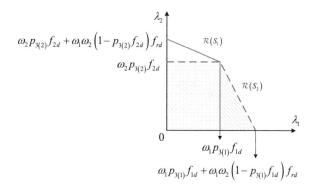

**Proof** See Sect. 4.9.

Similar to the system with a GEO relay satellite, $(\lambda_1, \lambda_2)$ in the stability region $\mathscr{R}(S_1)$ and $\mathscr{R}(S_2)$ assures the stability of the original system. Therefore, we get the stability region $\mathscr{R}(S_1) \cup \mathscr{R}(S_2)$ of the original system when the time sharing vector $[\omega_1, \omega_2]$ is given, as shown in Fig. 4.5.

Then we will find the whole stability region $\mathscr{R}(S)$ formulated by (4.19) for all feasible $W$ as (3.10) when there are two users and a LEO relay satellite in the system. We get the complete characterization of the stability region of the LEO relay system with the time slot allocation protocol in Theorem 4.2.

**Theorem 4.2** *With the time slot allocation protocol, the stability region of the whole system with two-use and an LEO relay is given by:*

$$\mathscr{R}(S) = \left\{ [\lambda_1, \lambda_2] \in R_+^2 \mid \lambda_2 < \max\left\{\Phi_1(\lambda_1), \Phi_2(\lambda_1)\right\} \right\}, \qquad (4.39)$$

*where $\Phi_1(\lambda_1)$ and $\Phi_2(\lambda_1)$ are defined as (4.40) and (4.41).*

$$\Phi_1(\lambda_1) = \begin{cases} \left(\frac{1}{2} + \frac{\lambda_1}{2p_{3(1)}f_{1d}} - \frac{p_{3(2)}f_{2d}}{2\psi_2}\right)^2 \psi_2 - \frac{\psi_2\lambda_1}{p_{3(1)}f_{1d}} + p_{3(2)}f_{2d}, & 0 \le \lambda_1 \le \lambda_{m1}, \\ -\frac{p_{3(2)}f_{2d}}{p_{3(1)}f_{1d}}\lambda_1 + p_{3(2)}f_{2d}, & \lambda_{m1} \le \lambda_1 \le \lambda_M, \end{cases}$$

$$(4.40)$$

$$\Phi_2(\lambda_1) = \begin{cases} -\frac{p_{3(2)}f_{2d}}{p_{3(1)}f_{1d}}\lambda_1 + p_{3(2)}f_{2d}, & 0 \le \lambda_1 \le \lambda_{m2}, \\ -2p_{3(2)}f_{2d}\sqrt{\frac{\lambda_1}{\psi_1}} + p_{3(2)}f_{2d} + \frac{p_{3(1)}p_{3(2)}f_{1d}f_{2d}}{\psi_1}, & \lambda_{m2} \le \lambda_1 \le \lambda_1^*, \end{cases}$$

$$(4.41)$$

*In (4.40) and (4.41), $\psi_i = \left(1 - p_{3(i)} f_{1d}\right) f_{rd}$ (i = 1, 2), $\lambda_{m1} = p_{3(1)} f_{1d} -$*

$\frac{p_{3(1)} p_{3(2)} f_{1d} f_{2d}}{\psi_2}$, $\lambda_M = p_{3(1)} f_{1d}$, $\lambda_{m2} = \frac{p_{3(1)}^2 f_{1d}^2}{(1 - p_{3(1)} f_{1d}) f_{rd}}$, *and $\lambda_1^*$ in (4.41) is defined as*

$$\lambda_1^* = \begin{cases} p_{3(1)} f_{1d}, & p_{3(1)} f_{1d} \geq \psi_1, \\ \frac{1}{4\psi_1} \left(p_{3(1)} f_{1d} + \psi_1\right)^2, & 0 \leq p_{3(1)} f_{1d} \leq \psi_1. \end{cases} \tag{4.42}$$

**Proof** First, for the system $S_1$, we establish the following optimization problem through (4.88) and (4.90) to maximize the arrival rate $\lambda_2$:

$$\max \ \lambda_2 = \omega_2 p_{3(2)} f_{2d} + \omega_1 \omega_2 \left(1 - \frac{\lambda_1}{\omega_1 p_{3(1)} f_{1d}}\right) \left(1 - p_{3(2)} f_{2d}\right) f_{rd}, \tag{4.43a}$$

$$\text{s.t. } \omega_1 + \omega_2 \leq 1, \tag{4.43b}$$

$$\lambda_1 \leq \omega_1 p_{3(1)} f_{1d}. \tag{4.43c}$$

Let $\psi_i = \left(1 - p_{3(i)} f_{1d}\right) f_{rd}$ (i = 1, 2), then the maximization problem is simplified into

$$\max \lambda_2 = (1 - \omega_1) p_{3(2)} f_{2d} + \omega_1 (1 - \omega_1) \left(1 - \frac{\lambda_1}{\omega_1 p_{3(1)} f_{1d}}\right) \psi_2. \tag{4.44}$$

Take the first derivative of (4.44) with respect to $\omega_1$:

$$\frac{\partial \lambda_2}{\partial \omega_1} = -p_{3(2)} f_{2d} + \psi_2 - 2\omega_1 \psi_2 + \frac{\lambda_1 \psi_2}{p_{3(1)} f_{1d}}. \tag{4.45}$$

Then the solution is given by

$$\omega_1' = \frac{1}{2\psi_2} \left(\psi_2 - p_{3(2)} f_{2d} + \frac{\lambda_1 \psi_2}{p_{3(1)} f_{1d}}\right). \tag{4.46}$$

According to (4.43c), $\omega_1 \geq \frac{\lambda_1}{p_{3(1)} f_{1d}}$. Then we discuss the optimal value of $\omega$ in difference cases.

(1) $\omega_1' \geq \frac{\lambda_1}{p_{3(1)} f_{1d}}$
we get

$$\lambda_1 \leq p_{3(1)} f_{1d} - \frac{p_{3(1)} p_{3(2)} f_{1d} f_{2d}}{\psi_2},$$

and the optimal solution $\omega_1^*$ is

$$\omega_1^* = \frac{1}{2\psi_2} \left(\psi_2 - p_{3(2)} f_{2d} + \frac{\lambda_1 \psi_2}{p_{3(1)} f_{1d}}\right). \tag{4.47}$$

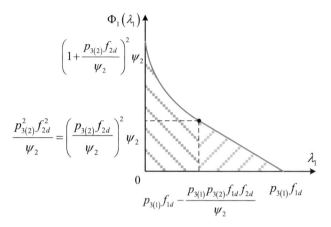

**Fig. 4.6** The first envelope for the stability region of the leo-relay system

(2) $\omega_1' \leq \dfrac{\lambda_1}{p_{3(1)} f_{1d}} \leq 1$
we get

$$p_{3(1)} f_{1d} - \frac{p_{3(1)} p_{3(2)} f_{1d} f_{2d}}{\psi_2} \leq \lambda_1 \leq p_{3(1)} f_{1d},$$

and the optimal solution $\omega_1^*$ is

$$\omega_1^* = \frac{\lambda_1}{p_{3(1)} f_{1d}}. \tag{4.48}$$

Then we get the optimal solution of $\omega$ for the optimization problem formulated by (4.43a)–(4.43c) as (4.49). Substituting (4.49) in (4.43a) as shown in Fig. 4.6, we get (4.50), which indicates the stability region of system $\mathscr{R}(S_1)$. In (4.50), $\psi_2 = \left(1 - p_{3(2)} f_{2d}\right) f_{rd}$.

$$\omega_1^* = \begin{cases} \frac{1}{2\psi_2}\left(\psi_2 - p_{3(2)} f_{2d} + \frac{\lambda_1 \psi_2}{p_{3(1)} f_{1d}}\right), & \lambda_1 \leq \lambda_m, \\ \frac{\lambda_1}{p_{3(1)} f_{1d}}, & \lambda_m \leq \lambda_1 \leq \lambda_M. \end{cases} \tag{4.49}$$

$$\Phi_1(\lambda_1) = \begin{cases} \left(\frac{1}{2} + \frac{\lambda_1}{2 p_{3(1)} f_{1d}} - \frac{p_{3(2)} f_{2d}}{2\psi_2}\right)^2 \psi_2 - \frac{\psi_2 \lambda_1}{p_{3(1)} f_{1d}} + p_{3(2)} f_{2d}, & \lambda_1 \leq \lambda_m, \\ -\frac{p_{3(2)} f_{2d}}{p_{3(1)} f_{1d}} \lambda_1 + p_{3(2)} f_{2d}, & \lambda_m \leq \lambda_1 \leq \lambda_M. \end{cases}$$

$$\tag{4.50}$$

In (4.49) and (4.50), $\lambda_m = p_{3(1)} f_{1d} - \frac{p_{3(1)} p_{3(2)} f_{1d} f_{2d}}{\psi_2}$, and $\lambda_M = p_{3(1)} f_{1d} - \frac{p_{3(1)} p_{3(2)} f_{1d} f_{2d}}{\psi_2}$.

Next, we consider the stability region $\mathscr{R}(S_2)$. Similar to the stability region branch of system $S_1$, let $\omega_1 + \omega_2 = 1$. According to (4.91), the maximum stabile rate of $\lambda_1$ can be achieved if $\lambda_2 = 0$. Then the arrival rate $\lambda_1$ can be given by

$$\max \lambda_1 = \omega_1 p_{3(1)} f_{1d} + \omega_1 (1 - \omega_1) \psi_1. \tag{4.51}$$

Take the first derivative of (4.51) with respect to $\omega_1$, and the extreme point is given by

$$\omega_1^* |_{\lambda_2 = 0} = \frac{1}{2\psi_1} \left( p_{3(1)} f_{1d} + \psi_1 \right). \tag{4.52}$$

Considering that $\omega_1 \leq 1$, we get $p_{3(1)} f_{1d} \leq \psi_1$. So the maximum of $\lambda_1$ under the condition that $\lambda_2 = 0$ is

$$\lambda_1^* |_{\lambda_2 = 0} = \begin{cases} p_{3(1)} f_{1d}, & p_{3(1)} f_{1d} \geq \psi_1, \\ \frac{1}{4\psi_1} \left( p_{3(1)} f_{1d} + \psi_1 \right)^2, & 0 \leq p_{3(1)} f_{1d} \leq \psi_1. \end{cases} \tag{4.53}$$

Next, we solve the maximum achievable tare $\lambda_2$ when $\lambda_1$ is fixed. According to (4.91), $\lambda_2$ can be write in terms of a function of $\lambda_1$ as

$$\lambda_2 = (1 - \omega_1) p_{3(2)} f_{2d} + \frac{p_{3(1)} p_{3(2)} f_{1d} f_{2d}}{\psi_1} - \frac{p_{3(2)} f_{2d}}{\omega_1 \psi_1} \lambda_1. \tag{4.54}$$

The second derivative of (4.54) with respect to $\omega_1$ is negative, which results that the extreme value calculated through the first derivative gets the maximum of $\lambda_2$ in (4.54), and the extreme value of $\omega$ is

$$\omega_1' = \sqrt{\frac{\lambda_1}{\psi_1}} = \sqrt{\frac{\lambda_1}{(1 - p_{3(1)} f_{1d}) f_{rd}}}. \tag{4.55}$$

On the other hand, $\omega_1 = 1 - \omega_2 \leq 1 - \frac{\lambda_2}{p_{3(2)} f_{2d}}$, i.e., $\lambda_2 \leq p_{3(2)} f_{2d} (1 - \omega_1)$. Then substitute $\lambda_2$ in (4.54) and after some manipulations, we get

$$\omega_1 \leq \frac{\lambda_1}{p_{3(1)} f_{1d}}. \tag{4.56}$$

According to (4.55) and (4.56), the optimum $\omega^*$ can be summarized as

$$\omega_1^* = \begin{cases} \frac{\lambda_1}{p_{3(1)} f_{1d}}, & \lambda_1 \leq \frac{p_{3(1)}^2 f_{1d}^2}{(1 - p_{3(1)} f_{1d}) f_{rd}}, \\ \sqrt{\frac{\lambda_1}{(1 - p_{3(1)} f_{1d}) f_{rd}}}, & \lambda_1 \geq \frac{p_{3(1)}^2 f_{1d}^2}{(1 - p_{3(1)} f_{1d}) f_{rd}}. \end{cases} \tag{4.57}$$

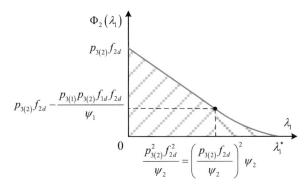

**Fig. 4.7** The second envelope for the stability region of the leo-relay system

Similar to (4.50), we substitute (4.57) in (4.54), the branch of the stability region $\mathcal{R}(S_2)$ can be specified as (4.58), which is shown in Fig. 4.7.

$$\Phi_2(\lambda_1) = \begin{cases} -\dfrac{p_{3(2)} f_{2d}}{p_{3(1)} f_{1d}} \lambda_1 + p_{3(2)} f_{2d}, & 0 \le \lambda_1 \le \lambda_m, \\ -2 p_{3(2)} f_{2d} \sqrt{\dfrac{\lambda_1}{\psi_1}} + p_{3(2)} f_{2d} + \dfrac{p_{3(1)} p_{3(2)} f_{1d} f_{2d}}{\psi_1}, & \lambda_m \le \lambda_1 \le \lambda_1^*. \end{cases}$$

$$(4.58)$$

In (4.58), $\psi_1 = \left(1 - p_{3(1)} f_{1d}\right) f_{rd}$, $\lambda_m = \dfrac{p_{3(1)}^2 f_{1d}^2}{(1 - p_{3(1)} f_{1d}) f_{rd}}$, and

$$\lambda_1^* = \begin{cases} p_{3(1)} f_{1d}, & p_{3(1)} f_{1d} \ge \psi_1, \\ \dfrac{1}{4\psi_1} \left(p_{3(1)} f_{1d} + \psi_1\right)^2, & 0 \le p_{3(1)} f_{1d} \le \psi_1. \end{cases}$$

Then we can get the stability region $R$ of the system with a LEO relay as

$$R = \left\{ (\lambda_1, \lambda_2) \in R_2^+ \mid \lambda_2 < \max \left\{ \Phi_1(\lambda_1), \Phi_2(\lambda_1) \right\} \right\}. \qquad (4.59)$$

This completes the proof of the Theorem 4.2.

*Delay Performance Analysis* In the LEO relay system, a data packet is not removed from the source satellite's queue until it is successfully transmitted to the destination. Then without regret to the transmission delay, the packet delay is the delay it waits in the source satellite's queue. Next, we analyze the delay performance of the symmetric two-user scenario with an LEO relay. The moment generating function of the joint queues' size processes $\left(Q_1^t, Q_2^t\right)$ can be defined as [29]

$$\Gamma(x, y) = \lim_{t \to \infty} E \left[ x^{Q_1^t} y^{Q_2^t} \right]. \qquad (4.60)$$

Assume that arrival processes are independent at different time slots. Then according to (4.73), we can get

$$E\left[x^{Q_1^{t+1}} y^{Q_2^{t+1}}\right] = E\left[x^{X_1^t} y^{X_2^t}\right] E\left[x^{(Q_1^t - Y_1^t)^+} y^{(Q_2^t - Y_2^t)^+}\right].$$ (4.61)

As assumed previously, the arrival processes follow Bernoulli random process. Then the first part of (4.61) can be written as

$$E\left[x^{X_1^t} y^{X_2^t}\right] = (x\lambda + 1 - \lambda)(y\lambda + 1 - \lambda),$$ (4.62)

where $\lambda$ is the arrival rate for the two symmetric source satellites. According to (4.88), the second part of (4.61) can be given by

$$\begin{aligned}
&E\left[x^{(Q_1^t - Y_1^t)^+} y^{(Q_2^t - Y_2^t)^+}\right] \\
&= E\left[I\left(Q_1^t = 0, Q_2^t = 0\right)\right] \\
&+ g_1(x) E\left[I\left(Q_1^t > 0, Q_2^t = 0\right) x^{Q_1^t}\right] \\
&+ g_1(y) E\left[I\left(Q_1^t = 0, Q_2^t > 0\right) y^{Q_2^t}\right] \\
&+ g_2(x, y) E\left[I\left(Q_1^t > 0, Q_2^t > 0\right) x^{Q_1^t} y^{Q_2^t}\right],
\end{aligned}$$ (4.63)

where

$$g_1(z) = 1 + \left(\frac{1}{z} - 1\right)\left[\omega p_{3(1)} f_{1d} + \omega^2 \left(1 - p_{3(1)} f_{1d}\right) f_{rd}\right],$$ (4.64a)

$$g_2(x, y) = \omega p_{3(1)} f_{1d} \left(\frac{1}{x} + \frac{1}{y}\right) + 2\omega \left(1 - p_{3(1)} f_{1d}\right),$$ (4.64b)

where $\omega$ is the symmetric resource sharing portion of each source satellite. Then (4.60) can be expressed as

$$\begin{aligned}
\Gamma(x, y) = (x\lambda + 1 - \lambda)(y\lambda + 1 - \lambda)\{\Gamma(0, 0) \\
+ g_1(x)[\Gamma(x, 0) - \Gamma(0, 0)] + g_1(y)[\Gamma(0, y) - \Gamma(0, 0)] \\
+ g_2(x, y)[\Gamma(x, y) - \Gamma(x, 0) - \Gamma(0, y) + \Gamma(0, 0)]\}.
\end{aligned}$$ (4.65)

Due to the symmetry, the average queue size can be given by $\Gamma_1(1, 1)$. To calculate $\Gamma_1(1, 1)$, the symmetry needs to be used. Applying L'Hopital limit theorem and through some straightforward calculations, we can get

$$\Gamma_1(1,1) = \frac{-\left[2\omega p_{3(1)} f_{1d} + \omega^2 \left(1 - p_{3(1)} f_{1d}\right) f_{rd}\right]\lambda^2 + 2\omega p_{3(1)} f_{1d}\lambda}{2\left[\omega p_{3(1)} f_{1d} + \omega^2 \left(1 - p_{3(1)} f_{1d}\right) f_{rd}\right]\left(\omega p_{3(1)} f_{1d} - \lambda\right)}.$$

(4.66)

Then the average queueing delay of the symmetric two-users system with an LEO relay is

$$\begin{aligned}
D &= \frac{1}{\lambda}\Gamma_1(1,1) \\
&= \frac{-\left[2\omega p_{3(1)} f_{1d} + \omega^2 \left(1 - p_{3(1)} f_{1d}\right) f_{rd}\right]\lambda + 2\omega p_{3(1)} f_{1d}}{2\left[\omega p_{3(1)} f_{1d} + \omega^2 \left(1 - p_{3(1)} f_{1d}\right) f_{rd}\right]\left(\omega p_{3(1)} f_{1d} - \lambda\right)}.
\end{aligned}$$

(4.67)

### 4.5.3   Multiple Users Case

The analytical method in this part can be extended to the general multi-user case. However, in such a case, the interactions between different users are much more complex, and the analysis tends to be very complicated. In [40], only the bounds on the stability region are given for ALOHA. In our work, we provide a feasible method for the symmetric scenario briefly. Consider there are $M$ user satellites can connect with the relay satellite (GEO/LEO). The dominant system, $S_M$, operates identically to the original $M$-user system, except that it does not help any of the LEO source satellites. Then the queue length in $S_M$ will never be shorter than that in the original system. In the symmetric scenario, all source satellites have the same success probability of transmission:

$$P\left[\left\{L_{id}^t = 1 \,\middle|\, L_{ir}^t = 1\right\} \cap C_{id}^t\right] = p_{3(1)} f_{1d}.$$

(4.68)

So the service rate per source satellite in the GEO relay system and LEO relay system are given by $\mu\left(S_M^{\text{GEO}}\right) = p_{3(1)} f_{1d}$, $\mu\left(S_M^{\text{LEO}}\right) = p_{3(1)} f_{1d}/M$, respectively, when the systems are symmetric. Applying Loynes theorem, the stability conditions for systems $S_M^{\text{GEO}}$ and $S_M^{\text{LEO}}$ are given by $\lambda^{\text{GEO}} < M p_{3(1)} f_{1d}$ and $\lambda^{\text{LEO}} < p_{3(1)} f_{1d}$, respectively, where $\lambda^{\text{GEO}}$ and $\lambda^{\text{LEO}}$ are the aggregate arrival rates of the systems. As discussed in Sect. 4.8, the stability condition of the dominant system can ensure the stability of original $M$-user system. Therefore, for the $M$-user symmetric systems applying cooperation protocols proposed, we can get the maximum stabile throughput as:

$$\lambda_{\text{MST}}^{\text{GEO}} = M p_{3(1)} f_{1d}, \quad \lambda_{\text{MST}}^{\text{LEO}} = p_{3(1)} f_{1d}.$$

(4.69)

## 4.6  Simulation Results

In this part, we perform simulation experiments to analyze performances of the two-user cooperative multiple access system with the proposed resource allocation protocol. In the bandwidth resource allocation system, the relay satellite is deployed in the geosynchronous orbit with the orbit radius of 42,164 km. In the time slot resource allocation system based on TDMA, the relay satellite is deployed in the low earth orbit with the height of 812 km. In both of the two systems above, the two users specifically indicate the two LEO satellites deployed in orbits with the height of 645 km and 785 km, respectively. The propagation path loss of ISL is given by $\gamma_1 = 2.1$, and the propagation path loss of the satellite-ground station link is given by $\gamma_2 = 2.8$, the transmit power $G = 10$ watt, and the average power of the Gaussian noise in ISL and the satellite-ground station link is $N_0 = 10^{-11}$ and $N_0 = 10^{-10}$, respectively. The ratio $K$ between the power in the LOS and the power in the other scattered paths of ISL and SGL is set as $K = 7.78$ dB and $K = 6.99$ dB [41], respectively, and the corresponding total power of the LOS and other scattered paths in ISL and SGL is set as $\Omega = 1 + K$.

First, we simulate the performance of the two proposed protocols under different settings of the SNR threshold $\beta$ and the maximum elevation angle of the ground-based radar $\alpha_2$. $\beta$ increases from 0 to 50 and $\alpha_2$ is selected as $30°$, $60°$ and $80°$. The inter-satellite link and the satellite-ground link connection status depends on the ON/OFF model and the physical channel model. Applying the bandwidth resource allocation protocol to the system of which the relay is a DRS on GEO, the aggregate maximum stable throughput of the two users is shown in Fig. 4.8. In this work, the aggregate throughput is defined as the ratio of the service to arrival in unit time. The corresponding results of the time resource allocation protocol are shown in Fig. 4.9, which indicates the aggregate maximum stable throughput of the two leo satellites in the cooperative transmission system, and in this system another LEO satellite operates as the relay. Numerical results indicate that the aggregate maximum stable throughput of source satellites decreases with the increase of $\beta$ and decrease of $\alpha_2$, and the effect of $\alpha_2$ is relatively weak, as shown in both Figs. 4.8 and 4.9. In both Figs. 4.8 and 4.9, the purple dotted lines indicate that under the ideal physical channel without noise ($f_{id} = 1$, $f_{ir} = 1$, $f_{rd} = 1$, $i = 1, 2$), the maximum aggregate throughput the GEO relay and the LEO relay systems can achieve according to the two resource allocation protocols, respectively. These two upper bounds are not much closed to 1, which results from discontinuous links between the source and the relay satellite and discontinuous links between the source satellite and the ground essentially according to (4.8)–(4.9). In addition, considering that the bandwidth allocation protocols allow both of the two source satellites to connect with the GEO relay in the same time slot, the upper bound of the GEO relay system is higher than the LEO relay system. Both of the bandwidth allocation protocol and the time slot allocation protocol perform well for low values of $\beta$.

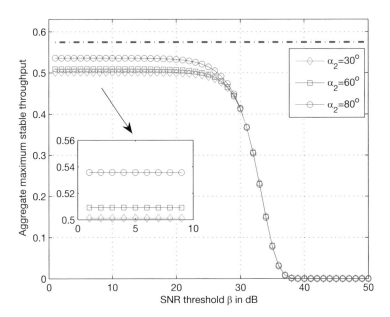

**Fig. 4.8**  Aggregate maximum stable throughput of geo-relay system versus SNR threshold $\beta$ and the maximum elevation angle $\alpha_2$ of the ground receiver

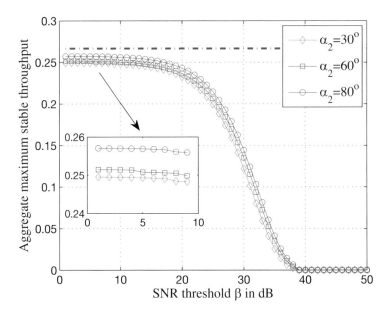

**Fig. 4.9**  Aggregate maximum stable throughput of leo-relay system versus SNR threshold $\beta$ and the maximum elevation angle $\alpha_2$ of the ground receiver

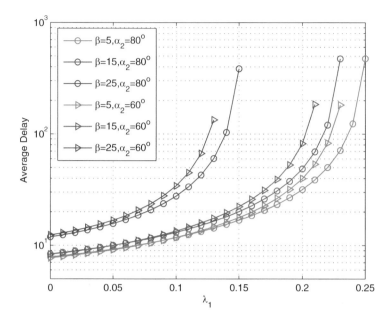

**Fig. 4.10** Average queueing delay per source satellite versus the arrival rate for different SNR threshold $\beta$ and the maximum elevation angle $\alpha_2$ of the ground receiver

For LEO relay time slot allocation protocol, the average queueing delay per LEO source satellite versus the arrival rate for different SNR threshold $\beta$ and the maximum elevation angle $\alpha_2$ is shown in Fig. 4.10. Results indicate that the delay increases with increasing $\beta$ and reducing $\alpha_2$. According to the premise that LEO source satellites can access the GEO relay, which can successfully forward all packet received, queueing delay of the GEO relay system is not discussed in this part.

Then we analyze the effects of the satellite orbit height on the aggregate throughput. We increase the two of the LEO satellite orbit height from 300 to 10,000 km, which is the typical range of the low earth orbit. Set the SNR threshold as $\beta = 10$, and other parameters are set as before. In Figs. 4.11 and 4.12, we get the aggregate maximum stable throughput of source satellites changed with different orbit height in the GEO relay system and LEO relay system, in which we use the bandwidth resource and time resource allocation protocols respectively. In the GEO relay system, the maximum throughput can be achieved when both of the satellites are deployed at the orbit height of about 2200 km. In the LEO relay system, when the orbit height is about 2800 km, the maximum throughput can be achieved.

Comparisons of the aggregate maximum stable throughput with existing cooperation protocols and no relay scene are shown in Fig. 4.13, in the case that both of the LEO source satellites can connect with the relay (GEO/LEO), and the link connection status depends on the ON/OFF model and the physical channel model. In the system without relays, we introduce the GEO relay to get the $p_{3(i)}$ ($i = 1, 2$) according to the previous premise. In the LEO relay system with the same parameter

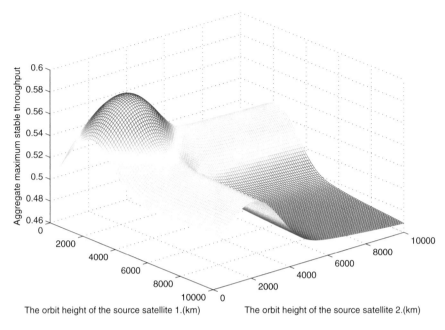

**Fig. 4.11** Aggregate maximum stable throughput of geo-relay system versus the orbit height of the two source satellites

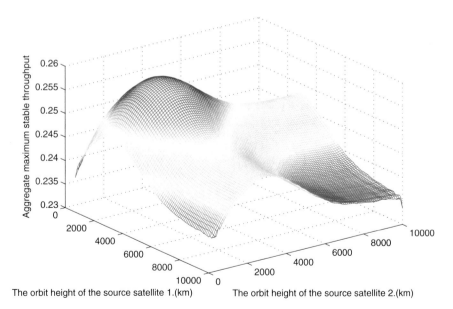

**Fig. 4.12** Aggregate maximum stable throughput of leo-relay system versus the orbit height of the two source satellites

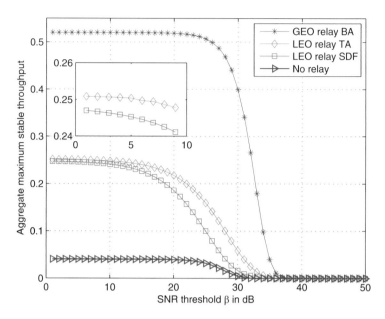

**Fig. 4.13** Aggregate maximum stable throughput versus different SNR threshold $\beta$

setting as the previous simulation, we compare the time slot allocation (TA) of this part with the selection decode-and-forward (SDF) protocol [42], the stability analysis of which are provided in [29] and the stability condition is specified as:

$$\frac{\lambda_1}{\text{Pr}_{1,\text{SDF}}\{C\}} + \frac{\lambda_2}{\text{Pr}_{2,\text{SDF}}\{C\}} < \frac{1}{2}, \tag{4.70}$$

where $\text{Pr}_{i,\text{SDF}}$ ($i = 1, 2$) represents the success transmission probability in the cooperative system with SDF, and $\text{Pr}_{i,\text{SDF}}$ can be calculated as:

$$\begin{aligned}
\text{Pr}_{i,\text{SDF}}\{C\} =& \left(1 - p_{3(i)}^2 f_{id}^2\right)^2 \left(1 - \left(1 - p_{3(i)}\right)^2 f_{ir}^2\right) \\
&+ \left(1 - p_{3(i)}^2 f_{id}^2\right)\left(1 - p_{3(i)}\right)^2 f_{ir}^2 \left(1 - f_{rd}^2\right).
\end{aligned} \tag{4.71}$$

In addition, the proposed bandwidth allocation (BA) protocol is used in the system with a GEO relay. Set $\alpha_2 = 80°$. The influence of the SNR threshold $\beta$ on the throughput is shown in Fig. 4.13. Without the relay, LEO can transmit data only when they are in the capture region of the ground station, and then the aggregate maximum stable throughput is rather low because of the low ratio of the connection. The maximum attained throughput for SDF is rather closed to the results of the LEO relay TA protocol because of the time slot repetition [29, 42], especially for the low values of $\beta$. But for high values of $\beta$, the performance of SDF is worse than the LEO

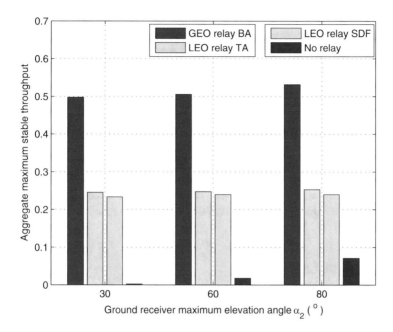

**Fig. 4.14** Aggregate maximum stable throughput versus the maximum elevation angle $\alpha_2$ of the ground receiver ($\beta = 10$)

relay TA protocol. GEO relay system with the BA protocol performs best, which results from the widespread cover of the GEO satellite for LEO source satellites and the time multiplex of the protocol.

Next, we test the effect of the ground receiver's capture region, which depends on $\alpha_2$, on the throughput for existing and proposed resource allocation protocols. As in previous simulation, $\alpha$ is selected as $30°$, $60°$ and $80°$, then we set $\beta = 10$ and leave the other arguments intact. Results are shown in Fig. 4.14, which indicates that the throughput increases with increasing $\alpha$ for the four testing scenes, although the effect is very weak. Moreover, the GEO relay system applied the BA protocol performs best, and the system without relay gets the worst result. The TA protocol and SDF for LEO relay system get similar results, and the former performs better. The comparatione results of these four scenes are same as the last simulation of $\beta$ influence.

## 4.7  Conclusion

In this part, we proposed two new multiple access communication systems with a GEO and a LEO satellite as the relay, respectively, to achieve cooperative transmission for space-based networks. According to the different network structures and the diversities of the transmission and store-and-forward capabilities of GEO and LEO relay satellites, two types of the resource allocation protocols were proposed for better resource utilization. By applying the proposed cooperation protocols, the idle bandwidth resource of GEO relay satellite and the time slot resource of LEO relay satellite can be allocated efficiently to improve the utilization of transmission resource. Considering the bandwidth and time resource allocation for GEO and LEO relay satellites, respectively, we analyzed the stability region for the two-user systems in detail and gave the stability region of the two types of cooperation systems. The system decomposition method introduced provides an efficient solution for the analytical difficulties of interacting queues. Simulation results indicate the performance of the proposed systems and corresponding protocols, and demonstrate that the cooperation protocols proposed can improve the utilization of transmission resource and increase the throughput of the multi-access systems effectively.

## 4.8  Proof of Lemma 4.1

(1)  $L_{1r} = L_{2r} = 0$:

In this case, neither of the LEO satellites can link with the relay to transmit data because of the Earth background. Therefore, the two arrival rates must satisfy

$$\lambda_1 = 0 \text{ and } \lambda_2 = 0. \tag{4.72}$$

(2)  $L_{ir} = 1 \cap L_{jr} = 0, \{i, j \in \{1, 2\} : i \neq j\}$:

In this case, LEO satellite $i$ can access the relay though it will switch the link to the ground station when $L_{id} = 1$ with the probability $p_{3(i)}$ according to (4.8) and the proposed protocol. Let $\omega_0$ $(0 < \omega_0 \leq 1)$ denote the relay transmission resource allocated to $i$ when $L_{ir} = 1 \cap L_{id} = 0$. We assume that the satellite-ground transmission resource is equal to the entire relay's, so when $L_{id} = 1$, the allocated resource is 1. Then according to the Loynes' theory, the arrival and service processes for each queue need to meet the queue stability condition. Let $Q_i^t(S)$ $(i = 1, 2)$ denote the queue size at the source node $i$ of the system $S$ at the time slot $t$. Then we can get

$$Q_i^{t+1}(S) = \max\left\{Q_i^t(S) - Y_i^t(S), 0\right\} + X_i^t(S), \tag{4.73}$$

where $X_i^t(S)$ is the number of arrivals to the source node $i$ of the system $S$ at the time slot $t$ and is a Poisson random arrival process with the finite mean $E\left[X_i^t(S)\right] = \lambda_i < \infty$, $Y_i^t(S)$ is the possible departure from node $i$ at time slot $t$. For source node $i$, $Y_i^t(S)$ can be modeled as

$$P\left[Y_i^t(S) = \omega_0\right] = P\left[\{L_{id}^t = 0 \,|\, L_{ir}^t = 1\} \cap C_{ir}^t\right], \tag{4.74a}$$

$$P\left[Y_i^t(S) = 1\right] = P\left[\{L_{id}^t = 1 \,|\, L_{ir}^t = 1\} \cap C_{id}^t\right]. \tag{4.74b}$$

So the statistical expectation of $Y_i^t(S)$ is

$$\lambda_i < \omega_0 \left(1 - p_{3(i)}\right) f_{ir} + p_{3(i)} f_{id}. \tag{4.75}$$

According to the Loynes' theory, stability of queues $i$ and $j$ in the system is achieved if the following conditions hold, respectively:

$$\lambda_i < \omega_0 \left(1 - p_{3(i)}\right) f_{ir} + p_{3(i)} f_{id}, \tag{4.76}$$

$$\lambda_j = 0. \tag{4.77}$$

**(3)** $L_{1r} = L_{2r} = 1$:

In this case, both of the LEO satellites have the opportunity of establishing the connection with ground stations with the probability $p_{3(i)}$ ($i = 1, 2$). Let $[\omega_1, \omega_2]$ $(0 < \omega_1, \omega_2 < \omega_0, \omega_1 + \omega_2 \le 1)$ denote the relay transmission resource allocated to the two accessed. We assume that the relay can sense the communication channel to detect the idle transmission resource as discussed in Sect. 4.4. As a result, there are interactions between the two queues of the source nodes. However, stability analysis of the interacting queues is a difficult problem. To solve this problem, we introduce the system decomposition method in [29] as follows.

We define $S_i$ ($i = 1, 2$) as a system that relay can only sense the arrival process of source node $i$ but cannot achieve the same sense for the other source node. Then we decompose the system into $S_1$ and $S_2$, of which the stability regions are $\mathscr{R}(S_1)$ and $\mathscr{R}(S_2)$, respectively. We notice that in system $S_1$, the length of queue 1 is always not shorter than the original system. This holds because satellite 1 can not send data when there is no data to be sent in the queue of satellite 2, as the defined rules in system $S_1$. So the stability conditions for system $S_1$ is sufficient for the original system. Similarly, the stability conditions for system $S_2$ can also ensure the stability of the original system. Consequently, we can get the stability region of our two-user system as $\mathscr{R}(S_1) \cup \mathscr{R}(S_2)$.

Firstly, we study the stability region of $S_1$. For the source node 1, $Y_1^t(S_1)$ can be modeled with the similar method in (4.74) as follows:

$$P\left[Y_1^t(S_1) = \omega_1\right] = P\left[\{L_{2d}^t = 0 \,|\, L_{2r}^t = 1\} \cap \{L_{1d}^t = 0 \,|\, L_{1r}^t = 1\} \cap C_{1r}^t\right], \tag{4.78a}$$

$$P\left[Y_1^t(S_1) = \omega_0\right] = P\left[\{L_{2d}^t = 1 \,|\, L_{2r}^t = 1\} \cap \{L_{1d}^t = 0 \,|\, L_{1r}^t = 1\} \cap C_{1r}^t\right], \tag{4.78b}$$

$$P\left[Y_1^t(S_1) = 1\right] = P\left[\{L_{1d}^t = 1 \,|\, L_{1r}^t = 1\} \cap C_{1d}^t\right]. \tag{4.78c}$$

Then according to Loynes' theory, the condition that promises the stability of the queue in source node 1 in the system $S_1$ can be achieved is formulated as

$$\lambda_1 < \left(1 - p_{3(1)}\right) \left[\omega_0 p_{3(2)} f_{1r} + \omega_1 \left(1 - p_{3(2)}\right) f_{1r}\right] + p_{3(1)} f_{1d}. \qquad (4.79)$$

Now we consider the queue 2 in system $S_1$. In $S_1$, the relay can also sense the arrival process of the source node 1, and when queue 1 has no packet to transmit at the time slot $t$, $\omega_0$ is allocated to the queue 2. When $\{L_{2d}^t = 0\} \cap \{L_{1d}^t = 0\}$ and there exists packets at the queue 1, $\omega_2$ is allocated to the queue 2. Moreover, when the node 2 can directly connect with the ground station, transmission resource allocated to it is 1. The above allocation strategies can be modeled as follows:

$$P\left[Y_2^t(S_1) = \omega_0\right]$$
$$= P\big[\{L_{2d}^t = 0 \,|\, L_{2r}^t = 1\} \qquad\qquad\qquad (4.80a)$$
$$\cap\{\{L_{1d}^t = 1 \,|\, L_{1r}^t = 1\} \cup \{Q_1^t(S_1) = 0\}\} \cap C_{2r}^t\big],$$

$$P\left[Y_2^t(S_1) = \omega_2\right]$$
$$= P\big[\{L_{2d}^t = 0 \,|\, L_{2r}^t = 1\} \cap \{L_{1d}^t = 0 \,|\, L_{1r}^t = 1\} \qquad (4.80b)$$
$$\cap\{Q_1^t(S_1) \neq 0\} \cap C_{2r}^t\big],$$

$$P\left[Y_2^t(S_1) = 1\right] = P\big[\{L_{2d}^t = 1 \,|\, L_{2r}^t = 1\} \cap C_{2d}^t\big], \qquad (4.80c)$$

where $\{Q_1^t(S_1) = 0\}$ denotes that the queue 1 has no packet to send at $t$. Under (4.79), we can get the probability of queue 1 empty as

$$\Pr\{Q_1^t(S_1) = 0\}$$
$$= 1 - \frac{\lambda_1}{\left(1 - p_{3(1)}\right)\left[\omega_0 p_{3(2)} + \omega_1 \left(1 - p_{3(2)}\right)\right] f_{1r} + p_{3(1)} f_{1d}}. \qquad (4.81)$$

Using the Loynes stability theory and given (4.15) and (4.81), the stability condition of queue 2 in system $S_1$ is given by

$$\lambda_2 < \omega_0 \left(1 - p_{3(2)}\right) f_{2r} + p_{3(2)} f_{2d}$$
$$- \frac{(\omega_0 - \omega_2)\left(1 - p_{3(1)}\right)\left(1 - p_{3(2)}\right) f_{2r} \lambda_1}{\left(1 - p_{3(1)}\right)\left[\omega_0 p_{3(2)} + \omega_1 \left(1 - p_{3(2)}\right)\right] f_{1r} + p_{3(1)} f_{1d}}. \qquad (4.82)$$

Both (4.79) and (4.82) constitute the conditions for the stability of system $S_1$ for a special sharing vector $[\omega_1, \omega_2]$. $\mathscr{R}(S_1)$ is exactly ranged by (4.79) and (4.82). Then we can get the stability region $\mathscr{R}(S_2)$ of dominant system $S_2$ by the parallel arguments, and $\mathscr{R}(S_2)$ can be specified as (4.83) and (4.84). We have

$$\lambda_1 < \omega_0 \left(1 - p_{3(1)}\right) f_{1r} + p_{3(1)} f_{1d}$$
$$- \frac{(\omega_0 - \omega_1)\left(1 - p_{3(1)}\right)\left(1 - p_{3(2)}\right) f_{1r} \lambda_2}{\left(1 - p_{3(2)}\right)\left[\omega_0 p_{3(1)} + \omega_2\left(1 - p_{3(1)}\right)\right] f_{2r} + p_{3(2)} f_{2d}}. \tag{4.83}$$

$$\lambda_2 < \left(1 - p_{3(2)}\right)\left[\omega_0 p_{3(1)} + \omega_2\left(1 - p_{3(1)}\right)\right] f_{2r} + p_{3(2)} f_{2d}. \tag{4.84}$$

According to above analysis, any point in region $\mathscr{R}(S_1)$ and $\mathscr{R}(S_2)$ can be achieved to the original system and the stability condition of $\mathscr{R}(S_1)$ and $\mathscr{R}(S_2)$ is sufficient for the stability of the system. This completes the proof of Lemma 4.1.

## 4.9   Proof of Lemma 4.2

According to the cooperation protocol discussed in the last section, when both of the two satellite can neither link with the ground station nor link with the available relay satellite, then the two arrival rates must satisfy

$$\lambda_1 = 0 \text{ and } \lambda_2 = 0. \tag{4.85}$$

Next, we focus on the situation that there are at least one LEO satellite can link with the relay that satisfying $L_{rd} = 1$. According to the time slot allocation protocol designed for the LEO relay system, the queues of the two source LEO source satellites are interactive in this case. Specifically, whether the packet sent by one source can be delivered to the ground successfully by the relay depends on the other source satellite has the idle time slot or not. To due with this interaction, we introduce the similar analysis as the last part of this section. The whole system is separated into two dominant systems $S_1$ and $S_2$. Different with the time slot allocation protocol for the LEO relay system, we define that in the system $S_1$, the queue 1 do not delete the packet even this packet is transmitted by the relay successfully, which means that the relay transmits the data sent from the queue 2, and the queue 1 acts as a node in TDMA systems. Then in the dominant system $S_1$, the length of the queue 1 will be never shorter than the original system, so the stability conditions of the queues 1 and 2 are sufficient for the original system [29]. The system $S_2$ is defined as well. As a result, the stability region of the LEO relay system is also formulated as $\mathscr{R}(S_1) \cup \mathscr{R}(S_2)$. Then, we analyze the station of queues in queues 1 and 2 through the Loynes' theorem. We define $Y_i^t(S_1)$ $(i = 1, 2)$ as the possible data send from the queue $i$ at the time slot $t$. Specifically, for source node 1, $Y_1^t(S_1)$ happens only under the circumstance that the time slot allocation and the ISL/SGL station due to the visibility and SNR support the data transmission at the same time $t$, which can be modeled as

$$Y_1^t(S_1) = I\left[A_1^t \cap \left\{L_{1d}^t = 1 \,\middle|\, L_{1r}^t = 1\right\} \cap C_{1d}^t\right], \tag{4.86}$$

where $A_1^t$ indicates that the time slot $t$ is allocated to the queue 1, $I[\cdot]$ is the indicator function. According to Loynes, the condition that promises the stability of the queue in source 1 in system $S_1$ can be achieved is

$$\lambda_1 < \omega_1 p_{3(1)} f_{1d}. \tag{4.87}$$

Then we analyze the queue 2 in the system $S_1$. When either of the following two events is set up, the packet from the queue 2 would be delivered to the ground from the satellite 2 or the relay:

*(1) At the time slot allocated to the queue 2:* the satellite 2 can link with the ground station directly, and the physical channel support the successful transmission;
*(2) At the time slot allocated to the queue 1:* the satellite cannot link with the relay or the queue 2 is empty. Meanwhile, the last time slot is assigned to the queue 2, which sends packet successfully to the relay and in the current time slot, the physical channel supports the successful transmission from the relay to the ground station.

The situation above can be modeled as

$$Y_2^t(S_1) = I\left[A_2^t \cap \left\{L_{2d}^t = 1 \,\middle|\, L_{2r}^t = 1\right\} \cap C_{2d}^t\right]$$
$$+ I\left[A_1^t \cap \left\{Q_1^t(S_1) = 0\right\} \cap A_2^{t-1} \cap \overline{\left\{L_{2d}^{t-1} = 1 \,\middle|\, L_{2r}^{t-1} = 1\right\}} \cap C_{2d}^{t-1} \cap C_{rd}^t\right], \tag{4.88}$$

where $Q_1^t(S_1) = 0$ denotes that the queue 1 has no packet to send at $t$. According to (4.88), the probability of queue 1 empty is

$$\Pr\left\{Q_1^t(S_1) = 0\right\} = 1 - \frac{\lambda_1}{\omega_1 p_{3(1)} f_{1d}}. \tag{4.89}$$

Introducing the Loynes stability and given (4.15) and (4.89), the stability condition of queue 2 in system $S_1$ is given by

$$\lambda_2 < \omega_2 p_{3(2)} f_{2d} + \omega_1 \omega_2 \left(1 - \frac{\lambda_1}{\omega_1 p_{3(1)} f_{1d}}\right)\left(1 - p_{3(2)} f_{2d}\right) f_{rd}. \tag{4.90}$$

Then (4.88) and (4.90) constitute the stability conditions of system $S_1$ for a special time sharing vector $[\omega_1, \omega_2]$. Similarly, we can get the stability region

$\mathscr{R}(S_2)$ of the system $S_2$, which can be ranged by (4.91) and (4.92). This proves the Lemma 4.2.

$$\lambda_1 < \omega_1 p_{3(1)} f_{1d} + \omega_1 \omega_2 \left(1 - \frac{\lambda_2}{\omega_2 p_{3(2)} f_{2d}}\right)\left(1 - p_{3(1)} f_{1d}\right) f_{rd}, \quad (4.91)$$

$$\lambda_2 < \omega_2 p_{3(2)} f_{2d}. \quad (4.92)$$

This completes the proof of Lemma 4.2.

# References

1. K. Bhasin and J. L. Hayden, "Space internet architectures and technologies for NASA enterprises," in *Aerospace Conference, IEEE Proceedings*, vol. 2.  Big Sky, MT, Mar. 2001, pp. 2–931.
2. J. Du, C. Jiang, Q. Guo, M. Guizani, and Y. Ren, "Cooperative earth observation through complex space information networks," *IEEE Wireless Commun.*, vol. 23, no. 2, pp. 136–144, May 2016.
3. Z. Xu, J. Du, J. Wang, C. Jiang, and Y. Ren, "Satellite image prediction relying on GAN and LSTM neural networks," in *IEEE Int. Conf. Commun. (ICC'19)*.  Shanghai, China, 20–24 May 2019.
4. J. Du, C. Jiang, X. Wang, Q. Guo, X. Wang, X. Zhu, and Y. Ren, "Detection and transmission resource configuration for space-based information network," in *IEEE Global Conf. Signal Inform. Process. (GlobalSIP'14)*.  Atlanta, GA, USA, 3–5 Dec. 2014.
5. B. Friedrichs, "Data processing for broadband relay satellite networks–digital architecture and performance evaluation," in *International Communications Satellite Systems Conferences, ICSSC*.  Florence, Italy, Oct. 2013.
6. R. M. Calvo, P. Becker, D. Giggenbach, F. Moll, M. Schwarzer, M. Hinz, and Z. Sodnik, "Transmitter diversity verification on artemis geostationary satellite," in *SPIE Proceeding*, vol. 8971.  International Society for Optics and Photonics, Mar. 2014, pp. 897 104.1–897 104.14.
7. C. Zhang, C. Jiang, J. Jin, S. Wu, L. Kuang, and S. Guo, "Spectrum sensing and recognition in satellite systems," *IEEE Trans. Veh. Tech.*, vol. 68, no. 3, pp. 2502–2516, Mar. 2019.
8. B. L. Edwards and D. J. Israel, "A geosynchronous orbit optical communications relay architecture," in *Aerospace Conference, IEEE*.  Big Sky, MT, Mar. 2014, pp. 1–7.
9. C. Jiang and Z. Li, "Decreasing big data application latency in satellite link by caching and peer selection," *IEEE Trans. Network Sci. Eng.*, vol. 7, no. 4, pp. 2555–2565, Oct. 2020.
10. L. Wang, C. Jiang, L. Kuang, S. Wu, L. Fei, and H. Huang, "Mission scheduling in space network with antenna dynamic setup times," *IEEE Trans. Aerosp. Electron. Syst.*, vol. 55, no. 1, pp. 31–45, Feb. 2019.
11. J.-P. Sheu, C.-C. Kao, S.-R. Yang, and L.-F. Chang, "A resource allocation scheme for scalable video multicast in WiMAX relay networks," *Mobile Computing, IEEE Transactions on*, vol. 12, no. 1, pp. 90–104, Jan. 2013.
12. A. Zafar, M. Shaqfeh, M. Alouini, and H. Alnuweiri, "Resource allocation for two source-destination pairs sharing a single relay with a buffer," *Communications, IEEE Transactions on*, vol. 62, no. 5, pp. 1444–1457, May. 2014.
13. R. O. Afolabi, A. Dadlani, and K. Kim, "Multicast scheduling and resource allocation algorithms for OFDMA-based systems: A survey," *Communications Surveys & Tutorials, IEEE*, vol. 15, no. 1, pp. 240–254, Feb. 2013.

14. S. Kandeepan, K. Gomez, L. Reynaud, and T. Rasheed, "Aerial-terrestrial communications: terrestrial cooperation and energy-efficient transmissions to aerial base stations," *Aerospace and Electronic Systems, IEEE Transactions on*, vol. 50, no. 4, pp. 2715–2735, Oct. 2014.
15. J. Du, C. Jiang, S. Yu, and Y. Ren, "Time cumulative complexity modeling and analysis for space-based networks," in *IEEE Int. Conf. Commun. (ICC'16)*.    Kuala Lumpur, Malaysia, 23–27 May 2016.
16. S. Morosi, S. Jayousi, and E. Del Re, "Cooperative strategies of integrated satellite/terrestrial systems for emergencies," in *Personal Satellite Services*.   Springer, Feb. 2010, pp. 409–424.
17. S. Morosi, E. Del Re, S. Jayousi, and R. Suffritti, "Hybrid satellite/terrestrial cooperative relaying strategies for DVB-SH based communication systems," in *Wireless Conference, European*.    IEEE, Aalborg, May. 2009, pp. 240–244.
18. I. del Portillo, E. Bou, E. Alarcon, M. Sanchez-Net, D. Selva, and A. Alvaro, "On scalability of fractionated satellite network architectures," in *Aerospace Conference, IEEE*.   Big Sky, MT, Mar. 2015, pp. 1–13.
19. S. K. Sharma, D. Christopoulos, S. Chatzinotas, and B. Ottersten, "New generation cooperative and cognitive dual satellite systems: Performance evaluation," in *32nd AIAA International Communications Satellite Systems Conference*.    San Diego, California, Aug. 2014.
20. A. Svigelj, M. Mohorcic, and G. Kandus, "Oscillation suppression for traffic class dependent routing in ISL network," *Aerospace and Electronic Systems, IEEE Transactions on*, vol. 43, no. 1, pp. 187–196, Jan. 2007.
21. J. Du, C. Jiang, Y. Qian, Z. Han, and Y. Ren, "Resource allocation with video traffic prediction in cloud-based space systems," *IEEE Transactions on Multimedia*, vol. 18, no. 5, pp. 820–830, Mar. 2016.
22. J. Du, C. Jiang, Y. Qian, Z. Han, and Y. Ren, "Traffic prediction based resource configuration in space-based systems," in *IEEE Int. Conf. Commun. (ICC'16)*.    Kuala Lumpur, Malaysia, 23–27 May 2016.
23. J. Wang, C. Jiang, H. Zhang, Y. Ren, and V. C. Leung, "Aggressive congestion control mechanism for space systems," *IEEE Aerospace and Electronic Systems Magazine*, vol. 31, no. 3, pp. 28–33, Jun. 2016.
24. C. Jiang, X. Wang, J. Wang, H.-H. Chen, and Y. Ren, "Security in space information networks," *IEEE Communications Magazine*, vol. 53, no. 8, pp. 82–88, Aug. 2015.
25. E. Blasch, K. Pham, G. Chen, G. Wang, C. Li, X. Tian, and D. Shen, "Distributed QoS awareness in satellite communication network with optimal routing (QuASOR)," in *Digital Avionics Systems Conference (DASC), 33rd AIAA*.    IEEE, Colorado Springs, CO, Oct. 2014, pp. 6C3–1–6C3–11.
26. W. Szpankowski, "Stability conditions for some distributed systems: Buffered random access systems," *Advances in Applied Probability*, vol. 26, no. 2, pp. 498–515, Jun. 1994.
27. G. D. Celik and E. Modiano, "Scheduling in networks with time-varying channels and reconfiguration delay," in *INFOCOM, 2012 Proceedings IEEE*.    IEEE, Orlando, FL, Mar. 2012, pp. 990–998.
28. R. Loynes, "The stability of a queue with non-independent inter-arrival and service times," in *Mathematical Proceedings of the Cambridge Philosophical Society*, vol. 58, no. 03. Cambridge Univ Press, 1962, pp. 497–520.
29. A. K. Sadek, K. R. Liu, and A. Ephremides, "Cognitive multiple access via cooperation: protocol design and performance analysis," *Information Theory, IEEE Transactions on*, vol. 53, no. 10, pp. 3677–3696, Oct. 2007.
30. A. A. El-Sherif and K. R. Liu, "Cooperation in random access networks: Protocol design and performance analysis," *Selected Areas in Communications, IEEE Journal on*, vol. 30, no. 9, pp. 1694–1702, Sept. 2012.
31. E. L. Lehmann and G. Casella, *Theory of point estimation*.    Springer Science & Business Media, 1998, vol. 31.
32. S. M. Kay, *Fundamentals of Statistical Signal Processing: Practical Algorithm Development*. Pearson Education, 2013, vol. 3.

33. A. Abdi, C. Tepedelenlioglu, M. Kaveh, and G. Giannakis, "On the estimation of the k parameter for the rice fading distribution," *Communications Letters, IEEE*, vol. 5, no. 3, pp. 92–94, Mar. 2001.

34. N. Benvenuto, S. G. Pupolin, and G. Guidotti, "Performance evaluation of multiple access spread spectrum systems in the presence of interference," *Vehicular Technology, IEEE Transactions on*, vol. 37, no. 2, pp. 73–77, May. 1988.

35. B. Hajek, A. Krishna, and R. O. LaMaire, "On the capture probability for a large number of stations," *Communications, IEEE Transactions on*, vol. 45, no. 2, pp. 254–260, Feb. 1997.

36. A. Dissanayake, A. Zarembowitch, K. Hogie, X. Yang, J. Lubelczyk, and H. Safavi, "TDRSS narrow-band simulator and test system," in *2016 IEEE Aerospace Conference*. IEEE, Big Sky, MT, Mar. 5–12, 2016, pp. 1–10.

37. M. Novak, E. Alwan, F. Miranda, and J. Volakis, "Conformal and spectrally agile ultra wideband phased array antenna for communication and sensing," NASA Glenn Research Center; Cleveland, OH United States, Tech. Rep. GRC-E-DAA-TN26916, Sept. 2015.

38. M. Xu, J. Wang, and N. Zhou, "Computational mission analysis and conceptual system design for super low altitude satellite," *Journal of Systems Engineering and Electronics*, vol. 25, no. 1, pp. 43–58, Feb. 2014.

39. B. Chapman, "Chinese military space power: US department of defense annual reports," *Astropolitics*, vol. 14, no. 1, pp. 71–89, 2016.

40. V. Naware, G. Mergen, and L. Tong, "Stability and delay of finite-user slotted ALOHA with multipacket reception," *Information Theory, IEEE Transactions on*, vol. 51, no. 7, pp. 2636–2656, Jul. 2005.

41. G. T. Irvine and P. J. McLane, "Symbol-aided plus decision-directed reception for PSK/TCM modulation on shadowed mobile satellite fading channels," *Selected Areas in Communications, IEEE Journal on*, vol. 10, no. 8, pp. 1289–1299, Oct. 1992.

42. J. N. Laneman, D. N. Tse, and G. W. Wornell, "Cooperative diversity in wireless networks: Efficient protocols and outage behavior," *Information Theory, IEEE Transactions on*, vol. 50, no. 12, pp. 3062–3080, Dec. 2004.

# Part III
# Cooperative Transmission in Integrated Satellite-Terrestrial Networks

# Chapter 5
# Introduction of Cooperative Transmission in Integrated Satellite-Terrestrial Networks

**Keywords** 6G · Integrated Satellite-terrestrial Networks · Cooperative Transmission

Cooperative networking of high, medium and low orbit satellites can achieve wide area or even global coverage, which can provide undifferentiated communication services for global users. At the same time, the current terrestrial fifth generation mobile communications (5G) will have a perfect industrial chain, a huge user group, flexible and efficient application service mode, etc. The integration of the satellite communication system and 5G, drawing on each other's strengths and complementing each other's weaknesses, together constitutes a comprehensive communication network with seamless global coverage of sea, land, air and space, which can meet the various service requirements of users everywhere. It is an important basis for the realization of the future 6G network integrating sea and space, and an important direction of future communication development. Focusing on the main problems of the dynamic optimal allocation of resources in the network integrating the satellite and the earth, this chapter studies the following three aspects of cooperative transmission problems in integrated satellite-terrestrial networks.

The first part proposes the auction-based traffic offloading scheme in integrated satellite-terrestrial networks. Aiming at the problem of important allocative externalities, i.e., other uncooperative transmission resource of satellites can benefit from the cooperation between terrestrial base stations and the transmission resource of satellites performing offloading, we design a second-priced auction-based traffic offloading and spectrum sharing mechanism. According to the alternative cooperative and competitive modes, high capacity transmission and co-channel interference control in the integrated satellite-terrestrial networks can be achieved. To solve the problem of secure transmission with interference control, the second part proposes a cooperative secure transmission beamforming and artificial noise designing scheme, which realizes the secure transmission and co-channel interference control in a coexistence system of the satellite-terrestrial network and cellular network, which is also referred to as the integrated satellite-terrestrial network. To solve the nonconvex optimization problems established, an iteration and convex quadratic approximation

J. Du, C. Jiang, *Cooperation and Integration in 6G Heterogeneous Networks*, Wireless Networks, https://doi.org/10.1007/978-981-19-7648-3_5

based genetic algorithm is designed in this part. By applying the designed scheme, the secrecy rate of the eavesdropped fixed satellite service terminal can be doubled, comparing with the beamforming scheme without cooperation. For the problem of adaptivity to traffic properties, the third part proposes a traffic prediction-based resource allocation mechanism for the transmission and service resource of the ground station when receiving data from multiple satellites. In this mechanism, a predictive Backpressure (PBP) based service mechanism is designed to minimize the time average cost of the multiple access system as well as the waiting time of packets after they enter the queue. Results validate that the delay of the SBIN queueing system can be reduced by the resource allocation mechanism that coordinates with traffic properties, and more than half packets in the queues do not need to wait for service, which means that they are pre-served before arriving to the system.

# Chapter 6
# Traffic Offloading in Satellite-Terrestrial Networks

**Abstract** Recently, hybrid satellite-terrestrial networks (H-STN) are expected to support extremely high data rates and exponentially increasing demands of data, which require new spectrum sharing and interference control technology paradigms. By achieving an efficient spectrum sharing among H-STN, traffic offloading is a promising solution for boosting the capacity of traditional cellular networks. In this chapter, a software-defined network (SDN) based spectrum sharing and traffic offloading mechanism is proposed to realize the cooperation and competition between the ground base stations (BSs) of the cellular network and beam groups of the satellite-terrestrial communication (STCom) system. Assume all BSs are operated by the same mobile network operator (MNO). Under the cooperation mode, all the BSs stop occupying a corresponding channel, and a selected beam group of the satellite helps offload the traffic from the BSs by exclusively using this channel. To facilitate the offloading negotiation between the MNO and satellite, we design a second-price auction mechanism, which presents positive allocative externalities, i.e., other uncooperative beam groups of the satellite can benefit from the cooperation between BSs and the beam group performing offloading. Meanwhile, the unique optimal biding strategies for different beam groups of the satellite to achieve the symmetric Bayesian equilibrium, as well as the expected utility of the MNO are derived and obtained in this part. The performance of the proposed traffic offloading mechanism is validated in the simulations, which also reveal that there exists the unique optimal offloading threshold for the MNO to achieve the maximum expected utility.

**Keywords** Integrated Satellite-terrestrial Networks · Software-defined Network (SDN) · Traffic Offloading · Spectrum Sharing · Auction

## 6.1 Introduction

Lately, driven by the exponentially increasing demands of multimedia data traffic, the Sixth generation (6G) network architecture, combining wireless mobile communications with satellite systems (telecommunication satellite, navigational satellite

© The Author(s), under exclusive license to Springer Nature Singapore Pte Ltd. 2023    91
J. Du, C. Jiang, *Cooperation and Integration in 6G Heterogeneous Networks*,
Wireless Networks, https://doi.org/10.1007/978-981-19-7648-3_6

and Earth observing satellite) [1–4], is expected to realize an ubiquitous information perception, and to increase the network capacity effectively by offloading the traffic from base stations (BSs) of cellular networks to satellite-terrestrial communication (STCom) systems. However, with more licensed and unlicensed spectrum shared by both ground cellular networks and STCom systems [5–7], the two systems will suffer more serious co-channel interference. In addition, implementing traffic offloading for BSs of cellular networks brings no capacity increase for STCom systems without cooperative and incentive spectrum management. Therefore, spectrum management is crucial in supporting of traffic offloading between cellular and satellite communications, and cooperative and efficient spectrum sharing has been confirmed as a promising technology that combats the co-interference and increases the capacity of the hybrid satellite-terrestrial network (H-STN). In this part, we will consider the scenario of multimedia multicast services in the H-STN [8, 9], and focus on the spectrum sharing and traffic offloading mechanism design in the H-STN to realize the cooperation between traditional cellular networks and STCom systems.

As space-based communication infrastructure, STCom systems are costly and time-consuming to be deployed and updated. To overcome the slow configuration, inflexible traffic engineering and other disadvantages resulting from the traditional design of STCom systems, software-defined network (SDN) has been considered as an effective and efficient network architecture to realize flexible resource management and system performance control [10, 11]. Leveraging the concept of SDN, the transmission resource and common spectrum occupied in traditional cellular networks and STCom systems can be controlled and managed in a central manner efficiently. In addition, by separating the control plane from the data plane of the system, the SDN promises a potential to revolutionize the H-STN design and transmission resource management to realize various applications and services. Therefore, in this work, an SDN-based architecture will be established for spectrum management and traffic offloading in the H-STN. With a centralized controller, the common spectrum used by the cellular network and STCom system simultaneously will be managed efficiently, and then an auction-based traffic offloading mechanism can be implemented to meet data requests as well as to control the co-channel interference in the H-STN.

We highlight the main contributions and our main ideas as follows:

- We establish an SDN framework for traffic offloading and spectrum sharing in the H-STN. With the aid of the SDN controller, the traffic requests and some public resource status can be obtained. According to the SDN architecture, the spectrum management and geo-distributed transmission resource can be separated.
- We design a second-priced auction based traffic offloading and spectrum sharing mechanism for the SDN-based H-STN. By applying the designed mechanism, the MNO operates its BSs to work in cooperative and competitive modes according to bids given by the satellite's beam groups. Then the common channels and spectrums shared in the H-STN can be managed efficiently to support the high capacity transmission and co-channel interference control.

- We analyze the performance of the designed auction based traffic offloading mechanism. Specifically, the optimal bidding strategy of the satellite is derived in this part, which demonstrates that the proposed auction-based traffic offloading presents positive allocative externalities. In addition, the expected utility of the MNO is also analyzed in this part. Moreover, simulation results validate the performance of the designed mechanism, and reveal that there exists a unique optimal offloading threshold for the MNO to achieve the maximum utility.

The remainder of this part is organized as follows. In Sect. 6.2, we discuss some related works. The architecture of SDN-based traffic offloading system for the H-STN is established in Sect. 6.3. Section 6.4 presents the system model. The second-price auction mechanism for traffic offloading in the H-STN is designed in Sect. 6.5. We provide satellite's equilibrium bidding strategies in Sect. 6.6 and utility of the MNO in Sect. 6.7. Simulations are shown in Sect. 6.8, and conclusions of this part are drawn in Sect. 6.9.

## 6.2  Related Works

The traffic offloading mechanism design for a better radio resource management has attracted researchers' great attention, especially in heterogeneous networks (HetNets) [12, 13]. However, research on traffic offloading between cellular and satellite communications is rather little. In addition, even in most of current studies for HetNets, interference management issues have not been taken into account [14–16]. Some recent work has started trying to introduce spectrum management to optimize the performance of traffic offloading. In [17], load balancing and interference management in HetNets were jointly optimized in the traffic offloading mechanism design to mitigate the inter-cell interference. Traffic management for an $M$-tier HetNet was investigated in [18], in which APs accessing the licensed spectrum opportunistically shared the unlicensed spectrum according to the Carrier Sense multiple Access/Collision Avoidance (CSMA/CA) protocol. Nevertheless, very little research has focused on the traffic offloading mechanism in the H-STN systems. In addition, incentive mechanisms that encourage BSs sharing the spectrum to participate in the offloading operation efficiently are little concerned in these studies above. To solve this problem, the cooperative and competitive modes based spectrum sharing proposed in this part can motivate the satellite to implement offloading for BSs of the MNO, and meanwhile motivate the MNO to attract satellite's offloading.

Consider that the beam groups of the satellite, which occupy different channels sharing with BSs of the MNO, provide multimedia multicast services with different transmission rates. These beam groups can provide traffic offloading for the MNO's users with the same service requests. For such different capabilities of transmission resource, auction theory becomes a feasible and effective tool in network economics to model and analyze the resource supply and demand. Therefore, to facilitate the traffic offloading negotiation between the MNO and satellite, auction-based mechanisms can be introduced in the H-STN. Recently, many research efforts have been devoted to the auction modeling for traffic offloading to improve the capacity and resource utilization of the system [19–22]. Double auction models were applied in HetNets for data offloading [23] and spectrum management [24], which improved the throughput and decreased the interference of the system. In [25] and [26], reverse auction was introduced to describe and model the traffic offloading mechanism in heterogeneous cellular networks. In the previous research above, double auction was utilized thanks to its ability of eliciting the hidden information between the resource provider and demanders, and the reverse auction model was applied to the scenario where there were multiple potential sellers (resource providers) and a single buyer (resource demander). However, important allocative externalities [27], i.e., other uncooperative beam groups of the satellite can benefit from the cooperation between BSs of the MNO and the beam group performing offloading, did not involve in the studies above. To solve this problem, a second-price auction mechanism will be proposed in this work. Different from traditional auction models, second-price auction specifies that the bidder submitting the highest (or lowest) bid will win the auction, and the auction item will be sold at the price determined by the second highest (or lowest) bid. Nevertheless, the analysis for such allocative externalities involved mechanism is difficult since that bidding truthfully is a weakly dominant strategy in the second-price auction [28, 29]. To reveal and confirm the performance of the designed mechanism by overcoming this difficulty, this work will analyze the optimal bidding strategy for beam groups of the satellite with different transmission rates, and provide their unique equilibrium bids.

The H-STN is expected to be integrated, heterogenous, efficient and intelligent, which will bring more difficulties to implement complicated resource management and system control. A novel architecture of satellite-terrestrial networks was initiatively proposed in [30], and a number of key technical challenges associated with such architecture were presented systematically and comprehensively. In addition, authors of [30] indicated that the SDN can be considered as a potential architecture which promises an increasing efficiency of resource management thanks to the separation of control and data transmission. Currently, different SDN-based architectures have been designed and proposed in many studies for H-STNs and other HetNets [19, 31]. In [32], a flexible network architecture was proposed for efficient integration of heterogeneous satellite-terrestrial networks. By synthesizing Locator/ID split and Information-Centric Networking, it is able to achieve routing scalability alleviation, mobility support, efficient content delivery, etc. To optimize the coverage probability, spectral and energy efficiency and other performance, an end-to-end H-STN with SDN was proposed in [33]. Leveraging the architecture of SDN, flexible resource management and system performance control can be achieved to support various applications and services. Therefore, this work will design an SDN-based architecture to realize an efficient and effective traffic offloading and spectrum sharing in the H-STN.

## 6.3   Architecture of SDN

The SDN is established as a network framework which separates the control plane from the data plane. The architecture of an SDN-based spectrum sharing and traffic offloading system in the hybrid satellite-terrestrial network (H-STN) is shown as Fig. 6.1, which consists of three parts: the service plane, control plane and management plane, the functions and operations of which will be introduced in detailed as follows.

### *6.3.1   Service Plane*

In the service plane, the H-STN provides multimedia multicast services with the satellite and distributed BSs, and BSs' mobile users (MUs) and satellite's users (SUs) are randomly distributed in the coverage of the satellite and BSs. All the BSs are operated by the same mobile network operator (MNO) who manages the spectrum and operating modes of these BSs to serve its MUs. Meanwhile, one

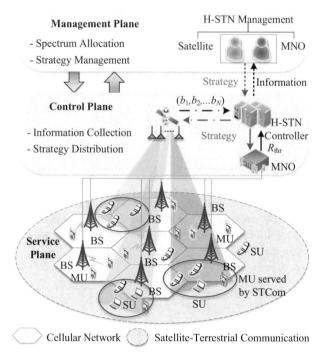

**Fig. 6.1** Architecture of SDN-based spectrum sharing and traffic offloading system in the hybrid STCom network

satellite can cooperatively provide services for its own SUs and BSs' MUs. In addition, the STCom system and the cellular network provide their multimedia multicast services by occupying the common spectrum, and their coverage areas are overlapped seriously. Through cooperative spectrum sharing and traffic offloading among the satellite and BSs, the throughput and other performance of the H-STN can be improved effectively.

## 6.3.2   Control Plane

As shown in Fig. 6.1, the SDN architecture separates spectrum resource management from the geo-distributed resource, which forms an auction-based market of spectrum and channel in the control plane. The control plane implements information collection and strategy distribution.

### 6.3.2.1   Information Collection

The centralized H-STN Controller receives the traffic offloading requests from the MNO and potential supplies of the traffic offloading service from the satellite, and then sends such received information to the management plane of the H-STN. To be specific, the MNO submits its required threshold of offloading rate $R_{thr}$ which denotes the minimum data rate that it is willing to be served by the satellite through the winning spectrum band. In addition, threshold $R_{thr}$ will be observed by the satellite by accessing to the H-STN Controller. In response, the satellite submits a bid vector denoted by $\mathbf{b} = [b_1, b_2, \cdots, b_N]$, which expresses a willingness to provide the traffic offloading service for the MNO.

### 6.3.2.2   Strategy Distribution

The H-STN controller receives the strategies of traffic offloading and spectrum sharing made by the management plane, and then distributes the corresponding strategies to the satellite and MNO to complete cooperative transmission.

## 6.3.3   Management Plane

After receiving the bidding and threshold information submitted by the satellite and MNO, respectively, the H-STN Management allocates the common spectrum and makes the traffic offloading decision for the satellite and MNO. Specifically, the H-STN Management will determine that which channel among the common channels will be occupied by the satellite exclusively and how much traffic of MNO's MUs to be onloaded to the satellite, according to the information obtained. Then these decisions will be returned back to the H-STN Controller in the control plane to guide the cooperative transmission between the satellite and MNO's BSs.

## 6.4   System Model of Traffic Offloading in H-STN

Consider the H-STN consists of one satellite and a set of BSs, and these BSs are operated by an MNO and distributed within the coverage of the satellite. Both the satellite and multiple BSs provide multimedia multicast services through the common spectrum bands occupying a set of channels, denoted by $\mathcal{N} = \{1, 2, \cdots, N\}$ ($N \geq 2$). Moreover, beams of the satellite are separated into $N$ groups according to transmission channels. Therefore, the beams transmitted from the satellite to SUs will cause co-channel interference to the corresponding MUs served by the BSs through the same channels, vice versa. In addition, consider that beams are transmitted from the satellite or BSs through different channels, so that there is no interference between them. Meanwhile, the interference between BSs are also ignored considering the architecture of cellular network. Before proceeding further, we summarize the main notations used throughout the following sections in Table 6.1 for convenience.

**Table 6.1**  List of main notations in H-STN

| Parameter | Definition |
| --- | --- |
| $\mathcal{N}$ | Set of channels occupied |
| $\mathcal{M}$ | Set of beam groups with the maximum bid |
| $R_{\text{thr}}$ | MNO's required threshold of offloading rate |
| $R_0$ | BS's original transmission rate |
| $\mathbf{b}$ | Satellite's bid vector |
| $b_n$ | Bid of beam group $n$ |
| $\mu_n$ | Throughput offered by beam group $n$ |
| $\mu_{\text{cost}}$ | Offloading rate of the winning beam group |
| $m$ | Index of the winning beam group |
| $\alpha$ | Gain-ratio of beam group's transmission rate |
| $\beta$ | Discounting factor of beam group's transmission rate |
| $\gamma$ | BS's increasing factor of transmission rate |
| $\pi_1/\pi_2$ | BS's benefit/discount |

### 6.4.1  Fully-Loaded Transmission

We assume that the traffic offloading system in the H-STN is time-slotted and quasi-static. To be specific, in each time slot, the system status remains constant, but changes over time slots. Assume that the satellite and all BSs work under a fully-loaded situation. In a real STCom system or cellular network, a single time slot is not enough to complete data transmission for the satellite or each BS because there is a lot of traffic to be served [34–36]. Thus the assumption of "fully-loaded" is reasonable as well as it can simplify system analysis.

### 6.4.2  Satellite's Transmission Rate Through Each Channel

We assume that $N$ groups of beams, occupying the number of $N$ channels and transmitted from the satellite, serve SUs with two rates, i.e., the rate achieved under the co-channel interference caused by BSs and the rate achieved without co-channel interference. In each time slot, the satellite senses the occupancy status of every channel, and selects the appropriate transmission rate through each of these channels determined by whether there exists co-channel interference.

#### 6.4.2.1  Transmission Rates Under Interference

In the interference system consisting of a satellite and multiple BSs, we use $\mu_n$ to denote the throughput that beam group $n$ of the satellite can achieve to serve SUs when this beam group shares channel $n$ with BSs within the coverage of the satellite. In this work, we assume that $\mu_n$ ($\forall n \in \mathcal{N}$) is a continuous random variable changing over interval $[\mu_{\min}, \mu_{\max}]$, and $0 \leq \mu_{\min} < \mu_{\max}$, and all $\mu_n$ ($\forall n \in \mathcal{N}$) obey the same probability distribution function (PDF) $f(\mu)$ and cumulative distribution function (CDF) $F(\mu)$. Then we notice that both $f(\mu)$ and $F(\mu)$ are independent of $n$. Moreover, the value of $\mu_n$ ($\forall n \in \mathcal{N}$) is considered as the local and private information of beam group $n$ of the satellite, so that BSs of the MNO and satellite's beams using other channels cannot obtain this information. However, we assume that the probability distribution of $\mu_n$ can be known by all BSs and the other $N-1$ beam groups of the satellite.[1]

---

[1] This assumption is reasonable for BSs since that they are operated by the MNO. On the other hand, the value of $\mu_n$ is a random variable changing over time, cognising its probability distribution is more feasible and efficient than its real-time value for other beam groups of the satellite. In addition, SUs served by different channels may belong to different third-party customers; in other words, these customers may rent different channels of the satellite to provide multicast services for their users. Therefore, the satellite may hide the information of transmission rate in each channel, and then consumers renting other channels cannot obtain the value of $\mu_n$.

#### 6.4.2.2 Transmission Rates Under Non-Interference

When beam group $n$ of the satellite has sensed that there is no other ground BS using channel $n$, this beam group will turn its transmission rate up to a higher value as much as $\alpha$ times of the rate under the interference, i.e., $\alpha\mu_n$ ($\alpha > 1$).

### 6.4.3 BSs' Cooperative and Competitive Modes

We consider that the MNO can operate all the BSs to work in one of the following modes:

#### 6.4.3.1 Cooperative Mode

The BSs can make a deal with one for the satellite's beam group, denoted by $m \in \mathcal{N}$, where all BSs turn off their channel $m$, stop transmitting in this channel, and then beam group $m$ of the satellite occupies this channel exclusively. In this case, there is no co-channel interference for beam group $m$, whose transmission rate is turned up to $\alpha\mu_m$ correspondingly, and remaining $N - 1$ beam groups of the satellite still work with their lower rates $\mu_n$ ($n \in \mathcal{N}$, $n \neq m$) due to the co-channel interference. As a compensation, the satellite will serve MUs of the MNO through channel $m$ with guaranteed offloading rate $\mu_{\text{cost}}$.

#### 6.4.3.2 Competitive Mode

In this mode, the satellite does not provide the traffic offloading service for the MNO through any of its beam groups, and each beam group still occupies its own channel to serve SUs. Therefore, all data requests from MUs have to be served by BSs. To ensure the total transmission rate under interference brought by the satellite and satisfy MUs' data requests, all BSs turn its competition mode on by selecting randomly from one of the $N$ channels and increasing the transmission power in this channel. As a result, the relative beam group of the satellite using this channel will suffer more serious interference, which decreases the transmission rate of this beam group. Let $\beta \in (0, 1)$ denote the transmission rate discounting factor.

## 6.5 Second-Price Auction Based Traffic Offloading Mechanism Design

### 6.5.1 Second-Price Auction

Next, we consider a second-price auction, where the MNO operating multiple BSs is the auctioneer (channel seller) and all beam groups of the satellite are performing as bidders (potential channel buyers) who can offload for BSs to get a probable better communication quality. The auction is operated at the beginning of each time slot. The "good" of the MNO to sell is the transmission channel occupied simultaneously by its BSs and the satellite. In addition, successful beam group $m$ will get a higher communication rate (bandwidth) by obtaining the channel, some parts of which will be contributed to the offloading service for MNO's BSs nevertheless. This transmission rate of traffic offloading for BSs is considered as the payment of beam group $m$ of the satellite for the transmission channel. When beam group $m$ makes a successful deal with the MNO, the MNO will operate all its BSs to work in the cooperative mode, stop using channel $m$, and onload its traffic to beam group $m$ of the satellite, which will transmit this traffic through channel $m$. Furthermore, different from traditional auctions, where the winner pays the auctioneer the winning price, the second-price auction introduced here requires the winning beam group to serve MUs with offloading rate $\mu_{\text{cost}}$ as the payment according to the second highest bid.

### 6.5.2 Auction Operation

The auction operation for traffic offloading in the H-STN can be decomposed into two steps, as shown in Fig. 6.1.

(1) **Stage I:** The MNO announces its required threshold of offloading rate $R_{\text{thr}} \in [0, +\infty)$, which denotes the minimum data rate that it is willing to be served by the satellite through the winning beam group.

(2) **Stage II:** After observing offloading rate threshold $R_{\text{thr}}$ required by the MNO, each beam group $n$ ($\forall n$) of the satellite submits its bid $b_n \in [R_{\text{thr}}, +\infty) \cup \{\varnothing\}$, where $b_n \in [R_{\text{thr}}, +\infty)$ indicates that beam group $n$ of the satellite expresses a willingness to provide the traffic offloading service for the MNO, and $b_n = \varnothing$ denotes the situation that beam group $n$ has no willingness to buy transmission channel $n$ from the MNO. In addition, the case of $b_n < R_{\text{thr}}$ is considered as $b_n = \varnothing$. Then denote the vector of satellite's bids as $\mathbf{b} = [b_1, b_2, \cdots, b_N]$.

### 6.5.3 Outcomes of Auction-Based Traffic Offloading

Next, we analyze the auction outcome of the traffic offloading system for different values of bid vector $\mathbf{b}$ and rate threshold $R_{\text{thr}}$. First, we define the set of beam groups with the maximum bid in a given time slot as

$$\mathscr{M} \triangleq \left\{ m \in \mathscr{N} : m = \arg\max_{n\in\mathscr{N}} b_n \right\}. \tag{6.1}$$

Denote the number of beam groups with the maximum bid by $|\mathscr{M}|$. Then the following three possible situations will happen:

- **Case I:** $|\mathscr{M}| = 1$.

In the case of $|\mathscr{M}| = 1$ after a successful auction, which means that among the bid vector of the satellite, only one beam group provides the highest bid for the channel, we consider beam group $m = \arg\max_{n\in\mathscr{N}} b_n$ as the winner. Then the MNO adjusts all its BSs to work in the cooperative mode and leaves channel $m$ to beam group $m$ of the satellite. Thus, winning beam group $m$ will exclusively use this channel to serve SUs, and meanwhile serve some of MUs with offloading rate $\mu_{\text{cost}}$. Based on the rule of second-price auction, when there is only one bid giving the highest price among the satellite's bid vector, the offloading rate for MUs provided by this winning beam group is given by

$$\mu_{\text{cost}} = \max \{ R_{\text{thr}}, b_1, \cdots, b_{m-1}, b_{m+1}, \cdots, b_N \}, \tag{6.2}$$

i.e., the highest bid among the rate threshold and all rate bids from other beam groups. One can notice that the transaction price (offloading rate) is smaller than the highest price of the satellite's bid vector $\mathbf{b}$.

- **Case II:** $|\mathscr{M}| \geq 2$.

In the case where $|\mathscr{M}| \geq 2$ and $\max_{n\in\mathscr{N}} b_n \in [R_{\text{thr}}, +\infty)$, the MNO adjusts all its BSs to work in the cooperative mode, selects one beam group $m$ randomly with the probability $1/|\mathscr{M}|$ from set $\mathscr{M}$, and then leaves channel $m$ to this winning beam group of the satellite. Then beam group $m$ exclusively occupies this channel to serve its own SUs and some parts of MUs. In this case, the offloading rate of beam group $m$ for the MNO is required to be the maximum bid:

$$\mu_{\text{cost}} = \max_{n\in\mathscr{N}} b_n, \tag{6.3}$$

which is equal to the winning beam group $m$'s bid.

- **Case III:** $|\mathscr{M}| = 0$.

The case of $|\mathscr{M}| = 0$ indicates that the satellite is not willing to buy MNO's transmission channel or provide traffic offloading service for BSs of the MNO through any of its beam group, and gives up bidding, i.e., $b_n = \emptyset$, $\forall n \in \mathcal{N}$. In this case, the MNO will work in the competitive mode, select one of $N$ channels randomly, increase its transmission power in this channel, and serve its users on its own. Then SUs of the satellite served by the beam group using this channel will suffer more serious co-channel interference.

Based on the analysis of three cases above, the offloading rate of the satellite through its selected beam group can be given by

$$
\mu_{\text{cost}}(\mathbf{b}, R_{\text{thr}}) = \begin{cases} \max\left\{R_{\text{thr}}, \; \max_{n \in \mathcal{N}\setminus\{m\}} b_n\right\}, & |\mathcal{M}| = 1, m = \arg\max_{n \in \mathcal{N}} b_n; \\ \max\{R_{\text{thr}}, b_n\}, & |\mathcal{M}| \geq 2, \max_{n \in \mathcal{N}} b_n \in [R_{\text{thr}}, +\infty); \\ 0, & \mathcal{M} = \emptyset. \end{cases} \quad (6.4)
$$

Then define the utility of MNO obtained from offloading as

$$
\pi_{\text{MNO}}(\mathbf{b}, R_{\text{thr}}) = \begin{cases} \mu_{\text{cost}}(\mathbf{b}, R_{\text{thr}}) + \pi_1, & \text{if } |\mathcal{M}| \geq 1; \\ \gamma R_0 - \pi_2, & \text{if } \mathcal{M} = \emptyset, \end{cases} \quad (6.5)
$$

where $R_0$ is the original transmission rate of each BS when sharing with the satellite's beams to use the same channel, $\gamma > 1$ captures the increasing rate of each BS when it works competitively, $\pi_1$ and $\pi_2$ reflect the rate benefit and discount for every BS resulting from the energy saving and cost under the cooperative and competitive modes, respectively.

Then the expected utility of the satellite obtained by beam group $n$ can be written by

$$
\pi_n(\mathbf{b}, R_{\text{thr}}) = \begin{cases} 2\mu_n - \max_{i \in \mathcal{N}\setminus\{n\}}\{R_{\text{thr}}, b_i\}, & b_n > \max_{i \in \mathcal{N}\setminus\{n\}} b_i; \\ \frac{1}{|\mathcal{M}|}\left[2\mu_n - \max_{i \in \mathcal{N}}\{R_{\text{thr}}, b_i\}\right] + \frac{|\mathcal{M}|-1}{|\mathcal{M}|}\mu_n, & b_n = \max_{i \in \mathcal{N}\setminus\{n\}} b_i; \\ \mu_n, & b_n < \max_{i \in \mathcal{N}} b_i; \\ \frac{1}{N}\beta\mu_n + \frac{N-1}{N}\mu_n, & \max_{i \in \mathcal{N}} b_i = \emptyset. \end{cases} \quad (6.6)
$$

According to (6.6), one can notice that beam group $n$ of the satellite fails to win the auction in the following three cases: (a) beam group $n$ submits the same highest bid $b_n = \max_{i \in \mathcal{N}, i \neq n} b_i$ with some other beam groups but is not selected by the MNO to provide the offloading service; (b) $b_n < \max_{i \in \mathcal{N}} b_i$, which indicates that there exists one beam group other than $b_n$ winning the auction; (c) $\max_{i \in \mathcal{N}} b_i = \emptyset$, which indicates that all beam groups fall to win the auction and the MNO adjusts all its BSs to work in the competitive mode. However, the utilities obtained by the failing beam groups are different in these three cases above: beam group $n$ will achieve a utility of $\mu_n$ when either of the first two cases above happens, and will obtain a small utility of $\beta\mu_n$ when $\max_{i \in \mathcal{N}} b_i = \emptyset$. In other words, rather than losing the auction, i.e., $\max_{i \in \mathcal{N}} b_i = \emptyset$, beam group $n$ is more willing to see other beam groups winning the auction even though it fails to win. This shows that the designed auction mechanism has the ability to motivate bidders (beam groups of the satellite) to participate in the offloading service and compete the channel resource actively.

In this work, we assume that all beam groups are rational to submit bidding strategies which may lead to potential maximum expected utilities for them, respectively. Based on the second-price auction designed above, the beam group $n$'s bidding strategy is affected by the distribution of other beam groups' communication quality under the interference. In other words, when beam group $n$ evaluates its utility when it fails to win the auction, it needs to consider that whether the other beam groups win or not. This characteristic reflects the positive allocative externalities [28], which lead to a delicate beam group bidding strategy to achieve the equilibrium. Next section will focus on finding satellite's equilibrium bidding strategies.

## 6.6   Satellite's Equilibrium Bidding Strategies

We first give the definition of *Symmetric Bayesian Equilibrium* in Definition 6.1.

**Definition 6.1** Given a offloading threshold $R_{thr}$, a bidding strategy function $b^*(\mu, R_{thr})$, $\mu \in [\mu_{min}, \mu_{max}]$, constitutes a Symmetric Bayesian Nash Equilibrium (SBNE) if $\forall s_n \in [R_{thr}, +\infty) \cup \{\varnothing\}$, $\forall \mu_n \in [\mu_{min}, \mu_{max}]$ and $\forall n \in \mathscr{N}$, it holds

$$E_{\boldsymbol{\mu}_{-n}} \left\{ \pi_n \left( \mathbf{b}^*, R_{thr} \right) | \mu_n \right\} \geq E_{\boldsymbol{\mu}_{-n}} \left\{ \pi_n \left( \mathbf{b}^*_{s_n}, R_{thr} \right) | \mu_n \right\}, \tag{6.7}$$

where vector $\boldsymbol{\mu}_{-n} \triangleq \{\mu_i : \forall i \in \mathscr{N}, i \neq n\}$ denotes all beam groups' communication quality except $n$, and is unknown to beam group $n$ of the satellite,

$$\mathbf{b}^* = \left( b^* (\mu_1, R_{thr}), \cdots, b^* (\mu_{n-1}, R_{thr}), b^* (\mu_n, R_{thr}), \right.$$
$$\left. b^* (\mu_{n+1}, R_{thr}), \cdots, b^* (\mu_N, R_{thr}) \right), \tag{6.8a}$$

$$\mathbf{b}^*_{s_n} = \left( b^* (\mu_1, R_{thr}), \cdots, b^* (\mu_{n-1}, R_{thr}), s_n, \right.$$
$$\left. b^* (\mu_{n+1}, R_{thr}), \cdots, b^* (\mu_N, R_{thr}) \right). \tag{6.8b}$$

Definition 6.1 implies that beam group $n$ cannot get a better expected utility if it changes its strategy from $b^*(\mu_n, R_{thr})$ to any $s_n \in [R_{thr}, +\infty) \cup \{\varnothing\}$.

Next, we analyze the optimal bidding strategy for all beam groups of the satellite considering different $\mu_n \in [\mu_{min}, \mu_{max}]$, $\forall n$. According to definitions in Sect. 6.5.3, the cost and benefit for beam group $m$ who wins the traffic offloading auction are denoted by $\mu_{cost}(\mathbf{b}, R_{thr})$ in (6.4) and $(\alpha - 1)\mu_m$, respectively. In addition, $\mu_{cost}(\mathbf{b}, R_{thr}) \leq \mu_m$ always holds. Consider that equilibrium bidding strategies will exist when the balance between the expected cost and benefit for beam groups can be achieved, i.e., $\alpha = 2$. In other words, the unique equilibrium bidding strategy can not be achieved when $\alpha$ is larger or small than 2. Therefore, in this work, we analyze the optimal equilibrium bidding strategies for beam groups in the case where $\alpha = 2$.

### 6.6.1   Bidding Strategy for $R_{thr} \in (\mu_{min}, \mu_{max}]$

We first analyze the beam groups' equilibrium bidding strategies when $R_{thr} \in (\mu_{min}, \mu_{max}]$ holds, which is the most complicated situation than the later ones. We first introduce Lemma 6.1 to provide the beam group $n$'s equilibrium strategy $b^* (\mu_n, R_{thr})$.

**Lemma 6.1** *There is at least one solution $\mu \in (\mu_{min}, R_{thr})$ holding the following equation:*

$$\sum_{n=1}^{N-1} \binom{N-1}{n} [F (R_{thr}) - F (\mu)]^n F (\mu)^{N-1-n} \frac{(\alpha - 1) \mu - R_{thr}}{n + 1}$$
$$+ F (\mu)^{N-1} \left[ \left( \alpha - 1 + \frac{1 - \beta}{N} \right) \mu - R_{thr} \right] = 0, \tag{6.9}$$

*when $\alpha = 2$, where $F (\mu)$ is the CFD of random variable $\mu \in (\mu_{min}, \mu_{max})$. Denote the solutions $\mu \in (\mu_{min}, R_{thr})$ of function (6.9) by $\tilde{\mu}_1 (R_{thr}), \tilde{\mu}_2 (R_{thr}), \cdots, \tilde{\mu}_K (R_{thr})$, where $K \in \mathbb{N}^+$ is the number of function (6.9)'s solutions in interval $(\mu_{min}, R_{thr})$.*

The proof of Lemma 6.1 is provided in Sect. 6.10. In addition, such uniqueness of the solution claimed in Lemma 6.1 will be validated by the simulation of this work.

Based on Lemma 6.1, we provide the SBNE bidding strategies in Theorem 6.1:

**Theorem 6.1** *Consider a $\tilde{\mu}_a (R_{thr}) \in (\mu_{min}, R_{thr})$ belonging to set $\{\tilde{\mu}_1 (R_{thr}), \tilde{\mu}_2 (R_{thr}), \cdots, \tilde{\mu}_K (R_{thr})\}$ defined in Lemma 6.1. For a given $R_{thr} \in (\mu_{min}, \mu_{max}]$, then the following bidding strategy $b^* (\mu_n, R_{thr})$ constitutes the unique form to achieve an SBNE for beam group n, $\forall n \in \mathcal{N}$:*

$$b^* (\mu_n, R_{thr}) = \begin{cases} \varnothing, & \mu_n \in [\mu_{min}, \tilde{\mu}_a (R_{thr})); \\ R_{thr} \text{ or } \varnothing, & \mu_n = \tilde{\mu}_a (R_{thr}); \\ R_{thr}, & \mu_n \in (\tilde{\mu}_a (R_{thr}), R_{thr}); \\ \mu_n, & \mu_n \in [R_{thr}, \mu_{max}); \\ any \text{ } value \in [\mu_{max}, +\infty), & \mu_n = \mu_{max}. \end{cases} \tag{6.10}$$

The proof of Theorem 6.1 is provided in Sect. 6.11. In addition, the structure of bidding strategy at SBNE summarized in Theorem 6.1 is shown in Fig. 6.2.

**Fig. 6.2** Bidding strategy at SBNE when $R_{thr} \in (\mu_{min}, \mu_{max}]$

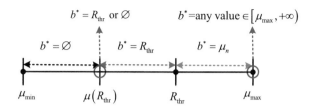

## 6.6.2 Bidding Strategy for $R_{thr} \in \left(\mu_{max}, \left(1 + \frac{1-\beta}{N}\right)\mu_{max}\right)$

Similar to the analysis of bidding strategy above, we first introduce Lemma 6.2 as follows.

**Lemma 6.2** *There is at least one solution $\mu \in (\mu_{min}, \mu_{max})$ holding the following equation:*

$$\sum_{n=1}^{N-1} \binom{N-1}{n} (1 - F(\mu))^n F(\mu)^{N-1-n} \frac{(\alpha - 1)\mu - R_{thr}}{n+1}$$

$$+ F(\mu)^{N-1}\left[\left(\alpha - 1 + \frac{1-\beta}{N}\right)\mu - R_{thr}\right] = 0, \tag{6.11}$$

*when $\alpha = 2$, where $F(\mu)$ is the CFD of random variable $\mu \in (\mu_{min}, \mu_{max})$. Denote the solutions $\mu \in (\mu_{min}, \mu_{max})$ of function (6.11) by $\tilde{\mu}_1(R_{thr}), \tilde{\mu}_2(R_{thr}), \cdots,$ $\tilde{\mu}_K(R_{thr})$, where $K \in \mathbb{N}^+$ is the number of function (6.11)'s solutions in interval $(\mu_{min}, \mu_{max})$.*

Such uniqueness of the solution claimed in Lemma 6.2 will be validated by the simulation of this work. Based on Lemma 6.2, we provide the SBNE bidding strategies of the satellite in Theorem 6.2:

**Theorem 6.2** *Consider a $\tilde{\mu}_b(R_{thr}) \in (\mu_{min}, \mu_{max})$ belonging to set $\{\tilde{\mu}_1(R_{thr}), \tilde{\mu}_2(R_{thr}), \cdots, \tilde{\mu}_K(R_{thr})\}$ defined in Lemma 6.2, then the following bidding strategy $b^*(\mu_n, R_{thr})$ constitutes the unique form of an SBNE for beam group n, $\forall n \in \mathcal{N}$:*

$$b^*(\mu_n, R_{thr}) = \begin{cases} \varnothing, & \mu_n \in [\mu_{min}, \tilde{\mu}_b(R_{thr})); \\ R_{thr} \text{ or } \varnothing, & \mu_n = \tilde{\mu}_b(R_{thr}); \\ R_{thr}, & \mu_n \in (\tilde{\mu}_b(R_{thr}), \mu_{max}]. \end{cases} \tag{6.12}$$

*In addition, in the case where there is no beam group with $\mu_n \in [\tilde{\mu}_b(R_{thr}), \mu_{max})$, then the optimal bid constituting an SBNE for the beam group or groups with $\mu_n = \mu_{max}$ is selecting any value $\in [R_{thr}, +\infty)$,*

**Fig. 6.3** Bidding strategy at SBNE when $R_{thr} \in$ $\left(\mu_{max}, \left(1 + \frac{1-\beta}{N}\right)\mu_{max}\right)$

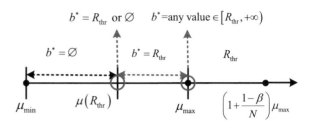

The proofs of Lemma 6.2 and Theorem 6.2 are similar to those of Lemma 6.1 and Theorem 6.1, respectively. Therefore, we omit the details of these proofs. In addition, the structure of bidding strategy at SBNE for $R_{thr} \in \left(\mu_{max}, \left(1 + \frac{1-\beta}{N}\right)\mu_{max}\right)$ is shown in Fig. 6.3.

*Remark 6.1* According to Theorems 6.1 and 6.2, one can notice that the SBNE bidding strategies of the satellite are determined by $R_{thr}$ and $\tilde{\mu}_a$ ($\tilde{\mu}_b$) obtained through Lemmas 6.1 and 6.2. In addition, the uniqueness of solutions of (6.9) and (6.11) is important for the feasibility and reasonability of Theorems 6.1 and 6.2. In Sect. 6.8.2, simulation results validates that there exists only one solution $\tilde{\mu}_a(R_{thr}) \in (\mu_{min}, R_{thr})$ and $\tilde{\mu}_b(R_{thr}) \in (\mu_{min}, \mu_{max})$ holding (6.9) and (6.11), respectively, when beam group's transmission rate $\mu_n$ ($\forall n \in \mathcal{N}$) follows the truncated normal distribution (TND) $N_T(\mu_T, \sigma_T^2)$ to interval $[\mu_{min}, \mu_{max}]$ or the uniform distribution (UD) $U(\mu_{min}, \mu_{max})$, where $\mu_T$ and $\sigma_T^2$ are the expectation and variance of the TND, respectively.

## 6.6.3  Bidding Strategy for $R_{thr} \in \left[\left(1 + \frac{1-\beta}{N}\right)\mu_{max}, +\infty\right)$

We assume that offloading rate threshold $R_{thr} \in \left(\left(1 + \frac{1-\beta}{N}\right)\mu_{max}, +\infty\right)$ is given. Then Theorem 6.3 provides the bidding strategy constituting an SBNE.

**Theorem 6.3** *For a given* $R_{thr} \in \left(\left(1 + \frac{1-\beta}{N}\right)\mu_{max}, +\infty\right)$, *bidding strategy* $b^*(\mu_n, R_{thr}) = \varnothing$ *constitutes an SBNE strategy for all* $\mu_n \in [\mu_{min}, \mu_{max}]$, $\forall n \in \mathcal{N}$. *Specifically, when* $R_{thr} = \left(1 + \frac{1-\beta}{N}\right)\mu_{max}$, *the unique form of SBNE bidding strategy for every beam group is given by*

$$b^*(\mu_n, R_{thr}) = \begin{cases} R_{thr} \cup \varnothing, & \mu_n = \mu_{max}; \\ \varnothing, & \mu_n = [\mu_{min}, \mu_{max}). \end{cases} \quad (6.13)$$

The proof of Theorem 6.3 is provided in Sect. 6.12.

**Fig. 6.4** Bidding strategy at SBNE when $R_{thr} \in \left[\left(1 + \frac{1-\beta}{N}\right)\mu_{\max}, +\infty\right)$

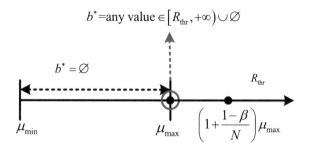

*Remarks* Strategy function (6.13) describes the situation where the MNO requires a large offloading rate but only promises a very limited improvement of transmission quality to the winning beam group of the satellite. Therefore, all beam groups submit the value of bid as $\varnothing$ with probability one. We show the structure of bidding strategy at SBNE in Fig. 6.4.

## 6.6.4 Bidding Strategy for $R_{thr} \in [0, \mu_{\min}]$

Considering the situation where the MNO announces a small value of offloading rate threshold satisfying $R_{thr} \in [0, \mu_{\min}]$, we provide the unique form of SBNE bidding strategy in Theorem 6.4.

**Theorem 6.4** *For a given $R_{thr} \in [0, \mu_{\min}]$, the SBNE bidding strategy of every beam group is given by a unique form as*

$$b^*(\mu_n, R_{thr}) = \begin{cases} any\ value \in [R_{thr}, \mu_{\min}], & \mu_n = \mu_{\min}; \\ \mu_n, & \mu_n \in (\mu_{\min}, \mu_{\max}); \\ any\ value \in [\mu_{\max}, +\infty), & \mu_n = \mu_{\max}. \end{cases} \quad (6.14)$$

We omit the proof of Theorem 6.4 due to its similarity with Theorem 6.1. In addition, the structure of bidding strategy at SBNE is shown in Fig. 6.5.

**Fig. 6.5** Bidding strategy at SBNE when $R_{thr} \in [0, \mu_{\min}]$

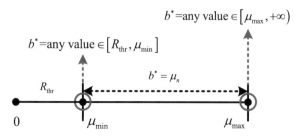

So far we have provided optimal bidding strategy $b^*(\mu_n, R_{thr})$ which constitutes the SBNE for every beam group of the satellite with $\mu_n \in [\mu_{min}, \mu_{max}]$, $\forall n \in \mathcal{N}$, when the MNO announces different values of offloading rate threshold $R_{thr} \geq 0$.

We notice that some beam groups select bidding rate threshold $R_{thr}$ conservatively for $R_{thr} \in \left(\mu_{min}, \left(1 + \frac{1-\beta}{N}\right)\mu_{max}\right)$ as their optimal strategy. Specifically, if there is at least one beam group bids rate from $[R_{thr}, +\infty)$ (situation analyzed in Sect. 6.6.1), then these beam groups prefer to not provide the offloading service to the MNO, which results from their relatively low $\mu_n$ and low increasing rates brought by the cooperation with MNO's BSs. This reason motivates satellite's beam groups with low transmission rates to reduce their chances of winning the auction. On the other hand, if none of the other beam groups of the satellite bids rate from $[R_{thr}, +\infty)$ (situation analyzed in Sect. 6.6.2), then these beam groups need to bid conservative rate $R_{thr}$ to facilitate the cooperation with the MNO, which can avoid the severe deterioration of transmission rate resulting from the competitive mode of the MNO. This reason can be considered as the motivation that encourages beam groups to bid from $[R_{thr}, +\infty)$. Therefore, bidding conservative rate $R_{thr}$ conservatively is the SBNE strategy for these beam groups of the satellite.

## 6.7  Expected Utility Analysis for MNO

In this section, we investigate the expected utility obtained by the MNO when it reports different $R_{thr}$, based on the bidding strategy analysis in Sect. 6.6. In addition, assume that transmission rate $\mu_n$ ($\forall n \in \mathcal{N}$) follows the TND $N_T\left(\mu_T, \sigma_T^2\right)$ to interval $[\mu_{min}, \mu_{max}]$ or the uniform distribution (UD) $U(\mu_{min}, \mu_{max})$. Therefore, there exists unique solution $\tilde{\mu}_a(R_{thr}) \in (\mu_{min}, R_{thr})$ and $\tilde{\mu}_b(R_{thr}) \in (\mu_{min}, \mu_{max})$ for equality (6.9) and equality (6.11), respectively, which will be validated by simulation results in Sect. 6.8.2.

### 6.7.1  Utility Analysis for $R_{thr} \in (\mu_{min}, \mu_{max}]$

We first provide the expected utility of the MNO when $R_{thr} \in (\mu_{min}, \mu_{max}]$ in Theorem 6.5.

**Theorem 6.5** *Consider that the distribution of satellite's transmission rate can be known by the MNO, and the MNO announces threshold of traffic offloading rate $R_{thr} \in (\mu_{min}, \mu_{max}]$. Assume that all beam groups of the satellite provide the offloading bids according to (6.10). In addition, the utility of the MNO is defined* as (6.5). Then the expected utility of the MNO is given by:

$$E\left(\pi_{MNO}\left(R_{thr}\right)\right)$$

$$= F(\tilde{\mu}_a\left(R_{thr}\right))^N\left(\gamma R_0 - \pi_2\right) + \left(1 - F(\tilde{\mu}_a\left(R_{thr}\right))^N\right)\pi_1 + \bar{\mu}_{cost}, \tag{6.15}$$

where $\tilde{\mu}_a\left(R_{thr}\right) \in \left(\mu_{\min}, R_{thr}\right)$ is obtained according to Lemma 6.1, $\bar{\mu}_{cost} = N\bar{\mu}_0$, and

$$\bar{\mu}_0 = (N - 1) \int_{R_{thr}}^{\mu_{\max}} \mu f\left(\mu\right) F(\mu)^{N-2}\left(1 - F\left(\mu\right)\right) d\mu$$

$$+ R_{thr}\left(1 - F\left(R_{thr}\right)\right) F(\mu)^{N-1} + \frac{R_{thr}}{N}\left[F(R_{thr})^N - F(\tilde{\mu}_a\left(R_{thr}\right))^N\right]. \tag{6.16}$$

**Proof** Denote the CDF of $b_{\max}^{-n}$ as $H\left(\mu\right)$, which is given by

$$H\left(\mu\right) = F(\mu)^{N-1}, \tag{6.17}$$

where $F\left(\mu\right)$ is the CDF of each beam group's transmission rate of the satellite. Then the PDF of $b_{\max}^{-n}$, denoted by $h\left(\mu\right)$, can be calculated as

$$h\left(\mu\right) = (N - 1) F(\mu)^{N-2} f\left(\mu\right), \tag{6.18}$$

where $f\left(\mu\right)$ is the PDF of $\mu_n$, $\forall n \in \mathcal{N}$. Consider $\mu_m = \max_{n\in\mathcal{N}}\mu_n$. Based on (6.4) and (6.5), and given $R_{thr} \in \left(\mu_{\min}, \mu_{\max}\right]$, the utility obtained by the MNO can be analyzed under three cases:

1. $\mu_m \in \left(R_{thr}, \mu_{\max}\right]$ and $b_{\max}^{-m} \in \left(R_{thr}, \mu_m\right]$, then the MNO can receive offloading rate $b_{\max}^{-m}$ from the satellite, and the corresponding utility received is $b_{\max}^{-m} + \pi_1$;
2. $\mu_m \in \left(R_{thr}, \mu_{\max}\right]$ and $b_{\max}^{-m} = R_{thr}$ or $\varnothing$, then the MNO can receive offloading rate $R_{thr}$ from the satellite, then the corresponding utility received is $R_{thr} + \pi_1$;
3. $\mu_m \in \left[\tilde{\mu}\left(R_{thr}\right), R_{thr}\right]$ and $b_{\max}^{-n} = R_{thr}$ or $\varnothing$, then the MNO might work in the cooperative mode and receive offloading rate $R_{thr}$ if there exists at least one of beam groups of the satellite giving the conservative bid $R_{thr}$; otherwise, the competitive mode will turned on if all beam groups of the satellite give bid as $\varnothing$. Therefore, in this case, the expected utility of the MNO is determined by the number of beam groups with bid $R_{thr}$.

In conclusion, if there exists at least on beam group of the satellite giving a bid no less than conservative bid $R_{thr}$, then the expected offloading rate received by the

MNO is given by

$$
\bar{\mu}_0 = \int_{R_{\text{thr}}}^{\mu_{\max}} \mu h\,(\mu)\,(1 - F\,(\mu))\,d\mu + R_{\text{thr}}\,(1 - F\,(R_{\text{thr}}))\,H\,(R_{\text{thr}})
$$

$$
+ \left[F\,(R_{\text{thr}}) - F\,(\tilde{\mu}\,(R_{\text{thr}}))\right] \sum_{n=1}^{N-1} \binom{N-1}{n}
$$

$$
\cdot \left[F\,(R_{\text{thr}}) - F\,(\tilde{\mu}\,(R_{\text{thr}}))\right]^n F\,(\tilde{\mu}\,(R_{\text{thr}}))^{N-1-n} \cdot \frac{R_{\text{thr}}}{n+1}
$$

$$
= (N - 1) \int_{R_{\text{thr}}}^{\mu_{\max}} \mu f\,(\mu)\,F\,(\mu)^{N-2}\,(1 - F\,(\mu))\,d\mu
$$

$$
+ R_{\text{thr}}\,(1 - F\,(R_{\text{thr}}))\,F\,(\mu)^{N-1} + \frac{R_{\text{thr}}}{N}\left[F\,(R_{\text{thr}})^N - F\,(\tilde{\mu}\,(R_{\text{thr}}))^N\right].
$$

Considering that there are $N$ beam groups generated by the satellite, then the expected offloading rate received by the MNO is $\bar{\mu}_{\text{cost}} = N\bar{\mu}_0$, and MNO's expected utility received can be given by

$$
E\,(\pi_{\text{MNO}}\,(R_{\text{thr}}))
$$
$$
= F(\tilde{\mu}\,(R_{\text{thr}}))^N\,(\gamma R_0 - \pi_2) + \left(1 - F(\tilde{\mu}\,(R_{\text{thr}}))^N\right)\pi_1 + \bar{\mu}_{\text{cost}}. \tag{6.19}
$$

This completes the proof of Theorem 6.5.

### 6.7.2   Utility Analysis for $R_{thr} \in \left(\mu_{\max}, \left(1 + \frac{1-\beta}{N}\right)\mu_{\max}\right)$

According to Theorem 6.2, one can notice that the MNO will obtain the conservative offloading rate $R_{\text{thr}}$ if and only if there exists at least one beam group of the satellite with $\mu_n \in (\tilde{\mu}_b\,(R_{\text{thr}}), \mu_{\max}]$. In addition, $R_{\text{thr}}$ is the maximum and the only possible offloading rate provided by the satellite. Otherwise, the MNO will work in the competitive mode. Therefore, when the MNO announces $R_{\text{thr}} \in \left(\mu_{\max}, \left(1 + \frac{1-\beta}{N}\right)\mu_{\max}\right)$ and all beam groups of the satellite give bids according to (6.12), the expected utility of the MNO can be computed by

$$
E\,(\pi_{\text{MNO}}\,(R_{\text{thr}}))
$$
$$
= F(\tilde{\mu}_b\,(R_{\text{thr}}))^N\,(\gamma R_0 - \pi_2) + \left[1 - F(\tilde{\mu}_b\,(R_{\text{thr}}))^N\right](R_{\text{thr}} + \pi_1), \tag{6.20}
$$

where $\tilde{\mu}_b\,(R_{\text{thr}}) \in (\mu_{\min}, \mu_{\max})$ is a solution of (6.11).

### 6.7.3   Utility Analysis for $R_{thr} \in \left[ \left( 1 + \frac{1-\beta}{N} \right) \mu_{\max}, +\infty \right)$

Consider that the probability for a beam group of the satellite to have the maximum transmission rate $\mu_{\max}$ is zero due to the continuous distribution of $\mu_n$. Therefore, as formulated in Theorem 6.3, all beam groups of the satellite will give up bidding for channels, i.e., $b_n = \varnothing, \forall n \in \mathcal{N}$. Thus, when the MNO announces $R_{\mathrm{thr}} \in \left[ \left( 1 + \frac{1-\beta}{N} \right) \mu_{\max}, +\infty \right)$, its expected utility obtained is $E\left( \pi_{\mathrm{MNO}}\left( R_{\mathrm{thr}} \right) \right) = \gamma R_0 - \pi_2$.

### 6.7.4   Utility Analysis for $R_{thr} \in [0, \mu_{\min}]$

Similar to the analysis in the proof of Theorem 6.5, when the MNO announces $R_{\mathrm{thr}} \in [0, \mu_{\min}]$ and all beam groups of the satellite give bids according to (6.14), then the expected offloading rate obtained by the MNO can be given by

$$
\begin{aligned}
\bar{\mu}_{\mathrm{cost}} &= N \int_{\mu_{\min}}^{\mu_{\max}} \mu g\left( \mu \right) \left( 1 - F\left( \mu \right) \right) d\mu \\
&= N\left( N - 1 \right) \int_{\mu_{\min}}^{\mu_{\max}} \mu f\left( \mu \right) F(\mu)^{N-2} \left( 1 - F\left( \mu \right) \right) d\mu.
\end{aligned}
\tag{6.21}
$$

and then the corresponding expected utility of the MNO is $E\left( \pi_{\mathrm{MNO}}\left( R_{\mathrm{thr}} \right) \right) = \bar{\mu}_{\mathrm{cost}} + \pi_1$.

For now, we have provided the expected utility of the MNO, as well as the optimal bidding strategies at SBNE for every beam group, when applying the designed auction based traffic offloading mechanism. These proposed mechanism, optimized bidding strategies and performance analysis for the SDN-based hybrid STCom network can be extended flexibly and will still stand in the further H-STN systems with more channels occupied by common spectrum bands, increasing throughput achieved by beam groups and other different system parameters.

## 6.8   Simulation Results

We simulate the auction-based traffic offloading system with one MNO and one satellite. In addition, the multiple beams of the satellite are divided into seven groups occupying seven channels separately, i.e., $N = 7$. Such number of channels using by the STCom system is in compliance with the actual satellite system [37–39]. Consider $\mu_n \in [50, 200]$ Mbps ($\forall n \in \mathcal{N}$) follows the TND $N_T \left( 125, 50^2 \right)$ to interval $[50, 200]$ Mbps, or the UD $U\left( 50, 200 \right)$.

### 6.8.1 Beam Group's Strategy of the Satellite

First, we investigate the impact of some system settings, i.e., the offloading threshold, transmission rate discounting factor and the distribution of beam groups' transmission rates, on the beam group' strategies. Consider that the probability for a beam group to have a transmission rate $\mu_{\min}$ or $\mu_{\max}$ is zero due to the distribution of $\mu_n$, $\forall n \in \mathcal{N}$. Then all beam groups of the satellite select the same bidding strategy, i.e., $\varnothing$ and $\mu_n$ for the situation $R_{\text{thr}} \in \left[\left(1 + \frac{1-\beta}{N}\right)\mu_{\max}, +\infty\right)$ and $R_{\text{thr}} \in [0, \mu_{\min}]$, respectively. Therefore, in this section, we only test the beam groups' strategies for the first two situations analyzed in Sect. 6.6, specifically, $R_{\text{thr}} \in (\mu_{\min}, \mu_{\max}]$ and $R_{\text{thr}} \in \left(\mu_{\max}, \left(1 + \frac{1-\beta}{N}\right)\mu_{\max}\right)$. Set $\beta \in \{0.01, 0.3, 0.7\}$ to denote different degeneration degrees of beam group's transmission rate due to MNO's competitive mode, and the small value of $\beta$ implies a serious degeneration.

For different distributions of $\mu_n$, TDN $N_T (125, 50^2)$ and UD $U (50, 200)$, the probabilities for a beam group to give up bidding, i.e., bidding $\varnothing$, when the MNO announces different $R_{\text{thr}} \in (\mu_{\min}, \mu_{\max}]$ and $R_{\text{thr}} \in \left(\mu_{\max}, \left(1 + \frac{1-\beta}{N}\right)\mu_{\max}\right)$ are shown in Figs. 6.6a, b, respectively. As shown in these two figures, more beam groups give up bidding with the increase of $R_{\text{thr}}$, which indicates that both $\tilde{\mu}_a (R_{\text{thr}})$ and $\tilde{\mu}_b (R_{\text{thr}})$ increase with increasing $R_{\text{thr}}$. In addition, the number of beam groups those give up bidding decreases with decreasing $\beta$, which means that the more serious degeneration for beam groups' transmission rates brought by the MNO's competitive mode, the more beam groups prefer to give bids in $[R_{\text{thr}}, +\infty)$ to facilitate the cooperation with the MNO. Moreover, results shown in Fig. 6.6b reveal that less beam groups bid $\varnothing$ when $\mu_n$ ($\forall n \in \mathcal{N}$) follows $U (50, 200)$ than $N_T (125, 50^2)$ for the same $R_{\text{thr}}$ reported by the MNO, before all beam groups bidding $\varnothing$. Such phenomenon reflects that there are less beam groups with transmission rates closed to $\mu_{\max}$ when $\mu_n$ follows the TDN than UD, and then according to the optimal bidding strategies formulated in Theorem 6.2, less beam groups will bid conservative offloading rate $R_{\text{thr}}$ and more beam groups bid $\varnothing$.

### 6.8.2 Expected Utility of the MNO

Next, we investigate the expected utility obtained by the MNO when it reports different $R_{\text{thr}}$. Sections 6.6.1 and 6.6.2 indicates that the bidding strategies of satellite's beam groups are determined by $\tilde{\mu}_a (R_{\text{thr}})$ and $\tilde{\mu}_b (R_{\text{thr}})$, the uniqueness of which is important for the feasibility and reasonability of Theorems 6.1 and 6.2, respectively. Therefore, we first validate that there exists the unique solution

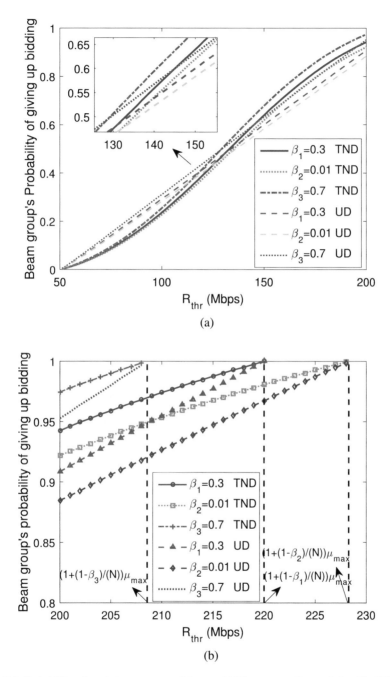

**Fig. 6.6** Probability of each beam group giving up bidding versus $R_{thr}$ and $\beta$ with different distributions of $\mu_n$. (**a**) $R_{thr} \in (\mu_{min}, \mu_{max})$. (**b**) $R_{thr} \in \left(\mu_{max}, \left(1 + \frac{1-\beta}{N}\right)\mu_{max}\right)$

for (6.9) and (6.11). Let $R_0 = 250$ Mbps [40], $\gamma = 1.2$, $\beta \in \{0.01, 0.3, 0.7\}$, and

$$
\Gamma_1(\mu) = F(\mu)^{N-1} \left[ \left( 1 + \frac{1-\beta}{N} \right) \mu - R_{\text{thr}} \right]
$$

$$
+ \sum_{n=1}^{N-1} \binom{N-1}{n} [F(R_{\text{thr}}) - F(\mu)]^n F(\mu)^{N-1-n} \frac{\mu - R_{\text{thr}}}{n+1}, \tag{6.22a}
$$

$$
\Gamma_2(\mu) = F(\mu)^{N-1} \left[ \left( 1 + \frac{1-\beta}{N} \right) \mu - R_{\text{thr}} \right]
$$

$$
+ \sum_{n=1}^{N-1} \binom{N-1}{n} (1 - F(\mu))^n F(\mu)^{N-1-n} \frac{\mu - R_{\text{thr}}}{n+1}, \tag{6.22b}
$$

where $\mu \sim N_T(125, 50^2)$ or $\mu \sim U(50, 200)$. Then Fig. 6.7 shows how $\Gamma_1(\mu)$ and $\Gamma_2(\mu)$ change over different values of $\mu$. Specifically, we first set $R_{\text{thr}} = 100$ Mbps $\in (\mu_{\min}, \mu_{\max})$, then Fig. 6.7a indicates that for different $\beta$ and probability distributions of $\mu$, $\Gamma_1(\mu)$ increases monotonically when $\mu$ increases from $\mu_{\min} = 50$ Mbps to $R_{\text{thr}} = 100$ Mbps, in addition, $\Gamma_1(\mu_{\min}) < 0$ and $\Gamma_1(R_{\text{thr}}) > 0$. Therefore, there exists only one $\tilde{\mu}_a(R_{\text{thr}}) \in (\mu_{\min}, R_{\text{thr}})$ which holds $\Gamma_1(\tilde{\mu}_a) = 0$. Similarly, we set $R_{\text{thr}} = \left(1 + \frac{1-\beta}{2N}\right) \mu_{\max} \in \left(\mu_{\max}, \left(1 + \frac{1-\beta}{N}\right) \mu_{\max}\right)$, and Fig. 6.7b indicates that $\Gamma_2(\mu)$ increases monotonically when $\mu$ increases around $\mu_{\max}$, and only one $\tilde{\mu}_b(R_{\text{thr}}) \in (\mu_{\min}, \mu_{\max})$ can hold $\Gamma_2(\tilde{\mu}_b) = 0$. Consequently, the conclusions in Theorems 6.1 and 6.2 are feasible and reasonable, which means that there exists the unique $\tilde{\mu}_a(R_{\text{thr}})$ and $\tilde{\mu}_b(R_{\text{thr}})$ for beam groups of the satellite to provide the unique SBNE bidding strategy when the MNO announces $R_{\text{thr}} \in (\mu_{\min}, \mu_{\max})$ and $R_{\text{thr}} \in \left(\mu_{\max}, \left(1 + \frac{1-\beta}{N}\right) \mu_{\max}\right)$, respectively.

Then we test the expected utility obtained by the MNO when it reports different $R_{\text{thr}}$. Let $\mu_n \sim N_T(125, 50^2)$ and $\beta = 0.3$ for instance. Set $(\pi_1, \pi_2)$ to select values from $\{(50, 50), (100, 50), (50, 10)\}$ to illustrate cases where $\mu_{\text{cost}}(\mathbf{b}, R_{\text{thr}}) + \pi_1 =$, $>$ and $< \gamma R_0 - \pi_2$. Then MNO's expected utilities for $R_{\text{thr}} \in [\mu_{\min}, \mu_{\max}]$ are shown in Fig. 6.8a. Results shown in Fig. 6.8a reflect that $E(\pi_{\text{MNO}}(R_{\text{thr}}))$ is strictly unimodal for $R_{\text{thr}} \in [\mu_{\min}, \mu_{\max}]$, which indicates that there exists the unique optimal announced threshold $R_{\text{thr}}^*$ for the MNO to achieve the maximum expected utility. In addition, when $\mu_{\text{cost}}(\mathbf{b}, R_{\text{thr}}) + \pi_1 > \gamma R_0 - \pi_2$, the MNO prefers to announce a smaller value of $R_{\text{thr}}$ to encourage beam groups of the satellite to bid from $[R_{\text{thr}}, +\infty)$, and then the corresponding optimal threshold $R_{\text{thr}}^*$ is smaller than that of the situation $\mu_{\text{cost}}(\mathbf{b}, R_{\text{thr}}) + \pi_1 = \gamma R_0 - \pi_2$. Similarly, $R_{\text{thr}}^*$ increases when $\mu_{\text{cost}}(\mathbf{b}, R_{\text{thr}}) + \pi_1 < \gamma R_0 - \pi_2$. Such properties are also validated in Fig. 6.8a.

The expected utilities of the MNO for different $R_{\text{thr}} \in \left(\mu_{\max}, \left(1 + \frac{1-\beta}{N}\right) \mu_{\max}\right)$ are shown in Fig. 6.8b. Similar to the results shown in Fig. 6.8a, the MNO's optimal announced threshold $R_{\text{thr}}^*$ is the smallest when $\pi_1 = 100$ and $\pi_2 = 50$, while when $\pi_1 = 50$ and $\pi_2 = 10$, the MNO needs to announce the largest $R_{\text{thr}}^*$ to get

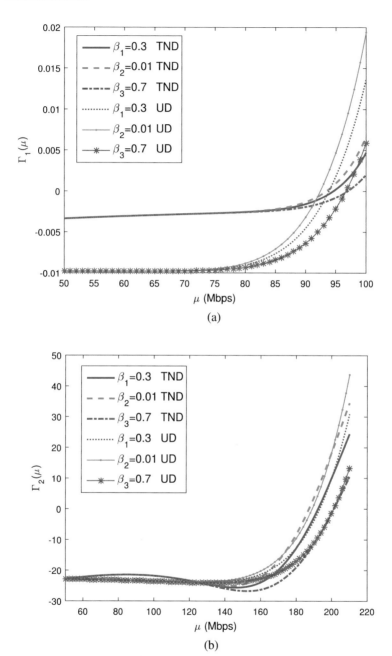

**Fig. 6.7** Monotonicity of $\Gamma_{1,2}(\mu)$ versus $\beta$ with different distributions of $\mu_n$. (**a**) $\Gamma_1(\mu)$, $R_{\text{thr}} \in (\mu_{\min}, \mu_{\max})$. (**b**) $\Gamma_2(\mu)$, $R_{\text{thr}} \in \left(\mu_{\max}, \left(1 + \frac{1-\beta}{N}\right)\mu_{\max}\right)$

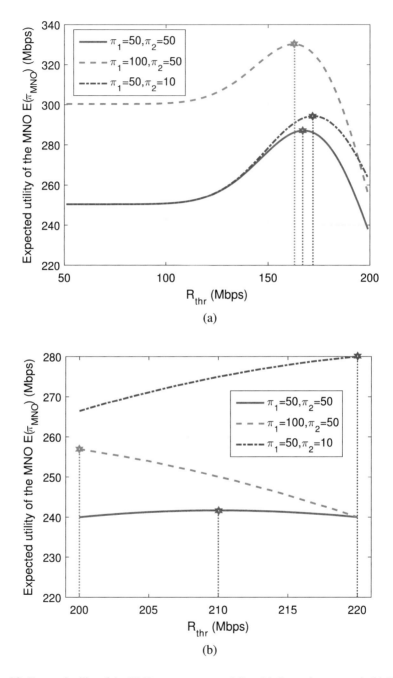

**Fig. 6.8** Expected utility of the MNO versus $\pi_1$, $\pi_2$ and $R_{thr}$. (**a**) $R_{thr} \in (\mu_{min}, \mu_{max})$. (**b**) $R_{thr} \in \left( \mu_{max}, \left( 1 + \frac{1-\beta}{N} \right) \mu_{max} \right)$

the maximum expected utility. In addition, when $R_{thr} \in \left( \mu_{max}, \left( 1 + \frac{1-\beta}{N} \right) \mu_{max} \right)$, beam groups of the satellite have only two optional bidding strategies, i.e., $R_{thr}^*$ and $\varnothing$, which will bring utilities $\pi_{MNO} = R_{thr} + \pi_1$ and $\pi_{MNO} = \gamma R_0 - \pi_2$, respectively, to the MNO. Therefore, when $R_{thr} + \pi_1 > \gamma R_0 - \pi_2$, the optimal threshold for the MNO should lead that the probability of all beam groups of the satellite giving up bidding is zero, i.e., $R_{thr}^* = \mu_{max}$. On the other hand, when $R_{thr} + \pi_1 < \gamma R_0 - \pi_2$, the MNO prefers to announce an $R_{thr}^*$ which can lead all beam groups of the satellite to bid $\varnothing$, i.e., $R_{thr}^* = \left( 1 + \frac{1-\beta}{N} \right) \mu_{max}$. Moreover, when $R_{thr} + \pi_1 = \gamma R_0 - \pi_2$, results shown in Fig. 6.8b reflect that $E\left( \pi_{MNO}\left( R_{thr} \right) \right)$ is strictly unimodal for $R_{thr} \in \left( \mu_{max}, \left( 1 + \frac{1-\beta}{N} \right) \mu_{max} \right)$, which indicates that there exists the unique optimal announced threshold $R_{thr}^*$ for the MNO to achieve the maximum expected utility.

Then we study that how the expected utility of the MNO changes for the changing number of channels occupied by BSs of the MNO and beam groups of the satellite. Considering that the optimal bidding strategies are the most complicated when $R_{thr} \in [\mu_{min}, \mu_{max}]$, we take this situation for instance. Set $R_{thr} = 100$ Mbps $\in (\mu_{min}, \mu_{max})$, $\mu_n \sim N_T \left( 125, 50^2 \right)$, and let the number of common channels $N$ in the H-STM traffic offloading system varies from 3 to 9, $\beta = 0.3$, and other parameters are set as before. The maximum expected utilities of the MNO versus $N$, $\pi_1$ and $\pi_2$ are shown in Fig. 6.9. Results in Fig. 6.9 indicate that for different values of $(\pi_1, \pi_2)$, the maximum $E\left( \pi_{MNO}\left( R_{thr} \right) \right)$ decreases when the number of common channels $N$ grows, which results from the increasing competition for the transmission resource.

**Fig. 6.9** Maximum Expected utility of the MNO versus $\pi_1$, $\pi_2$ and $N$. ($R_{thr} \in (\mu_{min}, \mu_{max})$)

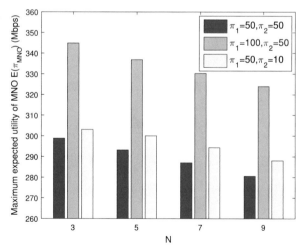

## 6.9   Conclusion

In this part, an SDN architecture was established for spectrum sharing and traffic offloading in the H-STN, which can support efficient resource management to realize high-capacity transmission and co-channel interference control. In addition, we proposed an auction-based mechanism to facilitate the traffic offloading negotiation between the MNO and satellite, in which the MNO announced its required threshold of offloading rate, and then each beam group of the satellite submitted the offloading bid according to the transmission rate. For different offloading rate threshold announced by the MNO, the optimal bidding rates for beam groups with diverse transmission rates have been derived in this work. In addition, the expected utility of the MNO was also analyzed for given threshold and satellite's optimal bids. Simulation results validated the performance of the proposed traffic offloading mechanism, including beam groups' bidding behaviors and MNO's expected utility, for different system parameter settings. Moreover, simulation results also revealed that there existed unique optimal threshold for the MNO to achieve the maximum utility by applying the proposed second-price auction mechanism.

## 6.10   Proof of Lemma 6.1

**Proof** Let

$$
\Gamma(\mu) = F(\mu)^{N-1} \left[ \left( \alpha - 1 + \frac{1-\beta}{N} \right) \mu - R_{\text{thr}} \right]
$$
$$
+ \sum_{n=1}^{N-1} \binom{N-1}{n} [F(R_{\text{thr}}) - F(\mu)]^n F(\mu)^{N-1-n} \frac{(\alpha-1)\mu - R_{\text{thr}}}{n+1},
\tag{6.23}
$$

where $\mu \in (\mu_{\min}, R_{\text{thr}})$ and $\alpha = 2$. Considering $F(\mu_{\min}) = 0$, then we have

$$
\Gamma(R_{\text{thr}}) = F(R_{\text{thr}})^{N-1} \left[ \left( 1 + \frac{1-\beta}{N} \right) R_{\text{thr}} - R_{\text{thr}} \right]
$$
$$
= F(R_{\text{thr}})^{N-1} \frac{1-\beta}{N} R_{\text{thr}},
\tag{6.24}
$$

and

$$\Gamma\left(\mu_{\min}\right) = F(\mu_{\min})^{N-1}\left[\left(1+\frac{1-\beta}{N}\right)\mu_{\min}-R_{\text{thr}}\right]$$
$$+\sum_{n=1}^{N-1}\binom{N-1}{n}\left[F\left(R_{\text{thr}}\right)-F\left(\mu_{\min}\right)\right]^{n}F(\mu_{\min})^{N-1-n}\frac{\mu_{\min}-R_{\text{thr}}}{n+1}$$
$$=\left[F\left(R_{\text{thr}}\right)-F\left(\mu_{\min}\right)\right]^{N-1}\frac{\mu_{\min}-R_{\text{thr}}}{n+1}.$$

Since $\mu_{\min} < R_{\text{thr}}$, $\beta < 1$ and $F\left(R_{\text{thr}}\right) > F\left(\mu_{\min}\right)$, then we conclude that

$$\Gamma\left(R_{\text{thr}}\right) > 0, \text{ and } \Gamma\left(\mu_{\min}\right) < 0. \tag{6.25}$$

Consider that $F(\mu)$ is the CDF of continuous random variable $\mu$ for all $\mu \in (-\infty, +\infty)$. Particularly, $F(\mu) = 0$ holds for $\mu \in (-\infty, \mu_{\min}]$ and $F(\mu) = 1$ holds for $\mu \in [\mu_{\max}, +\infty)$. In addition, $\Gamma(\mu)$ is a polynomial of $F(\mu)$. Therefore, function $\Gamma(\mu)$ is continuous for $\mu \in (-\infty, +\infty)$. According to (6.25), we conclude that there exists at least one solution $\mu \in (\mu_{\min}, R_{\text{thr}})$ holding (6.9) in Lemma 6.1. This completes the proof.

## 6.11   Proof of Theorem 6.1

***Proof*** We consider the strategy for beam group $n$ ($\forall n \in \mathcal{N}$). Assuming that all the other beam groups adopt strategy $b^*\left(\mu_i, R_{\text{thr}}\right), \forall i \in \mathcal{N}, i \neq n$ formulated in (6.10). Then we first proof that selecting strategy $b^*\left(\mu_n, R_{\text{thr}}\right)$ in (6.10) can maximize beam group $n$'s utility by four situations in following sections.

### 6.11.1   $\mu_n \in [R_{thr}, \mu_{max}]$

Assume that beam group $n$'s communication quality $\mu_n \in [R_{\text{thr}}, \mu_{\max}]$. There exist the following two cases:

#### 6.11.1.1   Case 1

$b_{\max}^{-n} \in [R_{\text{thr}}, +\infty)$. Then the MNO can always select one beam group of the satellite to offload its traffic. We compare bid $\mu_n$ with any bid larger or smaller than $\mu_n$, respectively, and show that bid $\mu_n$ is optimal for beam group $n$.

First, we compare bid $\mu_n$ with any bid $\hat{\mu}_1 \in [R_{thr}, \mu_n)$. It is easy to find that when $b_{max}^{-n} \in (\mu_n, +\infty)$ or $b_{max}^{-n} \in (R_{thr}, \hat{\mu}_1)$, bid $\mu_n$ and bid $\hat{\mu}_1$ generate the same utility for beam group $n$. So we only consider the situation in which $b_{max}^{-n} \in [\hat{\mu}_1, \mu_n]$.

- When $b_{max}^{-n} \in (\hat{\mu}_1, \mu_n)$, bidding $\hat{\mu}_1$ generates a utility $\mu_n$ for beam group $n$. Since $\mu_n > b_{max}^{-n}$ in this case, bidding $\mu_n$ brings a utility $2\mu_n - b_{max}^{-n} = \mu_n + \mu_n - b_{max}^{-n} > \mu_n$. Thus, bidding $\mu_n$ brings a higher utility than that under bidding $\hat{\mu}_1$.
- When $b_{max}^{-n} = \mu_n$, bidding $\hat{\mu}_1$ generates utility $\mu_n$ for beam group $n$. In this case, beam group $n$'s utility under bid $\mu_n$ equals $\varepsilon \left(2\mu_n - b_{max}^{-n}\right) + (1 - \varepsilon) \mu_n = \mu_n$, where $\varepsilon$ denotes the probability with which beam group $n$ is selected as the winning bidder ($\varepsilon = 1/|\mathcal{M}|$). Hence, in this case, bidding $\mu_n$ and $\hat{\mu}_1$ generate the same expected utility for beam group $n$.
- When $b_{max}^{-n} = \hat{\mu}_1$, the utility under bid $\mu_n$ equals $2\mu_n - \hat{\mu}_1$ since $\hat{\mu}_1 < \mu_n$. However, bidding $\hat{\mu}_1$ brings a utility of $\varepsilon \left(2\mu_n - \hat{\mu}_1\right) + (1 - \varepsilon) \mu_n < 2\mu_n - \hat{\mu}_1$, which results from $2\mu_n - \hat{\mu}_1 > \mu_n$. Therefore, bidding $\mu_n$ brings a higher utility than that under bidding $\hat{\mu}_1$.

To conclude, considering all cases above, bidding $\mu_n$ brings a higher utility than that under bidding $\hat{\mu}_1 < \mu_n$. Then we compare bid $\mu_n$ with any bid $\hat{\mu}_2 \in (\mu_n, +\infty)$. Similarly, we can notice that bid $\mu_n$ and bid $x_2$ generate a difference on beam group $n$'s utility only when $b_{max}^{-n} \in (\mu_n, \hat{\mu}_2]$. Next, we analyze situation $b_{max}^{-n} \in (\mu_n, \hat{\mu}_2)$ and $b_{max}^{-n} = \hat{\mu}_2$, separately:

- When $b_{max}^{-n} \in (\mu_n, \hat{\mu}_2)$, bidding $\mu_n$ generates a utility of $\mu_n$ for beam group $n$. Since $\mu_n < b_{max}^{-n}$ in this case, bidding $\hat{\mu}_2$ brings a utility $2\mu_n - b_{max}^{-n} < \mu_n$. Thus, bidding $\mu_n$ brings a higher utility than that under bidding $\hat{\mu}_2$.
- When $b_{max}^{-n} = \hat{\mu}_2$, the utility for beam group $n$ under bid $\mu_n$ equals $\mu_n$. However, bidding $\hat{\mu}_2$ brings a utility of $\varepsilon \left(2\mu_n - \hat{\mu}_2\right) + (1 - \varepsilon) \mu_n < \varepsilon\mu_n + (1 - \varepsilon) \mu_n = \mu_n$, which results from $2\mu_n - \hat{\mu}_2 < \mu_n$. Therefore, bidding $\mu_n$ brings a higher utility than that under bidding $\hat{\mu}_2$.

Summarily, bidding $\mu_n$ is beam group $n$'s optimal bidding strategy comparing with any value larger or smaller than $\mu_n$ when $b_{max}^{-n} \in [R_{thr}, +\infty)$.

Particularly, we analyze the optimal strategy for beam group $n$ with $\mu_n = \mu_{max}$. If all other beam groups select strategy $b^* (\mu_i, R_{thr})$ given by (6.10), one can notice that $b_{max}^{-k} < \mu_{max}$ always holds. As a result, bidding any value from $(\mu_{max}, +\infty)$ can bring the same utility for beam group $n$ as bidding $\mu_{max}$. In other words, if $\mu_n = \mu_{max}$, bidding any value from $[\mu_{max}, +\infty)$ is the optimal bidding strategy.

### 6.11.1.2  Case 2

$b_{max}^{-n} = \varnothing$. In this case, bidding any value from $[\mu_{max}, +\infty)$ generates the same utility, i.e., $2\mu_n - R_{thr}$ for beam group $n$ according to (6.6). On the contrary, if beam group $n$ gives bid $\varnothing$, its utility will fall to $\beta\mu_n$. Considering that $\beta\mu_n < \mu_n \leq 2\mu_n - R_{thr}$ (due to $\mu_n \geq R_{thr}$), bidding $\mu_n$ is one of optimal strategies for beam group $n$. In addition, if $\mu_n = \mu_{\max}$, bidding any value from $(\mu_{\max}, +\infty)$ is also beam group $n$'s optimal strategy as well as bidding $\mu_{\max}$.

Base on the analysis of Case 1 and Case 2, one can conclude that when all the other beam groups of the satellite submit their bids according to (6.10), the optimal bidding strategy for beam group with $\mu_n \in [R_{\text{thr}}, \mu_{\max}]$ is giving bid as (6.10).

## 6.11.2   $\mu_n \in \left(\tilde{\mu}_a\left(R_{thr}\right), R_{thr}\right)$

We consider the situation when $\mu_n \in (\tilde{\mu}_a(R_{\text{thr}}), R_{\text{thr}})$. Then we compare the utilities that can be obtained by beam group $n$ when it submits bid as $R_{\text{thr}}$, $\varnothing$ and any value from $(R_{\text{thr}}, +\infty)$, considering that all the other $N-1$ beam groups choose the optimal bidding strategies as (6.10).

### 6.11.2.1   $R_{thr}$ vs $\varnothing$

Denote $\tilde{\mu} \triangleq \tilde{\mu}_a(R_{\text{thr}})$, then the utility can be obtained by beam group $n$ when $b_n = R_{\text{thr}}$:

$$
\begin{aligned}
&\pi_n\left(\left(b_n = R_{\text{thr}}, \mathbf{b}^*_{-n}\right), R_{\text{thr}}\right) \\
&= \left[1 - F(R_{\text{thr}})^{N-1}\right]\mu_n + F(\tilde{\mu})^{N-1}\left(2\mu_n - R_{\text{thr}}\right) \\
&\quad + \sum_{n=1}^{N-1} \binom{N-1}{n} (F(R_{\text{thr}}) - F(\tilde{\mu}))^n (1 - F(\tilde{\mu}))^{N-1-n} h(\mu_n),
\end{aligned}
\tag{6.26}
$$

where $h(\mu_n) = \frac{n}{n+1}\mu_n + \frac{1}{n+1}(2\mu_n - R_{\text{thr}})$. Similarly, the utility can be obtained by beam group $n$ when $b_n = \varnothing$ is given by

$$
\begin{aligned}
&\pi_n\left(\left(b_n = \varnothing, \mathbf{b}^*_{-n}\right), R_{\text{thr}}\right) \\
&= \left[1 - F(R_{\text{thr}})^{N-1}\right]\mu_n + F(\tilde{\mu})^{N-1}\left[\frac{\beta\mu_n}{N} + \frac{(N-1)\mu_n}{N}\right] \\
&\quad + \sum_{n=1}^{N-1} \binom{N-1}{n} (F(R_{\text{thr}}) - F(\tilde{\mu}))^n (1 - F(\tilde{\mu}))^{N-1-n} \mu_n.
\end{aligned}
\tag{6.27}
$$

Then we have

$$
\begin{aligned}
&\pi_n\left(\left(b_n = R_{\mathrm{thr}}, \mathbf{b}^*_{-n}\right), R_{\mathrm{thr}}\right) - \pi_n\left(\left(b_n = \varnothing, \mathbf{b}^*_{-n}\right), R_{\mathrm{thr}}\right) \\
&= F(\tilde{\mu})^{N-1}\left[\left(1 + \frac{1-\beta}{N}\right)\mu_n - R_{\mathrm{thr}}\right] \\
&\quad + \sum_{n=1}^{N-1}\binom{N-1}{n}(F(R_{\mathrm{thr}}) - F(\tilde{\mu}))^n(1 - F(\tilde{\mu}))^{N-1-n}\frac{\mu_n - R_{\mathrm{thr}}}{n+1} \\
&\triangleq \delta(\mu_n).
\end{aligned} \tag{6.28}
$$

It is easy to find that $\delta(\mu_n)$ defined in (6.28) is a strictly increasing function with $\mu_n$. In addition, let $\mu_n = \tilde{\mu}$, based on the definition of $\tilde{\mu} = \tilde{\mu}_a(R_{\mathrm{thr}})$ in Lemma 6.1, we have

$$
\begin{aligned}
\delta(\tilde{\mu}) &= F(\tilde{\mu})^{N-1}\left[\left(1 + \frac{1-\beta}{N}\right)\tilde{\mu} - R_{\mathrm{thr}}\right] \\
&\quad + \sum_{n=1}^{N-1}\binom{N-1}{n}(F(R_{\mathrm{thr}}) - F(\tilde{\mu}))^n(1 - F(\tilde{\mu}))^{N-1-n}\frac{\tilde{\mu} - R_{\mathrm{thr}}}{n+1} \\
&= 0.
\end{aligned} \tag{6.29}
$$

Therefore, for all $\mu_n \in (\tilde{\mu}_a(R_{\mathrm{thr}}), R_{\mathrm{thr}}]$,

$$
\pi_n\left(\left(b_n = R_{\mathrm{thr}}, \mathbf{b}^*_{-n}\right), R_{\mathrm{thr}}\right) - \pi_n\left(\left(b_n = \varnothing, \mathbf{b}^*_{-n}\right), R_{\mathrm{thr}}\right) > 0 \tag{6.30}
$$

will always hold, which implies that in this case, bidding $R_{\mathrm{thr}}$ brings more utility to beam group $n$ than bidding $\varnothing$.

### 6.11.2.2 $R_{\mathrm{thr}}$ vs $\hat{\mu} \in [R_{\mathrm{thr}}, +\infty)$

We notice that either when $b^{-n}_{\max} = \varnothing$ or $b^{-n}_{\max} \in (\hat{\mu}, +\infty) (> R_{\mathrm{thr}})$, beam group $n$ obtains the same utility, i.e., $2\mu_n - R_{\mathrm{thr}}$ and $\mu_n$, respectively, no matter it submits bid as $b_n = R_{\mathrm{thr}}$ or $b_n = \hat{\mu}$. Hence we only analyze the situation when $b^{-n}_{\max} \in [R_{\mathrm{thr}}, \hat{\mu}]$:

- $b^{-n}_{\max} = R_{\mathrm{thr}}$: Considering that $\mu_n \in (\tilde{\mu}_a(R_{\mathrm{thr}}), R_{\mathrm{thr}})$, i.e, $2\mu_n - R_{\mathrm{thr}} < \mu_n$, then beam group $n$'s utility when bidding $R_{\mathrm{thr}}$ and $\hat{\mu}$ can be given by

$$
\begin{aligned}
\pi_n\left(\left(b_n = R_{\mathrm{thr}}, \mathbf{b}^*_{-n}\right), R_{\mathrm{thr}}\right) &= \varepsilon(2\mu_n - R_{\mathrm{thr}}) + (1 - \varepsilon)\mu_n \\
&> \varepsilon(2\mu_n - R_{\mathrm{thr}}) + (1 - \varepsilon)(2\mu_n - R_{\mathrm{thr}}) = 2\mu_n - R_{\mathrm{thr}},
\end{aligned} \tag{6.31a}
$$

$$
\pi_n\left(\left(b_n = \hat{\mu}, \mathbf{b}^*_{-n}\right), R_{\mathrm{thr}}\right) = 2\mu_n - R_{\mathrm{thr}}, \tag{6.31b}
$$

respectively. In (6.31a), $\varepsilon$ denotes the probability with which beam group $n$ is selected to provide the offloading service. Therefore, in this case, bidding $R_{\text{thr}}$ generates a higher utility for beam group $n$ than bidding $\hat{\mu}$.

- $b_{\text{max}}^{-n} \in (R_{\text{thr}}, \hat{\mu})$: Beam group $n$'s utility when bidding $R_{\text{thr}}$ and $\hat{\mu}$ can be given by

$$\pi_n \left( \left( b_n = R_{\text{thr}}, \mathbf{b}_{-n}^* \right), R_{\text{thr}} \right) = \mu_n, \tag{6.32a}$$

$$\pi_n \left( \left( b_n = \hat{\mu}, \mathbf{b}_{-n}^* \right), R_{\text{thr}} \right) = 2\mu_n - b_{\text{max}}^{-n} < 2\mu_n - R_{\text{thr}} \leq \mu_n, \tag{6.32b}$$

respectively, which indicates that bidding $R_{\text{thr}}$ generates a higher utility for beam group $n$ than bidding $\hat{\mu}$.

- $b_{\text{max}}^{-n} = \hat{\mu}$: In this case, bidding $R_{\text{thr}}$ generates a utility of $\mu_n$ for beam group $n$, and its utility under bid $\hat{\mu}$ equals $\varepsilon \left( 2\mu_n - \hat{\mu} \right) + (1 - \varepsilon) \mu_n < \mu_n$, where $\varepsilon$ denotes the probability with which beam group $n$ is selected to provide the offloading service. Thus, bidding $R_{\text{thr}}$ generates a higher utility than that under bid $\hat{\mu}$ in this case.

So in conclusion, when all the other $N - 1$ beam groups of the satellite adopt their bidding strategies according to (6.10), the optimal strategy for beam group $n$ with $\mu_n \in (\tilde{\mu}_a (R_{\text{thr}}), R_{\text{thr}})$ is $b^* (\mu_n, R_{\text{thr}}) = R_{\text{thr}}$.

### 6.11.3 $\mu_n = \tilde{\mu}_a (R_{thr})$

Similar to the analysis in Sects. 6.11.1 and 6.11.2, we can notice that when $\mu_n = \tilde{\mu}_a (R_{\text{thr}})$, bidding $R_{\text{thr}}$ and bidding $\varnothing$ generate the same expected utility to beam group $n$. In addition, these two bids weakly dominate any bid in $[R_{\text{thr}}, +\infty)$. Consequently, when other beam groups of the satellite choose bidding strategies in (6.10), the optimal bidding strategy for beam group $n$ with $\mu_n = \tilde{\mu}_a (R_{\text{thr}})$ is also adopting (6.10).

### 6.11.4 $\mu_n \in \left[ \mu_{min}, \tilde{\mu}_a (R_{thr}) \right)$

Now we consider the last situation where $\mu_n \in \left[ \mu_{\text{min}}, \tilde{\mu}_a (R_{\text{thr}}) \right)$. Based on the second-price auction designed in Sect. 6.5 and similar analysis in Sect. 6.11.2, bidding $R_{\text{thr}}$ weakly dominates any bidding strategy in $[R_{\text{thr}}, +\infty)$. So we only need to compare the obtained utility when bidding $R_{\text{thr}}$ and $\varnothing$. According to (6.28) and (6.29), we have

$$\pi_n \left( \left( b_n = R_{\text{thr}}, \mathbf{b}_{-n}^* \right), R_{\text{thr}} \right) - \pi_n \left( \left( b_n = \varnothing, \mathbf{b}_{-n}^* \right), R_{\text{thr}} \right) < 0 \tag{6.33}$$

when $\mu_n < \tilde{\mu}_a (R_{thr})$, which implies that bidding $\varnothing$ brings a larger utility than bidding $R_{thr}$ for beam group $n$ with $\mu_n < \tilde{\mu}_a (R_{thr})$.

Summarizing the four situations above, we can conclude that when other beam groups of the satellite choose bidding strategies in (6.10), the optimal bidding strategy for beam group $n$ is also adopting (6.10). This completes the proof of Theorem 6.1.

## 6.12   Proof of Theorem 6.3

***Proof*** Notice that in Sects. 6.11.2–6.11.4, we have analyzed the case where $\mu_n \in [\mu_{min}, R_{thr}] \subset [\mu_{min}, \mu_{max}]$, and have provided a two-part structure of optimal bidding strategy: some beam groups with $\mu_n$ closed to $R_{thr}$ bid rate threshold $R_{thr}$, and some bema groups with small $\mu_n$ bid $\varnothing$. This situation above is similar to that considered in Theorem 6.1, i.e., $\mu_n \in [\mu_{min}, \mu_{max}] \subset \left[\mu_{min}, \left(1 + \frac{1-\beta}{N}\right)\mu_{max}\right] \subset [\mu_{min}, R_{thr}], \forall\, n$. Therefore, the equilibrium analysis for the situation in Theorem 6.3 can be considered as a similar and special situation analyzed in Sect. 6.11.2. Let revisit (6.28):

$$\pi_n \left((b_n = R_{thr}, \mathbf{b}^*_{-n}), R_{thr}\right) - \pi_n \left((b_n = \varnothing, \mathbf{b}^*_{-n}), R_{thr}\right) \tag{6.34a}$$

$$= F(\tilde{\mu})^{N-1} \left[\left(1 + \frac{1-\beta}{N}\right)\mu_n - R_{thr}\right] \tag{6.34b}$$

$$+ \sum_{n=1}^{N-1} \binom{N-1}{n} (F(R_{thr}) - F(\tilde{\mu}))^n (1 - F(\tilde{\mu}))^{N-1-n} \frac{\mu_n - R_{thr}}{n+1}. \tag{6.34c}$$

It is easy to find that the term shown in (6.34c) is always negative since $\mu_n < R_{thr}, \forall\, n$. In addition, noticing that $R_{thr} > \left(1 + \frac{1-\beta}{N}\right)\mu_n$ holds for all $n \in \mathcal{N}$ since $R_{thr} > \left(1 + \frac{1-\beta}{N}\right)\mu_{max}$, the term shown in (6.34b) is also always negative. Therefore, inequality

$$\pi_n \left((b_n = R_{thr}, \mathbf{b}^*_{-n}), R_{thr}\right) < \pi_n \left((b_n = \varnothing, \mathbf{b}^*_{-n}), R_{thr}\right) \tag{6.35}$$

holds for all $n \in \mathcal{N}$ when $R_{thr} \in \left[\left(1 + \frac{1-\beta}{N}\right)\mu_{max}, +\infty\right)$. This implies that bid $\varnothing$ generates a higher utility than that under bid $R_{thr}$.

Specifically, if $R_{\text{thr}} = \left(1 + \frac{1-\beta}{N}\right)\mu_{\max}$, the expected utilities for beam group $n$ with $\mu_n = \mu_{\max}$ when it bids $R_{\text{thr}}$ and $\varnothing$, i.e.,

$$\pi_n\left(\left(b_n = R_{\text{thr}}, \mathbf{b}^*_{-n}\right), R_{\text{thr}}\right) = 2\mu_n - R_{\text{thr}}, \tag{6.36a}$$

$$\pi_n\left(\left(b_n = \varnothing, \mathbf{b}^*_{-n}\right), R_{\text{thr}}\right) = \frac{\beta\mu_n}{N} + \frac{N-1}{N}\mu_n, \tag{6.36b}$$

are equal. This completes the proof of Theorem 6.3.

# References

1. M. Alzenad, M. Z. Shakir, H. Yanikomeroglu, and M. S. Alouini, "FSO-based vertical backhaul/fronthaul framework for 5G+ wireless networks," *IEEE Commun. Mag.*, vol. 56, no. 1, pp. 218–224, Jan. 2018.
2. L. Jacob, A. G. Armada, O. A. Dobre, and A. Almuttiri, "Global communications newsletter," *IEEE Commun. Mag.*, vol. 56, no. 4, pp. 11–13, Apr. 2018.
3. J. Du, C. Jiang, H. Zhang, X. Wang, Y. Ren, and M. Debbah, "Secure satellite-terrestrial transmission over incumbent terrestrial networks via cooperative beamforming," *IEEE J. Sel. Areas Commun.*, vol. 36, no. 7, pp. 1367–1382, Jul. 2018.
4. P. Pawnkumar and R. T. Selvi, "An intermediate study of meta communication for wide cellular networks," *I-manager's J. Mobile Applicat. Technol.*, vol. 4, no. 1, pp. 30–38, Jan./Jun. 2017.
5. ERC/DEC/(00)07, "The shared use of the band 17.7-19.7 GHz by the fixed service and earth stations of the fixed-satellite service (space-to-earth)," in *ECC Report 241*. Electronic Commun. Committee, Copenhagen, Denmark, approved: 19 Oct. 2000, amended: 4 Mar. 2016.
6. E. Lagunas, S. K. Sharma, S. Maleki, S. Chatzinotas, and B. Ottersten, "Resource allocation for cognitive satellite communications with incumbent terrestrial networks," *IEEE Trans. Cognitive Commun. Networking*, vol. 1, no. 3, pp. 305–317, Sept. 2015.
7. X. Zhu, C. Jiang, L. Yin, L. Kuang, N. Ge, and J. Lu, "Cooperative multigroup multicast transmission in integrated terrestrial-satellite networks," *IEEE J. Sel. Areas Commun.*, vol. 36, no. 5, pp. 981–992, May 2018.
8. G. Araniti, M. Condoluci, P. Scopelliti, A. Molinaro, and A. Iera, "Multicasting over emerging 5G networks: Challenges and perspectives," *IEEE Network*, vol. 31, no. 2, pp. 80–89, Mar./Apr. 2017.
9. C. Jiang, X. Zhu, L. Kuang, Y. Qian, and J. Lu, "Multimedia multicast beamforming in integrated terrestrial-satellite networks," in *13th Int. Wireless Commun. Mobile Computing Conf. (IWCMC)*. Valencia, Spain, 26-30 Jun. 2017.
10. J. Liu, Y. Shi, L. Zhao, Y. Cao, W. Sun, and N. Kato, "Joint placement of controllers and gateways in SDN-enabled 5G-satellite integrated network," *IEEE J. Sel. Areas Commun.*, vol. 36, no. 2, pp. 221–232, Feb. 2018.
11. T. Li, H. Zhou, H. Luo, and S. Yu, "SERvICE: A software defined framework for integrated space-terrestrial satellite communication," *IEEE Trans. Mobile Computing*, vol. 17, no. 3, pp. 703–716, Mar. 2018.
12. M. G. Kibria, G. P. Villardi, W. S. Liao, K. Nguyen, K. Ishizu, and F. Kojima, "Outage analysis of offloading in heterogeneous networks: Composite fading channels," *IEEE Trans. Veh. Technol.*, vol. 66, no. 10, pp. 8990–9004, October. 2017.
13. Y. Zhong, T. Q. S. Quek, and X. Ge, "Heterogeneous cellular networks with spatio-temporal traffic: Delay analysis and scheduling," *IEEE J. Sel. Areas Commun.*, vol. 35, no. 6, pp. 1373–1386, Jun. 2017.

14. X. Chen, J. Wu, Y. Cai, H. Zhang, and T. Chen, "Energy-efficiency oriented traffic offloading in wireless networks: A brief survey and a learning approach for heterogeneous cellular networks," *IEEE J. Sel. Areas Commun.*, vol. 33, no. 4, pp. 627–640, Apr. 2015.

15. H. Ko, J. Lee, and S. Pack, "Performance optimization of delayed WiFi offloading in heterogeneous networks," *IEEE Trans. Veh. Technol.*, vol. PP, no. 99, pp. 1–1, Oct. 2017.

16. X. Li, X. Wang, K. Li, and V. C. M. Leung, "Collaborative hierarchical caching for traffic offloading in heterogeneous networks," in *IEEE Int. Conf. Commun. (ICC)*. Paris, France, 21–25 May 2017, pp. 1–6.

17. H. M. T. Ho, H. D. Tuan, D. Ngo, T. Q. Duong, and V. Poor, "Joint load balancing and interference management for small-cell heterogeneous networks with limited backhaul capacity," *IEEE Trans. Wireless Commun*, vol. 16, no. 2, pp. 872–884, Feb. 2017.

18. C. H. Liu and H. C. Tsai, "Traffic management for heterogeneous networks with opportunistic unlicensed spectrum sharing," *IEEE Trans. Wireless Commun.*, vol. 16, no. 9, pp. 5717–5731, Sept. 2017.

19. J. Du, C. Jiang, Z. Han, H. Zhang, S. Mumtaz, and Y. Ren, "Contract mechanism and performance analysis for data transaction in mobile social networks," *IEEE Trans. Network Sci. Eng.*, vol. 6, no. 2, pp. 103–115, Apr. –Jun. 2019.

20. A. Bousia, E. Kartsakli, A. Antonopoulos, L. Alonso, and C. Verikoukis, "Multiobjective auction-based switching-off scheme in heterogeneous networks: To bid or not to bid?" *IEEE Trans. Veh. Technol.*, vol. 65, no. 11, pp. 9168–9180, Nov. 2016.

21. J. Du, E. Gelenbe, C. Jiang, H. Zhang, Z. Han, and Y. Ren, "Data transaction modeling in mobile networks: Contract mechanism and performance analysis," in *IEEE Global Commun. Conf. (GLOBECOM)*. Singapore, Singapore, 4-8 Dec. 2017.

22. J. Du, C. Jiang, Z. Han, H. Zhang, S. Mumtaz, and Y. Ren, "Contract mechanism and performance analysis for data transaction in mobile social networks," *IEEE Trans. Network Sci. Eng.*, vol. 6, no. 2, pp. 103–115, Apr. - Jun. 2019.

23. P. Li, S. Guo, and I. Stojmenovic, "A truthful double auction for device-to-device communications in cellular networks," *IEEE J. Sel. Areas Commun.*, vol. 34, no. 1, pp. 71–81, Jan. 2016.

24. Y. Sun, Q. Wu, J. Wang, Y. Xu, and A. Anpalagan, "VERACITY: Overlapping coalition formation-based double auction for heterogeneous demand and spectrum reusability," *IEEE J. Sel. Areas Commun.*, vol. 34, no. 10, pp. 2690–2705, Oct. 2016.

25. X. Zhuo, W. Gao, G. Cao, and S. Hua, "An incentive framework for cellular traffic offloading," *IEEE Trans. Mobile Computing*, vol. 13, no. 3, pp. 541–555, Mar. 2014.

26. G. Athanasiou, P. C. Weeraddana, and C. Fischione, "Auction-based resource allocation in millimeterwave wireless access networks," *IEEE Commun. Lett.*, vol. 17, no. 11, pp. 2108–2111, Nov. 2013.

27. P. Jehiel and B. Moldovanu, "Efficient design with interdependent valuations," *Econometrica*, vol. 69, no. 5, p. 1237–1259, Dec. 2003.

28. H. Yu, G. Iosifidis, J. Huang, and L. Tassiulas, "Auction-based coopetition between LTE unlicensed and Wi-Fi," *IEEE J. Sel. Areas Commun.*, vol. 35, no. 1, pp. 79–90, Jan. 2017.

29. J. Du, C. Jiang, H. Zhang, Y. Ren, and V. C. Leung, "Second-price auction based cognitive traffic offloading in heterogeneous networks," in *IEEE Int. Wireless Commun. Mobile Comput. Conf. (IWCMC'19)*.

30. H. Yao, L. Wang, X. Wang, Z. Lu, and Y. Liu, "The space-terrestrial integrated network (STIN): An overview," *IEEE Commun. Mag.*, vol. PP, no. 99, pp. 2–9, 2018.

31. T. Ahmed, R. Ferrus, R. Fedrizzi, O. Sallent, N. Kuhn, E. Dubois, and P. Gelard, "Satellite gateway diversity in SDN/NFV-enabled satellite ground segment systems," in *IEEE Int. Conf. Commun. Workshops (ICC Workshops)*. Paris, France, 21-25 May 2017.

32. B. Feng, H. Zhou, H. Zhang, G. Li, H. Li, S. Yu, and H. Chao, "HetNet: A flexible architecture for heterogeneous satellite-terrestrial networks," *IEEE Network*, vol. 31, no. 6, pp. 86–92, Nov. 2017.

33. J. Zhang, X. Zhang, M. A. Imran, B. Evans, Y. Zhang, and W. Wang, "Energy efficient hybrid satellite terrestrial 5G networks with software defined features," *J. Commun. Networks*, vol. 19, no. 2, pp. 147–161, Apr. 2017.

34. J. Schloemann, H. S. Dhillon, and R. M. Buehrer, "Toward a tractable analysis of localization fundamentals in cellular networks," *IEEE Trans. Wireless Commun.*, vol. 15, no. 3, pp. 1768–1782, Mar. 2016.

35. M. Jia, X. Zhang, X. Gu, and Q. Guo, "Energy efficient cognitive spectrum sharing scheme based on inter-cell fairness for integrated satellite-terrestrial communication systems," in *IEEE 87th Veh. Tech. Conf. (VTC Spring)*. Porto, Portugal, 3-6 Jun. 2018.

36. K. Ng, C. H. Chan, and K. Luk, "Low-cost vertical patch antenna with wide axial-ratio beamwidth for handheld satellite communications terminals," *IEEE Trans. Antennas Propagation*, vol. 63, no. 4, pp. 1417–1424, Apr. 2015.

37. Y. Ma, X. Zou, and F. Weng, "Potential applications of small satellite microwave observations for monitoring and predicting global fast-evolving weathers," *IEEE J. Sel. Topics Applied Earth Observations Remote Sensing*, vol. 10, no. 6, pp. 2441–2451, Jun. 2017.

38. T. Rossi, M. D. Sanctis, M. Ruggieri, C. Riva, L. Luini, G. Codispoti, E. Russo, and G. Parca, "Satellite communication and propagation experiments through the alphasat Q/V band Aldo Paraboni technology demonstration payload," *IEEE Aerosp. Electron. Syst. Mag.*, vol. 31, no. 3, pp. 18–27, Mar. 2016.

39. D. V. Ionov, V. V. Kalinnikov, Y. M. Timofeyev, N. A. Zaitsev, Y. A. Virolainen, V. S. Kostsov, and A. V. Poberovskii, "Comparison of radiophysical and optical infrared ground-based methods for measuring integrated content of atmospheric water vapor in atmosphere," *Radiophysics Quantum Electron.*, vol. 60, no. 4, pp. 300–308, Sept. 2017.

40. C. Cano, D. Lopez-Perez, H. Claussen, and D. J. Leith, "Using LTE in unlicensed bands: Potential benefits and coexistence issues," *IEEE Commun. Mag.*, vol. 54, no. 12, pp. 116–123, Dec. 2016.

# Chapter 7
# Cooperative Beamforming for Secure Satellite-Terrestrial Transmission

**Abstract** In this part, we consider a scenario where the satellite-terrestrial network is overlaid over the legacy cellular network. The established communication system is operated in the millimeter wave (mmWave) frequencies, which enables the massive antennas arrays to be equipped on the satellite and terrestrial base stations (BSs). The secure communication in this coexistence system of the satellite-terrestrial network and cellular network through the physical layer security techniques is studied in this work. To maximize the achievable secrecy rate of the eavesdropped fixed satellite service (FSS), we design a cooperative secure transmission beamforming scheme, which is realized through the satellite's adaptive beamforming, artificial noise (AN) and BSs' cooperative beamforming implemented by terrestrial BSs. A non-cooperative beamforming scheme is also designed, according to which BSs implement the maximum ratio transmission (MRT) beamforming strategy. Applying the designed secure beamforming schemes to the coexistence system established, we formulate the secrecy rate maximization problems subjected to the power and transmission quality constraints. To solve the nonconvex optimization problems, we design an approximation and iteration based genetic algorithm, through which the original problems can be transformed into a series of convex quadratic problems. Simulation results show the impact of multiple antenna arrays at the mmWave on improving the secure communication. Our results also indicate that through the cooperative and adaptive beamforming, the secrecy rate can be greatly increased. In addition, the convergence and efficiency of the proposed iteration based approximation algorithm are verified by the simulations.

**Keywords** Integrated Satellite-terrestrial Networks · Physical Layer Security · Millimeter Wave (mmWave) Communications · Cooperative Beamforming

## 7.1 Introduction

Recently, the fifth generation (5G) of mobile communications is willing to bring an order of magnitude improvement for the network capacity, reliability, availability and security, and to satisfy the current dramatically increasing data traffic

© The Author(s), under exclusive license to Springer Nature Singapore Pte Ltd. 2023    129
J. Du, C. Jiang, *Cooperation and Integration in 6G Heterogeneous Networks*,
Wireless Networks, https://doi.org/10.1007/978-981-19-7648-3_7

demands. To achieve these performance improvements, millimeter wave (mmWave) communication becomes a potential technology for the future outdoor wireless networks [1, 2]. Many recent studies have demonstrated that the mmWave communication is feasible and effective by using massive antenna arrays in conjunction with the adaptive beamforming technique. Due to its physical properties, the mmWave techniques can solve many problems brought by the high speed broadcast wireless transmission, such as compensating the propagation loss at high frequencies. Specifically, with much smaller wavelengths of mmWave frequencies, the mmWave techniques can reduce the size of antenna array and enable the large arrays in a given area, and can support the directional beams to the receivers [3].

On the other hand, satellite communication (SatCom) has become an outgrowth of the continuing demand for higher capacity, real-time communication and wider coverage, due to its unique ability to provide seamless connectivity and high data rate [4]. In addition, SatCom is a more economical solution to provide a seamless and high speed connectivity than deploying other terrestrial networks, especially in some remote and sparsely populated locations. To support the higher data rate requirement, SatCom using the mmWave band, especially in Ka band (17.7–19.7 GHz for the downlink, and 27.5–29.5 GHz for the uplink), has been investigated for many years [5, 6]. However, the Ka band ranged above has been primarily assigned to the terrestrial fixed service (FS) microwave links, according to Decision ECC/DEC/(00)07 adopted by the European Conference of Postal and Telecommunications Administrations (ECPT) [9]. Therefore, in order to share this non-exclusive spectrum, it is necessary to investigate the co-channel interference, cooperation beamforming schemes and many other issues in the coexistence system with SatCom and incumbent terrestrial networks to improve the system performance and efficiency of spectrum utilization, and reduce the energy consumption.

Currently, the on-going development of 5G communication brings an opportunity for a seamless integration of SatCom with terrestrial networks. In addition, SatCom will play a vital role in the development and full realization of 5G [7, 8]. However, resulting from the immense and open coverage, the transmission security in SatCom with fixed satellite service (FSS) is confronted with an increasing serious challenge, especially for the military applications. Therefore, how to minimize the interference between the FSS terminals and terrestrial networks, meanwhile guarantee their transmission quality and security requirements, plays an important role to realize an efficient and secure transmission in the satellite terrestrial networks. In this work, we will consider the downlink communication in a coexistence system with FSS terminals and terrestrial cellular networks sharing the same Ka band. Subjected to the power and transmission quality constraints, we study the cooperation based beamforming schemes among the satellite and terrestrial based stations (BSs) to maximize the achievable secrecy rate of the wiretapped FSS terminals.

The main contributions in this part can be summarized as follows:

- We establish a coexistence system of FSS and cellular networks, in which one satellite communicating with multiple FSS terminals and multiple terrestrial BSs communicating with their own users are sharing the Ka band. Consider that the satellite and BSs carry multiple antennas, and FSS terminals and BSs' users are equipped with the single antenna. Then a multiple-input-single-output (MISO) channel in mmWave frequency band is modeled. The system model and related assumptions established are reasonable and can be applied to model the current coexistence system of SatCom and terrestrial cellular networks.

- To prevent the eavesdropper from wiretapping the FSS terminals, we analyze the physical layer security issues. Based on the establish security scenario, the non-cooperation based secure transmission beamforming and cooperation based security transmission beamforming schemes are designed to ensure the security of SatCom. Simulation results show that the cooperative beamforming scheme can improve the secrecy rate of the eavesdropped FSS terminal greatly by sacrificing the BS users' transmission quality.
- We formulate the physical layer security problem in the established MISO mmWave system. The objective of the security problem is to maximize the achievable secrecy rate of the eavesdropped FSS terminal, subjected to the power and SINR threshold constraints of FSS terminals and BSs' users. In the coexistence system, we consider that the communication of terrestrial network has higher priority and legacy right of protection. This precondition is in conformity with the current regulations and rules of the satellite terrestrial communication.
- To solve the formulated nonlinear and nonconvex optimization problems, we introduce an approximation and iteration based solution to transform the original problem into a series of convex quadratic problems. Our results show that the proposed algorithm can achieve high efficiency and fast convergence to solve the original nonconvex optimization problems.

The rest is arranged as follows. Section 7.2 retrospects the existing studies for security problems in satellite terrestrial networks. Section 7.3 sets up the system model. In Sect. 7.4, the secure transmission beamforming schemes are designed, and the corresponding secrecy rate maximization problems are formulated. The iteration based solution for the optimization problems is proposed in Sect. 7.5. Simulations are shown in Sect. 7.6, and conclusions are drawn in Sect. 7.7.

Notations: In this part, $(\cdot)^H$ and $(\cdot)^T$ demote conjugate transpose and transpose, respectively. $\|\mathbf{x}\|_2$ denotes the Euclidean norm of vector $\mathbf{x}$. $\mathbb{E}\{\cdot\}$ denotes the expectation, $\Re\{\cdot\}$ defines the real operator, and $\nabla$ defines the first order differential operator. Define $\langle \mathbf{x}, \mathbf{y} \rangle \triangleq \mathbf{x}^H \mathbf{y}$.

## 7.2 Related Works

### 7.2.1 Satellite Terrestrial Networks

Recently, many research efforts have been devoted to the analysis and improvement of the system performance in satellite terrestrial networks by spectrum sharing [10–12]. In [13], authors considered terrestrial users as the primary users, and studied the optimal power control schemes for real-time applications in cognitive satellite terrestrial networks. Without degrading the communication quality of the primary terrestrial users, the delay-limited capacity and outage capacity can be maximized through the designed power control schemes. Considering the multiple co-channel interferes at both the terrestrial relay and destination, a multiple-antenna hybrid satellite terrestrial relay network was analyzed in [14]. In [15], a multimedia multicast beamforming scheme was investigated for the integrated terrestrial satellite networks, in which the maximum ratio transmission (MRT) based beamforming scheme and the zero-forcing beamforming (ZFBF) scheme were applied by BSs and the satellite, respectively.

However, the transmission security issues are hardly investigated in current studies for satellite terrestrial networks. In [16], although an optimized power allocation strategy was designed to support the secure transmission only for the SatCom scenario, the terrestrial networks were not considered in the system. A secure beamforming scheme was proposed in [17] for a satellite terrestrial network, in which the terrestrial user's capacity was maximized subjected to the power and Signal-to-Interference Plus Noise Ratio (SINR) constraints. Nevertheless, the framework established in [17] has a limited ability to model the current complex and large-scale networks due to its simplified system structure, in which the satellite communicated with only one FSS terminal and there was one terrestrial BS with an associated user. In this work, we will establish a coexistence system of SatCom and terrestrial networks using the mmWave channels, in which the satellite and terrestrial BSs carry multiple antennas. Moreover, the multiple FSS terminals associated to the satellite and mobile users associated to the BSs are equipped with single antenna and are distributed among the terrestrial part of the system.

## 7.2.2  Physical Layer Security

Using an information theoretic point, physical layer security aims to enable the legitimate destinations to successfully receive the source information and prevent eavesdropping without upper layer data encryption [18, 19]. In the theoretical framework of physical layer security, "secrecy capacity" is defined as the maximum reliable rate of information transmitted from the source to the intended destination, while eavesdroppers are kept as ignorant of this information as possible. As first pioneered in [20], physical layer security has been generalized to the wireless fading channel and communication networks with multiple nodes [21]. In order to maximize the secrecy rate of destinations, cooperative jamming has been studied to increase the SINR at the intended destinations and decrease that at eavesdroppers, through power control, adaptive beamforming and other techniques. In [22], authors studied the secrecy transmission with the assistance of multiple wireless energy harvesting-enabled amplify-and-forward relays, who perform cooperative jamming to ensure the secure transmission of the wireless sensing network. Physical layer security theory was applied in [19, 23] to investigated the effect of outdated CSI on the secrecy performance of MIMO wiretap channels with multiple eavesdroppers in non-identical Nakagami-$m$ fading and Rayleigh fading, respectively. A physical layer security game framework was established in [24], in which the source was modeled as a buyer who want to optimize its secrecy capacity minus cost, meanwhile friendly jammers modified their jamming power to maximize their own utility. The study in [24] demonstrated the effectiveness of cooperative jamming on improving the secrecy capacity.

On the other hand, the artificial noise (AN) aided transmission strategy is another efficient method to improve the secrecy rate. In [25], authors introduced AN into multi-antenna wiretap channels, and demonstrated that jointly optimizing the precoder matrix and the portion of power allocated to AN can outperform the solutions which rely on optimizing the precoder only. The secure transmission in a cognitive satellite-terrestrial network with a multi-antennas eavesdropper was investigated in [26], whose objective was to minimize the transmit power by jointly optimizing the cooperative beamforming and AN. The power allocation problem was studied in [27] for AN secure precoding system in MISOSE (MISO, single-eavesdropper), MISOME (MISO, multiple-eavesdropper) and MIMOME (MIMO, multiple-eavesdropper) channels, and the secrecy rate was analyzed and its lower bounds were derived. In this work, we will consider the terrestrial BSs as friendly jammers who operate cooperative beamforming to improve the secrecy rate of FSS terminals. In addition, AN will be introduced into the system to further confuse the eavesdropper, who is located on the ground and wiretapping the information transmitted from the satellite to the FSS terminals.

However, it is not easy to implement an optimal beamforming design due to complicated and nonconvex optimization problems formulated. In [28], an extra penalty function optimization approach was introduced to solve the convex semidefinite programs (SDPs) together with rank-one constraints on the outer products. A Gaussian randomization method to construct the rank-one solutions from the non-rank one results was utilized in [29] when finding the optimal beamforming. In [30], authors proposed a suboptimal algorithm based on the Lagrange duality method to reduce the computational complexity. The optimized beamforming in [30] can enhance the physical layer security for information receivers and yet satisfy energy harvesting requirements for energy receivers. In this work, we will design a path-pursuit and iteration based approach to improve the efficiency and convergence rate when searching the optimal beamforming.

## 7.3  System Model

In a coexistence system of FSS and cellular networks in the mmWave bands, we consider a security scenario as shown in Fig. 7.1. In particular, in this system, the satellite, communicating with $N$ FSS terminals distributed within its coverage, is equipped with $N_s > N$ antennas to illustrate beams through beamforming coherently. Considering an interference mmWave scenario, there are $M$ multi-antenna BSs and their associated users within the coverage of the satellite. Assume that there are $N_p \geq M$ antennas at each BS, and BSs' users are equipped with single-

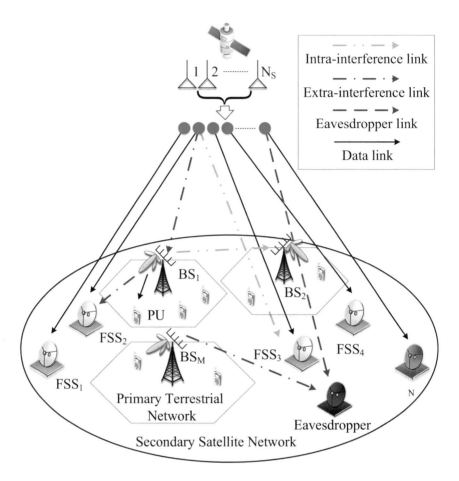

**Fig. 7.1** The coexistence system of SatCom and terrestrial cellular networks

antenna. One eavesdropper, located inside the satellite coverage, intends to wiretap the confidential message transmitted to one FSS terminal, named eavesdropped FSS terminal [31, 32]. Assume that both legitimate FSSs and the eavesdropper are equipped with a single antenna. Therefore, the communications from the sources, i.e., the satellite and terrestrial BSs, to the destinations, which refer to FSS terminals, BSs' users and the eavesdropper, can be considered as the MISO wiretap channels. In addition, we assume the system is time-slotted and quasi-static. Specifically, in each time slot, the system status remains constant.

In this work, we consider that the satellite downlinks and terrestrial BS down-links are both operating in the Ka band (17.7–19.7 GHz). According to Decision ECC/DEC/(00)07 adopted by the ECPT [9], the terrestrial BS links are incumbent links in the 17.7–19.7 GHz band, which means that BSs have the higher priority and legacy right to use this specific part of the spectrum. In other words, FSS terminals

**Table 7.1** List of Main Notations in Satellite-terrestrial Communications with Cellular Networks

| Parameter | Definition |
|---|---|
| $N$ | Number of FSSs |
| $M$ | Number of BSs |
| $N_s$ | Number of antennas equipped on Sat |
| $N_p$ | Number of antennas equipped on every BS |
| $P_s$ | Total transmit power of Sat |
| $P_p$ | Total transmit power of every BS |
| $\mathbf{h}_n$ | Channel vector between Sat and $FSS_n$ |
| $\mathbf{h}_e$ | Channel vector between Sat and EVE |
| $\mathbf{g}_m$ | Channel vector between $BS_m$ and $PU_m$ |
| $\mathbf{g}_{j,m}$ | Channel vector between $BS_j$ and $PU_m$ |
| $\mathbf{g}_{m,e}$ | Channel vector between $BS_m$ and EVE |
| $\mathbf{f}_m$ | Channel vector between Sat and $PU_m$ |
| $\mathbf{f}_{m,n}$ | Channel vector between $BS_m$ and $FSS_n$ |
| $\mathbf{w}_n$ / $\mathbf{w}$ | Beamforming vector of Sat for $FSS_n$ / FSSs |
| $\mathbf{u}_m$ / $\mathbf{u}$ | Beamforming vector of $BS_m$ / BSs |
| $\mathbf{v}$ | Artificial noise signal generated by Sat |
| $s_n$ | Transmitted data symbols from Sat to $FSS_n$ |
| $s_{ms}$ | Transmitted data symbols from $BS_m$ and $PU_m$ |
| $\gamma_n$ | SINR threshold of $FSS_n$ |
| $\gamma_{ms}$ | SINR threshold of $PU_m$ |

can be deployed without the right of protection, and their interference bringing to BSs and BS users needs to be limited [5]. Therefore, we define the transmission from BSs to their users as the primary link, while the satellite downlink to its FSS terminals as the secondary links in our work. Before proceeding further, we summarize the main notations used throughout the following sections in Table 7.1 for convenience.

## 7.3.1   Channel Model

The channel characteristics of satellite networks and terrestrial networks have been investigated and modeled in many current studies [33, 34]. The mmWave channels are expected to have limited scattering [35]. In addition, for the transmission from the satellite to FSS terminals, the line-of-sight (LOS) signal is much stronger than the others. Therefore, we consider a link with small number of paths to model the mmWave channel between the satellite and FSS terminals. Specifically, the channel vector $\mathbf{h}_n \in \mathbb{C}^{N_s \times 1}$ of FSS terminal $n$ ($n \in \mathcal{N} \triangleq \{1, 2, \cdots, N\}$) is given by Zhou et al. [36]:

$$\mathbf{h}_n = \sqrt{\frac{N_s}{L}} \sum_{l=1}^{L} \delta_{n,l} \boldsymbol{\alpha}\left(\theta_{n,l}\right), \ \forall n \in \mathcal{N}, \tag{7.1}$$

where $L$ is the number of scatters, $\delta_{n,l}$ and $\theta_{n,l}$ are the complex gain and normalized direction of the LOS path for FSS $n$, respectively. In addition, $\delta_{n,l}^2 \sim \mathcal{CN}\left(0, \sigma_0^2\right)$ is independent identically distributed (i.i.d.) complex Gaussian distribution with zero-mean, and covariance $\sigma_0^2 = 1$, which indicates Rician factor, and $\theta_{n,l} \sim U\left[-1, 1\right]$ is i.i.d. uniformly distributed. Moreover, when a uniform linear array (ULA) is adopted, the normalized array response $\boldsymbol{\alpha}\left(\theta\right)$ is given by

$$\boldsymbol{\alpha}\left(\theta\right) = \frac{1}{\sqrt{N_s}} \left[1, e^{-j\frac{2\pi}{\lambda} d \sin(\varphi)}, \cdots, e^{-j\frac{2\pi}{\lambda}(N_s-1)d \sin(\varphi)}\right]^T. \tag{7.2}$$

Here, normalized direction $\theta_n$ is related to the physical azimuth angle of departure (AoD) of $\varphi \in \left[-\pi/2, \pi/2\right]$ as $\theta = (2d/\lambda) \sin\left(\varphi\right)$, where $d$ is the antenna spacing (i.e., the distance between the two adjacent antennas), and $\lambda$ is the carrier wavelength. In this work, we assume the critically sampled environment, i.e, $d/\lambda = 0.5$, considering that the normalized AoD is the sine function of the actual AoD.

For the terrestrial cellular network, we adopt a multi-path channel with $L_m$ scatters to model the links between BSs and their users and interference links between BSs and FSS terminals [3, 6]. In this work, we assume that there is one associated user for each BS. Then the channel vector $\mathbf{g}_m \in \mathbb{C}^{N_p \times 1}$ ($m \in \mathcal{M} \triangleq \{1, 2, \cdots, M\}$) from BS $m$ to its user can be given by

$$\mathbf{g}_m = \sqrt{\frac{N_p}{L_m}} \sum_{l=1}^{L_m} \delta_{m,l} \boldsymbol{\alpha}\left(\theta_{m,l}\right), \ \forall m \in \mathcal{M}, \tag{7.3}$$

where $\delta_{m,l} \sim \mathcal{CN}\left(0, 1\right)$ and $\theta_{m,l} \sim U\left[-1, 1\right]$ are the path gain and AoD of the $l$th path of the channel vector $\mathbf{g}_m$, respectively. $L_m$ is the number of multi-path from BS $m$ to its user. Similar to (7.2), we have

$$\boldsymbol{\alpha}\left(\theta_{m,l}\right) = \frac{1}{\sqrt{N_p}} \left[1, e^{-j\frac{2\pi}{\lambda} d \sin(\varphi_{m,l})}, \cdots, e^{-j\frac{2\pi}{\lambda}(N_p-1)d \sin(\varphi_{m,l})}\right]^T \tag{7.4}$$

where normalized direction $\theta_{m,l}$ are related to the physical AoD $\varphi \in \left[-\pi/2, \pi/2\right]$ as $\theta = (2d/\lambda) \sin\left(\varphi\right)$.

### 7.3.2   Received Signal Model

Let $s_n$ be the transmitted data symbols to the $n$th FSS terminal denoted by $FSS_n$, and $s_{ms}$ be the transmitted data symbols from the $m$th BS, $BS_m$, to its user $PU_m$. The amplitude of the signal is normalized to one, i.e., $\mathbb{E}\left\{|s_n|^2\right\} = \mathbb{E}\left\{|s_{ms}|^2\right\} = 1$, $\forall n \in \mathcal{N}$, $m \in \mathcal{M}$. The transmit signals from the satellite and $BS_m$ are mapped onto the antenna array elements by the beamforming vectors $\mathbf{w}_n \in \mathbb{C}^{N_s \times 1}$, $\forall n$ and $\mathbf{u}_m \in \mathbb{C}^{N_p \times 1}$, $\forall m$, respectively. To confuse the eavesdropper, the satellite adds an AN signal, which is denoted by $\mathbf{v} \in \mathbb{C}^{N_s \times 1}$ [37, 38]. In this work, we consider that the interference AN $\mathbf{v}$ and all beamforming vectors $\mathbf{w}_n$ are controlled by the satellite-terrestrial system. In other words, for each FSS terminal, the beamforming vectors for the other FSS terminals and AN are known. Consequently, AN signal $\mathbf{v}$ is treated as interference at the eavesdropper, however $\mathbf{v}$ will not decrease the quality of legitimate communications in the system. Therefore, the interference of $\mathbf{v}$ is relatively weak for the legitimate FSS terminals of the SatCom system. Without loss of generality, assume that $\|\mathbf{w}_n\|_2 = P_n$, $\|\mathbf{u}_m\|_2 = P_{ms}$, $\forall n, m$, and $\|\mathbf{v}\|_2 = P_v$. The total transmit power of the satellite is $P_s$. Assume that the total transmit power of every BS is the same, i.e., $P_p$. Then $\sum_{n=1}^{N} \|\mathbf{w}_n\|_2 + \|\mathbf{v}\|_2 \leq P_s$ and $\|\mathbf{u}_m\|_2 \leq P_p$, $\forall m$.

Thus, after beamforming, the transmitted signal from the satellite is

$$\mathbf{x} = \sum_{n=1}^{N} \mathbf{w}_n s_n + \mathbf{v}, \tag{7.5}$$

and for each $FSS_n$, the received signal is

$$\mathbf{x}_n = \mathbf{w}_n s_n + \mathbf{v}, \ \forall n \in \mathcal{N}, \tag{7.6}$$

and the signal received by $FSS_n$ can be expressed

$$
\begin{aligned}
y_n =& \mathbf{h}_n^H \mathbf{w}_n s_n + \rho_{int} \sum_{i=1, i \neq n}^{N} \mathbf{h}_n^H \mathbf{w}_i s_i + \rho_{int} \mathbf{h}_n^H \mathbf{v} \\
&+ \rho_{ext} \sum_{m=1}^{M} \mathbf{f}_{m,n}^H \mathbf{u}_m s_{ms} + n_n, \ \forall n \in \mathcal{N},
\end{aligned}
\tag{7.7}
$$

where $\mathbf{f}_{m,n} \in \mathbb{C}^{N_p \times 1}$ is the channel vector between the $m$th BS and $n$th FSS. $0 \leq \rho_{int} < \rho_{ext} \leq 1$ are the interference coefficients of the inter-system and extra-system interference, respectively. In addition, $\rho_{int}$ is with a very small value due to that each FSS terminal knows the beamforming vectors of other FSS terminals and AN, as assumed above. $n_n \sim \mathcal{CN}\left(0, \sigma_s^2\right)$ is the i.i.d. noise with a zero-mean complex circular Gaussian distribution with variance $\sigma_s^2$.

The signal received by $PU_m$ can be given by

$$
\begin{aligned}
y_{ms} = & \mathbf{g}_m^H \mathbf{u}_m s_{ms} + \rho_{ext} \sum_{n=1}^{N} \mathbf{f}_m^H \mathbf{w}_n s_n + \rho_{ext} \mathbf{f}_m^H \mathbf{v} \\
& + \rho_{int} \sum_{j=1, j \neq m}^{M} \mathbf{g}_{j,m}^H \mathbf{u}_j s_{js} + n_{ms}, \quad \forall m \in \mathcal{M},
\end{aligned}
\tag{7.8}
$$

where $\mathbf{f}_m \in \mathbb{C}^{N_s \times 1}$ is the channel vector between the satellite and $PU_m$, $\mathbf{g}_{j,m} \in \mathbb{C}^{N_p \times 1}$ is the channel vector between $BS_j$ ($j \in \mathcal{M} \backslash m$) and $PU_m$, and $n_{ms} \sim \mathscr{CN}\left(0, \sigma_p^2\right)$ is the i.i.d. noise with a zero-mean complex circular Gaussian distribution with variance $\sigma_p^2$.

Without loss of generality, we assume that the eavesdropper is wiretapping $FSS_N$, which is similar to the model in [31]. Therefore, the received signal at the eavesdropper is given by

$$
\begin{aligned}
y_e = & \mathbf{h}_e^H \mathbf{w}_N s_N + \rho_e \sum_{i=1}^{N-1} \mathbf{h}_e^H \mathbf{w}_i s_i + \rho_e \mathbf{h}_e^H \mathbf{v} \\
& + \rho_e \sum_{m=1}^{M} \mathbf{g}_{m,e}^H \mathbf{u}_m s_{ms} + n_e,
\end{aligned}
\tag{7.9}
$$

where $\mathbf{h}_e \in \mathbb{C}^{N_s \times 1}$ and $\mathbf{g}_{m,e} \in \mathbb{C}^{N_p \times 1}$ are the channel vectors between the satellite and the eavesdropper and between $BS_m$ and the eavesdropper, respectively. $0 \leq \rho_e \leq 1$ is the interference coefficient, and $n_e \sim \mathscr{CN}\left(0, \sigma_e^2\right)$ is the i.i.d. noise at the eavesdropper. Comparing the expressions in (7.7) and (7.9), we can notice that the received signal at eavesdropped $FSS_N$ and the eavesdropper have the similar composition. However, for the eavesdropper, AN signal $\mathbf{v}$ and $\mathbf{w}_i$ ($i = 1, 2, \ldots, N-1$) can hardly known precisely. Therefore, the interference caused by AN and other transmitted signals can bring more serious reduction of the SINR for the eavesdropper.

### 7.3.3 Signal-to-Interference Plus Noise Ratio

Given the received signal formulated in (7.7), (7.8) and (7.9), the SINR of each FSS terminal, BS's user and the eavesdropper can be derived as

$$
\Gamma_n = \frac{\mathbf{w}_n^H \mathbf{R}_n \mathbf{w}_n}{\rho_{int} I_{int,n} + \rho_{ext} I_{ext,n} + \rho_{int} I_{AN,n} + \sigma_s^2}, \quad \forall n,
\tag{7.10a}
$$

$$\Gamma_{ms} = \frac{\mathbf{u}_m^H \mathbf{G}_m \mathbf{u}_m}{\rho_{ext} I_{ext,ms} + \rho_{int} I_{int,ms} + \rho_{ext} I_{AN,ms} + \sigma_p^2}, \forall m, \tag{7.10b}$$

$$\Gamma_{eN} = \frac{\mathbf{w}_N^H \mathbf{R}_e \mathbf{w}_N}{\rho_e I_{s,e} + \rho_e I_{p,e} + \rho_e I_{AN,e} + \sigma_e^2}, \tag{7.10c}$$

Respectively. In (7.10a), $I_{int,n} = \sum_{i=1,i\neq n}^N \mathbf{w}_i^H \mathbf{R}_n \mathbf{w}_i$, $I_{ext,n} = \sum_{m=1}^M \mathbf{u}_m^H \mathbf{F}_{m,n} \mathbf{u}_m$ and $I_{AN,n} = \mathbf{v}^H \mathbf{R}_n \mathbf{v}$, where $\mathbf{R}_n \triangleq \mathbf{h}_n \mathbf{h}_n^H$ and $\mathbf{F}_{m,n} \triangleq \mathbf{f}_{m,n} \mathbf{f}_{m,n}^H$. In (7.10b), $I_{ext,ms} = \sum_{n=1}^N \mathbf{w}_n^H \mathbf{F}_m \mathbf{w}_n$, $I_{int,ms} = \sum_{j=1,j\neq m}^M \mathbf{u}_j^H \mathbf{G}_{j,m} \mathbf{u}_j$ and $I_{AN,ms} = \mathbf{v}^H \mathbf{F}_m \mathbf{v}$, where $\mathbf{G}_m \triangleq \mathbf{g}_m \mathbf{g}_m^H$, $\mathbf{G}_{j,m} \triangleq \mathbf{g}_{j,m} \mathbf{g}_{j,m}^H$ and $\mathbf{F}_m \triangleq \mathbf{f}_m \mathbf{f}_m^H$. In (7.10c), $I_{s,e} = \sum_{n=1}^{N-1} \mathbf{w}_n^H \mathbf{R}_e \mathbf{w}_n$, $I_{p,e} = \sum_{m=1}^M \mathbf{u}_m^H \mathbf{G}_{m,e} \mathbf{u}_m$ and $I_{AN,e} = \mathbf{v}^H \mathbf{R}_e \mathbf{v}$, where $\mathbf{R}_e \triangleq \mathbf{h}_e \mathbf{h}_e^H$ and $\mathbf{G}_{m,e} \triangleq \mathbf{g}_{m,e} \mathbf{g}_{m,e}^H$.

### 7.3.4 Achievable Secrecy Rate

There have been several works that analyzed the MISO and MIMO wiretap channels. For the case of one eavesdropper, an achievable secrecy rate for the eavesdropped $FSS_N$ (the $N$th FSS) can be given by Lei et al. [16], Dong et al. [18]:

$$C_{sN} = \max\{C_N - C_{eN}, 0\}, \tag{7.11}$$

where

$$C_N = \log(1 + \Gamma_N), \quad C_{eN} = \log(1 + \Gamma_{eN}) \tag{7.12}$$

are the achievable rate of the link between the satellite and the eavesdropped $FSS_N$, and the achievable rate of the link between the satellite and the eavesdropper, respectively.

## 7.4 Secure Transmission Beamforming Schemes for Satellite Terrestrial Networks

In this section, we will design the secure transmission beamforming schemes by introducing AN. In addition, considering the terrestrial BSs distributed within the satellite coverage are performing as friendly jammers, we will also design a cooperative beamforming scheme to further increase the secrecy rate of the eavesdropped FSS terminal. Then we formulate the secrecy rate maximization problems for the designed beamforming schemes in this section.

### 7.4.1   Non-Cooperative Beamforming for Secure Transmission

Let us first discuss the beamforming and AN optimization for the satellite transmission without the cooperative jamming from the terrestrial cellular networks. As assumed previously, we consider that $FSS_N$ is wiretapped by the eavesdropper. The optimization goal is to maximize the achievable secrecy rate of $FSS_N$ by modifying the beamforming vectors and AN vector. Meanwhile, the required quality of service (QoS) of the system, i.e., the SINR requirements from both BSs' users and other legitimate FSS terminals, needs to be guaranteed. In addition, the beamforming scheme must meet the power constraint of the satellite. Thus, for the non-cooperative secure transmission beamforming (NCoSTB) scheme, the secrecy rate optimization problem can be formulated as

$$\max_{\mathbf{w}_n, \forall n, \mathbf{v}} \quad C_{sN}(\mathbf{w}, \mathbf{v}) = C_N(\mathbf{w}, \mathbf{v}) - C_{eN}(\mathbf{w}, \mathbf{v}), \tag{7.13a}$$

$$\text{s.t.} \quad \sum_{n=1}^{N} \|\mathbf{w}_n\|^2 + \|\mathbf{v}\|^2 \leq P_s, \tag{7.13b}$$

$$\Gamma_n(\mathbf{w}, \mathbf{v}) \geq \gamma_n, \ \forall n \in \mathcal{N}, \tag{7.13c}$$

$$\Gamma_{ms}(\mathbf{w}, \mathbf{v}) \geq \gamma_{ms}, \ \forall m \in \mathcal{M}, \tag{7.13d}$$

where $\mathbf{w} = \{\mathbf{w}_n\}_{n \in \mathcal{N}}$ and $\mathbf{v}$ are the optimization variables, $\gamma_n$ and $\gamma_{ms}$ are the SINR threshold required by $FSS_n$ and $PU_m$, respectively. The maximum ratio transmission (MRT) is a classic beamforming scheme applied in many current studies due to its low complexity and good performance [23, 39–41]. In this work, we consider that the BSs implement beamforming according to MRT for the NCoSTB scheme, i.e., for each BS,

$$\tilde{\mathbf{u}}_m = \sqrt{P_p} \frac{\mathbf{g}_m}{\|\mathbf{g}_m\|_2}, \quad m \in \mathcal{M}. \tag{7.14}$$

### 7.4.2   Cooperative Secure Beamforming for Secure Transmission

In the NCoSTB scheme, BSs implement fixed beaming determined by the channel states. In the coexistence system of the SatCom and terrestrial network when they are sharing the mmWave band, the BSs' transmitted signals after the beamforming can bring the noise and confuse the eavesdropper, which decreases the achievable rate at the eavesdropper according to (7.9). On the other hand, these signals from the terrestrial network can also influence the received rate at FSS terminals. Therefore, how to minimize the BSs' interference to the FSS terminals as well as to confuse the eavesdropper at the same time, has a significant effect on improving the security and capacity of the SatCom system.

In recent works that study the physical layer security, the cooperative jamming has been employed to reduce the eavesdropper's ability to decode the target receiver's information [42, 43]. Assume that the channel state information can be shared among the satellite terrestrial system. When the BSs transmit to their users, they implement their beamforming according to the channel state information not only of their own but also of the SatCom system. Specifically, BSs can perform as friendly jammers to minimize their interference to FSS terminals, meanwhile, improve the transmission security. Next, we will formulate the secrecy rate optimization problem for this cooperative secure transmission beamforming (CoSTB) scheme above.

Let $\mathbf{u} = \{\mathbf{u}_m\}_{m \in \mathcal{M}}$. The optimization problem aims to maximize the secrecy rate of the eavesdropped FSS terminal by jointly adjusting the beamforming of satellite and BSs, subjected to the power and SINR constraints of both the satellite and BSs. Thus, we formulate the optimization problem for the CoSTB scheme as

$$\max_{\substack{\mathbf{w}_n, \forall n, \mathbf{v} \\ \mathbf{u}_m, \forall m}} C_{sN}\left(\mathbf{w}, \mathbf{v}, \mathbf{u}\right) = C_N\left(\mathbf{w}, \mathbf{v}, \mathbf{u}\right) - C_{eN}\left(\mathbf{w}, \mathbf{v}, \mathbf{u}\right), \tag{7.15a}$$

$$\text{s.t.} \quad \sum_{n=1}^{N} \|\mathbf{w}_n\|^2 + \|\mathbf{v}\|^2 \leq P_s, \tag{7.15b}$$

$$\|\mathbf{u}_m\|^2 \leq P_p, \forall m \in \mathcal{M}, \tag{7.15c}$$

$$\Gamma_n\left(\mathbf{w}, \mathbf{v}, \mathbf{u}\right) \geq \gamma_n, \forall n \in \mathcal{N}, \tag{7.15d}$$

$$\Gamma_{ms}\left(\mathbf{w}, \mathbf{v}, \mathbf{u}\right) \geq \gamma_{ms}, \forall m \in \mathcal{M}. \tag{7.15e}$$

So far, we have formulated the secrecy rate optimization problems for the NCoSTB and CoSTB schemes. We can notice that in (7.13) and (7.15), the objective fuctions (7.13a) and (7.15a) are not concave. For constraints, (7.13b), (7.15b), and (7.15c) are convex. However, constraints (7.13c), (7.13d), (7.15d) and (7.15e) are not convex in their current forms. In the next section, we will focus on pursuing the solutions of such nonconvex optimization problems approximately but effectively and efficiently.

## 7.5   Solutions of the Optimization Problems

Currently, many works studying the beamforming design focus on solving such complicated and nonconvex optimization problems formulated in the previous section [44]. For instance, the tractable semidefinite technique is introduced to transform the nonconvex problems into a tractable SDP [45, 46]. However, when the total dimension of the optimization variables increase explosively in the scenarios where massive antennas are deployed for mmWave communications, the SDP approach will become computationally expensive. Concerning this issue, we will design a path-pursuit iteration based algorithm to solve the secrecy rate maximization problems (7.13) and (7.15) with high efficiency. Through the proposed algorithm, (7.13) and (7.15) will be decomposed into a series of iterative optimization problems, and each iteration can be formulated as a convex quadratic program in $(\mathbf{w}, \mathbf{v})$ and $(\mathbf{w}, \mathbf{v}, \mathbf{u})$, respectively. In this section, we will first provide a feasible solution to solve the optimization problems in the previous sections. To improve the efficiency and convergence rate of the introduced optimization algorithm, we will design a path-pursuit and iteration based approach later. Then we will prove the feasibility of the designed optimization.

### 7.5.1   Feasible Solution of the Optimization Problems

First, we introduce a classic optimization algorithm to solve the formulated optimization problems in the previous section. As discussed in the previous section, the objective functions of secrecy rate maximization problems (7.13) and (7.15) are nonconvex. To find out the approximate solutions, we introduce an efficient and effective stochastic and cooperation based optimization technique, called the cooperative particle swarm optimization (CPSO) algorithm [47]. CPSO was proposed based on the traditional particle swarm optimization (PSO). In PSO, the term of swarm indicates multiple particles, and there is only one swarm with many particles. Each of these particles refers to a possible solution of the optimization problem. PSO is operated with a series of iterations. In each iteration, every particle finds its own best solution and then accelerates in the direction of this position, as well as in the direction of the global best position having been found at present. However, the performance of PSO often deteriorates rapidly as the dimensionality of the problem increases. CPSO can be considered as an improvement of PSO, by expanding the single swarm, aiming to find the optimal $S$-dimensional vector, into $S$ swarms. Each of these swarms has many particles. Through the cooperative optimization of the one-dimensional vector operated by each of the $S$ swarms, CPSO can achieve a faster convergence to find the optimal solution than PSO.

In addition, CPSO is an effective and efficient approach to deal with a large range of optimization problems, such as nonconvex, nonsmooth and nonlinear high-dimensional optimization problems [48–50]. We summarize the main operation of CPSO proposed in [47] as Algorithm 1.

### 7.5.2   Path-Pursuit Iteration Based Approach

However, sometimes the CPSO algorithm may converge to a local optimal solution when applying it directly to deal with the optimization problems, which depends

---

**Algorithm 1** CPSO Algorithm [47]

---

**Initialization:**
1: Create and initialize $S$ one-dimensional PSOs: $P_j$, $j = 1, 2, \cdots, S$;
2: Define:
3: $g(j, z) \equiv \left( P_1 \cdot \hat{\mathbf{w}}, P_2 \cdot \hat{\mathbf{w}}, \cdots, P_{j-1} \cdot \hat{\mathbf{w}}, z, P_{j+1} \cdot \hat{\mathbf{w}}, \cdots, P_S \cdot \hat{\mathbf{w}} \right)$;
4: Iterations $T$.
5: **for** $t \leq T$ **do**
6:     **for** each swarm $j = 1, 2, \cdots, S$ **do**
7:         **for** each particle $i = 1, 2, \cdots, I$ **do**
8:             **if** $C_{sN} \left( g \left( j, P_j \cdot \mathbf{x}_i \right) \right) < C_{sN} \left( g \left( j, P_j \cdot \mathbf{w}_i \right) \right)$ **then**
9:                 $P_j \cdot \mathbf{w}_i = P_j \cdot \mathbf{x}_i$
10:             **end if**
11:             **if** $C_{sN} \left( g \left( j, P_j \cdot \mathbf{w}_i \right) \right) < C_{sN} \left( g \left( j, P_j \cdot \hat{\mathbf{w}} \right) \right)$ **then**
12:                 $P_j \cdot \hat{\mathbf{w}} = P_j \cdot \mathbf{w}_i$
13:             **end if**
14:         **end for**
15:         Update $P_j$ by PSO with:

$$u_{ij}(t+1) = w u_{ij}(t) + c_1 \zeta_{1i}(t) \left[ w_{ij}(t) - x_{ij}(t) \right] + c_2 \zeta_{2i}(t) \left[ \hat{w}_j(t) - x_{ij}(t) \right], (7.16)$$

$$\mathbf{x}_i(t+1) = \mathbf{x}_i(t) + \mathbf{u}_i(t+1), \tag{7.17}$$

16:         where $j = 1, 2, \cdots, S$, $S$: swarm size;
17:         $i = 1, 2, \cdots, I$, $I$: number of particles;
18:         $\mathbf{x}_i = [x_{i1} \, x_{i2} \cdots x_{iS}]$: current position in search space;
19:         $\mathbf{u}_i = [u_{i1} \, u_{i2} \cdots u_{iS}]$: current velocity;
20:         $\mathbf{w}_i = [w_{i1} \, w_{i2} \cdots w_{iS}]$: local best position;
21:         $c_1, c_2$: acceleration coefficients;
22:         $\zeta_1, \zeta_{2i} \sim U(0, 1)$: random sequences.
23:     **end for**
24: **end for**

---

on the initial feasible values selection. Especially when the objective function and constraints of the optimization problem are nonconvex, the genetic algorithms tend to converge much slower, and are much easier to converge to a local optimal solution. In response, we will design an iteration based CPSO (ICPSO) to improve the convergence speed and the reliability of the CPSO algorithm in this section.

### 7.5.2.1 Approximation of Optimization Problems

To solve the original optimization problems formulated in (7.13) and (7.15) with efficiency, we decompose them into a series of iterative optimization problems. In each iteration, the optimization problem will be approximatively formulated into a simple convex quadratic program in $(\mathbf{w}, \mathbf{v})$ or $(\mathbf{w}, \mathbf{v}, \mathbf{u})$. In addition, the solution of the current iterative optimization problem will be set as the initial values of the next iteration. Through the approximation and path-pursuit iteration process above, the optimal point will be evolved and optimized over the iterations.

　　The approximate and convex transformation mentioned above is the key operation to achieve a feasible and approximate optimal solution after a series of iterations. We can notice that although the objective functions shown in (7.13a) and (7.15a) are nonconvex, the components of them, i.e., $C_N$ and $C_{eN}$, can be transformed into the convex and concave functions, respectively. The proof of the convexity of $C_N$ and concavity of $C_{eN}$ can be found in Sects. 7.8 and 7.9, respectively. Additionally, we consider that the Taylor expansion can represent any differentiable nonlinear function as a polynomial with infinite terms, and the coefficient of each term is calculated from the value of this function's relevant order derivative at a given point. If the function is convex (concave), which means that its second derivative is positive (negative), then we can find the lower (upper) bound at a given point when only considering the terms of constant and the first derivative of the Taylor expansion. Furthermore, when the iterative algorithm is implemented, the given point for the Taylor expansion in every iteration can be set as the optimal solution obtained in the last iteration.

　　According to the analysis above, we can establish the approximate optimization problems of (7.13) and (7.15) for every iteration. Denote the approximate objective functions in the $t$th iteration of the NCoSTB and CoSTB as $C_{sN}^{(t)}(\mathbf{w}, \mathbf{v})$ and $C_{sN}^{(t)}(\mathbf{w}, \mathbf{v}, \mathbf{u})$, respectively, which can be given by

$$C_{sN}^{(t)}(\mathbf{w}, \mathbf{v}) = C_N^{(t)}(\mathbf{w}, \mathbf{v}) - C_{eN}^{(t)}(\mathbf{w}, \mathbf{v}), \tag{7.18a}$$

$$C_{sN}^{(t)}(\mathbf{w}, \mathbf{v}, \mathbf{u}) = C_N^{(t)}(\mathbf{w}, \mathbf{v}, \mathbf{u}) - C_{eN}^{(t)}(\mathbf{w}, \mathbf{v}, \mathbf{u}), \tag{7.18b}$$

where $C_N^{(t)}(\mathbf{w}, \mathbf{v})$ and $C_N^{(t)}(\mathbf{w}, \mathbf{v}, \mathbf{u})$ are the lower bounds of $C_N$ in the $t$th iteration, which will be provided in Theorem 7.1, and $C_{eN}^{(t)}(\mathbf{w}, \mathbf{v})$ and $C_{eN}^{(t)}(\mathbf{w}, \mathbf{v}, \mathbf{u})$ are the upper bounds of $C_{eN}$ in the $t$th iteration, which will be provided in Theorem 7.2.

**Theorem 7.1** *Let $\left(\mathbf{w}^{(t)}, \mathbf{v}^{(t)}\right)$ and $\left(\mathbf{w}^{(t)}, \mathbf{v}^{(t)}, \mathbf{u}^{(t)}\right)$ be the feasible solutions of (7.13) and (7.15), respectively, and be the datums in the $t$th iterative problems. Denote*

$$\psi_N(\mathbf{w}, \mathbf{v}) = \rho_{int}\sum_{i=1}^{N-1}\mathbf{w}_i^H\mathbf{R}_N\mathbf{w}_i + \rho_{int}\mathbf{v}^H\mathbf{R}_N\mathbf{v} + \rho_{ext}\sum_{m=1}^{M}\tilde{\mathbf{u}}_m^H\mathbf{F}_{m,N}\tilde{\mathbf{u}}_m + \sigma_s^2,$$
$$\tag{7.19a}$$

$$\psi_N(\mathbf{w}, \mathbf{v}, \mathbf{u}) = \rho_{int}\sum_{i=1}^{N-1}\mathbf{w}_i^H\mathbf{R}_N\mathbf{w}_i + \rho_{int}\mathbf{v}^H\mathbf{R}_N\mathbf{v} + \rho_{ext}\sum_{m=1}^{M}\mathbf{u}_m^H\mathbf{F}_{m,N}\mathbf{u}_m + \sigma_s^2,$$
$$\tag{7.19b}$$

where $\tilde{\mathbf{u}}_m$ in (7.19a) is obtained by the MRT strategy according to (7.14). For the NCoSTB scheme, the approximate lower bound of $C_N$ $(\mathbf{w}, \mathbf{v})$ can be given by

$$C_N (\mathbf{w}, \mathbf{v}) \geq C_N^{(t)} (\mathbf{w}, \mathbf{v})$$

$$\triangleq C_N \left( \mathbf{w}^{(t)}, \mathbf{v}^{(t)} \right) + \frac{2}{\ln 2} \frac{\Re \left\{ \left( \mathbf{w}_N^{(t)} \right)^H \mathbf{R}_N \mathbf{w}_N \right\}}{\psi_N \left( \mathbf{w}^{(t)}, \mathbf{v}^{(t)} \right)}$$

$$- \frac{1}{\ln 2} \frac{\left( \mathbf{w}_N^{(t)} \right)^H \mathbf{R}_N \mathbf{w}_N^{(t)} \left( \psi_N (\mathbf{w}, \mathbf{v}) + \mathbf{w}_N^H \mathbf{R}_N \mathbf{w}_N \right)}{\psi_N \left( \mathbf{w}^{(t)}, \mathbf{v}^{(t)} \right) \left[ \psi_N \left( \mathbf{w}^{(t)}, \mathbf{v}^{(t)} \right) + \left( \mathbf{w}_N^{(t)} \right)^H \mathbf{R}_N \mathbf{w}_N^{(t)} \right]} \qquad (7.20)$$

$$- \frac{1}{\ln 2} \frac{\left( \mathbf{w}_N^{(t)} \right)^H \mathbf{R}_N \mathbf{w}_N^{(t)}}{\psi_N \left( \mathbf{w}^{(t)}, \mathbf{v}^{(t)} \right)}.$$

Similarly, for the CoSTB scheme, the approximate lower bound of $C_N \left( \mathbf{w}_s, \mathbf{v}, \mathbf{w}_p \right)$ is given by

$$C_N (\mathbf{w}, \mathbf{v}, \mathbf{u}) \geq C_N^{(t)} (\mathbf{w}, \mathbf{v}, \mathbf{u})$$

$$\triangleq C_N \left( \mathbf{w}^{(t)}, \mathbf{v}^{(t)}, \mathbf{u}^{(t)} \right) + \frac{2}{\ln 2} \frac{\Re \left\{ \left( \mathbf{w}_N^{(t)} \right)^H \mathbf{R}_N \mathbf{w}_N \right\}}{\psi_N \left( \mathbf{w}^{(t)}, \mathbf{v}^{(t)}, \mathbf{u}^{(t)} \right)}$$

$$- \frac{1}{\ln 2} \frac{\left( \mathbf{w}_N^{(t)} \right)^H \mathbf{R}_N \mathbf{w}_N^{(t)} \left( \psi_N (\mathbf{w}, \mathbf{v}, \mathbf{u}) + \mathbf{w}_N^H \mathbf{R}_N \mathbf{w}_N \right)}{\psi_N \left( \mathbf{w}^{(t)}, \mathbf{v}^{(t)}, \mathbf{u}^{(t)} \right) \left[ \psi_N \left( \mathbf{w}^{(t)}, \mathbf{v}^{(t)}, \mathbf{u}^{(t)} \right) + \left( \mathbf{w}_N^{(t)} \right)^H \mathbf{R}_N \mathbf{w}_N^{(t)} \right]} \qquad (7.21)$$

$$- \frac{1}{\ln 2} \frac{\left( \mathbf{w}_N^{(t)} \right)^H \mathbf{R}_N \mathbf{w}_N^{(t)}}{\psi_N \left( \mathbf{w}^{(t)}, \mathbf{v}^{(t)}, \mathbf{u}^{(t)} \right)}.$$

**Proof** See Sect. 7.8.

*Remark* As defined in (7.20) and (7.21), $C_N^{(t)}$ $(\mathbf{w}, \mathbf{v})$ and $C_N^{(t)}$ $(\mathbf{w}, \mathbf{v}, \mathbf{u})$ are concave functions of $(\mathbf{w}, \mathbf{v})$ and $(\mathbf{w}, \mathbf{v}, \mathbf{u})$, respectively.

**Theorem 7.2** *Let*

$$\psi_e\left(\mathbf{w}, \mathbf{v}\right) = \rho_e \sum_{n=1}^{N-1} \mathbf{w}_n^H \mathbf{R}_e \mathbf{w}_n + \rho_e \mathbf{v}^H \mathbf{R}_e \mathbf{v} + \rho_e \sum_{m=1}^{M} \tilde{\mathbf{u}}_m^H \mathbf{G}_{m,e} \tilde{\mathbf{u}}_m + \sigma_e^2, \qquad (7.22a)$$

$$\psi_e\left(\mathbf{w}, \mathbf{v}, \mathbf{u}\right) = \rho_e \sum_{n=1}^{N-1} \mathbf{w}_n^H \mathbf{R}_e \mathbf{w}_n + \rho_e \mathbf{v}^H \mathbf{R}_e \mathbf{v} + \rho_e \sum_{m=1}^{M} \mathbf{u}_m^H \mathbf{G}_{m,e} \mathbf{u}_m + \sigma_e^2.$$

$$(7.22b)$$

*Then for the NCoSTB scheme, the approximate upper bound of $C_{eN}\left(\mathbf{w}, \mathbf{v}\right)$ can be given by*

$$C_{eN}\left(\mathbf{w}, \mathbf{v}\right) \le C_{eN}^{(t)}\left(\mathbf{w}, \mathbf{v}\right)$$

$$\triangleq C_{eN}\left(\mathbf{w}^{(t)}, \mathbf{v}^{(t)}\right) - \frac{1}{\ln 2}$$

$$(7.23)$$

$$+ \frac{1}{\ln 2} \frac{\psi_e\left(\mathbf{w}^{(t)}, \mathbf{v}^{(t)}\right)}{\psi_e\left(\mathbf{w}^{(t)}, \mathbf{v}^{(t)}\right) + \left(\mathbf{w}_N^{(t)}\right)^H \mathbf{R}_e \mathbf{w}_N^{(t)}} \left(\frac{\mathbf{w}_N^H \mathbf{R}_e \mathbf{w}_N}{\psi_e^{(t)}\left(\mathbf{w}, \mathbf{v}\right)} + 1\right),$$

*where*

$$\psi_e^{(t)}\left(\mathbf{w}, \mathbf{v}\right) = \rho_e \sum_{n=1}^{N-1} \Re\left\{\left\langle \mathbf{h}_e^H \mathbf{w}_n^{(t)}, 2\mathbf{h}_e^H \mathbf{w}_n - \mathbf{h}_e^H \mathbf{w}_n^{(t)}\right\rangle\right\}$$

$$+ \rho_e \Re\left\{\left\langle \mathbf{h}_e^H \mathbf{v}^{(t)}, 2\mathbf{h}_e^H \mathbf{v} - \mathbf{h}_e^H \mathbf{v}^{(t)}\right\rangle\right\} \qquad (7.24)$$

$$+ \rho_e \sum_{m=1}^{M} \tilde{\mathbf{u}}_m^H \mathbf{G}_{m,e}^H \tilde{\mathbf{u}}_m + \sigma_e^2.$$

*Similarly, for CoSTB, the approximate upper bound of $C_{eN}\left(\mathbf{w}, \mathbf{v}, \mathbf{u}\right)$ is given by*

$$C_{eN}\left(\mathbf{w}, \mathbf{v}, \mathbf{u}\right) \le C_{eN}^{(t)}\left(\mathbf{w}, \mathbf{v}, \mathbf{u}\right)$$

$$\triangleq C_{eN}\left(\mathbf{w}^{(t)}, \mathbf{v}^{(t)}, \mathbf{u}^{(t)}\right) - \frac{1}{\ln 2}$$

$$(7.25)$$

$$+ \frac{1}{\ln 2} \frac{\psi_e\left(\mathbf{w}^{(t)}, \mathbf{v}^{(t)}, \mathbf{u}^{(t)}\right)}{\psi_e\left(\mathbf{w}^{(t)}, \mathbf{v}^{(t)}, \mathbf{u}^{(t)}\right) + \left(\mathbf{w}_N^{(t)}\right)^H \mathbf{R}_e \mathbf{w}_N^{(t)}} \left[\frac{\mathbf{w}_N^H \mathbf{R}_e \mathbf{w}_N}{\psi_e^{(t)}\left(\mathbf{w}, \mathbf{v}, \mathbf{u}\right)} + 1\right],$$

*where*

$$\psi_e^{(t)}(\mathbf{w}, \mathbf{v}, \mathbf{u}) = \rho_e \sum_{n=1}^{N-1} \Re \left\{ \left\langle \mathbf{h}_e^H \mathbf{w}_n^{(t)}, 2\mathbf{h}_e^H \mathbf{w}_n - \mathbf{h}_e^H \mathbf{w}_n^{(t)} \right\rangle \right\}$$

$$+ \rho_e \Re \left\{ \left\langle \mathbf{h}_e^H \mathbf{v}^{(t)}, 2\mathbf{h}_e^H \mathbf{v} - \mathbf{h}_e^H \mathbf{v}^{(t)} \right\rangle \right\} \qquad (7.26)$$

$$+ \rho_e \sum_{m=1}^{M} \tilde{\mathbf{u}}_m^H \mathbf{G}_{m,e}^H \tilde{\mathbf{u}}_m + \sigma_e^2.$$

***Proof***  See Sect. 7.9.

***Remark***  As defined in (7.23) and (7.25), $C_{eN}^{(t)}(\mathbf{w}, \mathbf{v})$ and $C_{eN}^{(t)}(\mathbf{w}, \mathbf{v}, \mathbf{u})$ are convex functions of $(\mathbf{w}, \mathbf{v})$ and $(\mathbf{w}, \mathbf{v}, \mathbf{u})$, on domains

$$\psi_e^{(t)}(\mathbf{w}, \mathbf{v}) \geq 0, \qquad (7.27a)$$

$$\psi_e^{(t)}(\mathbf{w}, \mathbf{v}, \mathbf{u}) \geq 0, \qquad (7.27b)$$

respectively.

According to Theorems 7.1 and 7.2, the secrecy rate maximization problems formulated in (7.13) and (7.15) can be transformed into a series of convex quadratic problems, which can be solved and processed with low computational complexity and high efficiency. In order to avoid repeated and similar analysis, in the following parts of this section, we will take the CoSTB scheme as the example to introduce the operation of the path-pursuit iteration approach to find out the approximate solutions of problem (7.15).

Using (7.21) and (7.25), The $t$th iteration of optimization problem (7.15) can be approximated as an inner convex program as

$$\max_{\substack{\mathbf{w}_n, \forall n, \mathbf{v} \\ \mathbf{u}_m, \forall m}} \quad C_{sN}^{(t)}(\mathbf{w}, \mathbf{v}, \mathbf{u}), \qquad (7.28a)$$

$$\text{s.t.}  (7.15b), (7.15c), (7.15d), (7.15e) \text{ and } (7.27b), \qquad (7.28b)$$

where $C_{sN}^{(t)}(\mathbf{w}, \mathbf{v}, \mathbf{u})$ is obtained by applying (7.21) and (7.25).

### 7.5.2.2  Path-Pursuit Iteration Based Algorithm Design

Based on the approximate optimization problem established above, we design a path-pursuit based approach to maximize the secrecy rate of the eavesdropped FSS terminal, as summarized in Algorithm 2. In this part, we still only provide the algorithm for the CoSTB scheme as the example.

To achieve Step 7 in the repeated part of Algorithm 2, apply the CPSO algorithm introduced in Sect. 7.5.1 to obtain the current optimal solution for each

---

**Algorithm 2** Path-pursuit iteration based algorithm (ICPSO)

---

**Initialization:**
 1: Iterative index: $t = 1$;
 2: Maximun iterative number: $N_{\text{iter}}$;
 3: Caculate initial feasible point $\left(\mathbf{w}^{(1)}, \mathbf{v}^{(1)}, \mathbf{u}^{(1)}\right)$: Caculate $\tilde{\mathbf{w}}$ and $\tilde{\mathbf{u}}$ according to MRT, initialize
    $\tilde{\mathbf{v}} = \mathbf{0}$, and then adjust $\left(\tilde{\mathbf{w}}, \tilde{\mathbf{v}}, \tilde{\mathbf{u}}\right)$ to meet constraint (7.28b).
 4:
 5: **for** $t \leq N_{\text{iter}}$ **do**
 6:     Solve optimization problem in (7.28),
 7:     obtain the optimal solution $(\mathbf{w}^*, \mathbf{v}^*, \mathbf{u}^*)$,
 8:     $t = t + 1$,
 9:     $\mathbf{w}^{(t)} = \mathbf{w}^*, \mathbf{v}^{(t)} = \mathbf{v}^*, \mathbf{u}^{(t)} = \mathbf{u}^*$.
10: **end for**
**Output:**
11: Optimal solution: $(\mathbf{w}^*, \mathbf{v}^*, \mathbf{u}^*)$.

---

iterative optimization problem. After $N_{\text{iter}}$ times of iterations, the obtained $N_{\text{iter}}$th $(\mathbf{w}^*, \mathbf{v}^*, \mathbf{u}^*)$ will be considered as the optimal solution of the original optimization problem in (7.15). Similarly, the iteration based NCoSTB (INCoSTB) can be achieved.

### 7.5.3  Feasibility of Path-Pursuit Iteration Based Solution

So far, we have provided the path-pursuit iteration based solution to solve the original nonconvex problems by transforming them into a series of convex optimization problems approximately. Next, we will analyze the effectiveness and feasibility of the proposed algorithm, and proof that in the $t$th iteration, the new optimal point $\left(\mathbf{w}^{(t+1)}, \mathbf{v}^{(t+1)}\right)$ / $\left(\mathbf{w}^{(t+1)}, \mathbf{v}^{(t+1)}, \mathbf{u}^{(t+1)}\right)$ is a better point than $\left(\mathbf{w}^{(t)}, \mathbf{v}^{(t)}\right)$ / $\left(\mathbf{w}^{(t)}, \mathbf{v}^{(t)}, \mathbf{u}^{(t)}\right)$ to get a larger $C_{sN}$, and that

$$\lim_{t \to \infty} C_{sN}\left(\mathbf{w}^{(t)}, \mathbf{v}^{(t)}\right) / \lim_{t \to \infty} C_{sN}\left(\mathbf{w}^{(t)}, \mathbf{v}^{(t)}, \mathbf{u}^{(t)}\right)$$

is a Karush-Kuhn-Tucker point of the optimization problem.

We still take the CoSTB scheme as the example to analyze the feasibility of the iteration base approach for the optimization problems. According to the previous definitions in (7.21) and (7.25), for the $t$th iterative optimization problem, we have

$$C_{sN}\left(\mathbf{w}, \mathbf{v}, \mathbf{u}\right) \geq C_{sN}^{(t)}\left(\mathbf{w}, \mathbf{v}, \mathbf{u}\right), \tag{7.29a}$$

$$C_{sN}\left(\mathbf{w}^{(t)}, \mathbf{v}^{(t)}, \mathbf{u}^{(t)}\right) = C_{sN}^{(t)}\left(\mathbf{w}^{(t)}, \mathbf{v}^{(t)}, \mathbf{u}^{(t)}\right), \tag{7.29b}$$

$$C_{sN}\left(\mathbf{w}^{(t+1)}, \mathbf{v}^{(t+1)}, \mathbf{u}^{(t+1)}\right) \geq C_{sN}^{(t)}\left(\mathbf{w}^{(t+1)}, \mathbf{v}^{(t+1)}, \mathbf{u}^{(t+1)}\right), \tag{7.29c}$$

$\forall$ **w**, **v**, **u**. Moreover, consider that both $\left(\mathbf{w}^{(t)}, \mathbf{v}^{(t)}, \mathbf{u}^{(t)}\right)$ and $\left(\mathbf{w}^{(t+1)}, \mathbf{v}^{(t+1)}, \mathbf{u}^{(t+1)}\right)$ are feasible points of the $t$th iterative optimization problem. According to Algorithm 2, $\left(\mathbf{w}^{(t+1)}, \mathbf{v}^{(t+1)}, \mathbf{u}^{(t+1)}\right)$ is the optimal point of $t$th iterative optimization problem. Therefore, we have

$$C_{sN}^{(t)}\left(\mathbf{w}^{(t)}, \mathbf{v}^{(t)}, \mathbf{u}^{(t)}\right) \le C_{sN}^{(t)}\left(\mathbf{w}^{(t+1)}, \mathbf{v}^{(t+1)}, \mathbf{u}^{(t+1)}\right). \tag{7.30}$$

Consequently,

$$\begin{aligned}
&C_{sN}\left(\mathbf{w}^{(t+1)}, \mathbf{v}^{(t+1)}, \mathbf{u}^{(t+1)}\right) \\
&\ge C_{sN}^{(t)}\left(\mathbf{w}^{(t+1)}, \mathbf{v}^{(t+1)}, \mathbf{u}^{(t+1)}\right) \\
&> C_{sN}^{(t)}\left(\mathbf{w}^{(t)}, \mathbf{v}^{(t)}, \mathbf{u}^{(t)}\right) = C_{sN}\left(\mathbf{w}^{(t)}, \mathbf{v}^{(t)}, \mathbf{u}^{(t)}\right).
\end{aligned} \tag{7.31}$$

Therefore, solution $\left(\mathbf{w}^{(t+1)}, \mathbf{v}^{(t+1)}, \mathbf{u}^{(t+1)}\right)$ in $t$th optimization is a better point than $\left(\mathbf{w}^{(t)}, \mathbf{v}^{(t)}, \mathbf{u}^{(t)}\right)$ as it result to a larger $C_{sN}$ for the original optimization problem in (7.15).

Consider that sequence $\left\{\left(\mathbf{w}^{(t)}, \mathbf{v}^{(t)}, \mathbf{u}^{(t)}\right) | t = 1, 2, \cdots, T\right\}$ is constrained by (7.15b), (7.15c), (7.15d) and (7.15e). Therefore, there must exist a subsequence

$$\left\{\left(\mathbf{w}^{(t_\tau)}, \mathbf{v}^{(t_\tau)}, \mathbf{u}^{(t_\tau)}\right) | t_\tau \in \{1, 2, \cdots, T\}\right\}$$

converging to a limited point $(\mathbf{w}^*, \mathbf{v}^*, \mathbf{u}^*)$, i.e.,

$$\lim_{\tau \to \infty}\left[C_{sN}\left(\mathbf{w}^{(t_\tau)}, \mathbf{v}^{(t_\tau)}, \mathbf{u}^{(t_\tau)}\right) - C_{sN}\left(\mathbf{w}^*, \mathbf{v}^*, \mathbf{u}^*\right)\right] = 0. \tag{7.32}$$

Then for every $t$, there is $\tau$ that $t_\tau \le t \le t_{\tau+1}$,

$$\begin{aligned}
0 &= \lim_{\tau \to \infty}\left[C_{sN}\left(\mathbf{w}^{(t_\tau)}, \mathbf{v}^{(t_\tau)}, \mathbf{u}^{(t_\tau)}\right) - C_{sN}\left(\mathbf{w}^*, \mathbf{v}^*, \mathbf{u}^*\right)\right] \\
&\le \lim_{t \to \infty}\left[C_{sN}\left(\mathbf{w}^{(t)}, \mathbf{v}^{(t)}, \mathbf{u}^{(t)}\right) - C_{sN}\left(\mathbf{w}^*, \mathbf{v}^*, \mathbf{u}^*\right)\right] \\
&\le \lim_{\tau \to \infty}\left[C_{sN}\left(\mathbf{w}^{(t_\tau+1)}, \mathbf{v}^{(t_\tau+1)}, \mathbf{u}^{(t_\tau+1)}\right) - C_{sN}\left(\mathbf{w}^*, \mathbf{v}^*, \mathbf{u}^*\right)\right] = 0.
\end{aligned}$$

Therefore, we have

$$\lim_{t \to \infty} C_{sN}\left(\mathbf{w}^{(t)}, \mathbf{v}^{(t)}, \mathbf{u}^{(t)}\right) = C_{sN}\left(\mathbf{w}^*, \mathbf{v}^*, \mathbf{u}^*\right). \tag{7.33}$$

As a result, every improved point $(\mathbf{w}^*, \mathbf{v}^*, \mathbf{u}^*)$ is a Karush-Kuhn-Tucker point of sequence $\left\{\left(\mathbf{w}^{(t)}, \mathbf{v}^{(t)}, \mathbf{u}^{(t)}\right) | t = 1, 2, \cdots, T\right\}$.

### 7.5.4  Complexity Analysis

In this part we will analyze the computational complexity of CPSO and pro-posed ICPSO when applying the two designed secure beamforming schemes, i.e., NCoSTB and CoSTB. Considering that the ICPSO algorithm is designed based on CPSO, we first provide the complexity of CPSO. The complexity of CPSO consist of two parts: the complexity per iteration and the complexity introduced by iterations. For each iteration, the fitness value of each particle of all subswarms has to be evaluated and compared to the personal best position and the global best position of the swarm [51]. Therefore, the computational complexity per iteration is

$$\mathscr{C}^{\text{iter}}\,(\text{CPSO-NCoSTB}) = \mathcal{O}\,(I\,(NN_s + N_s))\,, \tag{7.34a}$$

$$\mathscr{C}^{\text{iter}}\,(\text{CPSO-CoSTB}) = \mathcal{O}\left(I\left(NN_s + MN_p + N_s\right)\right), \tag{7.34b}$$

where $I$ is the number of particles defined in Algorithm 1. Then the overall complexity of CPSO is thus given by

$$\mathscr{C}\,(\text{CPSO-NCoSTB}) = \mathcal{O}\,(TI\,(NN_s + N_s))\,, \tag{7.35a}$$

$$\mathscr{C}\,(\text{CPSO-CoSTB}) = \mathcal{O}\left(TI\left(NN_s + MN_p + N_s\right)\right), \tag{7.35b}$$

where $T$ corresponds to the maximum number of iterations required for conver-gence. Similarly, the computational complexity of ICPSO can be given by

$$\mathscr{C}\,(\text{ICPSO-NCoSTB}) = \mathcal{O}\,(N_{\text{iter}}T_0 I\,(NN_s + N_s))\,, \tag{7.36a}$$

$$\mathscr{C}\,(\text{ICPSO-CoSTB}) = \mathcal{O}\left(N_{\text{iter}}T_0 I\left(NN_s + MN_p + N_s\right)\right), \tag{7.36b}$$

where $N_{\text{iter}}$ is the maximun iterative number in Algorithm 2, and $T_0$ is the maximum number of iterations required when applying CPSO in Step 7.

## 7.6  Simulation Experiments and Analysis

This part provides numerical results to demonstrate and test the validity and effec-tiveness of designed secure beamforming schemes. In addition, the convergence and efficiency of the proposed iteration based solution for the optimization problem are also verified through the simulation.

First of all, we introduce the scenario setup for simulations. We consider a satellite terrestrial network consisted with one satellite, five FSS terminals and fifteen terrestrial BSs [16, 52]. Assume that the satellite carries fifteen antenna elements and each BS carries sixteen antenna elements [16] (Table 7.2).

**Table 7.2** Detailed system parameters

| Parameters | Value |
|---|---|
| Terrestrial spanning frequency | 17700 $\sim$ 18934 MHz [52, 53] |
| Satellite spanning frequency | 17700 $\sim$ 18895.2 MHz [52, 53] |
| Terrestrial transmit power | $-26 \sim -22$ dBW [52] |
| Satellite transmit power | 23.01 dBW [54] |
| Terrestrial bandwidth | 56 MHz [52] |
| Satellite bandwidth | 62.4 MHz [5, 52] |
| Terrestrial noise power $\sigma_p$ | $-121.52$ dBW [52] |
| FSS terminals noise power $\sigma_s$ | $-126.47$ dBW [5, 52] |
| Eavesdropper noise power $\sigma_e$ | $-121.52$ dBW |
| Number of scatters $L$ | 2 [36] |
| Number of scatters $L_m$ | 3 [13] |

First, we test the convergence of the CPSO algorithm and the proposed ICPSO algorithm when dealing with the optimization problems for the two designed secure beamforming schemes, i.e., NCoSTB and CoSTB. In addition, for the CPSO and ICPSO algorithms, the values of $(\mathbf{w}, \mathbf{v})$ and $(\mathbf{w}, \mathbf{v}, \mathbf{u})$ are initialized randomly and adjusted to satisfy the constraints if the random values are not feasible points of the optimization problems. Moreover, the maximum number of iteration when applying CPSO to solve (7.15) and (7.13) directly is set as 100. For ICPSO, let $N_{\text{iter}} = 5$ in Algorithm 2, and for each iterative optimization problem, the maximum number of iteration of CPSO is set as 20. Therefore, the 1st, 21st, 41st, 61st and 81st iterations are the beginnings of the new updated iterative optimization problems formulated in Sect. 7.5.2.1 and (7.28), by setting $t = 1, 2, \cdots, 5$. Thus, the total iteration number is 100, the same as that of the contrast experiment above applying CPSO directly. In addition, fix the SINR threshold as $\gamma_n = \gamma_{ms} = 0$ dB, $\forall n, m$ [17]. The achievable secrecy rate of $FSS_N$, the eavesdropped FSS terminal, updated in each iteration when applying the CPSO and ICPSO algorithms for the NCoSTB and CoSTB schemes are shown in Fig. 7.2a, b, respectively. For both NCoSTB and CoSTB, "CPSO 1" and "CPSO 2" in Fig. 7.2 indicate two different initial value settings of $(\mathbf{w}, \mathbf{v})$ and $(\mathbf{w}, \mathbf{v}, \mathbf{u})$. To present the influence of the eavesdropper, we test the achievable rates of $FSS_N$ when there is no eavesdropper in the system, and results are shown as the solid lines in Figs. 7.2a, b.

Results in Fig. 7.2 show that through ICPSO, the solutions of the optimization problem can converge to higher secrecy rates than through CPSO, no matter whether applying the NCoSTB or CoSTB scheme. In other words, for the same times of updating iteration, ICPSO tends to produce better beamforming and AN vectors and bring a higher secrecy rate than CPSO. For both of the secure beamforming schemes, the proposed ICPSO algorithm can achieve a faster convergence to reach the maximum secrecy rate, which results from its convex approximation operation of the original nonconvex objective function. In addition, Fig. 7.2 also indicates that when the optimization variables are initialized differently, the CPSO algorithm may converge to different optimal values, which might be the local optimal points.

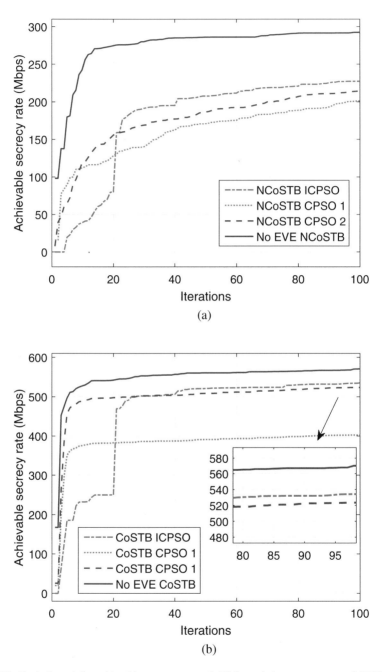

**Fig. 7.2** Evolution of the achievable secrecy rate of $FSS_N$ and the convergence of CPSO and ICPSO for the NCoSTB and CoSTB schemes

Moreover, results in Fig. 7.2 also reveal that with the assistance of cooperative beamforming from BSs, the achievable secrecy rate of the eavesdropped FSS terminal can be greatly improved, comparing with the beamforming scheme without the BSs' cooperation.

The results shown in Fig. 7.3 reveal the effect of the optimization variable's initialization on the convergence of the nonconvex optimization problems. In this experiment, the initial values $\left(\mathbf{w}^{(1)}, \mathbf{v}^{(1)}\right)$ and $\left(\mathbf{w}^{(1)}, \mathbf{v}^{(1)}, \mathbf{u}^{(1)}\right)$, of NCoSTB and CoSTB, respectively, are obtained by applying MRT and randomly (denoted by "Non-MRT" in Fig. 7.3). As shown in Fig. 7.3a, for the NCoSTB scheme, the ICPSO algorithm can achieve a faster convergence reaching to a larger secrecy rate than the traditional CPSO, when applying the same initialization strategy. Moreover, no matter whether to apply MRT or non-MRT based initialization, we can notice that although the secrecy rates obtained by ICPSO are relatively lower than by CPSO in the beginning of the iterations (from iteration 1 to 20), the rates increase more rapidly and reach higher values in the later iterations than that of CPSO. For the CoSTB scheme, results in Fig. 7.2b present a similar phenomenon. On the other hand, due to the nonconvex characteristic of original objective function and the drawback of CPSO, the convergence points sometimes are not the global optimal solutions, which depends much on the selection of the initial feasible point. Results in Fig. 7.3 indicate that the initialization obtained through the MRT can achieve better beamforming and AN vectors to get a higher secrecy rate. Even for the improved ICPSO algorithm, a random initialization may result to a weaker solution than the CPSO algorithm does with an MRT based initialization. Moreover, both results in Figs. 7.2 and 7.3 show that, some achievable secure rates at the beginning of iterative algorithms are zero. This phenomenon results from the initial value setting of iterative algorithms. Specifically, the initial value setting may lead to a negative value of $C_N (\mathbf{w}, \mathbf{v}) - C_{eN} (\mathbf{w}, \mathbf{v})$ or $C_N (\mathbf{w}, \mathbf{v}, \mathbf{u}) - C_{eN} (\mathbf{w}, \mathbf{v}, \mathbf{u})$. Then according to (7.11), the achievable secure rate is considered as zero when $C_N - C_{eN} < 0$. However, as iterations progress, achievable secure rates approach the positive optimal values.

Next, we show the achievable secrecy rate in Fig. 7.4 when the number of antennas carried on the satellite varies from 5 to 15 and the BSs' SINR threshold are set as $\gamma_{ms} = \gamma_p^1 = 0$ dB and $\gamma_{ms} = \gamma_p^2 = 6$ dB, $\forall m \in \mathcal{M}$ [55]. As we can see, as the number of antennas on the satellite increases, the secrecy rate of the eavesdropped FSS terminal increases, no matter whether the terrestrial BSs apply the cooperative beamforming and which optimization algorithm is applied. This result shows that thanks to the mmWave techniques, multiple antennas can greatly improve the transmission capacity and security of the communication network. Moreover, results in Fig. 7.4 also demonstrate that when the BSs require a higher SINR threshold, the secrecy rate of the eavesdropped FSS terminal will decrease. This dropping of performance results from the fact that the satellite has to lower its transmit power and adjust its beamforming and AN vectors to reduce its interference to BSs' users, which will sacrifice its own transmission rates and secrecy rates. However, with $N_s$ increasing, a higher achievable secrecy rate can be still achieved even when the system is constrained by a higher $\gamma_p$. In a real coexistence system of FSSs and

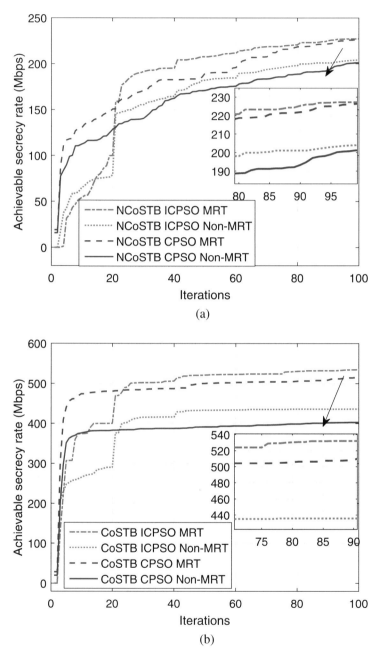

**Fig. 7.3** Achievable secrecy rate versus optimization variable's initialization. (**a**) NCoSTB. (**b**) CoSTB

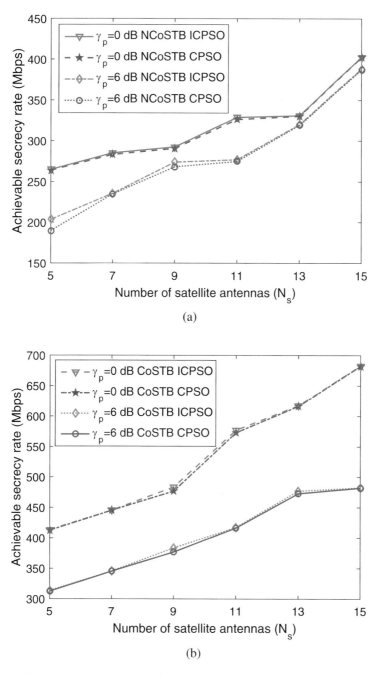

**Fig. 7.4** Achievable secrecy rate versus the number of antennas carried on the satellite $N_s$ and BSs' SINR threshold $\gamma_p$. (**a**) NCoSTB. (**b**) CoSTB

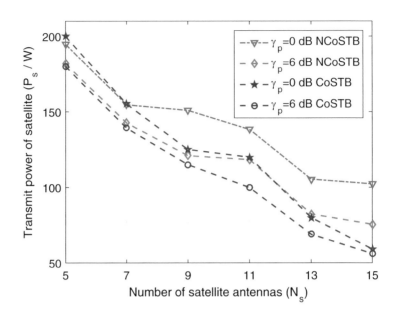

**Fig. 7.5** Transmit power of the satellite versus the number of antennas carried on the satellite $N_s$ and BSs' SINR threshold $\gamma_p$

cellular networks, the higher priority and legacy right of using some specific part of the spectrum for terrestrial BSs may reduce the capacity and security of the SatCom system. These results shown in Fig. 7.4 indicate that the communication quality of both FSS terminals and BSs' users can be guaranteed by the multiple antennas (such as MIMO/MISO), which can be realized when using the mmWave spectrum.

To further illustrate the effect of the number of antennas $N_s$ and BSs' SINR threshold $\gamma_{ms}$ on the system performance, we present the transmit power consumption of the satellite when maximizing the secrecy rate with power and SINR threshold constraints. As the results shown in Fig. 7.5, for the NCoSTB and CoSTB schemes, the transmit power of satellite decreases with $N_s$ increasing. In addition, when the BSs' SINR threshold is larger, i.e., $\gamma_{ms} = \gamma_p = 6$ dB, satellite will consume less power to guarantee the transmission quality of the BSs. Therefore, results shown in Figs. 7.4 and 7.5 reveal that the multiple antennas can contribute to improve the secure transmission capacity, meanwhile, to reduce the transmit power of the system.

Then we show the achievable secrecy rate in Fig. 7.6 when the number of antennas carried on each BS varies from 4 to 24 and the BSs' SINR threshold are set as $\gamma_{ms} = \gamma_p = 0$ dB $\forall m \in \mathcal{M}$. We can notice that the achievable secrecy rate does not change a lot when the NCoSTB scheme is applied. This phenomenon results from that the total transmission power of each BS remains unchanged, while these BSs do not perform their beamforming cooperatively. Therefore, BSs' interference for both FSSs and the eavesdropper presents a small change when $N_p$ increases.

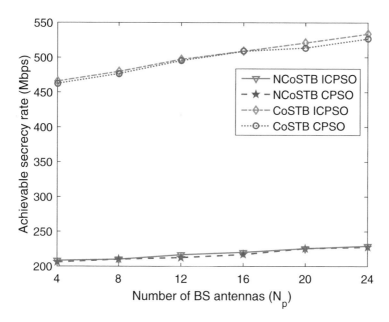

**Fig. 7.6** Achievable secrecy rate versus the number of antennas carried on each BS $N_p$

On the other hand, the achievable secrecy rate increases with increasing $N_p$ more distinctly when the CoSTB scheme is applied, comparing with that of NCoSTB, as shown in Fig. 7.6. These results indicate that when BSs perform as cooperators, their increasing number of antenna will bring more interference to the eavesdropper, meanwhile less interference to the satellite-terrestrial communication.

We also verify the complexity analysis by changing the number of antennas equipped on the satellite from 5 to 15. For both CoSTB and NCoSTB, when solving the beamforming optimization problems by CPSO, the maximum number of iterations is set as 100. On the other hand, when applying ICPSO, let $N_{iter} = 5$, and for each iterative optimization problem, the maximum number of iteration of CPSO is set as 20. According to this setting, the total number of iterations for both CPSO and ICPSO is 100. Then for the two beamforming strategies, Fig. 7.7 presents the time consumption on solving the beamforming optimization problems when introducing CPSO and ICPSO. Results in Fig. 7.7 indicate that the complexity grows near-linearly as $N_s$ increasing for the four cases. In addition, when the same beamforming strategy is applied, ICPSO costs less time than CPSO. Such efficiency of ICPSO results from the transformed convex quadratic problems it solves. Moreover, results in Fig. 7.7 also reveal the tradeoff between complexity and performance improvement when BSs participate in the cooperative beamforming. Specifically, according to results shown in Figs. 7.2, 7.3, and 7.4, CoSTB scheme brings higher achievable secrecy rate for the eavesdropped FSS terminal. However, the computational complexity for CoSTB is much higher than NCoSTB, which results from there are $NN_s + MN_p + N_s$ variables to be optimized for CoSTB,

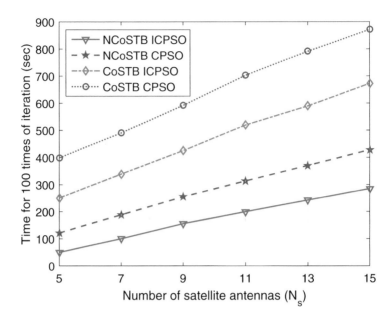

**Fig. 7.7** Time consumption on solving the beamforming optimization problems versus the number of antennas carried on the satellite $N_s$

while only $NN_s + N_s$ for NCoSTB. Such complexity properties indicated in (7.35) and (7.36) are also reflected in Fig. 7.7.

## 7.7   Conclusion

In this part, we have considered a mmWave and MISO channel based coexistence system of FSS and terrestrial cellular networks. The physical layer security problem is analyzed for the established scenario. To achieve the secure transmission, the adaptive beamforming and AN techniques are introduced to prevent the eavesdropper from receiving and decoding information successfully. We have proposed a non-cooperative beamforming scheme, through which the BSs process precoding through an MRT beamforming. On the other hand, to further improve the secrecy rate, the CoSTB scheme has been designed, according to which BSs implement the cooperative beamforming to decrease the SINR at the eavesdropper and increase the SINR at the eavesdropped FSS terminal, meanwhile ensure the SINR at BSs' users and other legitimate FSS terminals. An iteration based approximate genetic algorithm has been designed to solve the nonconvex secrecy rate maximization problem.

The simulation results show that multiple antenna arrays and the designed secure transmission beamforming schemes can improve the secrecy rate of the eavesdropped terminal, as well as guarantee the transmission quality of the incumbent communication in the coexistence system. In addition, the convergence and efficiency of the proposed iteration based approximation algorithm are verified by the simulation results.

## 7.8   Proof of Theorem 7.1

For the two security beamforming schemes, i.e., NCoSTB and CoSTB, the analysis and derivation of the lower bound of $C_N(\mathbf{w}, \mathbf{v})$ and $C_N(\mathbf{w}, \mathbf{v}, \mathbf{u})$ are similar, except that the beamforming strategies of BSs are fixed according to MRT under the non-cooperation scheme. In other words, for the NCoSTB, $\mathbf{u}$ is set as a constant vector by (7.14). Therefore, in this part, we only provide the derivation of the lower bound of $C_N(\mathbf{w}, \mathbf{v}, \mathbf{u})$ for simplification.

As defined in (7.12), we have

$$
\begin{aligned}
C_N(\mathbf{w}, \mathbf{v}, \mathbf{u}) &= \log_2\left(1 + \frac{\mathbf{w}_N^H \mathbf{R}_N \mathbf{w}_N}{\psi_N(\mathbf{w}, \mathbf{v}, \mathbf{u})}\right) \\
&= -\log_2\left(1 - \frac{\mathbf{w}_N^H \mathbf{R}_N \mathbf{w}_N}{\psi_N(\mathbf{w}, \mathbf{v}, \mathbf{u}) + \mathbf{w}_N^H \mathbf{R}_N \mathbf{w}_N}\right) \\
&\triangleq -\log_2\left(1 - \frac{g_1(\mathbf{w}_N)}{g_2(\mathbf{w}, \mathbf{v}, \mathbf{u})}\right),
\end{aligned}
\tag{7.37}
$$

where

$$
g_1(\mathbf{w}_N) = \mathbf{w}_N^H \mathbf{R}_N \mathbf{w}_N, \tag{7.38a}
$$

$$
g_2(\mathbf{w}, \mathbf{v}, \mathbf{u}) = \psi_N(\mathbf{w}, \mathbf{v}, \mathbf{u}) + \mathbf{w}_N^H \mathbf{R}_N \mathbf{w}_N > g_1(\mathbf{w}_N). \tag{7.38b}
$$

Consider that $f(x) = -\log_2(1 - x)$ is an increasing convex function of independent variable $x$ in the domain $\{x \mid x < 1\}$. Thus $f(g_1/g_2) = -\log_2(1 - g_1/g_2) \triangleq C_N(g_1, g_2)$ is convex in the domain $\{(g_1, g_2) \mid 0 < g_1 < g_2\}$ $(g_1/g_2 < 1)$, where $g_1 = g_1(\mathbf{w}_N)$ and $g_2 = g_2(\mathbf{w}, \mathbf{v}, \mathbf{u})$ are defined as (7.38). Considering the Taylor expansion and the convexity of $C_N(g_1, g_2)$ when $0 < g_1 < g_2$, we have

$$
\begin{aligned}
C_N(g_1, g_2) \geq {}& C_N\left(g_1^{(t)}, g_2^{(t)}\right) \\
&+ \left\langle \nabla C_N\left(g_1^{(t)}, g_2^{(t)}\right), (g_1, g_2) - \left(g_1^{(t)}, g_2^{(t)}\right)\right\rangle.
\end{aligned}
\tag{7.39}
$$

Denote $\mathbf{x}^{(t)} = \{\mathbf{w}^{(t)}, \mathbf{v}^{(t)}, \mathbf{u}^{(t)}\}$ and $\mathbf{x} = \{\mathbf{w}, \mathbf{v}, \mathbf{u}\}$ as simplified representations. Then in (7.39),

$$\left\langle \nabla C_N \left( g_1^{(t)}, g_2^{(t)} \right), (g_1, g_2) - \left( g_1^{(t)}, g_2^{(t)} \right) \right\rangle$$

$$= \frac{1}{\ln 2} \frac{g_2\left(\mathbf{x}^{(t)}\right)}{g_2\left(\mathbf{x}^{(t)}\right) - g_1\left(\mathbf{x}^{(t)}\right)} \left[ \frac{2\Re\left\{ \left(\mathbf{w}_N^{(t)}\right)^H \mathbf{R}_N \left(\mathbf{w}_N - \mathbf{w}_N^{(t)}\right) \right\}}{g_2\left(\mathbf{x}^{(t)}\right)} \right]$$

$$- \frac{1}{\ln 2} \frac{g_2\left(\mathbf{x}^{(t)}\right)}{g_2\left(\mathbf{x}^{(t)}\right) - g_1\left(\mathbf{x}^{(t)}\right)} \left( \frac{g_1\left(\mathbf{x}^{(t)}\right)}{g_2^2\left(\mathbf{x}^{(t)}\right)} \right) \left[ g_2\left(\mathbf{x}^{(t)}\right) - g_2\left(\mathbf{x}\right) \right]$$

$$= \frac{2}{\ln 2} \frac{\Re\left\{ \left(\mathbf{w}_N^{(t)}\right)^H \mathbf{R}_N \left(\mathbf{w}_N - \mathbf{w}_N^{(t)}\right) \right\}}{\psi_N\left(\mathbf{x}^{(t)}\right)}$$

$$- \frac{1}{\ln 2} \left[ \frac{1}{\psi_N\left(\mathbf{x}^{(t)}\right)} - \frac{1}{\psi_N\left(\mathbf{x}^{(t)}\right) + \left(\mathbf{w}_N^{(t)}\right)^H \mathbf{R}_N \mathbf{w}_N^{(t)}} \right] \left[ \psi_N\left(\mathbf{x}\right) \right.$$

$$\left. + \mathbf{w}_N^H \mathbf{R}_N \mathbf{w}_N - \psi_N\left(\mathbf{x}^{(t)}\right) - \left(\mathbf{w}_N^{(t)}\right)^H \mathbf{R}_N \mathbf{w}_N^{(t)} \right]$$

$$= \frac{1}{\ln 2} \frac{\Re\left\{ \left(\mathbf{w}_N^{(t)}\right)^H \mathbf{R}_N \left(\mathbf{w}_N - \mathbf{w}_N^{(t)}\right) \right\}}{\psi_N\left(\mathbf{x}^{(t)}\right)}$$

$$- \frac{1}{\ln 2} \frac{\left(\mathbf{w}_N^{(t)}\right)^H \mathbf{R}_N \mathbf{w}_N^{(t)} \left(\psi_N\left(\mathbf{x}\right) + \mathbf{w}_N^H \mathbf{R}_N \mathbf{w}_N\right)}{\psi_N\left(\mathbf{x}^{(t)}\right) \left[ \psi_N\left(\mathbf{x}^{(t)}\right) + \left(\mathbf{w}_N^{(t)}\right)^H \mathbf{R}_N \mathbf{w}_N^{(t)} \right]} + \frac{1}{\ln 2} \frac{\left(\mathbf{w}_N^{(t)}\right)^H \mathbf{R}_N \mathbf{w}_N^{(t)}}{\psi_N\left(\mathbf{x}^{(t)}\right)}$$

$$= \frac{2}{\ln 2} \frac{\Re\left\{ \left(\mathbf{w}_N^{(t)}\right)^H \mathbf{R}_N \mathbf{w}_N \right\}}{\psi_N\left(\mathbf{x}^{(t)}\right)} - \frac{1}{\ln 2} \frac{\left(\mathbf{w}_N^{(t)}\right)^H \mathbf{R}_N \mathbf{w}_N^{(t)}}{\psi_N\left(\mathbf{x}^{(t)}\right)}$$

$$- \frac{1}{\ln 2} \frac{\left(\mathbf{w}_N^{(t)}\right)^H \mathbf{R}_N \mathbf{w}_N^{(t)} \left(\psi_N\left(\mathbf{x}\right) + \mathbf{w}_N^H \mathbf{R}_N \mathbf{w}_N\right)}{\psi_N\left(\mathbf{x}^{(t)}\right) \left[ \psi_N\left(\mathbf{x}^{(t)}\right) + \left(\mathbf{w}_N^{(t)}\right)^H \mathbf{R}_N \mathbf{w}_N^{(t)} \right]}.$$

Substituting the result obtained above into (7.39), then (7.21) can be achieved. This completes the proof of Theorem 7.1.

## 7.9   Proof of Theorem 7.2

In this part, we will only derive the upper bound of $C_{eN}(\mathbf{w}, \mathbf{v}, \mathbf{u})$ for the CoSTB scheme. When applying the NCoSTB scheme, the derivation is similar to that of NCoSTB, by considering $\mathbf{u}$ as a constant vector.

According to the definition in (7.12), we have

$$C_{eN}(\mathbf{w}, \mathbf{v}, \mathbf{u}) = \ln\left(1 + \frac{\mathbf{w}_N^H \mathbf{R}_e \mathbf{w}_N}{\psi_e(\mathbf{w}, \mathbf{v}, \mathbf{u})}\right) \tag{7.40}$$

$$= \log_2(1 + \Gamma_e(\mathbf{w}, \mathbf{v}, \mathbf{u})) \triangleq C_{eN}(\Gamma_e(\mathbf{w}, \mathbf{v}, \mathbf{u})),$$

which is an increasing concave function of $\Gamma_e(\mathbf{w}, \mathbf{v}, \mathbf{u})$. Denote $\mathbf{x}^{(t)} = \{\mathbf{w}^{(t)}, \mathbf{v}^{(t)}, \mathbf{u}^{(t)}\}$ and $\mathbf{x} = \{\mathbf{w}, \mathbf{v}, \mathbf{u}\}$. Thus we have

$$\log_2(1 + \Gamma_e(\mathbf{x})) \leq \log_2\left(1 + \Gamma_e\left(\mathbf{x}^{(t)}\right)\right)$$
$$+ \left\langle \nabla C_{eN}\left(\Gamma_e\left(\mathbf{x}^{(t)}\right)\right), \Gamma_e^{(t)}(\mathbf{x}) - \Gamma_e\left(\mathbf{x}^{(t)}\right) \right\rangle, \tag{7.41}$$

where

$$\left\langle \nabla C_{eN}\left(\Gamma_e\left(\mathbf{x}^{(t)}\right)\right), \Gamma_e(\mathbf{x}) - \Gamma_e\left(\mathbf{x}^{(t)}\right)\right\rangle$$

$$= \frac{1}{\ln 2} \frac{\psi_e\left(\mathbf{x}^{(t)}\right)}{\psi_e\left(\mathbf{x}^{(t)}\right) + \left(\mathbf{w}_N^{(t)}\right)^H \mathbf{R}_e \mathbf{w}_N^{(t)}} \left[\frac{\mathbf{w}_N^H \mathbf{R}_e \mathbf{w}_N}{\psi_e^{(t)}(\mathbf{x})} - \frac{\left(\mathbf{w}_N^H\right)^{(t)} \mathbf{R}_e \mathbf{w}_N^{(t)}}{\psi_e\left(\mathbf{x}^{(t)}\right)}\right]$$

$$= \frac{1}{\ln 2} \frac{\psi_e\left(\mathbf{x}^{(t)}\right)}{\psi_e\left(\mathbf{x}^{(t)}\right) + \left(\mathbf{w}_N^{(t)}\right)^H \mathbf{R}_e \mathbf{w}_N^{(t)}} \left[\frac{\mathbf{w}_N^H \mathbf{R}_e \mathbf{w}_N}{\psi_e^{(t)}(\mathbf{x})} + 1 \frac{\left(\mathbf{w}_N^H\right)^{(t)} \mathbf{R}_e \mathbf{w}_N^{(t)}}{\psi_e\left(\mathbf{x}^{(t)}\right)} - 1\right]$$

$$= \frac{1}{\ln 2} \frac{\psi_e\left(\mathbf{x}^{(t)}\right)}{\psi_e\left(\mathbf{x}^{(t)}\right) + \left(\mathbf{w}_N^{(t)}\right)^H \mathbf{R}_e \mathbf{w}_N^{(t)}} \left[\frac{\mathbf{w}_N^H \mathbf{R}_e \mathbf{w}_N}{\psi_e^{(t)}(\mathbf{x})} + 1\right] - \frac{1}{\ln 2},$$

where $\psi_e^{(t)}(\mathbf{w}, \mathbf{v}, \mathbf{u})$ is defined by (7.26). Substituting the result obtained above into (7.41), then (7.25) can be achieved. This completes the proof of Theorem 7.2.

# References

1. J. Du, C. Jiang, Z. Han, H. Zhang, S. Mumtaz, and Y. Ren, "Contract mechanism and performance analysis for data transaction in mobile social networks," *IEEE Trans. Network Sci. Eng.*, vol. 6, no. 2, pp. 103–115, Apr. –Jun. 2019.
2. A. Osseiran, F. Boccardi, V. Braun, K. Kusume, P. Marsch, M. Maternia, O. Queseth, M. Schellmann, H. Schotten, H. Taoka, H. Tullberg, M. A. Uusitalo, B. Timus, and M. Fallgren, "Scenarios for 5G mobile and wireless communications: the vision of the METIS project," *IEEE Commun. Mag.*, vol. 52, no. 5, pp. 26–35, May 2014.
3. F. Guidolin, M. Nekovee, L. Badia, and M. Zorzi, "A study on the coexistence of fixed satellite service and cellular networks in a mmWave scenario," in *IEEE Int. Conf. on Commun. (ICC 2015)*. London, UK, 8-12 Jun. 2015, pp. 2444–2449.
4. J. Du, C. Jiang, Q. Guo, M. Guizani, and Y. Ren, "Cooperative earth observation through complex space information networks," *IEEE Wireless Commun.*, vol. 23, no. 2, pp. 136–144, May 2016.
5. E. Lagunas, S. K. Sharma, S. Maleki, S. Chatzinotas, and B. Ottersten, "Resource allocation for cognitive satellite communications with incumbent terrestrial networks," *IEEE Trans. Cognitive Commun. and Networking*, vol. 1, no. 3, pp. 305–317, Sept. 2015.
6. F. Guidolin, M. Nekovee, L. Badia, and M. Zorzi, "A cooperative scheduling algorithm for the coexistence of fixed satellite services and 5G cellular network," in *IEEE Int. Conf. on Commun. (ICC 2015)*. London, UK, 8-12 Jun. 2015, pp. 1322–1327.
7. M. Corici, A. Kapovits, S. Covaci, A. Geurtz, I.-D. Gheorghe-Pop, B. Riemer, and A. Weber, "Assessing satellite-terrestrial integration opportunities in the 5G environment," *[On-line] Available:* https://artes.esa.int/sites/default/files/Whitepaper, Sept. 2016.
8. J. Du, C. Jiang, Y. Qian, Z. Han, and Y. Ren, "Resource allocation with video traffic prediction in cloud-based space systems," *IEEE Trans. Multimedia*, vol. 18, no. 5, pp. 820–830, Mar. 2016.
9. ERC/DEC/(00)07, "The shared use of the band 17.7-19.7 GHz by the fixed service and earth stations of the fixed-satellite service (space-to-earth)," in *ECC Report 241*. Electronic Commun. Committee, Copenhagen, Denmark, approved: 19 Oct. 2000, amended: 4 Mar. 2016.
10. C. Niephaus, M. Kretschmer, and G. Ghinea, "QoS provisioning in converged satellite and terrestrial networks: A survey of the state-of-the-art," *IEEE Commun. Surveys & Tutorials*, vol. 18, no. 4, pp. 2415–2441, Apr. 2016.
11. C. Jiang, Y. Chen, Y. Gao, and K. J. R. Liu, "Joint spectrum sensing and access evolutionary game in cognitive radio networks," *IEEE Trans. Wireless Commun.*, vol. 12, no. 5, pp. 2470–2483, May 2013.
12. C. Jiang, Y. Chen, K. J. R. Liu, and Y. Ren, "Renewal-theoretical dynamic spectrum access in cognitive radio network with unknown primary behavior," *IEEE J. Sel. Areas Commun.*, vol. 31, no. 3, pp. 406–416, Mar. 2013.
13. S. Shi, G. Li, K. An, Z. Li, and G. Zheng, "Optimal power control for real-time applications in cognitive satellite terrestrial networks," *IEEE Commun. Lett.*, vol. 21, no. 8, pp. 1815–1818, Aug. 2017.
14. K. An, M. Lin, T. Liang, J.-B. Wang, J. Wang, Y. Huang, and A. L. Swindlehurst, "Performance analysis of multi-antenna hybrid satellite-terrestrial relay networks in the presence of interference," pp. 4390–4404, Nov. 2015.
15. C. Jiang, X. Zhu, L. Kuang, Y. Qian, and J. Lu, "Multimedia multicast beamforming in integrated terrestrial-satellite networks," in *13th Int. Wireless Commun. and Mobile Computing Conf. (IWCMC 2017)*. Valencia, Spain, 26-30 Jun. 2017, pp. 340–345.
16. J. Lei, Z. Han, M. Á. Vazquez-Castro, and A. Hjorungnes, "Secure satellite communication systems design with individual secrecy rate constraints," *IEEE Trans. Inf. Forens. Security*, vol. 6, no. 3, pp. 661–671, Sept. 2011.
17. K. An, M. Lin, J. Ouyang, and W.-P. Zhu, "Secure transmission in cognitive satellite terrestrial networks," *IEEE J. Sel. Areas Commun.*, vol. 34, no. 11, pp. 3025–3037, Nov. 2016.

18. L. Dong, Z. Han, A. P. Petropulu, and H. V. Poor, "Improving wireless physical layer security via cooperating relays," *IEEE Trans. Signal Process.*, vol. 58, no. 3, pp. 1875–1888, Mar. 2010.

19. Y. Huang, F. S. Al-Qahtani, T. Q. Duong, and J. Wang, "Secure transmission in MIMO wiretap channels using general-order transmit antenna selection with outdated CSI," *IEEE Trans. Commun.*, vol. 63, no. 8, pp. 2959–2971, Aug. 2015.

20. A. D. Wyner, "The wire-tap channel," *Bell Lab. Tech. J.*, vol. 54, no. 8, pp. 1355–1387, Oct. 1975.

21. Y. Liang, H. V. Poor, and S. Shamai, "Secure communication over fading channels," *IEEE Trans. Inf. Theory*, vol. 54, no. 6, pp. 2470–2492, May 2008.

22. H. Xing, K.-K. Wong, A. Nallanathan, and R. Zhang, "Wireless powered cooperative jamming for secrecy multi-AF relaying networks," *IEEE Trans. Wireless Commun.*, vol. 15, no. 12, pp. 7971–7984, Dec. 2016.

23. Y. Huang, J. Wang, C. Zhong, T. Q. Duong, and G. K. Karagiannidis, "Secure transmission in cooperative relaying networks with multiple antennas," *IEEE Trans. Wireless Commun.*, vol. 15, no. 10, pp. 6843–6856, Oct. 2016.

24. Z. Han, N. Marina, M. Debbah, and A. Hjørungnes, "Physical layer security game: interaction between source, eavesdropper, and friendly jammer," *EURASIP J. on Wireless Commun. and Networking*, vol. 2009, no. 452907, pp. 1–10, Jan. 2010.

25. S. R. Aghdam and T. M. Duman, "Joint precoder and artificial noise design for MIMO wiretap channels with finite-alphabet inputs based on the cut-off rate," *IEEE Trans. Wireless Commun.*, vol. 16, no. 6, pp. 3913–3923, Jun. 2017.

26. B. Li, Z. Fei, X. Xu, and Z. Chu, "Resource allocations for secure cognitive satellite-terrestrial networks," *IEEE Wireless Commun. Lett.*, vol. PP, no. 99, pp. 1–1, Sept. 2017.

27. S.-H. L. Tsai and H. V. Poor, "Power allocation for artificial-noise secure MIMO precoding systems," *IEEE Trans. Signal Process.*, vol. 62, no. 13, pp. 3479–493, Jul. 2014.

28. A. H. Phan, H. D. Tuan, H. H. Kha, and D. T. Ngo, "Nonsmooth optimization for efficient beamforming in cognitive radio multicast transmission," *IEEE Trans. Signal Process.*, vol. 60, no. 6, pp. 2941–2951, Jun. 2012.

29. M. Zhang, K. Cumanan, and A. Burr, "Secrecy rate maximization for MISO multicasting SWIPT system with power splitting scheme," in *2016 IEEE 17th Int. Workshop Signal Process. Advances Wireless Commun. (SPAWC)*.    Edinburgh, UK, 3–6 Jul. 2016.

30. M. Zhang, Y. Liu, and R. Zhang, "Artificial noise aided secrecy information and power transfer in OFDMA systems," *IEEE Trans. Wireless Commun.*, vol. 15, no. 4, pp. 3085–3096, Apr. 2016.

31. T. Lv, H. Gao, and S. Yang, "Secrecy transmit beamforming for heterogeneous networks," *IEEE J. Sel. Areas Commun.*, vol. 33, no. 6, pp. 1154–1170, Jun. 2015.

32. B. Li, Z. Fei, Z. Chu, and Y. Zhang, "Secure transmission for heterogeneous cellular networks with wireless information and power transfer," *IEEE Syst. J.*, vol. PP, no. 99, pp. 1–12, Jun. 2017.

33. J. Arnau, D. Christopoulos, S. Chatzinotas, C. Mosquera, and B. Ottersten, "Performance of the multibeam satellite return link with correlated rain attenuation," *IEEE Trans. Wireless Commun.*, vol. 13, no. 11, pp. 6286–6299, Nov. 2014.

34. A. Sayeed and J. Brady, "Beamspace mimo for high-dimensional multiuser communication at millimeter-wave frequencies," in *2013 IEEE Global Commun. Conf. (GLOBECOM)*. Atlanta, GA, USA, Dec. 2013, pp. 3679–3684.

35. Y. R. Ramadan, H. Minn, and A. S. Ibrahim, "Hybrid analog-digital precoding design for secrecy mmWave MISO-OFDM systems," *IEEE Trans. Commun.*, vol. 65, no. 11, pp. 5009–5026, Nov. 2017.

36. Z. Zhou, W. Feng, Y. Chen, and N. Ge, "Adaptive scheduling for millimeter wave multi-beam satellite communication systems," *J. Communi. Inf. Networks*, vol. 1, no. 3, pp. 42–55, Oct. 2016.

37. A. A. Nasir, H. D. Tuan, T. Q. Duong, and H. V. Poor, "Secrecy rate beamforming for multicell networks with information and energy harvesting," *IEEE Trans. Signal Process.*, vol. 65, no. 3, pp. 677–689, Feb. 2017.

38. F. Zhou, Z. Li, J. Cheng, Q. Li, and J. Si, "Robust AN-aided beamforming and power splitting design for secure MISO cognitive radio with SWIPT," *IEEE Trans. Wireless Commun.*, vol. 16, no. 4, pp. 2450–2464, Apr. 2017.

39. L. Wang, K. K. Wong, R. W. Heath, and J. Yuan, "Wireless powered dense cellular networks: How many small cells do we need?" *IEEE J. Sel. Areas Commun.*, vol. 35, no. 9, pp. 2010–2024, Sept. 2017.

40. J. Li, D. Wang, P. Zhu, J. Wang, and X. You, "Downlink spectral efficiency of distributed massive MIMO systems with linear beamforming under pilot contamination," *IEEE Trans. Veh. Technol.*, vol. PP, no. 99, pp. 1–1, Jul. 2017.

41. S. Abeywickrama, T. Samarasinghe, C. K. Ho, and C. Yuen, "Wireless energy beamforming using received signal strength indicator feedback," *IEEE Trans. Signal Process.*, vol. 66, no. 1, pp. 224–235, Jan. 2018.

42. H. Zhang, H. Xing, J. Cheng, A. Nallanathan, and V. C. Leung, "Secure resource allocation for OFDMA two-way relay wireless sensor networks without and with cooperative jamming," *IEEE Trans. Ind. Informat.*, vol. 12, no. 5, pp. 1714–1725, Oct. 2016.

43. T.-X. Zheng, H.-M. Wang, J. Yuan, Z. Han, and M. H. Lee, "Physical layer security in wireless Ad Hoc networks under a hybrid full-/half-duplex receiver deployment strategy," *IEEE Trans. Wireless Commun.*, vol. 16, no. 6, pp. 3827–3839, Jun. 2017.

44. Z. Fei, B. Li, S. Yang, C. Xing, H. Chen, and L. Hanzo, "A survey of multi-objective optimization in wireless sensor networks: Metrics, algorithms, and open problems," *IEEE Commun. Surveys Tutorials*, vol. 19, no. 1, pp. 550–586, Firstquarter 2017.

45. F. Wang, C. Xu, Y. Huang, X. Wang, and X. Gao, "REEL-BF design: Achieving the SDP bound for downlink beamforming with arbitrary shaping constraints," *IEEE Trans. Signal Process.*, vol. 65, no. 10, pp. 2672–2685, Feb. 2017.

46. F. Zhu and M. Yao, "Improving physical-layer security for CRNs using SINR-based cooperative beamforming," *IEEE Trans. Veh. Technol.*, vol. 65, no. 3, pp. 1835–1841, Mar. 2016.

47. F. Van den Bergh and A. P. Engelbrecht, "A cooperative approach to particle swarm optimization," *IEEE Trans. Evol. Comput.*, vol. 8, no. 3, pp. 225–239, Jun. 2004.

48. Y. Dong, L. Qiu, and X. Liang, "Energy efficiency maximization for uplink SCMA system using CCPSO," in *IEEE Global Commun. Conf. Workshops (GLOBECOM Wkshps 2016)*. Washington, DC, USA, 4-8 Dec. 2016, pp. 1–5.

49. M. R. Javan, N. Mokari, F. Alavi, and A. Rahmati, "Resource allocation in decode-and-forward cooperative communication networks with limited rate feedback channel," *IEEE Trans. Veh. Technol.*, vol. 66, no. 1, pp. 256–267, Jan. 2017.

50. A. Modiri, X. Gu, A. M. Hagan, and A. Sawant, "Radiotherapy planning using an improved search strategy in particle swarm optimization," *IEEE Trans. Biomed. Eng.*, vol. 64, no. 5, pp. 980–989, May 2017.

51. C. Knievel, M. Noemm, and P. A. Hoeher, "Low-complexity receiver for large-mimo space-time coded systems," in *2011 IEEE Veh. Technology Conf. (VTC Fall)*. San Francisco, CA, USA, 5-8 Sept. 2011.

52. E. Lagunas, S. Maleki, L. Lei, C. Tsinos, S. Chatzinotas, and B. Ottersten, "Carrier allocation for hybrid satellite-terrestrial backhaul networks," in *IEEE Int. Conf. on Commun. Workshop (ICC Wkshps 2017) on Satellite Commun.: Challenges and Integration in the 5G ecosystem*, 21–25 May 2017, pp. 1–6.

53. S. K. Sharma, S. Chatzinotas, J. Grotz, and B. Ottersten, "3D beamforming for spectral coexistence of satellite and terrestrial networks," in *82nd IEEE Veh. Technology Conf. (VTC Fall 2015)*. Boston, MA, USA, 6-9 Sept. 2015, pp. 1–5.

54. Y. Hong, A. Srinivasan, B. Cheng, L. Hartman, and P. Andreadis, "Optimal power allocation for multiple beam satellite systems," in *2008 IEEE Radio and Wireless Symp.* Orlando, FL, USA, Jan. 2008, pp. 823–826.

55. J. Lei, Z. Han, M. Vázquez-Castro, and A. Hjørungnes, "Multibeam SATCOM systems design with physical layer security," in *IEEE Int. Conf. on Ultra-Wideband (ICUWB 2011)*. Bologna, Italy, 14-16 Sept. 2011, pp. 555–559.

# Chapter 8
# Traffic Prediction Based Transmission in Satellite-Terrestrial Networks

**Abstract** This part considers the resource allocation problems for video transmission in space based information networks. The queueing system analyzed in this work is constituted by multiple users and a single server. The server is operated as a cloud that can sense the traffic arrivals to each user's queue, and then allocates the transmission resource and service rate for users. The objectives are to make configurations over time to minimize the time average cost of the system, and to minimize the waiting time of packets after they enter the queue. Meanwhile, the constraints on the queue stability of the system must be satisfied. In this part, we introduce a predictive backpressure algorithm, which considers the future arrivals with a certain prediction window size, into the consideration of resource allocation to make decision on which packets to be served first. In addition, this part designs a multi-resolution wavelet decomposition based backpropagation network for the prediction of video traffic, which exhibits the long-range dependence property. Simulation results indicate that the delay of the queueing system can be reduced through this prediction based resource allocation, and the prediction accuracy for the video traffic is improved according to the proposed prediction system.

**Keywords** Space-based Information Networks · Resource Allocation · Video Traffic Prediction · Cloud Service · Queueing Theory · Predictive Backpressure

## 8.1 Introduction

In recent years, the space-based information network (SBIN) is proposed to improve the detection and transmission capabilities of a single satellite or satellite system. Through the cooperation scheme of the SBIN, the real-time data acquisition and transfer can be achieved. Therefore, how to design appropriate cooperation mechanisms and achieve efficient network resource allocation to maximize the utility of the whole system become a key issue for SBIN operation. On the other hand, the demand of multimedia services for satellite communications has increased. Take typhoon tracking systems for instance, there are multiple satellites deployed in the low earth orbit (LEO) and geosynchronous orbit (GEO) getting video and image data of typhoons to monitor their trends. The obtained multimedia data from different satellites needs to be send to the earth as soon as possible. However, the communication resource, such as power and service rate, to receive data from different satellites is limited. This constraint results that satellites accessing to the ground station cannot send all of their obtained date simultaneously. Then resource allocation policies are needed to maximize the utilization of network resource and minimize the average transmission delay of every access satellite. This part will focus on the resource allocation for the SBIN with multiple satellite users and a single server deployed on the ground. We assume that the server performs as a cloud processing center, which can sense traffic arrivals to every access satellite and the channel state information, and serve packets sent from satellites according to the allocated power and service rate. The assumption that the cloud service can sense traffic arrivals is feasible. This sensing ability can be realized by many current traffic sensing technologies such as dynamic traffic monitor [1], packet sampling based on Kullback-Leibler Divergence (KLD) measure [2] and traffic estimation based on sensing order confidence [3], etc.

Resource allocation and optimal control of multi-access queueing systems and communication networks have been active research topics over past decades. However, most of current resource allocation mechanisms were operated depending on current traffic arrivals, while future arrivals based on the prediction were not considered. In other words, packets can only be served by servers after they have arrived into the queueing system according to works above. While in many current systems for multimedia services, the prediction for the future traffic is feasible. Moreover, learning and predicting the user behavior and then pre-serving the future traffic can improve the system performance significantly [4]. Considering that the transmission resource of the SBIN is much more limited and expensive, the resource allocation policy needs to maximize both the utilization of network resource and the network performance. Therefore, we design a resource allocation policy for video tasks in the SBIN based on the traffic prediction, which is obtained by learning and training the traffic characteristics. The main contributions of this part are summarized as follows.

- Proposing a cloud-based allocation resource system for the video business in the SBIN. In this system, the cloud server is designed to have the ability to sense the current traffic arrivals from different user satellites, predict the future video traffic and perform the prediction based resource allocation policy.
- Establishing a multi-level wavelet based backpropagation neural network for video traffic prediction according to properties of the video traffic. Specifically, the traffic sequence is first decomposed into levels with different resolutions, which will be trained by backpropagation networks. All of these networks constitute the prediction system for video traffic in the SBIN.
- Designing a resource allocation policy based on the future traffic prediction for video tasks in the SBIN. In this policy, the backpressure algorithm is introduced based on the prediction traffic information. Moreover, the power consumption and channel state of the SBIN is also considered in the resource allocation.

The remainder of this part is organized as follows. In Sect. 8.2, we review the related works associated with resource allocation and optimal control of multi-access queueing systems. In Sect. 7.3 the system model is described. The multi-resolution wavelet decomposition based backpropagation network for video traffic prediction is proposed in Sect. 8.4. Then we introduce the predictive backpressure scheme into the resource allocation for the SBIN in Sect. 8.5. Simulations are shown in Sects. 8.6, and 8.7 concludes this part.

## 8.2 Related Works

Resource allocation and optimal control of multi-access queueing systems and communication networks have been active research topics over past decades. A dynamic resource allocation scheme based on the prediction of packet loss probability and end-to-end distortion was proposed in [5] for video streaming over multi-hop networks. A quality-fair and Pareto optimal resource allocation for the multimedia system was proposed in [6], which jointly considered the available system resource and the video decoding task feature. In [7], a subcarrier and power allocation scheme was proposed in the context of orthogonal frequency division multiple access (OFDMA)-based cognitive radio (CR) video application systems. Over OFDMA wireless networks, the authors of [8] also designed a cross-layer resource allocation scheme to maximize the sum of the achievable rates and minimize the distortion difference among multiple videos. In [9], maximizing the video quality was optimized jointly with the time-domain resource partitioning, and a rate allocation algorithm was proposed for heterogonous cellular networks. In addition, game theory has also been widely utilized for modeling resource [10–13]. However, these works above did not consider the future traffic arrival, which can optimize the resource allocation and improve the system performance.

The future traffic information is needed for the current resource allocation. The capability to predict video traffic can significantly improve the effectiveness of the following dynamic resource allocation. There have been several studies on the prediction of video traffic. An adaptive traffic prediction method based on the identification of scene changes was proposed for variable-bit-rate (VBR) MPEG videos in [14]. To reduce the power consumption of wireless LAN infrastructure, a discrete autoregressive video prediction model was designed in [15]. In [16], a short-term bandwidth prediction of a video bit stream was performed for the dynamic resource allocation. The studies above predicted the video traffic based on the traffic flow characteristics, which plays a key role in improving the prediction precision. It has been demonstrated in numerous studies that the video traffic has the property of long-range dependence [17, 18]. The correlation structure that accompanies long-range dependence means that the traffic exhibits sustained burstiness, summarized in the related term self-similarity. These properties bring difficulties to the video traffic prediction. Multi-resolution wavelet decomposition can transform discrete sequences into different resolution levels, in each level the abundance of frequency components can be decreased. In other words, burstiness of the traffic arrivals can be reduced, which will be easier for training and prediction. Many wavelet based prediction schemes were proposed for long-range dependence discrete sequences, such as network traffic, river discharges, etc. [19, 20]. Moreover, many supervised learning approaches, such as support vector regression (SVR) and artificial neural networks (ANNs), were applied to non-linear training and forecasting systems [21, 22].

## 8.3  System Model

In this part, we propose a cloud-based space information system, in which control management capabilities of the cloud can greatly improve the scalability and flexibility of the system. The SBIN is operated under high dynamic circumstance. On the one hand, inter-satellite links (ISLs) and satellite-ground station links (SGLs) can hardly keep stable and continuous because of the rapid changing network topology. On the other hand, more satellites will be launched to increase the scale and capabilities of the SBIN, and the satellites that have been launched can be updated or replaced for enhancing functions. These changes of satellite infrastructure and capabilities can give rise to difficulties in network management and control. The cloud enables the ubiquitous and task-driven network access, and can provide global management and configuration of the network resource. Therefore, we introduce a cloud server to sense the traffic and access information and implement resource allocation, which can enhance the scalability and flexibility of the system.

We consider a general multiple queues system with a single server, as shown in Fig. 8.1. In this system, the server can obtain the traffic information of a finite number $N < \infty$ of users, and each of these users utilizes the service of the server.

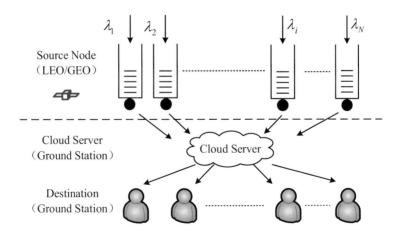

**Fig. 8.1** The multi-queue system with a cloud server serving workload for traffic arrivals from different users/satellites

The user is numbers by $1, 2, \cdots, N$, and the server is denoted as $d$. In our work, the users especially refer to Landsat satellites deployed in different earth orbits, and the server serving as a cloud processing center is deployed at the ground station, which is the destination of the satellite data, and performs traffic sensing, video traffic prediction and resource allocation. Moreover, to provide the time index for the analysis and real system operation, we assume that the system operates in slotted time, then slots are normalized to integral units $t \in \{0, 1, 2, \cdots\}$.

### 8.3.1  The Traffic Model

We consider the video traffic from different satellites as the arrivals of the system, and use $A_i(t)$ to denote the amount of new packets arriving to satellite $i$ ($i = 1, 2, \cdots, N$) at time slot $t$. Let $\mathbf{A}(t) = [A_1(t), A_2(t), \cdots, A_N(t)]$ denote the vector of arrivals at $t$. We assume that arrivals to each satellite are independent and identically distributes (i.i.d.) at different time slots, and $\lambda_i = E\{A_i(t)\}$ denotes the arrival rate at the queue of satellite $i$. In essence, the arrivals to different satellites can be arbitrarily correlated, and the correlation can be not considered and ignored when analyzing. We also assume that there exists $A_{\max}$ such that $0 \le A_i(t) \le A_{\max}$ for all $i$ and $t$.

### 8.3.2  Physical Channel Model

In the transmission of the SBIN, the line of sight (LOS) signal is much stronger than the others, which is different from ground networks. Therefore, we consider the wireless channels for satellites and ground stations as a Rician fading channel model with additive Gaussian noise. The signal received at the destination $d$ at time slot $t$ is modeled as

$$y_i^t = \sqrt{G l_i^{-\gamma}} h_i^t x_i^t + n_i^t, \tag{8.1}$$

where $i$ ($i = 1, 2, \cdots, N$) is the index for the source satellite, $d$ represents the destination (cloud server), $x_i$ is the data transmit from satellite $i$, $G$ is the transmitting power, $l_i$ is the distance between satellite $i$ and destination $d$, $\gamma$ denotes the path loss exponent, and $n_i^t$ is i.i.d. additive Gaussian noise between $i$ and $d$ at time slot $t$ with zero-mean and variance $N_0$. In (8.1), $h_i = X_1 + jX_2$ is the channel fading coefficient modeled as a circularly symmetric complex Gaussian random variable for $i$, in which $X_1 \sim \mathcal{N}(\mu_1, \sigma^2/2)$ and $X_2 \sim \mathcal{N}(\mu_2, \sigma^2/2)$ are Gaussian random variables. Then the distribution of $|h_i|$ is given by the Rician probability density function (PDF)

$$f_{|h_i|}(h) = \frac{2h}{\sigma^2} \exp\left\{\frac{-(h^2+s^2)}{\sigma^2}\right\} I_0\left(\frac{2sh}{\sigma^2}\right), \tag{8.2}$$

where $s^2 = v_1^2 + v_2^2$ is the power due to the Line of Sight (LOS) signal, and $I_0(\cdot)$ is the 0th order modified Bessel function of the first kind [23, 24]. Then $SNR_i$, the Signal-to-Noise Ratio (SNR) between node $i$ and the destination, can be specified as

$$\text{SNR}_i = \frac{|h_i|^2 l_i^{-\gamma} G}{N_0}, \tag{8.3}$$

where $|h_i|^2$ follows the non-central chi-squares ($\mathcal{X}^2$) distribution with the PDF as

$$f_{|h_i|^2}(h) = \frac{K+1}{\Omega} \exp\left\{-K - \frac{(K+1)h}{\Omega}\right\} \cdot I_0\left(2\sqrt{\frac{K(K+1)h}{\Omega}}\right). \tag{8.4}$$

In (8.4), $\Omega = s^2 + \sigma^2$ is the total power of the LOS and scattering signal, $K = s^2/\sigma^2$ is the ratio between the power in the direct path and the power in the other scattered paths [24, 25]. Next, we introduce the outage event and outage probability to characterize the success and failure of the packet transmission and reception. The condition of outrage is defined as that the SNR is less than the given SNR threshold $\beta$. Then outage event can be expressed as (8.5).

$$\{h_i : \text{SNR}_i < \beta\} = \left\{h_i : |h_i|^2 < \frac{\beta N_0 l_i^{\gamma}}{G}\right\}. \tag{8.5}$$

Since the success probability of the packet between $i$ and destination $d$ at SNR threshold $\beta$ is

$$
\begin{aligned}
f_i &\triangleq \Pr\{C_i\} = \Pr\left\{|h_i|^2 \geq \frac{\beta N_0 l_i^{\gamma}}{G}\right\} \\
&= \int_{\frac{\beta N_0 l_i^{\gamma}}{G}}^{+\infty} \frac{K+1}{\Omega} \exp\left\{-K - \frac{(K+1)h}{\Omega}\right\} \cdot I_0\left(2\sqrt{\frac{K(K+1)h}{\Omega}}\right) dh,
\end{aligned}
\tag{8.6}
$$

where $C_i$ denotes the success transmission between node $i$ and the destination.

### 8.3.3 The Cloud-Based Predictive Service Model

We consider that the cloud server allocates power for transmitting data packets at each time slot $t$. Let $P_i(t)$ denote the power allocated to serving packets from satellite $i$ at time slot $t$. Then we get the power allocation vector of the cloud server as $\mathbf{P}(t) = [P_1(t), P_2(t), \cdots, P_N(t)]^{\text{T}}$. Next, we discuss the link state between satellites and the destination. The SBIN is a kind of dynamic system, which results from the change of the channel fading coefficients $h_i$, various service requests from different users at different time slots, etc. This situation may generate different power and other resource consumption and requests different service rate for the server. To distinguish the change of connection and transmission, we use $S_i(t)$ to denote the link state between satellite node $i$ and the destination at time slot $t$. Then we get the whole link state of the system as the vector $\mathbf{S}(t) = [S_1(t), S_2(t), \cdots, S_N(t)]^T$. We assume that $\mathbf{S}(t)$ takes values in $\{s_j\}_{j=1}^{K}$, and let $p_{s_j}^i(t) = \Pr\left[S_i(t) = s_j\right]$. Under a link state $s_j$, power allocation vector $\mathbf{P}(t)$ chooses values in power allocation set $\mathscr{P}^{(s_j)}$, which is compact and there exists the constraint $0 \leq P_i(t) \leq P_{\max}$ for each $i = 1, 2, \cdots, N$ [4]. We assume that the cloud server can sense the link state, and allocate the power from the appropriate power allocation set. Then given link state $\mathbf{S}(t)$ and power allocation vector $\mathbf{P}(t)$, we define the amount of data packets served by the server from each queue of satellites as

$$\mu_i(t) = \mu_i(S_i(t), P_i(t)), \quad \forall i = 1, 2, \cdots, N, \tag{8.7}$$

where $\mu_i(S_i(t), P_i(t))$ is the continuous function of $S_i(t)$ and $P_i(t)$. We assume that $0 \leq \mu_i(S_i(t), P_i(t)) \leq \mu_{\max}$ for all $i$, $\mathbf{S}(t)$ and $\mathbf{P}(t)$ at any time slot $t$. In our work, we set $S_i(t) \in \{1, 2\}$. $S_i(t) = 1$ denotes that there does not exist a link between node $i$ and the destination, and the lower service rate will be allocated. $S_i(t) = 2$ denotes that the link exists, and the service rate allocated for $i$ will be higher than the case of $S_i(t) = 1$. According to Sect. 8.3.2, let

$$p_1^i(t) = 1 - f_i(t), \quad p_2^i(t) = f_i(t), \tag{8.8}$$

where $f_i(t)$ is calculated through (8.6). In this part, we assume that the service rate is given by

$$\mu_i(t) = \lfloor \log(1 + S_i(t) P_i(t)) \rfloor, \tag{8.9}$$

where $\lfloor x \rfloor$ is the floor function mapping the largest integer not greater than $x$ [4].

Most previous works studying multi-queue systems allocate transmitting resource according to the current or past arrivals to different queues. This kind of resource allocation mechanism can lead to the serious delay resulting from the stochastic and burst arrivals, especially for video business. To reduce the delay and improve the quality of service, a predictive scheduling, i.e., predicting and pre-serving arrivals, was proposed in [4]. Next, we introduce the predictive service mechanism to the video transmission in the SBIN. We assume that the cloud server at the ground station can predict and serve the future packet arrivals to each queue of satellites, and allocate power for these queues according to its prediction. Let $D_i \geq 1$ as the prediction window size of satellite $i$. Then at each time slot $t$ and for each $i = 1, 2, \cdots, N$, the cloud server can predict arrival information in the lookahead window $\{A_i(t), A_i(t+1), \cdots, A_i(t+D_i-1)\}$, where $A_i(t+1), \cdots, A_i(t+D_i-1)$ are considered as the future arrivals. We assume that the packets arriving at a time slot can only be served in following time slots. At each time slot $t$, let $\mu_i^{(\tau)}(t)$ ($\tau = 0, 1, \cdots, D_i - 1$) denote the service rate allocated to arrival packet $A_i(t + \tau)$, and let $\mu_i^{(-1)}(t)$ denote the service rate allocated for the arrival packets that are already in the queue. For each $\mu_i(t)$, there always has

$$\sum_{\tau=-1}^{D_i-1} \mu_i^{(\tau)}(t) \le \mu_i(t).$$ (8.10)

### 8.3.4 The Queueing Model

Let $Q_i(t)$ be the amount of packets queued at the cloud server from satellite $i$ at time slot $t$. Then the dynamic of the queue is given by

$$Q_i(t+1) = \max\left\{Q_i(t) - \mu_i^{(-1)}(t), 0\right\} + A_i^{(-1)}(t),$$ (8.11)

where $A_i^{(-1)}(t)$ is the amount of packets that actually enter the queue after experiencing a series of predictive service processing [4]. Specifically, the packets arriving to the queue are served by the pre-allocated rate in previous time slots, and the processing of which can be formulated as

1. $-1 \le \tau \le D_i - 2$:

$$A_i^{(\tau)}(t) = \max\left\{A_i^{(\tau+1)}(t) - \mu_i^{(\tau+1)}(t - \tau - 1), 0\right\}.$$ (8.12)

2. $\tau = D_i - 1$:

$$A_i^{(\tau)}(t) = A_i(t).$$ (8.13)

In this part, we consider the system stability as the finiteness of the time average queue size. Let $E\{Q_i(t)\}$ be the mean length of queue $i$. The system is stable, if

$$\bar{Q} \triangleq \limsup_{t \to \infty} \frac{1}{t} \sum_{\tau=0}^{t-1} \sum_{i=1}^{N} E\{Q_i(\tau)\} < \infty.$$ (8.14)

The optimization target for the system is to find a power allocation and scheduling scheme to minimize the time average cost, which subjects to the constraint that queues of the system must keep stable. Then establish the optimization problem as

$$\min \ \bar{f_c} = \limsup_{t \to \infty} \frac{1}{t} \sum_{\tau=0}^{t-1} E\left\{f_c\left(\mathbf{S}\left(\tau\right), \mathbf{P}\left(\tau\right)\right)\right\}, \tag{8.15a}$$

$$\text{s.t.} \ \limsup_{t \to \infty} \frac{1}{t} \sum_{\tau=0}^{t-1} \sum_{i=1}^{N} E\left\{Q_i\left(\tau\right)\right\} < \infty, \tag{8.15b}$$

where $f_c\left(\mathbf{S}\left(t\right), \mathbf{P}\left(t\right)\right)$ denotes the cost of the server resulting from the power consumption in every time slot, for given $\mathbf{S}\left(t\right)$ and $\mathbf{P}\left(t\right)$. In this work, we set the total power consumption

$$f_c\left(\mathbf{S}\left(t\right), \mathbf{P}\left(t\right)\right) = \sum_{i=1}^{N} P_i\left(t\right) \tag{8.16}$$

as the cost of the server. We assume that there always has $f_c\left(\mathbf{S}\left(t\right), \mathbf{P}\left(t\right)\right) \le f_{c\,\text{max}}$, $\forall t, \mathbf{S}\left(t\right)$ and $\mathbf{P}\left(t\right)$.

## 8.4 Wavelet Based Backpropagation Prediction for Traffic

As discussed in the literature review of this part, the video traffic is burst and corresponds to the existence of clusters of occurrences. Moreover, the autocorrelation of video traffic decays hyperbolically, which indicates the characteristic of long-range dependence, so the video traffic is a type of self-similar stochastic process. Therefore, the prediction for the video traffic is very difficult and complicated because of its characteristics different with other business.

The discrete wavelet transform (DWT) decomposes a signal sequence at different dilations to get multiple dimensions and resolutions of approximation coefficients and detail coefficients. The approximation coefficients represent the high-scale and low-frequency components of the signal, and the detail coefficients represent the low-scale and high-frequency components. DWT provides both spatial and frequency description of signals and is very useful for processing of non-stationary signals. The long-range dependence feature of the video traffic can be well extracted through this kind of multi-dimension decomposition. For the prediction of the traffic according to extracted multi-dimension features, artificial neural networks are adopted. First, artificial neural networks are non-linear methods that learn from patterns and obtain hidden functional relationships, which can be unknown or difficult to identify. In addition, artificial neural networks can model non-stationary and dynamic system, which will deal with the feature of strong dependence of the video traffic. In this section, we will design a backpropagation neural network prediction system based on the multi-resolution wavelet decomposition to predict the video traffic. This prediction system is a part of the cloud processing center operated at the cloud server, and provides prediction information for the power and service rate allocation in the next stage.

### 8.4.1  Multi-Level Wavelet Decomposition

In this part, the signal decomposed by DWT in our work is the video traffic, specifically, the amount sequence of the arriving packets. Wavelet transforms decompose signals through a basic mother wavelet by adjusting the time-shifting and time-dilation parameters. DWT is derived from a continuous Wavelet transform (CWT), the definition of which for a given continuous signal $x(t)$ is defined by

$$W(a, b) = \frac{1}{\sqrt{a}} \int_{-\infty}^{+\infty} x(t)\, \phi\left(\frac{t-b}{a}\right) dt, \tag{8.17}$$

where $a > 0$ is the scaling parameter that determines the wavelet spread, $b > 0$ is the translation parameter that determines the central position, and $\phi(t)$ is the mother wavelet function. DWT of the discrete signal $x(k)$ is defined as

$$W(m, n) = \frac{1}{\sqrt{2^m}} \sum_{k=0}^{T-1} x(k)\, \phi\left(\frac{k - n \cdot 2^m}{2^m}\right), \tag{8.18}$$

where $T$ is the length of the signal, $k$ is the discrete time index, integer variable $m$ and $n$ are the scaling and translation parameter ($a = 2^m$, $b = n \cdot 2^m$). Then we can analyze the signal at different frequencies with different resolutions through an efficient filtering algorithm proposed by Mallat [26]. By using complementary low-pass and high-pass filters, the algorithm decomposed signal $x(k)$ into the approximation and detail components, which represent the low and high frequency components of the signal, respectively. The process for signal $x(k)$ is formulated as

$$y_{\text{low}}(k) = \sum_{n=-\infty}^{+\infty} x(n)\, h(2k - n), \tag{8.19a}$$

$$y_{\text{high}}(k) = \sum_{n=-\infty}^{+\infty} x(n)\, g(2k - n), \tag{8.19b}$$

where $h(k)$ is the high-pass filter, and $g(k)$ is the low-pass filter. The approximation components reflect the general trend of the signal, and can be decomposed into multiple levels through a series of processing as previous. A two-layer filter bank implementation of DWT is shown in Fig. 8.2, where $A1$, $A2$ and $D1$, $D2$ are the approximation and detail components of Levels 1 and 2, respectively. Through the

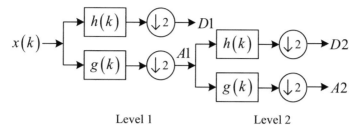

Level 1                                    Level 2

**Fig. 8.2** A 2-layer wavelet decomposition, $A$ and $D$ denote the approximate and detail components, respectively. $h(k)$ is the high-pass filter, $g(k)$ is the low-pass filter, and $\downarrow 2$ denotes down-sampling

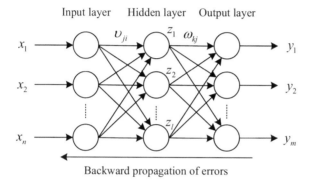

**Fig. 8.3** A typical 3-layer backpropagation neural network structure

down-sampling, which is represented by ($\downarrow 2$), the length of the signal can remain unchanged.

## 8.4.2 Backpropagation Neural Network Prediction

Backpropagation (BP) is an important method for supervised learning with feed-forward artificial neural networks. The BP neural network is composed of the input level, output level and one or more hidden layers. The structure of a typical BP neural network with only one hidden layer is shown in Fig. 8.3. The input layer receives the external vector, which passes through a weighted connection to the neurons in the hidden layer. Then neurons in the hidden layer compute their activations, which will be passed to neurons in the output or next hidden layer. This process can be considered as the given input vector is propagated forward through the network, and the activations vector is formed in the output layer finally. In Fig. 8.3, the number of neurons in the input, hidden and output layer is $n$, $l$ and $m$, respectively. $x_i$ ($i = 1, 2, \cdots, n$), $z_j$ ($j = 1, 2, \cdots, l$) and $y_k$ ($k = 1, 2, \cdots, m$) indicate the activations in the related layer. $\upsilon_{ji}$ denotes the connection weight from neuron $i$ in the input layer to neuron $j$ in the hidden layer, and $\omega_{kj}$ denotes the connection weight from neuron $j$ in the hidden layer to neuron $k$ in the output layer. Activations of neurons in the input and hidden layers are computed through a non-linear activation function, and then we get activation outputs of the hidden and output layers as

$$z_j = f\left(net_j\right) = f\left(\sum_{i=1}^{n} v_{ji}x_i - \theta_j\right), \quad j = 1, 2, \cdots, l, \quad (8.20)$$

$$y_k = f\left(net_k\right) = f\left(\sum_{j=1}^{l} \omega_{kj}z_j - \vartheta_k\right), \quad k = 1, 2, \cdots, m, \quad (8.21)$$

where $\theta_j$ and $\vartheta_k$ are the unit bias values, which can be treated as weights. In (8.20) and (8.21), $f(\cdot)$ is the non-linear activation function, and the standard sigmoid logistic function as (8.22) is mostly selected because of its nice property as (8.23).

$$f_{\text{sig}}(net) = \left(1 + e^{-net}\right)^{-1}, \quad (8.22)$$

$$\frac{\partial f_{\text{sig}}(net)}{\partial net} = f_{\text{sig}}(net)\left(1 - f_{\text{sig}}(net)\right). \quad (8.23)$$

The basic idea of the backpropagation learning algorithm is the repeat application of the chain rule to compute the contribution of each weight, and getting a desired mapping of input to output activations. The mapping constitutes the pattern set $\mathscr{P}$. Through training the weights, the resulting output activations of the network should equal or approach to ideal outputs. The distance between the resulting and ideal outputs is measured by an error or cost function as (8.24), which can be considered as a fitness index of weights.

$$err = \frac{1}{2}\sum_{p\in\mathscr{P}}\sum_{k=1}^{m}\left(t_k^p - y_k^p\right)^2, \quad (8.24)$$

where $\mathbf{t}^p = \left[t_1^p, t_2^p, \cdots, t_m^p\right]^{\text{T}}$ is the target or ideal activation vector, $\mathbf{y}^p = \left[y_1^p, y_2^p, \cdots, y_m^p\right]^{\text{T}}$ is the resulting output vector of the network. The training goal is to get the global minimum of $err$. So the backpropagation algorithm can be divided into the following two stages:

1. *Feed-forward stage:* activations in each layer are calculated using the weights, the activation function and the activations in the previous layer.
2. *Backpropagation stage:* the algorithm checks whether the distance between the resulting and ideal outputs $err$ is within a given threshold. If not, all weights are modified through (8.25a) and (8.25b), and the feed-forward stage is repeated.

When *err* is within the threshold or the maximum number of iterations is exceeded, the learning stops, and the current weights are used to generate the learned prediction network. We have

$$\Delta \omega_{kj} = -\alpha \sum_{p \in \mathscr{P}} \left( y_k^p - t_k^p \right) y_k^p \left( 1 - y_k^p \right) z_j^p, \tag{8.25a}$$

$$\Delta \upsilon_{ji} = -\alpha \sum_{p \in \mathscr{P}} \sum_{k=1}^{m} \delta_k^p \omega_{kj} z_j^p \left( 1 - z_j^p \right) x_i^p, \tag{8.25b}$$

where $\alpha$ is the learning rate, and $\delta_k^p$ in (8.25b) is given by

$$\delta_k^p = \sum_{p \in \mathscr{P}} \left( y_k^p - t_k^p \right) y_k^p \left( 1 - y_k^p \right). \tag{8.26}$$

### 8.4.3   Wavelet Based Backpropagation Prediction

Through the multi-resolution wavelet transform, the decomposed traffic has less frequency components and is more stationary than the original traffic. The accuracy of prediction for these different resolutions' components can be improved compared with the non-stationary and burst original traffic. Therefore, we establish the prediction system with multiple backpropagation neural networks, of which the inputs are approximation or detail coefficients obtained through the multi-resolution wavelet decomposition, and the outputs are the corresponding predictive coefficients of the future traffic. The structure of the proposed prediction system is shown as Fig. 8.4. Specifically, according to Sect. 8.3, for a given video traffic time series $\{A_i(0), A_i(1), \cdots, A_i(t)\}$ of user $i$ ($i = 1, 2, \cdots, N$), the prediction process for the traffic of the next $D_i - 1$ time slots consists of the following steps:

- The original video traffic time series $\{A_i(0), A_i(1), \cdots, A_i(t)\}$ are processed with the $L$-resolution discrete wavelet decomposition. Then we get approximation coefficients of the $L$th level $a_L(t)$ and detail coefficients of all $L$ levels as $d(t) = \{d_1(t), d_2(t), \cdots, d_L(t)\}$.
- For each of approximation and detail coefficients $a_L(t)$ and $d(t)$, we establish $L + 1$ backpropagation network to predict each of the next value at time slot $t + 1$. Take $a_L(t)$ as an example, the training input matrix of the backpropagation network is

$$I_{\text{Tr}} = \begin{bmatrix} a_L(0) & a_L(1) & \cdots a_L(\tau_0 - 1) \\ a_L(1) & a_L(2) & \cdots & a_L(\tau_0) \\ \vdots & \vdots & \ddots & \vdots \\ a_L(t - \tau_0) & a_L(t - \tau_0 + 1) & \cdots & a_L(t - 1) \end{bmatrix}, \tag{8.27}$$

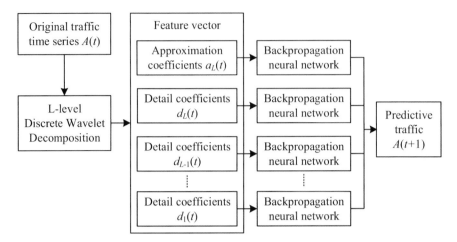

**Fig. 8.4** The multi-level Wavelet decomposition based backpropagation neural network traffic prediction system

where $\tau_0$ is the length of each training input vector, and each row vector is an input of the backpropagation network. The corresponding outputs of each row constitute the training output vector of the backpropagation network: $\mathbf{O}_{\mathrm{Tr}} = [a_L(\tau_0), a_L(\tau_0 + 1), \cdots, a_L(t)]^{\mathrm{T}}$.

Through the backward propagation of error operated by the training network, we get the prediction network for approximation coefficients $a_L(t)$.

- Set vector $[a_L(t - \tau_0 + 1), a_L(t - \tau_0 + 2), \cdots, a_L(t)]^{\mathrm{T}}$ as the input of the prediction network obtained in the last step, and the output is the predictive approximation coefficient $a_L(t + 1)$.
- $a_L(t + 1)$ is combined with $a_L(t - \tau_0 + 2)$, $a_L(t - \tau_0 + 3)$, $\cdots$, $a_L(t)$ to predict $a_L(t + 2)$. Repeat this process and we can get the all following $D_i - 1$ approximation coefficients. The detail coefficients in the following $D_i - 1$ time slots are predicted through similar processes.
- Reestablish the predictive $A(t + 1)$ to $A(t + D_n - 1)$ through $a_L(\tau)$ and $d(\tau)$ $(\tau = 0, 1, \cdots, D_n - 1)$.

The generated backpropagation prediction networks are modified in every time slot by adding the new arrival traffic as the training data. Then at each time slot, we can get future arrivals $A_i(t + 1)$, $A_i(t + 2)$, $\cdots$, $A_i(t + D_i - 1)$ to constitute the look-ahead window, which is discussed in the Sect. 8.3.3.

## 8.5   Resource Allocation Based on the Predictive Backpressure

According to the last section, the future traffic arrival information of satellites in the cloud-based SBIN from time slot $t+1$ to $t+D_i-1$ can be predicted and available in the current time slot $t$. In this section, we design a resource allocation and scheduling scheme based on the predictive traffic information, and the backpressure scheme [4, 27, 28] is introduced into the prediction based resource allocation to reduce the queue length and delay of the system.

### 8.5.1   Dynamic Evolution of Queues

First, we introduce the conception of prediction queue [4]. As discussed in the previous sections, future arrivals can be predicted and pre-served in the current time slot. Let $\mu_i^{(\tau)}(t)$ $(\tau = 0, 1, \cdots, D_i - 1)$ denote the pre-allocated service rate for the future arrivals $A_i(t+\tau)$. We use $Q_i^{(\tau)}(t)$ $(\tau = 0, 1, \cdots, D_i - 1)$ to denote the amount of remaining arrivals currently of queue $i$ in future time slot $t+\tau$. Notice that $Q_i^{(\tau)}(t)$ just records the residual number of arrivals in the queue over the time line, but is not the real number of arrivals. Let $Q_i^{(-1)}(t)$ denote the number of arrival packets already at queue $i$ in time slot $t$, which is the same as $Q_i(t)$ in (8.11). Different from $Q_i^{(0)}(t)$, $Q_i^{(1)}(t)$, $\cdots$, $Q_i^{(D_i-1)}(t)$, we notice that $Q_i^{(-1)}(t)$ records the real backlog in the queue, and is the only real queue in $\left\{ Q_i^{(\tau)}(t) \right\}_{\tau=-1}^{D_i-1}$. Therefore, the system is stable if $Q_i^{(-1)}(t)$ is stable. As the same assumption as mentioned in Sect. 8.3.3, packets arriving at a time slot can only be served in following time slots. Then the dynamic evolement process of queue $i$ can be modeled as following conditions:

1. $\tau = D_i - 1$:

$$Q_i^{(\tau)}(t+1) = A_i(t+D_i).  \tag{8.28}$$

2. $0 \le \tau \le D_i - 2$:

$$Q_i^{(\tau)}(t+1) = \max\left\{ Q_i^{(\tau+1)}(t) - \mu_i^{(\tau+1)}(t), 0 \right\}.  \tag{8.29}$$

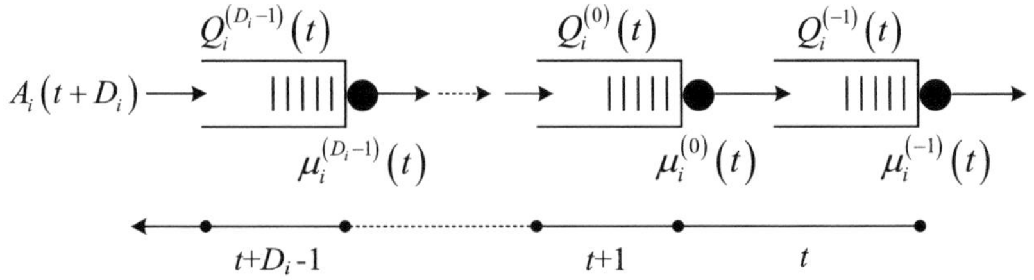

**Fig. 8.5** The dynamic evolvement process of the prediction queue

3. $\tau = -1$:

$$Q_i^{(\tau)} (t+1) = \max \left\{ Q_i^{(\tau)} (t) - \mu_i^{(\tau)} (t), 0 \right\}$$
$$+ \max \left\{ Q_i^{(0)} (t) - \mu_i^{(0)} (t), 0 \right\},$$

(8.30)

where $Q_i^{(-1)} (0) = 0$.

The dynamic evolvement process of the prediction queue is shown in Fig. 8.5. Then for each user of the system, we get the amount of remaining arrivals at every time slot of the prediction window, which can be considered for making decision of the resource allocation.

### 8.5.2  Prediction Based Backpressure

Backpressure is originally designed to make decisions that minimize the amount of queue backlogs in the network from one time slot to the next, and is mathematically analyzed via the theory of Lyapunov drift [27–30]. The algorithmic mechanism is similar to how water flows through a network of pipes via pressure gradients. However, the future arrival information of the predictive window is available for the current time slot, which presents difficulties in analysis and prohibits the application of Lyapunov theory. Fortunately, [4] provided a queue-equivalence between the predictive queueing network with a fully efficient scheduling and an equivalent queueing network without any prediction. First, we introduce the definition of the fully efficient predictive scheduling scheme and the equivalence between the predictive and non-predictive queueing system proposed in [4].

**Definition 8.1**  A predictive scheduling scheme is fully efficient when the following conditions is satisfied for any user $i$:

1. $\sum_{\tau} \mu_i^{(\tau)} (t) = \mu_i (t)$;
2. $\mu_i^{(\tau)} (t) > Q_i^{(\tau)} (t), \mu_i^{(\tau')} (t) \geq Q_i^{(\tau')} (t), \forall -1 \leq \tau \leq D_i - 1, \tau' \neq \tau$.

Definition 8.1 indicates that for a queueing system with the fully efficient predictive scheduling, all service opportunities and resource will be in full use and not be wasted. Moreover, any queue of the system will not be allocated extra service resource unless other queues are accomplished with serves.

**Theorem 8.1** *In a single-queue system, if:*

1. $\tilde{Q}_i (0) = \sum_{t=0}^{D_i-1} A_i (t)$, *where $\tilde{Q}_i (t)$ denotes the length of queue $i$,*
2. *arrivals of the single-queue system satisfy $\tilde{A}_i (t) = A_i (t + D_i)$,*
3. *service of the system satisfies $\tilde{\mu}_i (t) = \sum_{\tau=-1}^{D_i-1} \mu_i^{(\tau)} (t)$,*

*and we get the queue evolution as*

$$\tilde{Q}_i (t + 1) = \max \left\{ \tilde{Q}_i (t) - \tilde{\mu}_i (t), 0 \right\} + \tilde{A}_i (t). \tag{8.31}$$

*Then for a predictive system with the fully efficient predictive scheduling scheme and $Q_i^{(-1)} (0) = 0$, it has*

$$\sum_{\tau=-1}^{D_i-1} Q_i^{(\tau)} (t) = \tilde{Q}_i (t), \ \forall i, \ t. \tag{8.32}$$

Theorem 8.1 provides a queue equivalence between the predictive queueing system with a fully efficient scheduling policy and a queueing system without any prediction. The complexity of analysis for the pre-serve queueing system with predictive window can be reduced by this equivalence. According to Theorem 8.1, the delay distribution characteristics of the non-predictive system can be considered as *shifted to the left by $D_i$ time slots* in the fully efficient predictive system with $D_i$-slots predictive window. Detailed proofs of the delay characteristics are addressed in [4].

Next, we introduce the resource allocation policy based on the predictive backpressure (PBP) for the cloud-based SBIN. In this policy, prediction queue length of every satellite in the SBIN is considered. The first object of the resource allocation is to find a power allocation and scheduling scheme to minimize the average cost, which is modeled as (8.15). The other optimization target is to minimize the amount of backlogs of the entire network, in other words, to minimize the delay of the network. This object results in that queues with or will with more backlogs are tend to be allocated the higher service rate. The two optimization targets of the resource allocation above can be modeled as (8.33a), where the control parameter $V \geq 1$ is defined to tradeoff the system cost and the delay. Specifically, the resource allocation tends to give more consideration to the power consumption with the increase of $V$. Then we present the predictive backpressure policy as following steps.

1. For the SBIN with $N$ satellites, compute $\sum_{\tau=-1}^{D_i-1} Q_i^{(\tau)} (t)$ for each satellite's traffic queue $i$ in every time slot $t$.
2. Observe the current link state vector $\mathbf{S}(t)$.

3. Then choose the power allocation vector of the server $\mathbf{P}(t)$ to solve the following optimization problem:

$$\min \ V f_c\left(\mathbf{S}(t), \mathbf{P}(t)\right) - \sum_{i=1}^{N} \sum_{\tau=-1}^{D_i-1} Q_i^{(\tau)}(t) \, \mu_i\left(\mathbf{S}(t), \mathbf{P}(t)\right), \qquad (8.33a)$$

$$\text{s.t. } \mathbf{P}(t) \in \mathscr{P}^{(\mathbf{S}(t))}, \qquad\qquad\qquad (8.33b)$$

where $\mathscr{P}^{(\mathbf{S}(t))}$ is the feasible power allocation set.

4. For each prediction queue $Q_i^{(\tau)}(t)$ ($\tau = -1, 0, \cdots, D_i - 1$), allocate the service rates $\mu_i^{(\tau)}(t)$ by the fully efficient scheme according to any queueing discipline.

5. Update the queues' length by (8.28)–(8.30).

The queueing discipline mentioned in Step 4 is how to select packets to serve and transmit from $\left\{Q_i^{(\tau)}(t)\right\}_{\tau=-1}^{D_i-1}$. Two typical queueing disciplines are FIFO (first input first output) and LIFO (last input first output).

## 8.6  Simulation Results and Analysis

In this part, we perform simulation experiments to analyze performances of the multi-level wavelet based backpropagation prediction system and the predictive backpressure based resource allocation. We simulate the SBIN service system with fifteen satellite users, i.e., $N = 15$. In the resource allocation system, the fifteen satellites are deployed around the geosynchronous orbit with the orbit radius from 42,154 km to 42,182 km with equally spaced distance of 2 km, respectively. The propagation path loss of channel between satellites and the ground cloud server is given by $\gamma = 2.8$, and the average power of the Gaussian noise in the channel is $N_0 = 10^{-10}$. The ratio $K$ between the power in the LOS and the power in the other scattered paths of the link is set as $K = 6.99$ dB, and the corresponding total power of the LOS and other scattered path is set as $\Omega = 1 + K$. Set the SNR threshold as $\beta = 5$. For each satellite user, we assume that the link state $\mathbf{S}(t)$ selects values in $\{1, 2\}$ with probabilities $1 - f_i$ and $f_i$ according to (8.6). The power allocated to serving packets from satellite node $i$ is set as $P_i(t) \in \mathscr{P}^{(S_i)} \triangleq \{0, 5, 10\}$. We assume that it allows more than one users can be allocated non-zero power in any time slot. The service rate is set as (8.9), and we set the total power consumption as the cost function according to (8.16). To simplify the analysis, we notice the performance of the first satellite and eleventh satellite in the following simulations.

### 8.6.1   Video Traffic Model

Video traffic arrivals $A_i(t)$ $(i = 1, 2, \cdots, 15)$ of the users are independent processes. First, we generate $A_i$ through the traffic model. Long-range dependence is the inherent property of the video traffic. Therefore, simulations in this work use appropriate stochastic time series model to generate the sequence with self-similarity characteristic. There are two classical stochastic models that can account for long-range dependent: the increment processes of self-similar models and the fractional autoregressive integrated moving average (FARIMA) processes. FARIMA models have been widely used in the modeling for the burst and long-range dependence traffic [31, 32]. In our simulations, we use the FARIMA model to generate the sequence of the video traffic. The FARIMA process evolved from the standard ARIMA $(p, d, q)$ model by allowing the degree of differencing $d$ to take non-integral values. Specifically, the FARIMA $(p, d, q)$ $(p, q \in \mathbb{N}^+,$ $d \in \mathbb{R})$ process is defined as a stochastic process $X = \{X_i \,|\, i = 0, 1, \cdots\}$ with a representation given by

$$\Phi\left(z^{-1}\right) \nabla^d X_i = \Theta\left(z^{-1}\right) n_i, \tag{8.34}$$

where $\Phi\left(z^{-1}\right)$ and $\Theta\left(z^{-1}\right)$ are the autoregressive (AR) and moving average (MA) polynomials of order $p$ and $q$, respectively, in the backward shift operator $z^{-1}\{X_i\} = X_{i-1}$. Then the differential operator can be expressed as $\nabla = 1 - z^{-1}$, and $\nabla^d$ denotes the fractional differential operator with order $d$, which is defined as the usual binomial expansion. $n_i$ is an i.i.d. non-Gaussian noise with finite mean and variance [31]. The important property of the FARIMA $(p, d, q)$ model is that, stochastic sequence $X = \{X_i \,|\, i = 0, 1, \cdots\}$ performs the long-range dependence feature when $0 < d < 0.5$.

Set $d = 0.1$, we generate fifteen stochastic series with the length of 1000, which present video traffic arrivals in 1000 time slots of the fifteen satellite users. Figure 8.6 shows the traffic of the first user and the eleventh user and the their autocorrelations, which can express the burst and the long-range dependence property of the video traffic.

### 8.6.2   Performance of Wavelet Based Backpropagation Prediction

As shown in Fig. 8.6, the generated video traffic is burst and has quite abundant frequency components, which is difficult to be trained and learnt for prediction. In this part, we apply the multi-resolution wavelet decomposition, and transform the original traffic into 7 levels of approximation and detail components. Take the traffic of user 1 for instance, the approximation components of Level 7 and detail

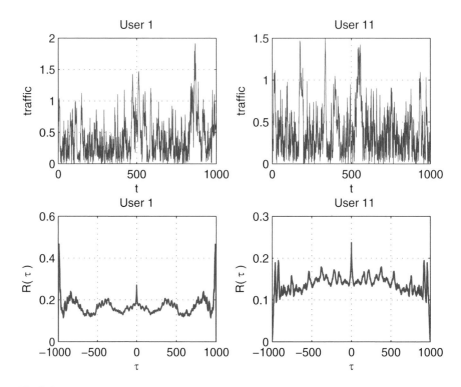

**Fig. 8.6** The traffic generated through the FARIMA model for the two sample users and their long range dependence characteristics

components of Level 7, 5 and 1 are shown in Fig. 8.7. Figure 8.7 indicates that the abundance of frequency components in each level trend to be reduced, which will increase the prediction accuracy.

To establish the backpropagation neural network for the traffic prediction, we generate the training traffic through the same FARIMA model with the same parameter settings. The length of the training traffic is set as 2000. Decompose the training traffic into 7 levels by multi-resolution wavelet transform. Then we get the training approximation and detail components, by which we establish the wavelet based backpropagation prediction networks introduced in Sect. 8.4.3. We set the length of each training input vector $\tau_0 = 5$ and $\tau_0 = 50$ in (8.27) for approximation and detail components, respectively. The threshold of $err$ is set as $10^{-4}$, and the maximum number of iterations is set as 1000. Through training, we get 8 backpropagation prediction networks. Take the first $\tau_0$ approximation and detail components of the fifteen users' traffic as the inputs of the corresponding prediction network, the outputs are the predictive components of the original traffic.

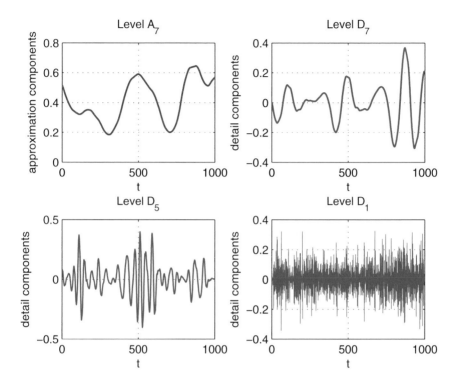

**Fig. 8.7** Approximation components and detail components of the 7 levels wavelet decomposition for the traffic of user 1

**Table 8.1** Mean square error of BP prediction for 7-Level wavelet decomposition

|                     | Level A7              | Level D7              | Level D6              | Level D5              |
| ------------------- | --------------------- | --------------------- | --------------------- | --------------------- |
| Traffic of User 1   | $2.970 \times 10^{-4}$ | $3.755 \times 10^{-5}$ | $8.341 \times 10^{-5}$ | $7.220 \times 10^{-5}$ |
| Traffic of User 11  | $8.560 \times 10^{-4}$ | $7.322 \times 10^{-5}$ | $4.620 \times 10^{-5}$ | $4.235 \times 10^{-5}$ |
|                     | Level D4              | Level D3              | Level D2              | Level D1              |
| Traffic of User 1   | $4.710 \times 10^{-5}$ | $8.739 \times 10^{-5}$ | $5.491 \times 10^{-5}$ | $8.423 \times 10^{-5}$ |
| Traffic of User 11  | $4.732 \times 10^{-5}$ | $8.252 \times 10^{-5}$ | $9.240 \times 10^{-5}$ | $9.223 \times 10^{-5}$ |

For the two sample users (User 1 and User 11), the prediction accuracy of each level of the components is shown in Table 8.1. The real traffic and the predictive traffic by compositing predictive approximation and detail components of the two sample users is shown in Fig. 8.8. Results indicate that the proposed method can accomplish precise prediction of the traffic with burst and long-range dependence property.

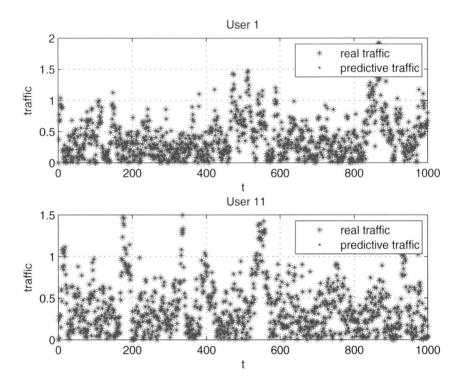

**Fig. 8.8** The real traffic and the predictive traffic through the wavelet decomposition based backpropagation prediction for the two sample users

## 8.6.3   *Performance of Predictive Backpressure*

Based on the SBIN queueing system, this subsection analyzes the performance of the resource allocation based on the predictive backpressure. The predictive traffic arrivals obtained in the previous simulations provide the prediction information needed by the predictive backpressure. Through the service resource allocation, we simulate the power consumption, queue length and the delay of the queueing system by the original real traffic. LIFO is selected as the queueing discipline.

First, we test the power consumption of the cloud sever over the different control parameter $V$ and the size of the prediction window $D_i$ ($i = 1, 2, \cdots, 15$). We set $V \in \{1, 5, 10, 25, 50, 100, 150\}$ and $D_1 = D_2 = \cdots = D_{15} \in \{5, 20\}$. Simulation results are shown in Fig. 8.9, in which *accurate prediction* denotes that the real traffic are used to provide predictive traffic information for backpressure. Results indicate that the power consumption decreases with the increase of $V$, which denotes the tradeoff between the system cost and delay.

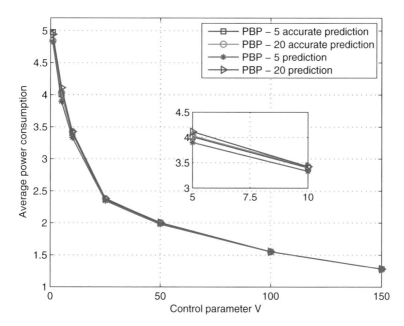

**Fig. 8.9** Average power consumption versus the control parameter $V$ and prediction window size $D_i$

Figure 8.10 shows the average queue length of the two sample users changes with the control parameter $V$ and the size of the prediction window $D_i$ ($i = 1, 2, \cdots, 15$). In Fig. 8.10, *0 prediction* expresses that the backpressure scheme is operated in the resource allocation without any prediction information. Simulation results indicate that the average queue length increases with the increase of $V$, and decreases with the prediction window size. Moreover, the average queue length will increase when the future arrivals are not considered, i.e., $D_i = 0$. Then we can conclude that large prediction window size can lead to shorter average queue length, which means shorter delay or waiting time. However, large prediction window size requires more storage and process capacities for the server to achieve the prediction of the future arrivals. So the setting of $D_i$ is a tradeoff between the service performance and operational costs of the cloud-based system.

Then we test the delay distributions of packets in the two sample users' queues. We set control parameter $V = 50$, and the prediction window size $D_1 = D_2 = \cdots = D_{15} = 20$. Results shown in Fig. 8.11 indicate that about 72.32% of the packets in queue 1 and 72.07% of the packets in queue 11 do not need to wait for service, which means that they are pre-served before they arrival to the system. These results demonstrate that the delay of the SBIN queueing system can be reduced by the resource allocation based on the predictive backpressure.

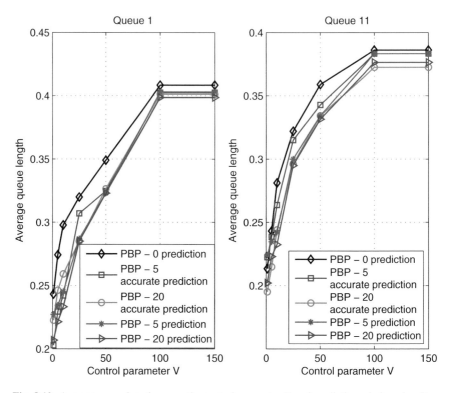

**Fig. 8.10**  Average queue length versus the control parameter $V$ and prediction window size $D_i$

## 8.7   Conclusion

In this part, we proposed a multi-resolution wavelet based backpropagation prediction system for the video traffic in the cloud-based SBIN. The predictive backpressure policy is introduced into the resource allocation of the SBIN with multiple satellite users and a single cloud server. Simulation results indicate that the wavelet backpropagation network can predict video traffic precisely. The performance of the predictive backpressure is simulated, and results demonstrate that the prediction information based resource allocation can reduce the queue length and delay of the queueing system.

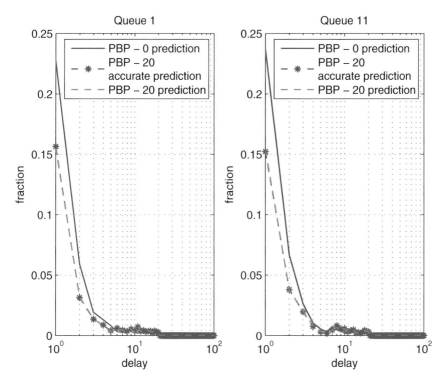

**Fig. 8.11** Packet delay distribution under PBP with LIFO scheduling when $V = 50$ and $D_1 = D_2 = \cdots = D_{15} = 20$

# References

1. P. Lalbakhsh, B. Zaeri, and M. N. Fesharaki, "Improving shared awareness and qos factors in antnet algorithm using fuzzy reinforcement and traffic sensing," in *Future Computer and Communication, 2009. ICFCC 2009. International Conference on*. Kuala Lumpar, Malaysia, Apr. 2009, pp. 47–51.
2. B. R. Tamma, N. Baldo, B. Manoj, and R. R. Rao, "Multi-channel wireless traffic sensing and characterization for cognitive networking," in *Communications, 2009. ICC'09. IEEE International Conference on*. Dresden, Germany, Jun. 2009, pp. 1–5.
3. C.-H. Liu, J. Tran, P. Pawelczak, D. Cabric *et al.*, "Traffic-aware channel sensing order in dynamic spectrum access networks," *Selected Areas in Communications, IEEE Journal on*, vol. 31, no. 11, pp. 2312–2323, Nov. 2013.
4. L. Huang, S. Zhang, M. Chen, and X. Liu, "When backpressure meets predictive scheduling," in *Proceedings of the 15th ACM international symposium on Mobile ad hoc networking and computing*. ACM, Philadelphia, USA, Aug. 2014, pp. 33–42.
5. Y. Zhang, Y. Zhang, S. Sun, S. Qin, and Z. He, "Multihop packet delay bound violation modeling for resource allocation in video streaming over mesh networks," *Multimedia, IEEE Transactions on*, vol. 12, no. 8, pp. 886–900, Aug. 2010.
6. N. H. Mastronarde and M. van der Schaar, "A bargaining theoretic approach to quality-fair system resource allocation for multiple decoding tasks," *Circuits and Systems for Video Technology, IEEE Transactions on*, vol. 18, no. 4, pp. 453–466, Apr. 2008.

7. H. Saki and M. Shikh-Bahaei, "Cross-layer resource allocation for video streaming over OFDMA cognitive radio networks," *Multimedia, IEEE Transactions on*, vol. 17, no. 3, pp. 333–345, Jan. 2015.

8. S. Cicalo and V. Tralli, "Distortion-fair cross-layer resource allocation for scalable video transmission in OFDMA wireless networks," *Multimedia, IEEE Transactions on*, vol. 16, no. 3, pp. 848–863, Mar. 2014.

9. A. Argyriou, D. Kosmanos, and L. Tassiulas, "Joint time-domain resource partitioning, rate allocation, and video quality adaptation in heterogeneous cellular networks," *Multimedia, IEEE Transactions on*, vol. 17, no. 5, pp. 736–745, Apr. 2015.

10. C. Jiang, Y. Chen, Y. Yang, C. Wang, and K. J. R. Liu, "Dynamic chinese restaurant game: theory and application to cognitive radio networks," *IEEE Trans. Wireless Commun.*, vol. 13, no. 4, pp. 1960–1973, Mar. 2014.

11. C. Jiang, Y. Chen, K. J. R. Liu, and Y. Ren, "Renewal-theoretical dynamic spectrum access in cognitive radio network with unknown primary behavior," *IEEE J. Sel. Areas Commun.*, vol. 31, no. 3, pp. 406–416, Mar. 2013.

12. C. Jiang, Y. Chen, Y. Gao, and K. J. R. Liu, "Joint spectrum sensing and access evolutionary game in cognitive radio networks," *IEEE Trans. Wireless Commun.*, vol. 12, no. 5, pp. 2470–2483, Mar. 2013.

13. C. Jiang, Y. Chen, and K. J. R. Liu, "Multi-channel sensing and access game: Bayesian social learning with negative network externality," *IEEE Trans. Wireless Commun.*, vol. 13, no. 4, pp. 2176–2188, Feb. 2014.

14. S.-J. Yoo, "Efficient traffic prediction scheme for real-time VBR MPEG video transmission over high-speed networks," *Broadcasting, IEEE Transactions on*, vol. 48, no. 1, pp. 10–18, Mar. 2002.

15. A. M. Kholaif, T. D. Todd, P. Koutsakis, and A. Lazaris, "Energy efficient h. 263 video transmission in power saving wireless lan infrastructure," *Multimedia, IEEE Transactions on*, vol. 12, no. 2, pp. 142–153, Jan. 2010.

16. M. Wu, R. Joyce, H.-S. Wong, L. Guan, S.-Y. Kung *et al.*, "Dynamic resource allocation via video content and short-term traffic statistics," *Multimedia, IEEE Transactions on*, vol. 3, no. 2, pp. 186–199, Jun. 2001.

17. S. H. Kang and A. Zakhor, "Effective bandwidth based scheduling for streaming media," *Multimedia, IEEE Transactions on*, vol. 7, no. 6, pp. 1139–1148, Dec. 2005.

18. M. M. Krunz and A. M. Ramasamy, "The correlation structure for a class of scene-based video models and its impact on the dimensioning of video buffers," *Multimedia, IEEE Transactions on*, vol. 2, no. 1, pp. 27–36, Mar. 2000.

19. E. Szolgayová, J. Arlt, G. Blöschl, and J. Szolgay, "Wavelet based deseasonalization for modelling and forecasting of daily discharge series considering long range dependence," *Journal of Hydrology and Hydromechanics*, vol. 62, no. 1, pp. 24–32, Feb. 2014.

20. S. A. Östring, H. R. Sirisena, and I. Hudson, "Rate control of elastic connections competing with long-range dependent network traffic," *Communications, IEEE Transactions on*, vol. 49, no. 6, pp. 1092–1101, Jun. 2001.

21. M. T. Asif, J. Dauwels, C. Y. Goh, A. Oran, E. Fathi, M. Xu, M. M. Dhanya, N. Mitrovic, and P. Jaillet, "Spatiotemporal patterns in large-scale traffic speed prediction," *Intelligent Transportation Systems, IEEE Transactions on*, vol. 15, no. 2, pp. 794–804, Apr. 2014.

22. L. Vanajakshi and L. R. Rilett, "A comparison of the performance of artificial neural networks and support vector machines for the prediction of traffic speed," in *IEEE Intelligent Vehicles Symposium*. Parma, Italy, Jun. 2004, pp. 194–199.

23. E. L. Lehmann and G. Casella, *Theory of point estimation*. Springer Science & Business Media, 1998, vol. 31.

24. S. M. Kay, *Fundamentals of Statistical Signal Processing: Practical Algorithm Development*. Pearson Education, 2013, vol. 3.

25. A. Abdi, C. Tepedelenlioglu, M. Kaveh, and G. Giannakis, "On the estimation of the k parameter for the rice fading distribution," *Communications Letters, IEEE*, vol. 5, no. 3, pp. 92–94, Mar. 2001.

26. S. Mallat, *A wavelet tour of signal processing*. New York: Academic press, 1999.

27. M. J. Neely and R. Urgaonkar, "Optimal backpressure routing for wireless networks with multi-receiver diversity," *Ad Hoc Networks*, vol. 7, no. 5, pp. 862–881, Jul. 2009.

28. M. J. Neely, "Energy optimal control for time-varying wireless networks," *Information Theory, IEEE Transactions on*, vol. 52, no. 7, pp. 2915–2934, Jul. 2006.

29. L. Huang, S. Moeller, M. J. Neely, and B. Krishnamachari, "LIFO-backpressure achieves near-optimal utility-delay tradeoff," *Networking, IEEE/ACM Transactions on*, vol. 21, no. 3, pp. 831–844, Jun. 2013.

30. L. Tassiulas and A. Ephremides, "Stability properties of constrained queueing systems and scheduling policies for maximum throughput in multihop radio networks," *Automatic Control, IEEE Transactions on*, vol. 37, no. 12, pp. 1936–1948, Dec. 1992.

31. J. Ilow, "Forecasting network traffic using FARIMA models with heavy tailed innovations," in *Acoustics, Speech, and Signal Processing, 2000. ICASSP'00. Proceedings. 2000 IEEE International Conference on*, vol. 6. IEEE, Istanbul, Turkey, Jun. 2000, pp. 3814–3817.

32. S. Ma and C. Ji, "Modeling heterogeneous network traffic in wavelet domain," *IEEE/ACM Transactions on Networking (TON)*, vol. 9, no. 5, pp. 634–649, Oct. 2001.

# Part IV
# Cooperative Computation and Caching in Heterogeneous Networks

# Chapter 9
# Introduction of Cooperative Computation and Caching

**Abstract** As discussed in Chap. 1, the allocated resource can be divided into communication resource, which includes channels and bandwidth, and computing resource, such as memory and processing power. In the two parts above, we have introduced some cooperation mechanisms for heterogeneous communication resource. Then this part will focus on cooperative computing and caching mechanisms for exploding computation-intensive and rich-media applications and on-demand services in 5G/6G.

**Keywords** 6G · Heterogeneous Networks · Cooperative Computation · Cooperative Caching

As discussed in Chap. 1, the allocated resource can be divided into communication resource, which includes channels and bandwidth, and computing resource, such as memory and processing power. In the two parts above, we have introduced some cooperation mechanisms for heterogeneous communication resource. Then this part will focus on cooperative computing and caching mechanisms for exploding computation-intensive and rich-media applications and on-demand services in 5G/6G.

The tremendous increase of computation-heavy applications has posed great challenges in terms of enhanced service coverage and high-speed data processing. In response, edge computing and caching have been expected as efficient approaches to support low-latency and on-demand services, which have significant impacts on the developments of heterogenous networks. To be specific, mobile edge computing (MEC) is a promising paradigm that brings computing resources to mobile devices at the network edge, which allows applications and contents to be processed quickly by edge servers instead of remote cloud servers to meet the requirements of mobile users via computation offloading. Similarly, vehicular edge computing (VEC) enables task execution and analysis in close proximity to the data sources, which eases the burden of backhaul links, and reduces the response latency drastically compared with vehicular cloud computing. On the other hand, distributed content caching is regarded as one of the most effective techniques to alleviate the transmission delay and improve the quality and experience of multimedia services through caching and forwarding contents at the edge of networks. Aiming at the realizing ultra-reliable and low-latency services through edge intelligence in heterogeneous networks, we introduce four cooperative computation and caching mechanisms in this chapter.

(1) Hierarchical edge-cloud computing: In a fog and cloud computing system operated by one cloud computing service provider (CCP) and multiple fog computing service providers (FCPs), a computing resource market can be considered, in which the CCP shares its cloud computing resource among FCPs and itself to serve users with computational tasks. To facilitate the resource trading between the CCP and FCPs, the first part proposes a Stackelberg differential game based resource sharing mechanism. In this mechanism, performance discrepancy is introduced as a penalty factor to denote the mismatch between the resource supply and demand, which will encourage all computing providers (CPs) to make their trading decisions that can truthfully reflect their resource capacity and requirements. In addition, an evolutionary game based replicator dynamics is established to analyze the users' service selection among CPs. Based on the established hierarchical game framework, we also investigate the interactions between the user selection and computing resource sharing.

(2) Caching resource allocation: The second part investigates a small-cell based caching system composed of one mobile network operator (MNO) and multiple video service providers (VSPs). Considering different video popularities and mobile users' (MUs') preferences of VSPs, we propose a caching mechanism based on double auction, which can encourage both the MNO and VSPs to truthfully report their acceptances and requirements of caching resource, respectively. Moreover, the proposed caching mechanism ensures the efficient operation of market by maximizing the social welfare.

(3) Priority-aware computational resource allocation: The third part investigates the priority-aware task offloading mechanism in vehicular fog computing based on deep reinforcement learning (DRL), in which vehicles are incentivized to share their idle computing resource by dynamic pricing, which comprehensively considers the mobility of vehicles, the task priority, and the service availability of vehicles.

(4) Hybrid decision based DRL approach for energy-aware computation offloading: Most current reinforcement learning (RL) or DRL based resource allocation approaches were modeled in a discrete action space, which restricts the optimization of offloading decisions in a limited action space. Such model assumption is unreasonable in practice, where the action space of offloading decision is often continuous-discrete hybrid. To be specific, in a task offloading enabled 6G network, the strategies of which node should be selected to implement traffic/computation offloading or caching constitute a discrete action space. On the other hand, the possible resource volume should be provided by the selected node for offloading is a continuous value usually. Such resource allocation problem with continuous-discrete hybrid decision spaces tends to be

extremely complex, especially when time-varying tasks, energy harvesting and security issues are also considered. To solve these problems above, the fourth part focuses on the hybrid decision of computation offloading in 6G networks based on DRL, in which the energy harvesting enabled devices can offload their computational tasks to edge computing servers. The sever selection problem is modeled in a discrete action space, and meanwhile the decision spaces of offloading ratio and local computation capacity are continuous. To validate the efficiency and superiority of our proposed hybrid-action-critic (Hybrid-AC) based computation offloading approach, we also test the average rewards received and execution time, comparing with deep Q-learning based offloading (DQLO), server execution and device execution mechanisms.

Experimental results validate that the mechanisms above have good performance of latency and resource efficiency, and effectively guarantee differentiated quality and experience of multimedia services.

# Chapter 10
# QoS-Aware Computational Resource Allocation

**Abstract** Recently, the boosting growth of computation-heavy applications raises great challenges for the Fifth Generation (5G) and future wireless networks. As responding, the hybrid edge and cloud computing (ECC) system has been expected as a promising solution to handle the increasing computational applications with low-latency and on-demand services of computation offloading, which requires new computing resource sharing and access control technology paradigms. This work establishes a software-defined networking (SDN) based architecture for edge/cloud computing services in 5G heterogeneous networks (HetNets), which can support efficient and on-demand computing resource management to optimize resource utilization and satisfy the time-varying computational tasks uploaded by user devices. In addition, resulting from the information incompleteness, we design an evolutionary game based service selection for users, which can model the replicator dynamics of service subscription. Based on this dynamic access model, a Stackelberg differential game based cloud computing resource sharing mechanism is proposed to facilitate the resource trading between the cloud computing service provider (CCP) and different edge computing service providers (ECPs). Then we derive the optimal pricing and allocation strategies of cloud computing resource based on the replicator dynamics of users' service selection. These strategies can promise the maximum integral utilities to all computing service providers (CPs), meanwhile the user distribution can reach the evolutionary stable state at this Stackelberg equilibrium. Furthermore, simulation results validate the performance of the designed resource sharing mechanism, and reveal the convergence and equilibrium states of user selection, and computing resource pricing and allocation.

**Keywords** Edge/Cloud computing · Software-defined Networking (SDN) · Resource Pricing and Allocation · Evolutionary Game · Stackelberg Differential Game

## 10.1  Introduction

Recently, computation-heavy applications are experiencing a dramatic increasing over the Fifth Generation (5G) and future wireless networks. There is evidence that such applications, including mining process for Proof-of-Work (PoW) in blockchain, interactive gaming, virtual reality, video services, etc., have become premier drivers of the exponential computing task growth [1–3]. To handle such increasing computing requirements, hybrid edge and cloud computing (ECC) systems have been expected to provide low-latency and on-demand computing services to users [4–6]. In ECC systems, cloud computing, as the traditional solution of computation offloading for user devices, is usually implemented at cloud nodes physically located far from users, which results in a long latency service response. Aiming at this problem, edge computing has been proposed as the complement of cloud computing by enabling users to upload computational tasks to the edge of networks [7, 8], which can eliminate the latency and enhance the reliability of services. However, with the growing amount of computational task requirements, computational power limited edge servers might be overwhelmed with severe performance degradation. A feasible solution for this problem is forwarding these tasks at edge nodes to the remote cloud center [9, 10], which can be considered as computation offloading between edge computing service providers (ECPs) and the cloud computing service provider (CCP). Therefore, to achieve the optimal and stable performance of CCP systems, an efficient cloud computing resource sharing mechanism plays an important role resulting from the constrained resource equipped by the CCP and time-varying user requirements among the CCP system. In addition, such mechanism is more challenging when the dynamic service subscription of users is taken into account [11]. This work will establish a hybrid ECC system, in which users can upload their computational tasks to nearby ECPs or the remote CCP dynamically. In addition, by considering the dynamic service subscription of users among the CCP and ECPs, this work will focus on the computing resource sharing and computation offloading mechanism design in the ECC system to realize an efficient utilization of computing resource and satisfy the service requirements of users.

As mentioned previously, the mobility and time varying service selection of users may bring difficulties to efficient resource sharing mechanism designs. In addition, there always exist bidirectional data interactions, including service subscription, task uploading, service response, etc., between end users and computing servers located at either the CCP or ECPs in the ECC system. These frequent interactions may lead to congestions at different computing providers (CPs) [12]. To solve these problems, an appropriate network architecture is necessary to realize an effective management of the hybrid ECC system. In recent years, software-defined networking (SDN) has been considered as an advanced network architecture to achieve flexible resource management and system performance control [13, 14], which can mitigate challenges above. Moreover, taking advantage of the available and accurate information of global system status collected by the SDN controller, the system can make optimal decisions to improve resource utilization and service quality [15]. On the other hand, latency problems, fault and Disruption tolerance, and scalability issues brought by the SDN-based fully centralized control architecture can be well solved by the integrated cloud and edge computing mechanism. Therefore, in this work, an SDN-based architecture will be established for computing resource sharing and computation offloading in the ECC system. With a centralized controller, SDN will help CPs to dynamically adjust the resource sharing and computation offloading strategies, which can match time-varying demands of users by observing their dynamic service selection.

Considering that the SDN-based fully centralized control architecture established in Sect. 10.3 will suffer from latency problems, fault and Disruption tolerance, and scalability issues, this section will introduce an ECC system to realize

Main contributions of this part are summarized as follows.

1. We establish an SDN-based architecture for computing resource sharing in the ECC system. Taking advantage of SDN controllers which separate the distributed infrastructure and resource management, the dynamic optimal pricing and allocation strategies can be obtained.
2. We design a Stackelberg differential game based cloud computing resource sharing, which determines the optimal resource pricing and allocation/request strategies dynamically. Then an open-loop Stackelberg equilibrium is derived as the optimal solution. Comparing with traditional static strategies, the proposed mechanism can achieve higher integral utilities in a time horizon and faster convergence speed of decision making.
3. We propose a hierarchical dynamic game framework composed of evolutionary game in the user layer and Stackelberg differential game in the edge and cloud layer, which can incentive the cooperation of cloud computing resource sharing. Different from the traditional separated control outside the user layer, this work considers the dynamic service selections of users among edge and cloud resource. Based on this consideration, the user service requirements can be satisfied as well as the edge/cloud computing resource can be utilized efficiently.
4. We analyze the performance of the designed computing resource sharing mechanism based on the hierarchical dynamic game. Specifically, the existence and uniqueness of equilibrium of user selections, as well as their evolutionary stable states, are analyzed. In addition, the optimal dynamic pricing and allocation of cloud computing resource are derived based on the replicator dynamics of users' service selection. Furthermore, simulation results validate the performance of the designed resource sharing mechanism, and reveal the convergence and stable states of user selection, resource pricing and resource allocation.

The roadmap of this thesis is as follows. In Sect. 10.2, we revisit the related works. The SDN-based architecture for the ECC system is established in Sect. 10.3. Section 10.4 presents the system model and proposed hierarchical game framework. An evolutionary game for service selection of user devices is designed in Sect. 10.5, and the Stackelberg differential game based computing resource pricing and allocation schemes are proposed in Sect. 10.6. Simulations are shown in Sect. 10.7, and conclusions are drawn in Sect. 10.8.

## 10.2   Related Works

For the integrated ECC system, the computation offloading mechanism plays a crucial role in improving resource utilization and service quality. Such computation offloading involves two aspects. Specifically, in the aspect of users, both the CCP and ECPs can offload users' computational tasks with different processing latency and transfer latency. On the other hand, resulting from the limited computational power equipped, ECPs are not qualified for providing services of heavy-computation tasks processing. Then ECPs have to forward some of these tasks to the remote CCP, which has powerful and dedicated computing resource and can provide services on demand. Such process above can be also considered as computation offloading between the CCP and ECPs. According to such two-layered resource sharing among different CPs and users, how to allocate cloud computing resource among ECPs and users selecting the CCP will influence the resource utilization and service quality significantly.

Driven by the supply and demand of computing resource among the CCP, ECPs and users, the resource trading can be formed and facilitated, which needs to satisfy the demands of users selecting different CPs, and meanwhile maximize the utility of each CP. For these purposes, many researches have focused on effect and efficient resource allocation and sharing mechanisms in edge/cloud systems, by introducing different economic models based on auction [16, 17], contract [18, 19], Stackelberg game [20–22], etc.. Among these studies, auction and contract based trading mechanisms were designed to motivate participants to report their service requirements or capacities truthfully, which can deal with trustworthiness and information asymmetric issues in the system. On the other hand, Stackelberg game provides a suitable framework to model the interactions of trading strategies made in supply and demand sides, including resource pricing, requests and proving for communications [23], storage [24], energy [25], etc., which can facilitate the resource trading efficiently and dynamically. In [20], a multi-leader multi-follower Stackelberg game was studied to provide cost-effective migrations of data centers in edge-cloud environment. To optimize resource allocation of all cloud and fog computing nodes, a Stackelberg game was formulated in [21], in which fog computing relied on a set of low-power fog nodes that were located close to the end users to offload the services originally targeting at cloud centers. A two-stage Stackelberg game was introduced into the blockchain consensus process in order to incentive the cooperation between the edge/cloud providers and the miners in a PoW-based blockchain network [22]. Similarly, in some current studies, Stackelberg game frameworks were also formulated to model the interaction between the edge/cloud nodes and users [26, 27]. However, all these studies above only considered the computing resource trading between users and CPs, or between the CCP and ECPs, while interactions and influences among the three levels were hardly investigated. In fact, users' service subscription will impact computation offloading between the CCP and ECPs. In addition, computing service qualities received by users selecting different CPs will vary with different resource sharing strategies made by CPs. Then users will change their selection strategies for better services, considering that users are rational. It is difficult to model and analyze these interactions above, since that the strategies made by the three parties will impact and be impacted by each other. To solve such interactive issues, this work will establish an evolutionary game based model to analyze the users' dynamic service selection among CPs. In addition, we will propose a Stackelberg differential game based cloud computing resource sharing mechanism, which will dynamically determine the optimal resource pricing and allocation/request strategies for the CCP and ECPs. In this Stackelberg differential game, the differential equation is introduced based on the replicator dynamics of user selections, which can establish the connection between evolutionary game operated among users and Stackelberg differential game operated among CPs. According to such hierarchical control and optimization, computing resource utilization and user service quality can be both improved.

As mentioned previously, dynamic user selections will bring challenges to the optimization of resource sharing. Such joint optimization for users and hetero-geneous CPs can be implemented and programmed efficiently by an SDN-based architecture. The SDN-based architecture design for the ECC systems has attracted researchers' great attention, especially in the 5G heterogeneous environment and various Internet of Things (IoT) applications [28–30]. To realize efficient and secure resource management, data processing and access control, different SDN-based architectures have been investigated. In [31], authors introduced a tunnel-less SDN scheme for scalable realization of virtual tenant networks across the 5G heterogeneous infrastructure, which could support migrations of software instances among geo-distributed computing resources. To meet requirements of various applications and improve the end-to-end system performance efficiently, a novel integrated framework including SDN, computing, and caching was designed in [32]. In [33], the cooperation among edge computing nodes was investigated, and their interactions were realized by establishing an SDN related mechanism. An SDN-based control scheme was designed in [34] for a multi-edge-cloud environment involves huge data migrations to realize an efficient traffic flow scheduling. In addi-tion, different SDN-based distributed and layered network architectures were also investigated to operate edge/cloud computing systems with blockchain techniques, which can deal with problems brought by limited bandwidth, high latency, large volume of data, and real-time analysis requirements [35, 36]. In summary, leverag-ing the SDN-based architecture, flexible management of heterogonous resource and optimal control of system performance can be implemented to support computation-heavy applications. Therefore, this work will design an SDN-based architecture to optimize the computing resource utilization by considering the dynamic users' service subscription.

## 10.3  SDN Architecture Design for Edge/Cloud Computing Systems

This section will propose a model of layered edge/cloud computing service providing system which takes advantage of SDN paradigms, and show that how to implement such cross-layer computing service for users in the designed SDN-based architecture in detail. The SDN-based architecture, which consists of three levels, i.e., the infrastructure plane, control plane and management plane, is established based on the infrastructure in 5G wireless heterogeneous networks (HetNets), as shown in Fig. 10.1. According to such SDN-based management, CPs will provide edge and cloud computing services to users. In this architecture, the cloud computing center and edge computing nodes are operated by the CCP and different ECPs, respectively. These CPs constitute the infrastructure plane and take charge of providing computing services to different types of user devices. In

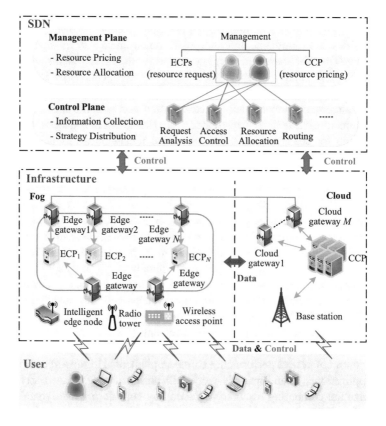

**Fig. 10.1** Architecture of SDN-based resource pricing, sharing and user scheduling in an ECC system

addition, we consider a user layer outside the SDN architecture, in which user devices such as mobile phones, tablets, etc., can receive computing services by subscribing to different CPs, according to their computational tasks' requirements. Such architecture can realize a dynamic and real-time information collection of task requirements and system workload, and decision making of computation offloading. Next, we will design the function and operation of the three planes in order to satisfy the computing requirements of users, and meanwhile manage the computing resource sharing and providing efficiently.

### 10.3.1   Infrastructure Plane

In the infrastructure plane, the CCP and ECPs provide remote and edge computing services for users, respectively, who can access these CCP and ECPs via edge nodes, such as wireless access points, radio towers, and macro-cell base station in 5G HetNets. In addition, surveillance cameras and servers at the edge of Radio Access Networks (RANs) can be also operated as edge servers to provide ubiquitous computing services. In current designed mobile computing systems, edge nodes can connect to different ECPs through edge gateways and Device-to-Device (D2D) communication without effecting the backhaul network [12]. By subscribing to these ECPs, users can receive fast response with respect to computing services. Alternatively, users can also select the remote CCP, who usually possesses richer computational power to provide higher-speed computing services than ECPs.

Considering the limited computational power of ECPs, the CCP can share parts of its computational power with ECPs through wireline connections between the edge gateways and cloud gateways. Resulting from the dynamic service selection of users, how to allocate the cloud computing resource among ECPs and users selecting the CCP will influence the efficiency of ECC system. Therefore, an efficient and dynamic control on computing sharing between the CCP and ECPs plays an important role on optimizing the resource utilization and satisfying the resource requirements of users with fast response.

### 10.3.2   Control Plane

As shown in Fig. 10.1, the SDN-based architecture separates computing resource management from the infrastructure, which forms a hierarchical game based market of cloud computing resource in the control plane. In this plane, information collection of users' subscriptions and strategy distribution for CPs in the infrastructure plane will be implemented through the data information interaction between the infrastructure plane and control plane.

#### 10.3.2.1   Information Collection

Through the controller of request analysis, the control plane collects the information of users' subscriptions among different CPs, as well as the local computational power of each CP, and then sends such received information to the management plane in real time. Specifically, the controllers of request analysis and access control are able to communicate with the CCP and ECPs through access points, and then call for the time-varying number of users selecting the corresponding CP.

#### 10.3.2.2   Strategy Distribution

The control plane receives pricing and request strategies of cloud computing resource made by the upper management plane, and then distributes these strategies to the CCP and ECPs through the controller of resource allocation. In addition, the access controller and core controller are responsible for resource sharing of cloud resource and access control between edge gateways and cloud gateways, i.e., setting approach network paths between gateways, managing computation offloading and so on.

### 10.3.3   Management Plane

After receiving the information of users' subscription and computational power equipped at each CP, the optimal pricing and request strategies of cloud resource will be determined at the management plane. To be specific, the management will help the CCP to make decisions on dynamic resource pricing, meanwhile help ECPs to determine how much computational power should be requested. Then these decisions will be returned back to the control plane, and then guide the cloud computing resource sharing between the CCP and ECPs. Such service response above can be implemented fast considering the wire connections between SDN controllers and edge and cloud gateways, and the powerful processing capacities of SDN servers.

## 10.4   System Model and Hierarchical Game Framework

Considering that the SDN-based fully centralized control architecture established in Sect. 10.3 will suffer from latency problems, scalability issues, etc., this section will introduce a hierarchical game based computation offloading mechanism to realize real-time and latency-sensitive services close to users.

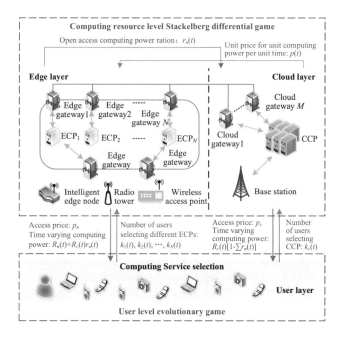

**Fig. 10.2** Hierarchical game based resource pricing and sharing in ECC systems

## 10.4.1   System Model

Consider an ECC system with a finite set $\mathcal{N} = \{1, 2, \cdots, N\}$ ECPs overlaying with one remote CCP and providing edge computing service to user devices, including mobile phones, wearable devices, tablets, etc., as shown in Fig. 10.2. In this work, we consider that the CCP and ECPs can be operated by the mobile network operators (MNOs) or provide the computing service as third parties. These user devices can access and send computational tasks to different ECPs by communicating with edge access points, such as intelligent edge nodes, radio towers, etc., which can upload these computational tasks or service requests to ECPs through edge gateways. In addition, to fulfil some high-complex computational tasks, users can also request the cloud computing resource through base stations and edge access points, which will upload their computational tasks to the CCP. Accordingly, the CCP and ECPs will respond to these computing requests on demand. As compensation, each user needs to pay the CCP or ECPs for accessing fee. Denote $p_n$ as the price charged by ECP $n$ ($n \in \mathcal{N}$) and $p_c$ as the price charged by the CCP, which are fixed access fees per device per unit of time paid by users, considering current charging models set by mobile service operators.[1]

---

[1] Price $p_n$ and $p_c$ can be set by the MNO or the third-party service providers, depending on who operate these CCP and ECPs.

In this system, ECPs, performing as light-weight computing servers, might be deployed at base stations, wireless access points, etc., and then providing computing services with shortened latency. In addition, the CCP, which can promise powerful and stationary computing services, connects to the $N$ ECPs via cloud gateways in a wireline manner. When ECPs cannot fulfill the received computational tasks under their constrained computing resource, they have to forward some part of computational tasks to the remote CCP through wireline backhaul links. Such computation offloading means that ECPs needs to request the CCP for additional computational power. For the shared computing resource and possible energy cost resulting from resource allocation and information exchange for the CCP, ECPs need to pay the CCP with uniform price $p$ in monetized payment unit computational power per unit time. In this work, $p$ is a function of time and can be denoted by $p(t)$. Once announced by the CCP, $p(t)$ can be observed by all ECPs in the system through the SDN architecture.

After observing price $p(t)$, ECP $n$ $(n \in \mathcal{N})$ decides to request $r_n(t) \in [0, 1)$ proportion of CCPs computational power[2] during a continuous observation period $[0, T]$. Then the computing resource allocation state of the system can be described by vector $\mathbf{r} = [r_1, r_2, \cdots, r_N]^T$. Let $r_c(t)$ denote the current remaining computational power level of the CCP at time $t$. Then we have $\sum_{n=1}^{N} r_n(t) + r_c(t) = 1$ for $t \in [0, T]$. In this resource transaction between the CCP and ECPs, the CCP, as the computing resource provider, needs to optimize unit price $p(t)$ to make its resource bring into maximal efficacy; on the other hand, the ECPs, who are the resource receivers and buyers, will make the optimal decision on how many resources to buy to create a tradeoff between its quality of service and cost.

Consider that the local computational power of ECP $n$ is $R_n$ for all $n \in \mathcal{N}$, the total computational power of CCP is $R_c$ $(R_c > R_n, \forall n \in \mathcal{N})$, and the number of all user devices in the system is denoted by $K$. In addition, let $k_n(t)$ and $k_c(t)$ be the numbers of user devices subscribing to ECP $n$ and the CCP at time $t$, respectively. Then $x_n(t) = k_n(t)/K$ $(x_n \in [0, 1])$ and $x_c(t) = k_c(t)/K$ represent the population share of ECP $n$ and the CCP accordingly, and we have $\sum_{n=1}^{N} x_n(t) + x_c(t) = 1$[3] Based on these definitions above, the computational power allocated to each user selecting ECP $n$ and the CCP can be calculated as

$$\omega_n(\mathbf{x}, \mathbf{r}) = \frac{R_n + R_c r_n(t)}{K x_n(t)}, n \in \mathcal{N}, \tag{10.1a}$$

$$\omega_c(\mathbf{x}, \mathbf{r}) = \frac{R_c \left(1 - \sum_{n=1}^{N} r_n(t)\right)}{K x_c(t)}, \tag{10.1b}$$

---

[2] In this work, the computational power is considered as the computing frequency or speed, which can be measured by the computing times per unit time. In blockchain applications, the unit of computational power can be defined as $H/s$.

[3] In this work, we only consider the users having the requirement of uploading their computational tasks to the CCP or ECPs. Therefore, there exist at least one computing service provider to be selected by users.

respectively, where vector $\mathbf{x} = [x_1, x_2, \cdots, x_N, x_c]^T$ is the population distribution state of the ECC system.

## 10.4.2   Hierarchical Game Framework

Based on the computing resource market model above, the service selection of user devices, pricing strategy of the CCP and computational power requests of ECPs are time-varying and interact with each other. To be specific, if too many users select the same CP, the received computational power for each of these users will decrease according to (10.1). Then these users tend to leave to other CPs with high average computational power. In addition, CPs (including the CCP and ECPs) with large number of user devices will expect to obtain much computational power to satisfy the computing requirements from accessing user devices. However, such resource request behavior also depends on and will further influence the computing resource price decided by the CCP. To model interactions among users, ECPs and the CCP analyzed above, this work designs a hierarchical dynamic game based scheme to improve the computing service quality and facilitate the computing resource sharing among CPs in the ECC system.

There are two levels in the hierarchical dynamic game designed for the computing service selection, pricing and sharing system, i.e., the user level and the computing resource level. In the user level, the behavior of users' dynamic service selection is formulated and analyzed through an evolutionary game model. Then to model the computing service providing and requirement between the CCP and ECPs, a Stackelberg differential game will be designed to optimize the pricing and sharing strategies for the limited computational power. Such hierarchical dynamic game framework is shown in Fig. 10.2.

### 10.4.2.1   Evolutionary Game in User Level

According to access price $p_c$ / $p_n$ released by the CCP or different ECPs, as well as the received computational power, each user makes its computing service selection among these CPs to improve its utility. For user devices, the access prices are fixed and stay the same over time. On the other hand, one can notice that the received computational power for users is time varying, which depends not only on the number of user devices accessing the same CP currently, but also the dynamic computing resource sharing of CCP among different CPs. Therefore, without complete information, users can hardly make the optimized service selection globally, i.e., among all CPs and over all time duration. As a response, each user device will learn and adapt its selection strategy gradually. To model this learning and adaptation process, an evolutionary game can be designed to describe and analyze the dynamic user behavior. Through replicator dynamics, all user devices in the ECC system will reach the same individual utility at the equilibrium [37, 38].

### 10.4.2.2   Stackelberg Differential Game in Resource Level

To respond to users' computational task requests on demand, computational power limited ECPs need to buy more computing resource from the CCP. In addition, to compensate the potential loss of accessing users and cost resulting from resource sharing, the CCP will charge corresponding ECP with time-varying unit price $p(t)$ for the provided computational power. Such pricing strategy is dynamic according to the dynamic computational power sharing/requests and number of accessing user devices at the CCP. Accordingly, given unit price $p(t)$, ECPs control their resource requests dynamically to maximize their own utilities. To analyze such dynamic computing resource pricing of CCP and dynamic resource requests of ECPs, this work establishes a non-cooperate Stackelberg differential game, in the two levels of which, all CPs optimize their own strategies to receive maximized utilities. In this Stackelberg game, the CCP performs as the leader and ECPs are followers. These players in the computing resource level makes their own optimal decisions dynamically according to the time-varying service selection decisions of user devices.

## 10.5   Evolutionary Game for Service Selection of User Devices

In the ECC system, online users with heavy computational tasks will compete for the limited computing resource by selecting and accessing different CPs. Initially, every user selects a candidate CP randomly or by experience. Then to achieve a better service quality, i.e., large computational power and/or low access price, these users will adapt their selection decisions periodically based on the dynamic received computational power, access price, and the population distribution of all users among different CPs. During this adaptation process, users cannot optimize their selection strategies globally, resulting from the asymmetry of information. Therefore, to improve their utilities, users will learn by imitating those selection strategies with high utilities gradually.

   Evolutionary game can be expected as a suitable tool for modeling such learning and imitating process. Thus this section will first formulate the evolutionary game framework and evolutionary strategy adaptation for user selection dynamics. Then the evolutionary equilibrium and evolutionary stable state (ESS) will be investigated for the established model.

### 10.5.1  Evolutionary Game Based Service Selection

We first formulate the computing service selection of users among different CPs as an evolutionary game model.

#### 10.5.1.1  Players

The set of $K$ user devices in the service area are the players of the evolutionary game.

#### 10.5.1.2  Strategy

$\mathscr{S} = \{1, 2, \cdots, n, \cdots, N, c\}$, where $n \in \mathscr{N}$ indicates selecting ECP $n$ for computing service, and $s = c$ means selecting the CCP directly.

#### 10.5.1.3  Population States

The population shares of all ECPs constitute the population distribution state denoted by vector $\mathbf{x} = [x_1, x_2, \cdots, x_N, x_c]^T \in \mathbb{X}$, where $\mathbb{X}$ represents the state space which contains all possible population distributions.

#### 10.5.1.4  Utility

Given computing resource allocation state $\mathbf{r} = [r_1, r_2, \cdots, r_N]^T$ and population distribution state $\mathbf{x} = [x_1, x_2, \cdots, x_N, x_c]^T$, the utility function of user device selecting ECP $n$ and the CCP are defined by

$$\pi_n(n, \mathbf{x}, \mathbf{r}) = \frac{\beta \omega_n(\mathbf{x}, \mathbf{r})}{p_n} = \frac{\beta(R_n + R_c r_n(t))}{K p_n x_n(t)}, \tag{10.2a}$$

$$\pi_c(c, \mathbf{x}, \mathbf{r}) = \frac{\beta \omega_c(\mathbf{x}, \mathbf{r})}{p_c} = \frac{\beta R_c \left(1 - \sum_{n=1}^{N} r_n(t)\right)}{K p_c x_c(t)}, \tag{10.2b}$$

respectively, $\beta > 0$ is a constant denoting the mapping factor.

#### 10.5.1.5  Replicator Dynamic

The replicator dynamic reflects the evolutionary behavior of the population among different strategies, i.e., selecting different CPs over time. In this work, we introduce the definition in [37, 39, 40], and then give the replicator dynamic as follows,

$$\dot{x}_n (t) = \delta x_n (t) \left[ \pi (n, \mathbf{x} (t), \mathbf{r} (t)) - \pi (\mathbf{x} (t), \mathbf{x} (t), \mathbf{r} (t)) \right], n \in \mathcal{N}; \quad (10.3a)$$

$$\dot{x}_c (t) = \delta x_c (t) \left[ \pi (c, \mathbf{x} (t), \mathbf{r} (t)) - \pi (\mathbf{x} (t), \mathbf{x} (t), \mathbf{r} (t)) \right], \quad (10.3b)$$

with initial population distribution state $\mathbf{x} (0) = \mathbf{x}_0 \in \mathbb{X}$ ($\forall n \in \mathcal{S}$), where constant $\delta > 0$ is the learning rate of the population which controls the frequency of strategy adaptation for service selection. Moreover, in (10.3),

$$\pi (\mathbf{x} (t), \mathbf{x} (t), \mathbf{r} (t)) = \sum_{n=1}^{N} x_n (t) \pi (n, \mathbf{x} (t), \mathbf{r} (t)) + x_c (t) \pi (c, \mathbf{x} (t), \mathbf{r} (t))$$

$$(10.4)$$

is the expected utility of the population given population distribution state $\mathbf{x} (t)$ and computing resource allocation state $\mathbf{r} (t)$. Based on the definitions above, we have

$$\dot{x}_n (t) = \frac{\delta \beta}{K} \left[ \frac{R_n + R_c r_n (t)}{p_n} \right.$$

$$\left. -x_n (t) \left( \sum_{m=1}^{N} \frac{R_m + R_c r_m (t)}{p_m} + \frac{R_c \left( 1 - \sum_{m=1}^{N} r_m (t) \right)}{p_c} \right) \right], \quad (10.5a)$$

$$\dot{x}_c (t) = \frac{\delta \beta}{K} \left[ \left( \frac{R_c}{p_c} - \sum_{m=1}^{N} \frac{R_c r_m (t)}{p_c} \right) \right.$$

$$\left. -x_c (t) \left( \sum_{m=1}^{N} \frac{R_m + R_c r_m (t)}{p_m} + \frac{R_c \left( 1 - \sum_{m=1}^{N} r_m (t) \right)}{p_c} \right) \right]. \quad (10.5b)$$

According to this replicator dynamics defined above, the number of user devices selecting ECP $n$ will increase when $\pi (n, \mathbf{x} (t), \mathbf{r} (t)) > \pi (\mathbf{x} (t), \mathbf{x} (t), \mathbf{r} (t))$, and vice versa.

## 10.5.2  Existence and Uniqueness of Equilibrium

According to the population replicator dynamic formulated in (10.3)–(10.5) and the established hierarchical game framework, there exists interactions between decisions of computing service selection made by users and computing resource pricing/allocation controls of CPs. In other words, the evolution of population distribution state defined in (10.3) is controlled by the pricing strategy of the CCP and resource requests of ECPs. Next, Theorem 10.1 presents that under these controls, there exists the unique population distribution state $\mathbf{x}(t)$ that constitutes the solution of (10.5).

**Theorem 10.1** *Consider a dynamic service selection system with a fixed population. For the evolutionary behavior of users among different strategies defined as (10.3) with initial condition $\mathbf{x}(0) = \mathbf{x}_0$, if resource allocation vector $\mathbf{r}(t)$ is a vector of measurable functions on $[0, \infty)$, then there exists the unique population distribution state $\mathbf{x}(t)$ constitute the solution of (10.5) for all $t \in [0, \infty)$.*

***Proof*** Given population distribution state $\mathbf{x}(t)$ and computing resource allocation state $\mathbf{r}(t)$, let $f_n(\mathbf{x}(t), \mathbf{r}(t))$ be the right side of (10.3), i.e.,

$$f_n(\mathbf{x}(t), \mathbf{r}(t)) \triangleq \delta x_n(t) [\pi(n, \mathbf{x}(t), \mathbf{r}(t)) - \pi(\mathbf{x}(t), \mathbf{x}(t), \mathbf{r}(t))]. \quad (10.6)$$

Given a fixed $t$, the partial derivative of $f_n(\mathbf{x}(t), \mathbf{r}(t))$ with respect to $\mathbf{x}(t)$ is continuous. In addition, if $\mathbf{r}(t)$ is measurable on $[0, \infty)$, then $f_n(\mathbf{x}(t), \mathbf{r}(t))$ is also measurable for fixed $x_n(t)$ on the same interval. Furthermore, when given any closed bounded set $\Delta \in \mathbb{X}$ and closed interval $[a, b] \in [0, \infty)$, there always exists a positive $I$ to construct an integrable function [37, 41, 42] by

$$I_n(t) = \left| \frac{\delta \beta (R_n + R_c r_n(t))}{K p_n} \right| + I |\Theta|, \quad (10.7)$$

where

$$\Theta = \frac{\delta \beta}{K} \left[ \sum_{n=1}^{N} \frac{R_n + R_c r_n(t)}{p_n} + \frac{R_c \left(1 - \sum_{n=1}^{N} r_n(t)\right)}{p_c} \right]. \quad (10.8)$$

Obviously, it holds that $|f_n(\mathbf{x}(t), \mathbf{r}(t))| \leq I_n(t)$ and $|\partial f_n(\mathbf{x}(t), \mathbf{r}(t))/\partial x_n(t)| \leq I_n(t)$, for all $(\mathbf{x}, t) \in \Delta \times [a, b]$. Therefore, we have $|f_n(\mathbf{x}^*(t), \mathbf{r}(t)) - f_n(\mathbf{x}(t), \mathbf{r}(t))| = \Theta |\mathbf{x}^*(t) - \mathbf{x}(t)|$. Denote $\Theta_m = \max\{\Theta\}$, then we can further derive that

$$\left| f_n(\mathbf{x}^*(t), \mathbf{r}(t)) - f_n(\mathbf{x}(t), \mathbf{r}(t)) \right| \leq \Theta_m \left| \mathbf{x}^*(t) - \mathbf{x}(t) \right|, \quad (10.9)$$

which implies that $f_n(\mathbf{x}(t), \mathbf{r}(t))$ satisfies the global Lipschitz condition. According to the analysis above, we can conclude that the solution to this dynamical population evolutionary system under controls of the CCP and ECPs is unique and exists globally [37, 43–45]. This completes the proof of Theorem 10.1.

### 10.5.3   Analysis of Evolutionary Stable State (ESS)

Consider a situation where a small proportion of user devices switching to a different mixed strategy $\mathbf{y} \neq \mathbf{x}$. Then these user devices can be regarded as mutants of the population. Denote the size of these mutants by a normalized value $\varepsilon \in (0, 1)$. Then the population state after mutation can be given by $(1 - \varepsilon)\mathbf{x} + \varepsilon\mathbf{y}$ [37]. According to definitions above, we first give the definition of *Evolutionary Stable Strategy (ESS)* in Definition 10.1.

**Definition 10.1 (Evolutionary Stable Strategy)** A strategy $\mathbf{x}^*$ is an ESS, if $\forall \mathbf{x} \neq \mathbf{x}^*$, there exist some $\varepsilon_{\mathbf{x}} \in (0, 1)^4$ such that $\forall \varepsilon \in (0, \varepsilon_{\mathbf{x}})$, the following inequality holds.

$$\pi\left(\mathbf{x}^*, (1 - \varepsilon)\mathbf{x}^* + \varepsilon\mathbf{x}, \mathbf{r}\right) > \pi\left(\mathbf{x}, (1 - \varepsilon)\mathbf{x}^* + \varepsilon\mathbf{x}, \mathbf{r}\right), \qquad (10.10)$$

where $\pi(\mathbf{x}^*, (1 - \varepsilon)\mathbf{x}^* + \varepsilon\mathbf{x}, \mathbf{r})$ and $\pi(\mathbf{x}, (1 - \varepsilon)\mathbf{x}^* + \varepsilon\mathbf{x}, \mathbf{r})$ are the expected utilities of non-mutants and mutants, respectively.

Considering that the ESS is the best response to the evolutionary system, then an ESS is also a Nash Equilibrium (NE). In addition, the evolutionary stability of ESS provides a string refinement of the NE. Moreover, in the NE, a single user cannot benefit through deviating from the equilibrium strategy. On the contrary, the ESS can avoid the deviation behavior of a set of players. Next, we summarize that the evolutionary service selection of users presents the globally asymptotical stability converging to the ESS in Theorem 10.2.

**Theorem 10.2** *Consider a dynamic computing resource selection system with a fixed population. For the evolutionary behavior of the population among different strategies defined as (10.3) with any initial condition $\mathbf{x}(0) = \mathbf{x}_0$ $(x_n, x_c \in (0, 1)$, $\forall n \in \mathscr{S})$, the replicator dynamics for resource selection is globally asymptotically stable and converges to the ESS of game.*

**Proof** Considering $\mathbf{x} = [x_1, x_2, \cdots, x_N, x_c]^T$ and the replicator dynamic derived in (10.5), we have $\dot{\mathbf{x}} = \mathbf{\Pi}\mathbf{x} + \boldsymbol{\pi}_o$, where $\mathbf{\Pi}$ is a matrix with dimensional $(N + 1) \times (N + 1)$, and

$$\boldsymbol{\pi}_o = \left[ \frac{\delta\beta\,(R_1 + R_c r_1\,(t))}{K\,p_1}, \cdots, \frac{\delta\beta\,(R_N + R_c r_N\,(t))}{K\,p_N}, \right.$$
$$\left. \frac{\delta\beta R_c\left(1 - \sum_{m=1}^{N} r_m\,(t)\right)}{K\,p_c} \right]^T . \qquad (10.11)$$

---

[4] $\varepsilon_{\mathbf{x}}$ represents the maximum proportion of users selecting mutant strategies that can be resisted by the ESS. A large $\varepsilon_{\mathbf{x}}$ indicates that the ESS is robust.

Therefore, the characteristic function of (10.5) can be derived as $\det(\gamma \mathbf{I} - \boldsymbol{\Pi}) = (\gamma + \Theta)^{N+1} = 0$, where $\mathbf{I}$ is the identity matrix and $\Theta$ is determined by (10.8). In addition, (10.8) implies that $\Theta > 0$ for all resource allocation states $\mathbf{r}(t)$. Consequently, $\boldsymbol{\Pi}$ always has $N+1$ negative eigenvalues, which means that the replicator dynamics for service selection is globally asymptotically stable and converges to the ESS of evolutionary game. This completes the proof of Theorem 10.2.

## 10.6   Stackelberg Differential Game Based Dynamic Computational Power Pricing and Allocation

The CCP and ECPs need to make the optimal decisions on computational power pricing and the amount of computational power requests, respectively, considering the dynamic service selection of user devices $\mathbf{x}(t)$. For the CCP, decreasing price $p(t)$ might incentivize ECPs to request and buy more remote computing resource, which will increase the sharing of CCP resource. However, the user devices accessing the CCP might then leave for other ECPs since that the amount of their received computational power decreases. On the other hand, for ECPs, increasing the amount of computational power requests will improve the utilities obtained by user devices according to (10.2), which will attract more users' selection according to the replicator dynamics as (10.3). Then further increasing number of users assessing will reduce the utility obtained by each user device. To analyze this dynamic and interactive decision making problem and then facilitate the computing resource trading between hierarchical CPs, we formulate a Stackelberg differential game, in which the CCP and ECPs perform as the leader and followers, respectively. To search the optimal strategies, an open-loop Stackelberg equilibrium is analyzed as the solution of the game.

### 10.6.1   Formulation of Stackelberg Differential Game

As shown in Fig. 10.2, the single CCP and $N$ ECPs perform as the players of the Stackelberg game. Specifically, the CCP, as the game leader, first announces its unit computing resource price $p(t)$, according to which ECPs, who are the followers of the game, then make their responding decisions of resource requests $r_n(t)$. In this work, we assume that both the CCP and ECPs are rational so that they can make the best response to the system states and strategies of other players, and follow the strategies made by SDN controllers. In addition, in the established Stackelberg differential game, the CCP and ECPs are willing to optimize their integral utilities over the time horizon $[0, T]$, but not the current utilities, by dynamically controlling their pricing and request strategies, responsibility.

As the followers of Stackelberg game, ECPs can optimize the amount of requesting computational power when observing the unit price released by the CCP. However, the time-varying strategies of all ECPs during the time horizon cannot be observed by the CCP in the present moment. Therefore, this work considers that the CCP is able to learn and predict the expected best response of ECPs and then make its pricing strategy dynamically. Next, we formulate the maximization problems of integral utility for both the CCP and ECPs.

### 10.6.1.1  Maximization of Integral Utility for ECPs

Consider that the utility of each ECP is composed of economic profits and penalty of resource sharing performance. To be specific, by setting accessing price $p_n$ ($\forall n \in \mathcal{N}$), ECP $n$ obtains the revenue from users selecting to it, which is depends on the number of subscribed users, i.e., $K p_n x_n(t)$. In addition, when requesting the CCP for proportion of could computational power $r_n(t)$, ECP $n$ will be charged $R_c p(t) r_n(t)$ by the CCP. Moreover, the costs resulting from the mismatch between resource supply and demand are also taken account of by ECP $n$ when optimizing its resource request strategy $r_n(t)$, from the performance aspect. This mismatch can be modeled as the distance between the current computational power requirements from all subscribing user devices and the current total computing resource can be provided after receiving the CCP's sharing resource. In the follower layer of Stackelberg game, each ECP is trying to maximize its own profits while minimize the costs resulting from the mismatch between resource supply and demand. Consequently, the instantaneous utility of ECP $n$ can be given by

$$u_n(r_n(t), \mathbf{x}(t), p(t)) = \eta_1 p_n N x_n(t) - \eta_2 R_c p(t) r_n(t)$$
$$- \eta_3 [K \varphi x_n(t) - (R_n + R_c r_n(t))]^2, \quad (10.12)$$

where $\varphi > 0$ is defined as a nominal value of accessible computing rate for all user devices, and $\eta_1$, $\eta_2$ and $\eta_3$ are positive weight factors. In addition, the third term in (10.12) reflects the matching between the computing resource requirement and available service capacity.

According to the instantaneous utility function (10.12), the ECP utility depends not only on the received computational power shared by the CCP, but also the population distribution of user devices among different CPs. Therefore, given the pricing strategy of the CCP, the integral utility maximization problem for ECPs can be established as an optimal control problem subject to the population state of evolutionary game operated in the user-level, which is given by

$$\max_{r_n(t)} \ U_n^{\text{int}} \left( r_n(t), \mathbf{x}(t), p(t) \right)$$

$$= \int_0^T e^{-\rho t} \left[ \eta_1 p_n N x_n(t) - \eta_2 R_c p(t) r_n(t) \right. \tag{10.13a}$$

$$\left. - \eta_3 [K \varphi x_n(t) - (R_n + R_c r_n(t))]^2 \right] dt; $$

$$\text{s.t.} \ \dot{x}_n(t) = \delta x_n(t) \left[ \pi(n, \mathbf{x}(t), \mathbf{r}(t)) \right.$$
$$\left. - \pi(\mathbf{x}(t), \mathbf{x}(t), \mathbf{r}(t)) \right], \forall n \in \mathcal{N}, \tag{10.13b}$$

$$\dot{x}_c(t) = \delta x_c(t) \left[ \pi(n, \mathbf{x}(t), \mathbf{r}(t)) \right.$$
$$\left. - \pi(\mathbf{x}(t), \mathbf{x}(t), \mathbf{r}(t)) \right], \tag{10.13c}$$

$$\mathbf{x}(0) = \mathbf{x}_0, \tag{10.13d}$$

$$r_n(t) \in \mathcal{R}, \forall n \in \mathcal{N}. \tag{10.13e}$$

In (10.13), $\rho > 0$ denotes the discount rate influencing the discount value of future utilities.

### 10.6.1.2   Maximization of Integral Utility for CCP

Similarly, the CCP optimizes its pricing strategy to maximize the profits paid by the subscribed users and the ECPs receiving the CCP's computational power, while minimize the costs resulting from the performance discrepancy. Then we have the instantaneous utility of CCP as follows,

$$u_c(p(t), r_c(t), \mathbf{r}(t)) = \xi_1 p_c N x_c(t) + \xi_2 R_c \sum_{n=1}^{N} p(t) r_n(t)$$

$$- \xi_3 \left[ K \varphi x_c(t) - R_c \left( 1 - \sum_{n=1}^{N} r_n(t) \right) \right]^2, \tag{10.14}$$

where $\xi_1$, $\xi_2$ and $\xi_3$ are positive weight factors. Therefore, the integral utility maximization problem for the CCP can be also established as an optimal control problem subject to the population state of evolutionary game operated in the user-level, which can be formulated as

$$
\max_{p(t)} \quad U_c^{\text{int}}\left(p\left(t\right), \mathbf{x}\left(t\right), \mathbf{r}\left(t\right)\right)
$$

$$
= \int_0^T e^{-\rho t} \left[ \xi_1 p_c N x_c\left(t\right) + \xi_2 R_c \sum_{n=1}^N p\left(t\right) r_n\left(t\right) \right. \tag{10.15a}
$$

$$
\left. - \eta_3 [K \varphi x_n\left(t\right) - (R_n + R_c r_n\left(t\right))]^2 \right] dt;
$$

$$
\text{s.t.} \quad \dot{x}_n\left(t\right) = \delta x_n\left(t\right) \left[ \pi\left(n, \mathbf{x}\left(t\right), \mathbf{r}\left(t\right)\right) \right.
$$
$$
\left. - \pi\left(\mathbf{x}\left(t\right), \mathbf{x}\left(t\right), \mathbf{r}\left(t\right)\right) \right], \forall n \in \mathcal{N}, \tag{10.15b}
$$

$$
\dot{x}_c\left(t\right) = \delta x_c\left(t\right) \left[ \pi\left(n, \mathbf{x}\left(t\right), \mathbf{r}\left(t\right)\right) \right.
$$
$$
\left. - \pi\left(\mathbf{x}\left(t\right), \mathbf{x}\left(t\right), \mathbf{r}\left(t\right)\right) \right], \tag{10.15c}
$$

$$
\mathbf{x}\left(0\right) = \mathbf{x}_0, \tag{10.15d}
$$

$$
p\left(t\right) \in \mathcal{R}. \tag{10.15e}
$$

## 10.6.2   Open-Loop Stackelberg Equilibrium Solutions

In this part, we will analyze the open-loop solutions to the optimal computing resource pricing and requesting problems established above in (10.15) and (10.13) for the CCP and ECPs, respectively. For these optimization problems, if the CCP and ECPs choose to commit their strategies from outset, their information structure can be seen as an open-loop pattern, and their strategies become functions of the initial state $\mathbf{x}_0$, $\mathbf{r}_0$ and time $t$, for both the CCP and ECPs. Considering the Stackelberg differential game operation, it needs to search the optimal solution for each ECP first for the given CCP's pricing strategy, and then the CCP can make the decision on the computing price based on solutions of resource request strategies. Next, we will first analyze the optimal resource request problem for each ECP (follower) in a finite time period $[0, T]$, and then the optimal pricing strategy for the CCP (leader) will be obtained based on the ECPs' strategies.

In a Stackelberg differential game, an open-loop Stackelberg equilibrium is regarded as the optimal solution [46, 47]. So we first introduce the definitions of optimal control strategies for the CCP and ECPs.

**Definition 10.2 (Optimal Control Strategy)** For the CCP, pricing strategy $p^*\left(t\right)$ is optimal if the following inequality holds for all feasible control paths $p\left(t\right) \neq p^*\left(t\right)$.

$$
U_c^{\text{int}}\left(p^*\left(t\right), \mathbf{x}\left(t\right), \mathbf{r}\left(t\right)\right) \geq U_c^{\text{int}}\left(p\left(t\right), \mathbf{x}\left(t\right), \mathbf{r}\left(t\right)\right). \tag{10.16}
$$

Similarly, for ECP $n$ ($\forall n \in \mathcal{N}$), the proportion of computational power request $r_n^*(t)$ is optimal if inequality (10.17) holds for all feasible control paths $r_n(t) \neq r_n^*(t)$.

$$U_n^{\text{int}}\left(r_n^*(t), \mathbf{x}(t), p(t)\right) \geq U_n^{\text{int}}\left(r_n(t), \mathbf{x}(t), p(t)\right). \tag{10.17}$$

Based on the definition of the optimal control strategy, we give the definition of open-loop Stackelberg game equilibrium in Definition 10.3.

**Definition 10.3 (Open-Loop Stackelberg Equilibrium)** Strategy profile $\Phi^*(t) \triangleq \{p^*(t), \mathbf{r}^*(t)\}$ constitutes an open-loop Stackelberg equilibrium if $p^*(t)$ and $\mathbf{r}^*(t)$ are the optimal control strategies for the CCP and ECPs, respectively, given others' strategies.

### 10.6.2.1   Open-Loop Stackelberg Equilibrium of ECPs

In order to get equilibrium solutions of the optimization problem formulated in (10.13), we need to establish the Hamiltonian system for each ECP. Then the open-loop equilibrium solutions of optimization problem can be characterized as the Pontryagin's Maximum Principle, which is the necessary conditions to find the candidate optimal strategies. First, we summarize the Pontryagin's Maximum Principle in Definition 10.4.

**Definition 10.4 (Pontryagin's Maximum Principle for ECPs)** A set of controls $\{r_n^*(t)\}$ constitutes an open-loop equilibrium to the optimization problem formulated in (10.13), and $\mathbf{x}_f^*(t)$ is the corresponding population distribution state trajectory, if there exists a set of costate functions

$$\mathbf{\Lambda}_n(t) = [\lambda_{n1}(t) \; \lambda_{n2}(t) \; \cdots \; \lambda_{nm}(t) \; \cdots \; \lambda_{nN}(t) \; \lambda_{nc}(t)] \tag{10.18}$$

such that the following relations are satisfied.[5]

$$r_n^*(t) = \arg\max_{r_n(t)} \left\{ u_n\left(r_n(t), \mathbf{x}^*(t), p(t)\right) + \mathbf{\Lambda}_n(t)\dot{\mathbf{x}}^*(t) \right\}, \tag{10.19a}$$

$$\dot{x}_n^*(t) = \delta x_n^*(t) \left[ \pi\left(n, \mathbf{x}^*(t), \mathbf{r}^*(t)\right) - \pi\left(\mathbf{x}^*(t), \mathbf{x}^*(t), \mathbf{r}^*(t)\right) \right], \tag{10.19b}$$

$$\dot{x}_c^*(t) = \delta x_c^*(t) \left[ \pi\left(c, \mathbf{x}^*(t), \mathbf{r}^*(t)\right) - \pi\left(\mathbf{x}^*(t), \mathbf{x}^*(t), \mathbf{r}^*(t)\right) \right], \tag{10.19c}$$

$$\mathbf{x}^*(0) = \mathbf{x}_0^*, \tag{10.19d}$$

$$\dot{\mathbf{\Lambda}}_n(t) = \rho\mathbf{\Lambda}_n(t) - \frac{\partial\left[u_n\left(r_n(t), \mathbf{x}(t), p(t)\right) + \mathbf{\Lambda}_n(t)\dot{\mathbf{x}}(t)\right]}{\partial\mathbf{x}^*(t)}. \tag{10.19e}$$

---

[5] Considering that $r_c(t) = 1 - \sum_{n=1}^{N} r_n(t)$, then element $\lambda_{nc}(t)$ in $\Lambda_n(t)$ can be eliminated.

In this system, the equilibrium solutions for ECPs are the solutions of the differential game. Therefore, these solutions also constitute the Stackelberg equilibrium for ECPs. Then we first introduce the Hamiltonian system for each ECP as follows. Based on the Pontryagin's Maximum Principle, the Hamiltonian system of ECP $n$ can be given by

$$
\begin{aligned}
& H_n\left(r_n(t), p(t), \mathbf{x}(t), \boldsymbol{\Lambda}_n(t), t\right) \\
& \triangleq u_n\left(r_n(t), \mathbf{x}(t), p(t)\right) + \boldsymbol{\Lambda}_n(t)\, \dot{\mathbf{x}}(t),
\end{aligned}
\tag{10.20}
$$

where costate function $\boldsymbol{\Lambda}_n(t)$ is a function associate with population state $\mathbf{x}(t)$, and is defined by (10.19e). In addition, each element of costate function $\boldsymbol{\Lambda}_n(t)$, i.e., $\boldsymbol{\Lambda}_{nm}(t)$, is the costate variable of ECP $n$ associated with state $x_m$. Based on the Hamiltonian function defined in (10.20), the corresponding maximized Hamiltonian function is defined as follows:

$$
\begin{aligned}
& H_n^*\left(\mathbf{x}(t), \boldsymbol{\Lambda}_n(t), t\right) \\
& \triangleq \max_{r_n(t)}\left\{H_n\left(r_n(t), p(t), \mathbf{x}(t), \boldsymbol{\Lambda}_n(t), t\right) \mid r_n(t) \in \mathscr{R}\right\}.
\end{aligned}
\tag{10.21}
$$

**Lemma 10.1** *The optimal computational power rate solutions for ECP* $n \in \mathcal{N}$ *is*

$$
r_n^*(t) = -\frac{\eta_2}{2\eta_3 R_c} p(t) + \frac{K\varphi x_n(t) - R_n}{R_c} + \frac{1}{2\eta_3 R_c}\frac{\delta\beta}{K}\boldsymbol{\Lambda}_n \mathbf{q}_n(\mathbf{x}),
\tag{10.22}
$$

*which also constitutes an open-loop Stackelberg equilibrium for ECP* $n$. *In (10.22), $\mathbf{q}_n(\mathbf{x})$ is an $N$-dimension vector which is given by*

$$
\mathbf{q}_n(\mathbf{x}) = \frac{1}{p_n}\mathbf{i}_n - \left(\frac{1}{p_n} - \frac{1}{p_c}\right)\left[x_1\; x_2\; \cdots\; x_N\right]^T,
\tag{10.23}
$$

*where $N$-dimension vector $\mathbf{i}_n$ is a standard basis, i.e., its $n$-th element is $1$ and other elements are $0$.*

***Proof*** According to the Pontryagin's Maximum Principle for ECPs, the optimal control strategy of optimization problem (10.13) must also maximize the corresponding Hamiltonian function. Therefore, all candidates' optimal strategies have to satisfy the following necessary optimality conditions:

$$
\frac{\partial H_n\left(r_n(t), p(t), \mathbf{x}(t), \boldsymbol{\Lambda}_n(t), t\right)}{\partial r_n(t)} = 0.
\tag{10.24}
$$

Then plug (10.2) into (10.24), and the optimal computing resource request can be deduced as

$$
r_n^*(t) = -\frac{\eta_2}{2\eta_3 R_c} p(t) + \frac{K\varphi x_n(t) - R_n}{R_c} + \frac{1}{2\eta_3 R_c}\frac{\delta\beta}{K}\boldsymbol{\Lambda}_n \mathbf{q}_n(\mathbf{x}).
\tag{10.25}
$$

Furthermore, according to (10.19e), we can calculate all elements of $\Lambda_n(t)$, which can be given by

$$\dot{\lambda}_{nm} = \lambda_{nm} (\rho + \Theta(\mathbf{r}(t))), \ m \neq n, \tag{10.26a}$$

$$\dot{\lambda}_{nn} = \lambda_{nn} (\rho + \Theta(\mathbf{r}(t))) - \eta_1 p_n K, \tag{10.26b}$$

where $\Theta(\mathbf{r}(t))$ is defined in (10.8). This completes the proof of Lemma 10.1.

According to the optimal solutions summarized in Lemma 10.1, we can observe that the optimal computational power requests and allocation for ECPs is an decreasing function of pricing $p(t)$ determined by the CCP.

### 10.6.2.2  Open-Loop Stackelberg Equilibrium of CCP

Similarly, we can obtain the open-loop equilibrium solutions of (10.15) for the CCP based on the dynamic optimal control. In particular, with the definition of optimal strategy for the CCP as (10.16) in Definition 10.2, the open-loop equilibrium solutions for the CCP can be characterized as the Pontryagin's Maximum Principle for CCP, as summarized in following Definition 10.5.

**Definition 10.5 (Pontryagin's Maximum Principle for CCP)** A set of controls $\{p^*(t)\}$ constitutes an open-loop equilibrium to the optimization problem formulated in (10.15), and $\mathbf{x}^*(t)$ is the corresponding population distribution state trajectory, if there exist costate functions $\mathbf{M}(t) = \begin{bmatrix} \mu_{c1}(t) \ \mu_{c2}(t) \cdots \mu_{cN}(t) \end{bmatrix}^T$ and $\boldsymbol{\Psi}(t) = \begin{bmatrix} \boldsymbol{\Psi}_1(t) \ \boldsymbol{\Psi}_2(t) \cdots \boldsymbol{\Psi}_N(t) \end{bmatrix}^T$ such that the following relations are satisfied.

$$p^*(t) = \arg\max_{\rho(t)} \{ H_c(p(t), \mathbf{x}(t), \mathbf{r}(t), \boldsymbol{\Lambda}(t), \mathbf{M}(t), \boldsymbol{\Psi}(t)) \}, \tag{10.27a}$$

$$\dot{x}_n^*(t) = \delta x_n^*(t) \left[ \pi\left(n, \mathbf{x}^*(t), \mathbf{r}^*(t)\right) - \pi\left(\mathbf{x}^*(t), \mathbf{x}^*(t), \mathbf{r}^*(t)\right) \right], \tag{10.27b}$$

$$\dot{x}_c^*(t) = \delta x_c^*(t) \left[ \pi\left(c, \mathbf{x}^*(t), \mathbf{r}^*(t)\right) - \pi\left(\mathbf{x}^*(t), \mathbf{x}^*(t), \mathbf{r}^*(t)\right) \right], \tag{10.27c}$$

$$\mathbf{x}^*(0) = \mathbf{x}_0^*, \tag{10.27d}$$

$$\dot{\mathbf{M}}(t) = \rho \mathbf{M}(t) - \frac{\partial H_c(p(t), \mathbf{x}(t), \mathbf{r}(t), \boldsymbol{\Lambda}(t), \mathbf{M}(t), \boldsymbol{\Psi}(t))}{\partial \mathbf{x}^*(t)}, \tag{10.27e}$$

$$\dot{\boldsymbol{\Psi}}_n(t) = \rho \boldsymbol{\Psi}_n(t) - \frac{\partial H_c(p(t), \mathbf{x}(t), \mathbf{r}(t), \boldsymbol{\Lambda}(t), \mathbf{M}(t), \boldsymbol{\Psi}(t))}{\partial \boldsymbol{\Lambda}_n(t)}, \tag{10.27f}$$

where the Hamiltonian function of the CCP is given by

$$H_c \left( p \left( t \right), \mathbf{x} \left( t \right), \mathbf{r} \left( t \right), \boldsymbol{\Lambda} \left( t \right), \mathbf{M} \left( t \right), \boldsymbol{\Psi} \left( t \right) \right)$$

$$= \xi_1 \, p_c K x_c \left( t \right) + \xi_2 R_c p \left( t \right) \sum_{n=1}^{N} r_n \left( t \right)$$

$$- \xi_3 \left[ K \varphi x_c \left( t \right) - R_c \left( 1 - \sum_{n=1}^{N} r_n \left( t \right) \right) \right]^2 \tag{10.28}$$

$$+ \sum_{n=1}^{N} \mu_{cn} \left( t \right) \dot{x}_n \left( t \right) + \sum_{n=1}^{N} \left( \sum_{m=1}^{N} \theta_{nm} \left( t \right) \dot{\lambda}_{nm} \left( t \right) \right),$$

$\boldsymbol{\Lambda} \left( t \right) = [\boldsymbol{\Lambda}_1 \left( t \right) \boldsymbol{\Lambda}_2 \left( t \right) \cdots \boldsymbol{\Lambda}_N \left( t \right)]^T$ determined by (10.19e), $\mathbf{M} \left( t \right) = [\mu_{c1} \left( t \right) \mu_{c2} \left( t \right) \cdots \mu_{cN} \left( t \right)]^T$ and $\boldsymbol{\Psi} \left( t \right) = [\boldsymbol{\Psi}_1 \left( t \right) \boldsymbol{\Psi}_2 \left( t \right) \cdots \boldsymbol{\Psi}_N \left( t \right)]^T$, where $\boldsymbol{\Psi}_n \left( t \right) = [\theta_{n1} \left( t \right) \theta_{n2} \left( t \right) \cdots \theta_{nm} \left( t \right) \cdots \theta_{nN} \left( t \right)]^T$, are costate functions for the CCP.

By solving optimization problem (10.27a) based on Hamiltonian function (10.28), we provide the optimal pricing strategy in Lemma 10.2

**Lemma 10.2** *The optimal computational power pricing solutions for the CCP is*

$$p^* \left( t \right) \triangleq f_p \left( \mathbf{x} \left( t \right), \boldsymbol{\Lambda} \left( t \right), \mathbf{M} \left( t \right), \boldsymbol{\Psi} \left( t \right), t \right), \tag{10.29}$$

*where*

$$f_p \left( \mathbf{x} \left( t \right), \boldsymbol{\Lambda} \left( t \right), \mathbf{M} \left( t \right), \boldsymbol{\Lambda} \left( t \right), t \right)$$

$$\triangleq \frac{1}{2NB \left( \xi_2 + \xi_3 R_c N B \right)} \left\{ \xi_2 \sum_{n=1}^{N} A_n \right.$$

$$+ 2\xi_3 N B \left[ K \varphi \left( 1 - \sum_{n=1}^{N} x_n \right) - R_c \left( 1 - \sum_{n=1}^{N} A_n \right) \right] \tag{10.30}$$

$$+ \frac{\delta \beta B}{K} \sum_{n=1}^{N} \mu_{cn} \left[ -\frac{1}{p_n} - x_n \left( -\sum_{n=1}^{N} \frac{1}{p_n} + \frac{N}{p_c} \right) \right]$$

$$+ \frac{\delta \beta B}{K} \sum_{n=1}^{N} \sum_{m=1}^{N} \theta_{nm} \lambda_{nm} \left( -\sum_{n=1}^{N} \frac{1}{p_n} + \frac{N}{p_c} \right) \right\}.$$

*This optimal pricing $p^* \left( t \right)$ also constitutes an open-loop Stackelberg equilibrium for the CCP.*

**Proof** According to Lemma 10.1, optimal response $r_n^*(t)$ of ECP $n$ can be expressed as

$$r_n^*(t) = A_n(\mathbf{x}(t)) - Bp(x),\tag{10.31}$$

where

$$A_n(\mathbf{x}(t), \mathbf{\Lambda}_n(t)) = \frac{K\varphi x_n(t) - R_n}{R_c} + \frac{1}{2\eta_3 R_c}\frac{\delta\beta}{K}\mathbf{\Lambda}_n(t)\mathbf{q}_n(\mathbf{x}),\tag{10.32a}$$

$$B = \frac{\eta_2}{2\eta_3 R_c}.\tag{10.32b}$$

As assumed previously, the CCP can learn and predict the optimal response $r_n^*(t)$ of ECP $n$, $\forall n \in \mathcal{N}$. Therefore, plugging (10.31) into the Hamiltonian function of the CCP (10.28), then the Hamiltonian function of the CCP become a concave function with respect to $p(t)$. Thus the optimal pricing strategy $p^*(t)$ is unique for the CCP, which has to satisfy the following necessary optimality conditions

$$\frac{\partial H_c(p(t), \mathbf{x}(t), \mathbf{r}^*(t), \mathbf{\Lambda}(t), \mathbf{M}(t), \mathbf{\Psi}(t), t)}{\partial p(t)}$$

$$\triangleq \frac{\partial H_c(p(t), \mathbf{x}(t), \mathbf{\Lambda}(t), \mathbf{M}(t), \mathbf{\Psi}(t), t)}{\partial p(t)} = 0.\tag{10.33}$$

Taking the first derivative of $H_c(t)$ with respect to $p(t)$ and then we have

$$\left(2N\xi_2 B + 2\xi_3 R_c N^2 B^2\right) p^*(t)$$

$$= \xi_2 \sum_{n=1}^{N} A_n + 2\xi_3 N B \left[ K\varphi\left(1 - \sum_{n=1}^{N} x_n\right) - R_c\left(1 - \sum_{n=1}^{N} A_n\right)\right]$$

$$+ \frac{\delta\beta B}{K} \sum_{n=1}^{N} \mu_{cn}\left[-\frac{1}{p_n} - x_n\left(-\sum_{n=1}^{N}\frac{1}{p_n} + \frac{N}{p_c}\right)\right]$$

$$+ \frac{\delta\beta B}{K} \sum_{n=1}^{N}\sum_{m=1}^{N} \theta_{nm}\lambda_{nm}\left(-\sum_{n=1}^{N}\frac{1}{p_n} + \frac{N}{p_c}\right).\tag{10.34}$$

Therefore, the optimal pricing strategy denoted by (10.29) and (10.30) can be obtained.

Furthermore, according to (10.27e) and (10.27f), we can calculate all elements of $\mathbf{M}(t)$ and $\boldsymbol{\Psi}(t)$, and then we obtain

$$\dot{\mu}_{cn} = \mu_{cn}\left(\rho + \Theta\left(\mathbf{r}(t)\right)\right) - \xi_1 p_c K, \ \forall n \in \mathcal{N}, \tag{10.35a}$$

$$\dot{\theta}_{nm} = \theta_{nm}\Theta\left(\mathbf{r}(t)\right), \ \forall n, m \in \mathcal{N}, \tag{10.35b}$$

where $\Theta\left(\mathbf{r}(t)\right)$ is defined through (10.8). This completes the proof of Lemma 10.2.

### 10.6.2.3  Open-Loop Stackelberg Equilibrium Solutions

According to the optimal resource pricing and allocation strategies described in Lemmas 10.1 and 10.2, $p^*(t)$ and $r_n^*(n)$ ($\forall n \in \mathcal{N}$) can be denoted by

$$p^*(t) \triangleq f_p\left(\mathbf{x}(t), \boldsymbol{\Lambda}(t), \mathbf{M}(t), \boldsymbol{\Psi}(t), t\right), \tag{10.36a}$$

$$r_n^*(t) \triangleq f_r\left(\mathbf{x}(t), \boldsymbol{\Lambda}(t), \mathbf{M}(t), \boldsymbol{\Psi}(t), t\right), \tag{10.36b}$$

$$r_c^*(t) = 1 - \sum_{n=1}^{N} r_n^*(t). \tag{10.36c}$$

Then substituting (10.36) into (10.2), (10.26), (10.35a) and (10.35b), and a dynamic control system composed of population distribution state $\mathbf{x}(t)$ and all costate variables $\boldsymbol{\Lambda}_n(t)$, $\mathbf{M}(t)$ and $\boldsymbol{\Psi}(t)$ can be provided as follows.

$$\mathbf{x}^*(t) \triangleq f_x\left(\mathbf{x}(t), \boldsymbol{\Lambda}(t), \mathbf{M}(t), \boldsymbol{\Psi}(t), t\right), \tag{10.37a}$$

$$\boldsymbol{\Lambda}_n^*(t) \triangleq f_{\Lambda,n}\left(\mathbf{x}(t), \boldsymbol{\Lambda}(t), \mathbf{M}(t), \boldsymbol{\Psi}(t), t\right), \tag{10.37b}$$

$$\mathbf{M}^*(t) \triangleq f_M\left(\mathbf{x}(t), \boldsymbol{\Lambda}(t), \mathbf{M}(t), \boldsymbol{\Psi}(t), t\right), \tag{10.37c}$$

$$\boldsymbol{\Psi}^*(t) \triangleq f_{\Psi}\left(\mathbf{x}(t), \boldsymbol{\Lambda}(t), \mathbf{M}(t), \boldsymbol{\Psi}(t), t\right). \tag{10.37d}$$

The dynamic control system formulated above is a typical two-point boundary value problem (TPBVP) [48, 49]. By solving this problem, optimal controls $\mathbf{x}^*(t)$, $\boldsymbol{\Lambda}_n^*(t)$, $\mathbf{M}^*(t)$ and $\boldsymbol{\Psi}^*(t)$ can be obtained. Based on the optimal solutions of TPBVP, the open-loop Stackelberg game equilibrium $\Phi^*(t) \triangleq \{p^*(t), \mathbf{r}^*(t)\}$ can be further derived.

## 10.7  Simulation Results

In this part, we will analyze the service selection behavior based on the evolutionary game, and then use MATLAB2019b to evaluate the performance of proposed computing resource pricing and allocation mechanisms based on the Stackelberg differential game. First of all, we introduce the scenario setup of the simulations. In the following simulations, we assume a typical ECC system, in which there are a single CCP and multiple ECPs who can access the computing resource of the CCP. These CPs provide edge and cloud computing services to $K = 100$ user devices randomly distributed within the coverage of an ECC system.

### 10.7.1 Evolution of Population Distribution

For the numerical analysis, we first consider the situation of two ECPs, i.e., ECP 1 and ECP 2. The local available computational power of the two ECPs are set as $R_1 = 2$ kH/s and $R_2 = 1$ kH/s, and the fixed access prices of the two ECPs are given by $p_1 = 0.3$ and $p_2 = 0.2$, respectively [37]. In addition, the initial population distribution state is set as $\mathbf{x}_0 = [x_1(t), x_2(t), x_c(t)] = [0.3, 0.3, 0.4]$, and the initial computing resource request state of ECPs is set as $\mathbf{r}_0 = [r_1(0), r_2(0)] = [0, 0]$, which means that each ECP serves its users with its own computing resource at the beginning of the time horizon. Consider different sharable cloud computational power of the CCP, i.e., $R_c = 5$ kH/s indicating a service quality with high-computational power and $R_c = 2$ kH/s indicating a service quality with low-computational power. Then the CCP fixes its access price $p_c$ by selecting values in $\{0.5 > \max\{p_1, p_2\}, 0.2 = \min\{p_1, p_2\}\}$, which can reflect different cost performance of CCP for users. Moreover, set the learning rate of users as $\delta = 1$.

Then we first investigate the dynamics of population distribution state and the evolution process of service selection from initial state $\mathbf{x}_0$, which is subject to the control of resource pricing and allocation strategies. Considering that the dynamic change of population distribution indicates the service selection adaptation of users, we record the population distribution state $\mathbf{x}(t) = [x_1(t), x_2(t), x_c(t)]$ over time, and the results of which are shown in Fig. 10.3. Results in Fig. 10.3 validate that the proportion of users selecting every CP converges to an equilibrium state at which there is no user willing to change its service selection strategy.

Then we analyze the influence of cloud computing capacity on user selection. As presented in Fig. 10.3a, when $R_c = 5$ kH/s, the CCP setting a lower access price ($p_c = 0.2$) tends to attract more users to select its cloud resource directly, meanwhile share less computing resource to ECPs, although it possesses a larger computing capacity. On the contrary, when setting a rather high access price ($p_c = 0.5$), the CCP will share all of its computing resource to ECPs and then drive its subscribed users away to ECPs. In this case, the utility of CCP mainly comes from its sharing resource to ECPs. Then we analyze the resource selection evolution trajectory when CCP is limited with computational power, i.e., $R_c = 2$ kH/s, which are shown in Fig. 10.3b. In this case, one can notice that the proportion of users selecting the CCP at equilibrium when $p_c = 0.2$ is larger than that when $p_c = 0.5$, which reflects the fact that the lower price will attract more users. Moreover, results in Fig. 10.3b also imply that when the CCP has limited computing resource, the

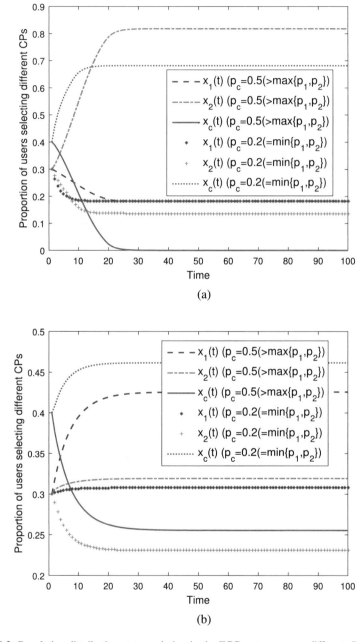

**Fig. 10.3** Population distribution state evolution in the ECC system versus different $R_c$ and $p_c$.
(a) $R_c = 5$. (b) $R_c = 2$

optimal pricing strategy for the CCP is reserving its resource to serve users directly by setting relatively higher unit price $p(t)$. These results can reveal the interaction and influence between the evolutionary game in the user layer and the Stackelberg game in computing resource layer, and validate the rational user behaviors and market rules.

We also test the population distribution dynamic when there are many ECPs in the ECC system. In particular, consider there are $N = 6$ different ECPs selecting the value of their computational power in $\{1, 2\}$ (kH/s) and the value of access price in $\{0.1, 0.2, 0.3\}$. For the CCP, set $R_c = 10$ kH/s and $p_c = 0.1$. In addition, the initial population distribution is set as $\mathbf{x} = [0.05, 0.1, 0.15, 0.05, 0.1, 0.15, 0.4]$. Then we get the evolution of population distribution, cloud resource allocation and user utilities, as shown in Fig. 10.4. As shown in Fig. 10.4a, one can observe that the proportions of users selecting ECPs with the same $R_n$ and $p_n$ simultaneously converge to the same equilibrium from different initial distributions. Moreover, results in Fig. 10.4a also present that the proportion of users selecting ECP 5 and ECP 6 are the highest among all ECPs, which indicates that users are more willing to select the ECPs with lower access price. Next, we investigate how the local computing capacity and access price affect the cloud resource allocation in the ECC system. As shown in Fig. 10.4b, ECP 5 and ECP 6 request and receive the most cloud computing shares among the six ECPs, and ECP 1 and ECP 2 are allocated the least. Combining the results in Fig. 10.4a, results in Fig. 10.4b indicate that ECPs with more population shares tend to request and receive more computing resource form the CCP, which can increase the utilities obtained by the users selecting these ECPs, as formulated in (10.2), and meanwhile boost the utilities of both the CCP and ECPs. Furthermore, results in Fig. 10.4c validate that through the replicator dynamics, all user devices will reach the same individual utility at the equilibrium.

Figure 10.5 illustrates the influence of user learning rate $\delta$ on the convergence speed of evolutionary game and Stackelberg differential game towards the equilibrium. As defined in (10.3), learning rate $\delta$ controls the frequency of strategy adaption of all users, which will further control the speed of convergence from initial states towards equilibrium. Results shown in Fig. 10.5 validate that the convergence speed of replicator dynamics grows with the learning rate increasing. In this part of simulation, we also introduce a classic static Stackelberg equilibrium control (SSEC) proposed in [50] and [51] to optimize the resource pricing and allocation strategies. In SSEC, the CCP and ECPs make their decisions only based on the users' selection strategies, but without the considering of dynamic pricing and allocation strategies among CPs. Then results in Fig. 10.5 indicate that the open-loop Stackelberg equilibrium control (OLSEC) applied in this work can receive a faster convergence speed than SSEC, resulting from the dynamic learning and prediction of all CPs' strategies.

**Fig. 10.4** Evolutions of
population distribution,
resource allocation and user
utilities versus different $R_n$
and $p_n$ when the number of
ECPs is $N = 6$. (**a**)
Population distribution $x_n$, $x_c$.
(**b**) Resource allocation $r_n$, $r_c$.
(**c**) User utilities $\pi_n$, $\pi_c$

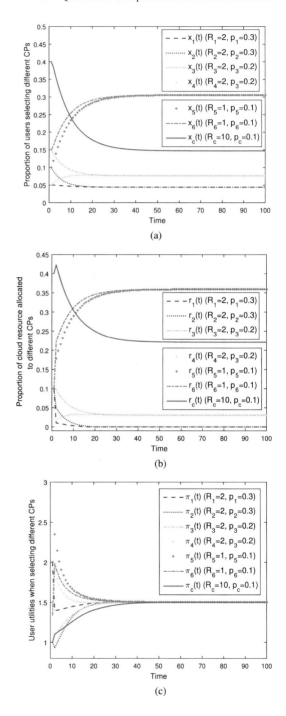

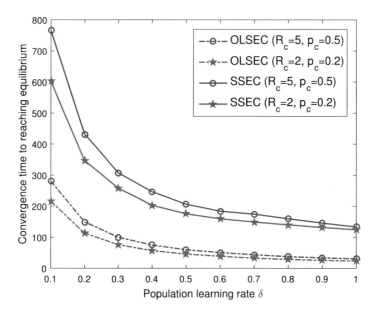

**Fig. 10.5** The convergence time versus increasing learning rate $\delta$ and different Stackelberg game equilibrium control schemes

## 10.7.2   Dynamic Pricing and Allocation of Computing Resource

To validate the performance of proposed Stackelberg differential game based resource pricing and allocation strategies, and investigate the impact of $R_c$ and $p_c$ on these strategies made in the computing resource level, we still consider the situation where there are two ECPs in the ECC system. Set $R_1 = 2\,\text{kH/s}$, $R_2 = 1\,\text{kH/s}$, $p_1 = 0.3$ and $p_2 = 0.2$, which are the same as the simulation in Sect. 10.7.1. Let $R_c$ select values in $\{5, 6, 7\}$ (kH/s) and $p_c$ choose values in $\{0.5 > \max\{p_1, p_2\}, 0.2 = \min\{p_1, p_2\}\}$.

By applying the evolutionary game based service selection and the Stackelberg differential game based resource pricing and allocation, we obtain the unit price of cloud computing resource and the proportion of could computing resource remaining to the CCP, which are shown in Fig. 10.6a and b, respectively. In Fig. 10.6a, results illustrate that the optimal price at equilibrium decreases with increasing total computational power of CCP. Meanwhile, the proportion of cloud computing resource remaining to the CCP at equilibrium increases with growing $R_c$. In addition, results in Fig. 10.6a also indicate that with the same $R_c$, the optimal price at equilibrium set by the CCP with lower access price $p_c$ is lower than that with high $p_c$. Combining the results in Fig. 10.6b, this trend implies that the utility of CCP with lower access price $p_c$ can be optimized by remaining more cloud computing, which will attract more users selecting the CCP, meanwhile setting a

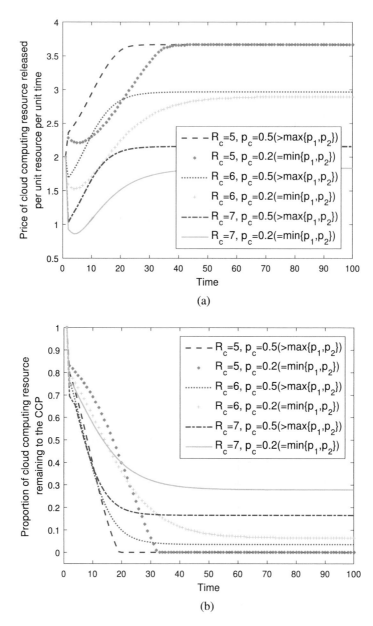

**Fig. 10.6** Dynamic computing resource pricing and allocation in the ECC system versus different $R_c$ and $p_c$. (**a**) Price $p(t)$. (**b**) Resource reservation $r_c(t)$

high $p_c$ to reduce the ECPs' willingness of purchasing cloud computing resource. In addition, results in Fig. 10.6 can also validate that the strategies of computing resource pricing and allocation will converge to the Stackelberg equilibrium.

### 10.7.3 Influence of Delay in Replicator Dynamics

Next, we study that how the population distribution states change with delay in replicator dynamics. As defined in (10.2), the utilities of user devices obtained by selecting different CPs are depends on the service selection strategies of all users in the evolutionary game. However, the information of population distribution state is always delayed resulting from the communication latency. Let $\tau_x \geq 0$ denote the delay of population information. Then the delayed replicator dynamics based on (10.3) can be given by

$$
\begin{aligned}
\dot{x}_{n/c}(t) = {} & \delta x_{n/c}(t - \tau_x)\left[\pi\left(n/c, \mathbf{x}(t - \tau_x), \mathbf{r}(t)\right)\right. \\
& \left. - \pi\left(\mathbf{x}(t), \mathbf{x}(t - \tau_x), \mathbf{r}(t)\right)\right],
\end{aligned}
\tag{10.38}
$$

where $n \in \mathcal{N}$. Considering that delayed replicator dynamics (10.38) can be rewritten as $\dot{\mathbf{x}}(t) = \mathbf{A}\mathbf{x}(t - \tau_x) + \mathbf{b}$, then its characteristic equation can be given by $\Theta e^{-\gamma \tau_x} + \gamma = 0$, where $\Theta$ has been defined in (10.8). Here we introduce the necessary and sufficient condition for the stability of delayed replicator dynamics proposed in [52], which can be given by $\tau_x < \pi/2\Theta$. Therefore, the stable ESS can be guaranteed with a small population delay. In this simulation, we set $\tau_x = 0.7$ and $\tau_x = 1.7$ to test different levels of population delay. Other parameters are set as $R_1 = 2\,\text{kH/s}$, $R_2 = 1\,\text{kH/s}$, $R_c = 2$, $p_1 = 0.3$, $p_2 = 0.2$ and $p_c = 0.2$. By applying the delayed replicator dynamics, the proportions of users selecting the CCP are shown in Fig. 10.7. Results in Fig. 10.7 validate that the population distribution state can still converge to the equilibrium after dynamic fluctuation, when $\tau_x$ is small. On the contrary, when $\tau_x$ is large, the equilibrium cannot be reached.

## 10.8 Conclusion

In this part, an SDN-based architecture has been established for edge and cloud computing services in 5G wireless HetNets, which can support efficient and on-demand computing resource management to optimize resource utilization and complete the time-varying computational tasks uploaded by user devices. In addition, considering the incompleteness of information, an evolutionary game based service selection was designed for users, which can model users' replicator dynamics of service subscription when they request the CCP or ECPs for computing resource. To complete these time-varying computational tasks from users, a Stackelberg differential game based cloud computing resource sharing mechanism was proposed to facilitate the resource trading between the CCP and different ECPs. Moreover, open-loop Stackelberg equilibrium solutions for the CCP (leader) and ECPs (followers), i.e., the optimal resource pricing and allocation strategies, were derived and obtained, which can promise the maximum integral utilities of the leader and followers over the time horizon, respectively. Simulation results have validated the performance of

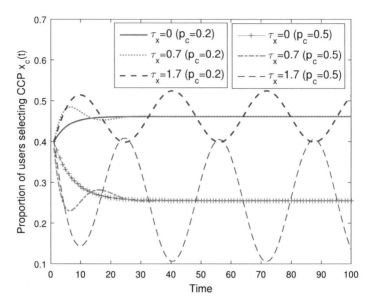

**Fig. 10.7** Proportions of users selecting the CCP under different population delays $\tau_x$ and access price of CCP $p_c$

the designed resource sharing mechanism, and revealed the convergence and stable states of user selection, resource pricing, and resource allocation in the ECC system.

# References

1. C. Jiang, H. Zhang, Y. Ren, Z. Han, K.-C. Chen, and L. Hanzo, "Machine learning paradigms for next-generation wireless networks," *IEEE Wireless Commun.*, vol. 24, no. 2, pp. 98–105, Apr. 2017.
2. H. Yao, M. Li, J. Du, P. Zhang, C. Jiang, and Z. Han, "Artificial intelligence for information-centric networks," *IEEE Commun. Mag.*, vol. 57, no. 6, pp. 47–53, Jun. 2019.
3. F. Li, H. Yao, J. Du, C. Jiang, and Y. Qian, "Stackelberg game-based computation offloading in social and cognitive industrial internet of things," *IEEE Trans. Ind. Inform.*, vol. 16, no. 8, pp. 5444–5455, Aug. 2020.
4. Y. Jiao, P. Wang, D. Niyato, and K. Suankaewmanee, "Auction mechanisms in cloud/fog computing resource allocation for public blockchain networks," *IEEE Trans. Parallel Distrib. Syst.*, vol. 30, no. 9, pp. 1975–1989, Sept. 2019.
5. J. Du, C. Jiang, J. Wang, Y. Ren, and M. Debbah, "Machine learning for 6G wireless networks: Carry-forward-enhanced bandwidth, massive access, and ultrareliable/low latency," *IEEE Veh. Technol. Mag*, vol. 15, no. 4, pp. 123–134, Dec. 2020.
6. J. Shi, J. Du, J. Wang, and J. Yuan, "Deep reinforcement learning-based V2V partial computation offloading in vehicular fog computing," in *IEEE Wireless Commun. Networking Conf. (WCNC'21).*   Nanjing, China, 29 Mar.-1 Apr. 2021.
7. G. Premsankar, M. Di Francesco, and T. Taleb, "Edge computing for the internet of things: A case study," *IEEE Internet Things J.*, vol. 5, no. 2, pp. 1275–1284, Apr. 2018.

8. B. Jaing, J. Du, C. Jiang, Y. Shi, and Z. Han, "Communication-efficient device scheduling via over-the-air computation for federated learning," in *IEEE Global Commun. Conf. (GLOBE-COM'22).* Rio de Janeiro, Brazil, 4–8 Dec. 2022.

9. Y. Liu, F. R. Yu, X. Li, H. Ji, and V. C. Leung, "Distributed resource allocation and computation offloading in fog and cloud networks with non-orthogonal multiple access," *IEEE Trans. Veh. Technol.*, vol. 67, no. 12, pp. 12 137–12 151, Dec. 2018.

10. J. Du, J. Song, Y. Ren, and J. Wang, "Convergence of broadband and broadcast/multicast in maritime information networks," *Tsinghua Sci. Techno.*, vol. 26, no. 5, pp. 592–607, 2021.

11. Z. Laaroussi, R. Morabito, and T. Taleb, "Service provisioning in vehicular networks through edge and cloud: an empirical analysis," in *2018 IEEE Conf. Standards Commun. Networking (CSCN).* Paris France, 29-31 Oct. 2018.

12. R. Chaudhary, N. Kumar, and S. Zeadally, "Network service chaining in fog and cloud computing for the 5G environment: data management and security challenges," *IEEE Commun. Mag.*, vol. 55, no. 11, pp. 114–122, Nov. 2017.

13. Z. Cao, S. S. Panwar, M. Kodialam, and T. Lakshman, "Enhancing mobile networks with software defined networking and cloud computing," *IEEE/ACM Trans. Networking*, vol. 25, no. 3, pp. 1431–1444, Jun. 2017.

14. J. Du, C. Jiang, H. Zhang, Y. Ren, and M. Guizani, "Auction design and analysis for SDN-based traffic offloading in hybrid satellite-terrestrial networks," *IEEE J. Sel. Areas Commun.*, vol. 36, no. 10, pp. 2202–2217, Oct. 2018.

15. C. Liang, Y. He, F. R. Yu, and N. Zhao, "Enhancing video rate adaptation with mobile edge computing and caching in software-defined mobile networks," *IEEE Trans. Wireless Commun.*, no. 10, pp. 7013–7026, Oct. 2018.

16. A. Kiani and N. Ansari, "Toward hierarchical mobile edge computing: An auction-based profit maximization approach," *IEEE Internet Things J.*, vol. 4, no. 6, pp. 2082–2091, Dec. 2017.

17. Y. Zhang, C.-Y. Wang, and H.-Y. Wei, "Parking reservation auction for parked vehicle assistance in vehicular fog computing," *IEEE Trans. Veh. Technol.*, vol. PP, no. 99, pp. 1–1, Feb. 2019.

18. R. Wang, J. Yan, D. Wu, H. Wang, and Q. Yang, "Knowledge-centric edge computing based on virtualized D2D communication systems," *IEEE Communications Magazine*, vol. 56, no. 5, pp. 32–38, May 2018.

19. J. Du, E. Gelenbe, C. Jiang, H. Zhang, and Y. Ren, "Contract design for traffic offloading and resource allocation in heterogeneous ultra-dense networks," *IEEE J. Sel. Areas Commun.*, vol. 35, no. 11, pp. 2457–2467, Nov. 2017.

20. G. S. Aujla, N. Kumar, A. Y. Zomaya, and R. Rajan, "Optimal decision making for big data processing at edge-cloud environment: An SDN perspective," *IEEE Trans. Ind. Inform.*, vol. 14, no. 2, pp. 778–789, Feb. 2018.

21. H. Zhang, X. Yong, S. Bu, D. Niyato, R. Yu, and H. Zhu, "Computing resource allocation in three-tier IoT fog networks: a joint optimization approach combining stackelberg game and matching," *IEEE Internet Things J.*, vol. 4, no. 5, pp. 1204–1215, Oct. 2017.

22. Z. Xiong, S. Feng, D. Niyato, W. Ping, and H. Zhu, "Edge computing resource management and pricing for mobile blockchain," *IEEE Internet Things J.*, vol. Early Access, 24 Sept. 2018.

23. G. S. Aujla, R. Chaudhary, N. Kumar, J. J. P. C. Rodrigues, and A. Vinel, "Data offloading in 5G-enabled software-defined vehicular networks: A stackelberg-game-based approach," *IEEE Commun. Mag.*, vol. 55, no. 8, pp. 100–108, Aug. 2017.

24. T. Sanguanpuak, D. Niyato, N. Rajatheva, and M. Latva-aho, "Radio resource sharing and edge caching with latency constraint for local 5G operator: Geometric programming meets stackelberg game," *IEEE Trans. Mobile Computing*, pp. 1–1, Early Access, Oct. 2019.

25. G. S. Aujla, M. Singh, N. Kumar, and A. Y. Zomaya, "Stackelberg game for energy-aware resource allocation to sustain data centers using RES," *IEEE Transactions on Cloud Computing*, vol. 7, no. 4, pp. 1109–1123, Oct.-Dec. 2019.

26. S. Feng, Z. Xiong, D. Niyato, and P. Wang, "Dynamic resource management to defend against advanced persistent threats in fog computing: A game theoretic approach," *IEEE Trans. Cloud Computing*, vol. Early Access, 31 Jan. 2019.

27. S. H. Kim, S. Park, C. Min, and C. H. Youn, "An optimal pricing scheme for the energy efficient mobile edge computation offloading with OFDMA," *IEEE Commun. Lett.*, vol. 22, no. 9, pp. 1922–1925, Sept. 2018.

28. A. Molina Zarca, J. Bernal Bernabe, I. Farris, Y. Khettab, T. Taleb, and A. Skarmeta, "Enhancing IoT security through network softwarization and virtual security appliances," *Int. J. Network Manage.*, vol. 28, no. 5, p. e2038, Jul. 2018.

29. I. Farris, T. Taleb, Y. Khettab, and J. Song, "A survey on emerging SDN and NFV security mechanisms for IoT systems," *IEEE Commun. Surveys & Tutorials*, vol. 21, no. 1, pp. 812–837, Firstquarter 2019.

30. G. S. Aujla and N. Kumar, "MEnSuS: An efficient scheme for energy management with sustainability of cloud data centers in edge–cloud environment," *Future Generation Computer Syst.*, vol. 86, pp. 1279–1300, Sept. 2018.

31. R. Bruschi, F. Davoli, P. Lago, and J. F. Pajo, "A multi-clustering approach to scale distributed tenant networks for mobile edge computing," *IEEE J. Sel. Areas Commun.*, vol. 37, no. 3, pp. 499–514, Mar. 2019.

32. Q. Chen, F. R. Yu, T. Huang, R. Xie, J. Liu, and Y. Liu, "Joint resource allocation for software-defined networking, caching, and computing," *IEEE/ACM Trans. Networking*, vol. 26, no. 1, pp. 274–287, Feb. 2018.

33. A. C. Baktir, A. Ozgovde, and C. Ersoy, "How can edge computing benefit from software-defined networking: A survey, use cases & future directions," *IEEE Commun. Surveys & Tutorials*, vol. 19, no. 4, pp. 2359–2391, Jun. 2017.

34. G. S. Aujla, N. Kumar, A. Y. Zomaya, and R. Ranjan, "Optimal decision making for big data processing at edge-cloud environment: An sdn perspective," *IEEE Transactions on Industrial Informatics*, vol. 14, no. 2, pp. 778–789, Feb. 2018.

35. P. K. Sharma, S. Singh, Y. S. Jeong, and J. H. Park, "Distblocknet: A distributed blockchains-based secure SDN architecture for IoT networks," *IEEE Commun. Mag.*, vol. 55, no. 9, pp. 78–85, Sept. 2017.

36. P. K. Sharma, S. Rathore, Y. S. Jeong, and J. H. Park, "SoftEdgeNet: SDN based energy-efficient distributed network architecture for edge computing," *IEEE Commun. Mag.*, vol. 56, no. 12, pp. 104–111, Dec. 2018.

37. K. Zhu, E. Hossain, and D. Niyato, "Pricing, spectrum sharing, and service selection in two-tier small cell networks: A hierarchical dynamic game approach," *IEEE Trans. Mobile Computing*, vol. 13, no. 8, pp. 1843–1856, Aug. 2014.

38. J. Du, C. Jiang, K.-C. Chen, Y. Ren, and H. V. Poor, "Community-structured evolutionary game for privacy protection in social networks," *IEEE Trans. Inf. Forens. Security*, vol. 13, no. 3, pp. 574–589, Mar. 2018.

39. P. D. Taylor and L. B. Jonker, "Evolutionary stable strategies and game dynamics," *Math. Biosci.*, vol. 40, no. 1-2, pp. 145–156, Jul. 1978.

40. J. Romero, A. Sanchis-Cano, and L. Guijarro, "Dynamic price competition between a macrocell operator and a small cell operator: A differential game model," *Wireless Commun. Mobile Computing*, vol. 2018, May 2018.

41. T. L. Friesz, "Dynamic optimization and differential games," *Alkalmaz.mat.lapok*, vol. 135, no. 1-2, pp. 203–209, 2007.

42. Y. C. Ho, "Differential games, dynamic optimization, and generalized control theory," *J. Optimization Theory & Applicati.*, vol. 6, no. 3, pp. 179–209, Sept. 1970.

43. C. Jiang, Y. Chen, Y. Gao, and K. R. Liu, "Joint spectrum sensing and access evolutionary game in cognitive radio networks," *IEEE Trans. Wireless Commun.*, vol. 12, no. 5, pp. 2470–2483, May 2013.

44. F. Li, H. Yao, J. Du, C. Jiang, and Y. Qian, "Stackelberg game-based computation offloading in social and cognitive industrial internet of things," *IEEE Trans. Ind. Inform.*, vol. 16, no. 8, pp. 5444–5455, Aug. 2020.

45. C. Jiang, Y. Chen, K. R. Liu, and Y. Ren, "Renewal-theoretical dynamic spectrum access in cognitive radio network with unknown primary behavior," *IEEE J. Sel. Areas Commun.*, vol. 31, no. 3, pp. 406–416, Mar. 2013.

46. M. Simaan and J. B. Cruz, "On the stackelberg strategy in nonzero-sum games," *J. Optimization Theory Applicat.*, vol. 11, no. 5, pp. 533–555, May 1973.

47. G. Freiling, G. Jank, and S. R. Lee, "Existence and uniqueness of open-loop stackelberg equilibria in linear-quadratic differential games," *J. Optimization Theory Applicat.*, vol. 110, no. 3, pp. 515–544, Sept. 2001.

48. G. M. Anderson, "Comparison of optimal control and differential game intercept missile guidance laws," *J. Guidance, Control, Dynamics*, vol. 4, no. 2, pp. 109–115, Mar. – Apr. 1981.

49. M. Pachter, E. Garcia, and D. W. Casbeer, "Toward a solution of the active target defense differential game," *Dynamic Games Applicat.*, vol. 9, no. 1, pp. 165–216, Mar. 2019.

50. Y. Chen, Z. Jin, and Z. Qian, "Utility-aware refunding framework for hybrid access femtocell network," *IEEE Trans. Wireless Commun.*, vol. 11, no. 5, pp. 1688–1697, Jun. 2012.

51. J. B. Clempner and A. S. Poznyak, "Solving transfer pricing involving collaborative and non-cooperative equilibria in nash and stackelberg games: Centralized–decentralized decision making," *Computational Econ.*, no. 1, pp. 1–29, Jul. 2018.

52. K. Gopalsamy, *Stability and Oscillations in Delay Differential Equations of Population Dynamics*. Berlin, Germany. Springer, Mar. 1992.

# Chapter 11
# QoS-Aware Caching Resource Allocation

**Abstract** Recently, wireless streaming of on-demand videos of mobile users (MUs) has become the major form of data traffic over cellular networks. As responding, caching popular videos in the storage of small base stations (SBSs) has been regarded as an efficient approach to reduce the transmission latency and alleviate the data traffic loaded over backhaul channels. This work considers a small-cell based caching market composed of one mobile network operator (MNO) and multiple video service providers (VSPs). In this system, the MNO manages and operates its SBSs, and assigns these SBSs' storage to different VSPs, who have caching requirements. However, videos have different popularities and MUs present different preferences to these VSPs when they request videos. In addition, the caching service brings different utilities to different VSPs, as well as that providing caching service to different VSPs causes distinct costs to the MNO. Such privacy information cannot be aware of among VSPs and the MNO. Therefore, to elicit this hidden information, this chapter designs a double auction based caching mechanism, which ensures the efficient operation of the market by maximizing the social welfare, i.e., the gap between VSPs' caching utilities and MNO's caching costs. Moreover, the chapter demonstrated economic properties of the designed caching mechanism, which are also validated by simulation results.

**Keywords** Video Caching · Double Auction · Heterogeneous Networks · Economic Property · Information Hidden

## 11.1 Introduction

Recently, mobile data traffic is experiencing a dramatic increasing over cellular networks. There is evidence that content distribution services, such as video on demand (VoD), catch-up TV, internet video streaming, etc., have become premier drivers of the exponential traffic growth [1–3]. A key feature of such type of video services is asynchronous content reuse [4]. Specifically, a relatively small number of popular video files provided by a certain part of video service providers (VSPs) account for the most of data traffic, and are requested frequently by mobile users (MUs). To deal with this fact of highly redundant video demands from MUs, caching techniques, e.g., storing video files in MUs' devices or potential helper nodes disseminated in the network, have been developed to avoid the high-throughput backhaul to the core network, which is too costly and constitutes a major bottleneck [5, 6]. In additions, by applying caching techniques, popular video files transmitted in backbone networks are dispersed into local caches of network nodes located at the edge of wireless networks, e.g., femto-cells and pico-cells, and the distance between video files and requesters is also shortened. Therefore, introducing wireless caching into networks brings low power consumption and latency [7].

In recent years, the small-cell based architecture has dominated in ultra-dense heterogeneous networks (HetNets). In HetNets, mobile network operates (MNOs) deploy multiple small base stations (SBSs) which work in conjunction with micro base stations (MBSs). For the MNO, the cost for long-distant transmission can be saved by this architecture. On the other hand, MUs receive their requested data through low-power consumption, low-latency and better-quality communications [8]. Based on this architecture, video caching relying on SBSs constitutes a feasible and low-cost solution to further cope with the increasing video data traffic over backhaul channels with assistance of the small-cell based architecture.

Small-cell based video caching generally consists of two stage: data placement and data delivery. In the data placement stage, popular videos are cached in the SBS storage during the off-peak time [9]. Then in the data delivery stage, if the requested video file of an MU has been pre-cached in an SBS whose communication range covers this MU, then the requested video is delivered from this SBS directly to this MU [10]. Otherwise, the MBS associated by this MU will request the video file to the related server through the backbone network via backhaul channels, and then deliver it to the MU.

Given a set of VSPs who can provide the same set of video files and are with different requested preferences, i.e., requested probabilities by MUs, the challenge in this context is to find an optimal caching policy. Specifically, in a real-world small-cell based caching system, VSPs' utilities obtained from receiving the caching service and the MNO's cost caused by providing caching service to different VSPs are local and privacy information, which means that such information is unknowable for anyone except themselves. Consequently, with this insufficient and asymmetric information, how to allocate SBSs' storage to different VSPs to place their video files, so as to maximize the social welfare, i.e., the sum of utilities obtained by the MNO and VSPs, becomes a difficult problem for system optimization. Therefore, in this part, we consider a small-cell based video caching system with hidden information. To elicit the hidden information and achieve the maximum social welfare, we study the caching resource allocation mechanism under an economic framework, i.e., double auction. This economics based caching mechanism processes all the following four economic properties, although these properties concluded in [11–13] have been demonstrated that they cannot be satisfied at the same time:

1. **Economic Efficiency (EE):** The designed caching mechanism is able to get the optimal solution that leads to the maximum social welfare.
2. **Individually Rationality (IR):** The service provider and requesters, i.e., referring to the MNO and VSPs, respectively, will never get worse or negative utilities

by participating than those obtained by not participating, which will bring zero utilities for participants.

3. **Incentive Compatibility (IC):** Under the designed mechanism, service requesters are induced to report their truthful requirements or private information directly or indirectly.

4. **(Weakly) Budge Balance (BB):** The broker does not have to invest additional "money" to make the mechanism go round. In other words, the negotiated payments from service requesters/buyers (VSPs) to the broker should not be less than those from the broker to the service provider/seller (MNO).

The main contributions of this part can be summarized as follows:

1. We establish a small-cell based video caching system in ultra-dense HetNets, in which the MNO operates a set of SBSs and leases SBS storage to multiple VSPs to placing their video files. Based on different VSP preferences, VSP utility functions, the MNO cost function and a social welfare maximization problem are formulated in this work.

2. In order to elicit the hidden information among VSPs and the MNO, i.e., VSP utility functions and the MNO cost function, a double auction model is introduced to solve the caching problem. Based on the designed bidding rules, resource allocation schemes for the SBS storage and pricing rules are designed to promote VSPs and the MNO to consciously provide bids reflecting their truthful caching requirements and admission, respectively. In addition, the designed pricing rules not only reflect the resource allocation constraint, but also the caching cost of the MNO.

3. We formulate an alternative optimization problem which has the same optimal solution as the social welfare optimization problem. By solving this problem, and applying the designed allocation schemes and pricing rules, the maximum social welfare can be achieved although there exists the hidden information.

4. We provide the detailed proof of convergence and economic properties of the designed double auction based caching mechanism, which are also validated by simulation results.

The remainder of the part is organized as follows. In Sect. 11.2, we briefly review the related works that associated with caching mechanisms in heterogenous networks. Section 11.3 sets up the system model. In Sect. 11.4, the caching problem is formulated and the system economic benefits are analyzed. A double auction mechanism for video caching in small-cell based networks is proposed in Sect. 11.5, and its implementation and characteristics are provided in Sect. 11.6. Simulations are shown in Sect. 11.7, and conclusions are drawn in Sect. 11.8.

## 11.2  Related Works

Small-cell based caching mechanisms can help offload data traffic from the MBSs and bring contents closer to the MUs, which will reduce the power consumption, shorten the transmission latency and offloading delay. Due to its significant performance on releasing the increasing mobile data traffic, video caching has received considerable attention in the wireless communications, and many researches have focused on effect and efficient video caching mechanism design.

Authors of [14] considered the problem of joint content placement and routing in HetNets that supported in-network caching and also provided a separate (uncached) path to a back-end content server. In [15], the caching problem was analyzed in a distributed HetNet, assuming that popularity profiles of cached data were unknown. In [16], cache-based content delivery in a three-tier HetNet, where the radio access network (RAN) caching and device-to-device (D2D) caching coexist, was proposed and analyzed. The secrecy caching capacity was investigated in [17], which derived the maximum amount of information which can be stored in the caching network such that there was no leakage of information during a partial repair process. In addition, caching techniques were also applied in D2D communication networks to reduce the downloading delay and power consumption [18–21].

Among caching techniques, video files placement and helper nodes' storage management also attract much recent research attention. In [22], the hit performance of cache systems that receive file requests with general arrival distributions and different popularities was analyzed by considering time policies. The established bounds on the number of objects cached by the optimal policy in [22] was first defined in [23]. Particularly, the author of [23] originally obtained formally the result that, under independent reference modeled traffic, the performance of RANDOM and FIFO (First Input First Output) in terms of hit probability, are the same. The mathematic model, replacement problem formulated and important results in [23] have become important theoretical foundations to investigate storage management, caching mechanism design and performance optimization in caching systems and content delivery networks [24–27].

As summarized above, most current research above on wireless caching mainly focused on video placement problems optimized to reduce the downloading delay and transmission power and latency. However, besides content placement, there are many issues involved in the video caching mechanism design, such as caching market operation and commercial property analysis. In a market of caching resource, the MNO, operating and managing SBSs and MBSs, plays the role of resource owner and caching service provider. On the other side, VSPs, who provide videos to MUs, are caching resource requesters. Caching resource in this market can be considered as the right of using SBSs, SBSs' storage resource and so on. Considering different popularities of videos, preference of VSPs and limited SBS storage, benefits of different VSPs obtained by assigned different amount of SBS storage to place their videos, as well as the cost of MNO caused by leasing different amount of SBS storage to different VSPs will be different from a commercial perspective. Therefore, an efficient caching resource allocation scheme working in conjunction with a seasonable and proper pricing rule will ensure a caching system operation with high efficiency [28]. So in this work, we consider a caching resource market where the MNO leases the storage of SBSs to multiple VSPs, and analyze and optimize the performance of the caching system by establishing an economic model, i.e., automatic auction.

Auction theory, as an effective theory in network economics, has been studied to model and analyze the process of resource requesting and providing [29, 30], especially for networks with heterogeneous and limited resource. In [29], different auction models based on Markov process were proposed and analyzed, which provided classic mathematical models for networked resource supply and demand. To achieve efficient data transaction and traffic offloading among mobile users, auction models were introduced in [30, 31] and [32]. Authors of [33] and [34] established a multi-object auction model to describe and analyze the storage competition for small-cell based caching systems. However, economic properties of auction models, such as social welfare, incentive compatibility, etc., were less considered in these studies above. Specifically, the heterogeneous characteristic in small-cell based caching system means that benefits for different VSPs who are assigned the same SBS resource, as well as costs for the MNO when it assigns the same resource to different VSPs tends to be different resulting from different preferences of VSPs. Through an auction based caching mechanism, the SBS storage can be assigned to VSPs with optimal utility of the entire system, i.e., the social welfare. On the other hand, considering the hidden information in the caching system, it is hard to achieve a globally and socially efficient solution. To cope with this difficulty, a double auction (DA) mechanism, which introduces a broker as a centralized controller and requests both service providers and requesters to submit bids for resource allocation [35, 36], is introduced in this work. With the design of resource allocation scheme and pricing rules, the DA based caching mechanism can make sure that the maximum social welfare can be gradually reached, without any prior knowledge for the market.

## 11.3   System Model

Consider a small-cell based caching system consisting of a set of VSPs $\mathcal{N} = \{1, 2, \cdots, N\}$, a number of MUs and a set of SBSs operated by a monopolist MNO, who also operates a micro-cell base station (MBS), as shown in Fig. 11.1. In such caching system, VSPs are willing to rent the SBSs' storage from the MNO to place their videos. Next, we define the caching problem as a market design, and the goal of which is maximizing the social welfare considering both the MNO and VSPs. Before proceeding further, we summarize the main notations used throughout the following sections in Table 11.1 for convenience.

### 11.3.1   Network Model

In the small-cell network composed of SBSs and MUs, we assume that SBSs are equipped with the same transmission power $P$ and the same storage of $Q$ video files. Consider that SBSs and MUs are spatially distributed in the coverage of an MBS

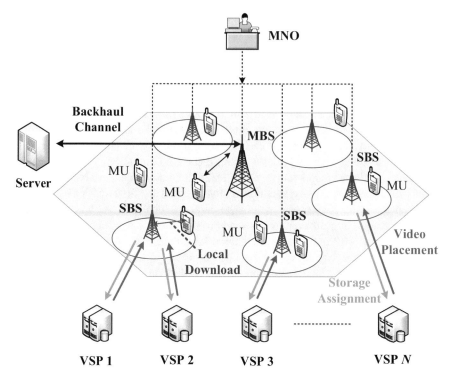

**Fig. 11.1** Small-cell based video caching system in HetUDNs

according to two Homogeneous Poisson Point Processes (HPPPs) $\Phi$ with density $\varphi$, and $\Psi$ with density $\psi$, which represent the number of SBSs and MUs within a unit area, respectively. In addition, the wireless downlink channels from SBSs to MUs are modeled as independent and identically distributed (i.i.d.) combination of path loss and Rayleigh fading [37, 38]. Considering a typical MU located at the original, then the path loss between this MU and the SBS located at $d$ can be presented by $\|d\|^{-\gamma}$, where $\gamma$ is the path loss exponent, and denote $h_d$, where $h_d \sim \exp(1)$, to be the channel power of Rayleigh fading between them. Furthermore, assume that SBSs transmit over channels those are orthogonal to that of the MBS. Therefore, there is no interference between SBSs and the MBS.

In this work, we consider the steady-state of a saturated network, where every SBS in the system keeps on transmitting in the entire frequency band allocated to it [39–42]. Therefore, the received signal-to-interference-plus-noise ratio (SINR) at the original-located MU from the SBS located at $d$ can be given by

$$SINR(d) = \frac{Ph_d\|d\|^{-\gamma}}{\sum_{d'\in\Phi\backslash d} Ph_{d'}\|d'\|^{-\gamma} + \sigma^2}, \tag{11.1}$$

**Table 11.1**  List of main notations in small-cell based caching system

| Parameter | Definition |
|---|---|
| $N$ | Number of VSPs |
| $P$ | Transmission power of each SBS |
| $Q$ | Number of videos can be stored at each SBS |
| $F$ | Number of video files |
| $K$ | Average number of video requests per MU per unit time |
| $\varphi$ | Density of SBS (number per unit area) |
| $\psi$ | Density of MU (number per unit area) |
| $p_f$ | Popularity of video $f = 1, 2, \cdots, F$ |
| $q_n$ | Preference of VSP $n = 1, 2, \cdots, N$ |
| $\pi_n$ | Fraction of SBSs assigned to VSP $n = 1, 2, \cdots, N$ |
| $\alpha$ | Exponent of video popularity |
| $\beta$ | Exponent of VSP preference |
| $d$ | SBS location |
| $\gamma$ | Path loss exponent |
| $h$ | Channel power of Rayleigh fading |
| $\sigma^2$ | Variance of the AWGN |
| $\delta$ | SINR threshold |
| $r_{ld}$ | Local downloading surcharging (LDS) |
| $r_{bh}$ | Average backhaul cost (ABC) |

where $\sigma^2$ is the variance of the i.i.d. additive white Gaussian distributed noise with zero mean at MUs. Notice that when $SINR(d)$ is no less than SINR threshold $\delta$,[1] i.e., $SINR(d) \geq \delta$, the original-located MU is considered to be within the coverage of the SBS located at $d$. In addition, an MU can be covered by multiple SBSs, generally.

## 11.3.2   Video Popularity

Consider a set of $F$ video files denoted by $\mathscr{F} = \{1, 2, \cdots, F\}$. Since each SBS can only storage at most $Q$ files, we assume that $F \geq Q$. In addition, we assume that all VSPs provide the same video set $\mathscr{F}$.[2] When a VSP rents an SBS, the first $Q$ video files with the most popularities will be cached in this SBS, since this is the most efficient way to place videos. Every file in set $\mathscr{F}$ can be a popular movie

---

[1] SINR threshold $\delta$ is defined as the highest delay of downloading a video file.

[2] This assumption is feasible since that although VSPs provide some different videos, the most popular videos tend to be the same. On other words, those different videos are usually not the popular ones, then they are not worthy to be cached at SBSs. So different videos do not affect the caching procedure design.

or video clip, and will be requested by MUs frequently. Then we first define the popularity distribution of video files among popular videos to be cached at SBSs as their probabilities of been requested. Consider that the popularity distribution of video file set $\mathscr{F}$, denoted by vector $\mathbf{p} = [p_1, p_2, \cdots, p_F]$, complies with a Zipf distribution [43],[3] and thus $p_f$ is defined by

$$p_f = \frac{1/f^\alpha}{\sum_{i=1}^{F} 1/i^\alpha}, \quad \forall f \in \mathscr{F}, \tag{11.2}$$

where exponent $\alpha > 0$ characterizes the video popularity. A larger $\alpha$ implies a frequent video reuse or request, and means that the most popular videos account for the majority of video requests. In addition, according to (11.2), the popularities of videos in set $\mathscr{F}$ decrease with increasing $f$.

### 11.3.3   VSP Preference

Practically, MUs present different preferences to different VSPs. For instance, most MUs in a certain area prefer YouTube to download videos. In this work, preferences of the $N$ VSPs, denoted by $\mathbf{q} = [q_1, q_2, \cdots, q_F]$, are also modeled as a Zipf distribution [43], and we define $q_n$ by

$$q_n = \frac{1/n^\beta}{\sum_{j=1}^{N} 1/j^\beta}, \quad \forall n \in \mathscr{N} \tag{11.3}$$

as the request probability that an MU prefers to download videos from VSP $n$. In (11.3), exponent $\beta > 0$ characterizes the VSP preference, and a larger $\beta$ implies that the most popular VSPs account for the majority of download request.

## 11.4   Caching Problem Formulation and Profit Analysis

In this section, we first introduce how to implement the video caching in the small-cell based network. Then we analyze the economic utility can be obtained by VSPs from receiving the caching service, and the cost of the MNO when providing the caching service to VSPs.

---

[3] The dataset analysed in [43] consisted of meta-information about user-generated videos from YouTube and Daum UGC services.

## 11.4.1 Caching Procedure

We first introduce the caching procedure in the small-cell based caching system. This procedure falls into three steps.

### 11.4.1.1 SBS Assignment

In this step, each VSP rents a certain fraction of the SBSs operated by the MNO for placing its video files. The fraction for every VSP is denoted by vector $\pi = \{\pi_1, \pi_2, \cdots, \pi_N\}$, where $\pi_n$ represents the fraction of SBSs released to VSP $n$, $\forall n \in \mathcal{N}$. Constitutionally, it always holds that $\pi_n \geq 0$ and $\sum_{n=1}^{N} \pi_n \leq 1$, and the case where $\pi_n = 0$ indicates that there is no SBS assigned to VSP $n$ for caching. Assume that the SBSs assigned to each VSP are randomly and uniformly distributed, and then the number of SBSs assigned to VSP $n$ ($\forall n \in \mathcal{N}$) is distributed as a thinned HPPP $\Phi_n$ with intensity $\pi_n \varphi$ [39].

### 11.4.1.2 Video File Placing

In this step, each VSP accesses and places the most popular video files at its assigned SBSs to achieve an efficient caching. This step can be completed during the off-peak time after the first step being finished and every VSP having been assigned a certain fraction of SBSs for caching.

### 11.4.1.3 MU Video Requests

In this step, an MU sends a request of video $f \in \mathcal{F}$ to VSP $n$. This request is first sent to this MU's nearest SBS in $\Phi_n$ which caches this video file. If there exists an SBS in $\Phi_n$ that has cached this requested video and its coverage can cover this MU, then this MU will download the video directly from this SBS. This successful event is denoted by $I_{n,f}$, and brings a local downloading saving (LDS) for VSP $n$. When the MNO leases fraction $\pi_n$ of SBSs to VSP $n$, the probability of evert $I_{n,f}$, denoted by $\Pr\{I_{n,f}\}$ is given by:

$$
\Pr\{I_{n,f}(\pi_n)\}
$$

$$
= \frac{\pi_n}{\pi_n G_1(\delta, \gamma) + (1 - \pi_n) G_2(\delta, \gamma) + \pi_n}, \quad \forall n = 1, 2, \cdots, Q. \tag{11.4}
$$

In (11.4), $G_1(\delta, \gamma) = \frac{2\delta}{\gamma-2} {}_2F_1\left(1, 1 - \frac{2}{\gamma}; 2 - \frac{2}{\gamma}, -\delta\right)$, and $G_2(\delta, \gamma) = \frac{2}{\gamma}\delta^{\frac{2}{\gamma}} B\left(\frac{2}{\gamma}, 1 - \frac{2}{\gamma}\right)$, where ${}_2F_1(a, b; c, z)$ is a Gaussian hypergeometric function, and

$$B(x, y) \triangleq \int_0^1 t^{x-1}(1 - t)^{y-1}dt$$

is Beta function. The detailed derivation of (11.4) can be found in [39]. Otherwise, the requested video will be transmitted to the MU by the MBS through MNO's backhaul channel remotely, which will give rise to extra cost for the MNO.

## 11.4.2  Benefit Analysis

In this section, we will formulate utilities of VSPs and the cost of MNO brought by the caching system. In this work, we focus on the average utility based on the stochastically geometrical distribution of network nodes in term of per unit area times unit period ($/UAP$), e.g., $/\text{second} \cdot \text{meter}^2$ or $/\text{month} \cdot \text{kilometer}^2$.

### 11.4.2.1  VSP Utility

As discussed previously, VSPs can provide videos to MUs ether through memories of assigned SBSs directly, or from their own servers via backhaul channels. Denoted the LDS by $r_{ld}$. Consider that there are average $K$ video requests from each MU within a unit period. Then the average utility ($/UAP$) obtained by VSP $n$ can be calculated by

$$u_n^{\text{VSP}} = \sum_{f=1}^{Q} K\psi p_f q_n \Pr\left\{I_{n,f}\right\} r_{ld}. \tag{11.5}$$

*Remark 11.1*  According to the definition in (11.4), $u_n^{\text{VSP}}$ can be rewritten as

$$u_n^{\text{VSP}} = \frac{\pi_n \sum_{f=1}^{Q} K\psi p_f q_n r_{ld}}{[G_1(\delta, \gamma) - G_2(\delta, \gamma) + 1]\pi_n + G_2(\delta, \gamma)}. \tag{11.6}$$

We can notice that every utility function of VSP, i.e., $u_n^{\text{VSP}}$, $\forall n \in \mathcal{N}$, is a positive, increasing and strictly concave function of SBS fraction $\pi_n$. Therefore, this increasing utility will capture VSPs' preferences of seeking more SBSs for caching. The solid line in Fig. 11.2 illustrates the utility function of VSPs.

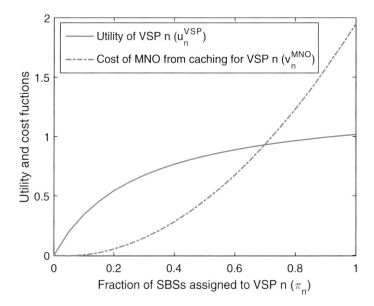

**Fig. 11.2** An example of the concave function of VSP utility ($u_n^{VSP}$) and convex function of MNO cost ($v_n^{MNO}$) when system parameters are set as $N = 3$, $F = 100$, $Q = 40$, $K = 50$, $n = 1$, $\gamma = 4$, $P = 10$, $\delta = 0.02$, $\alpha = 0.2$, $\beta = 1$, $\sigma^2 = 10^{-10}$, $\varphi = 20$, $\psi = 80$, $r_{ld} = 10^{-3}$ and $r_{bh} = 5 \times 10^{-5}$

### 11.4.2.2   MNO Cost

The utility obtained from the caching procedure consists two parts: the first part is the income from reducing the cost of backhaul by leasing SBSs to VSPs, and the second part, which is negative, is the cost for maintaining these SBSs. Denote the average backhaul cost (ABC) for SBSs to transmit video files by $r_{bh}$. Then the first part of utility obtained by placing VSP $n$'s video is given by

$$u_n^{bh} = \sum_{f=1}^{Q} K \psi p_f q_n \Pr \{I_{n,f}\} r_{bh}.$$  (11.7)

The cost caused by maintaining the leased SBSs to VSP $n$ is defined as $c_n = c\varphi\pi_n$, where $c > 0$ is the unit cost of the MNO to maintain the leased SBSs. Then for the MNO, the total utility obtained by placing VSP $n$'s video is given by

$$u_n^{MNO} = u_n^{bh} - c_n.$$  (11.8)

*Remark 11.2* Consider the total cost function of the MNO when placing VSP $n$'s video as:

$$v_n^{MNO} = -u_n^{MNO}.$$  (11.9)

In this work, we consider that the cost of maintaining SBSs is dominant, and the utility obtained by avoiding backhaul transmission $u_n^{bh}$ is relatively small. Then we can consider the total cost function $v_n^{MNO}$ is a positive, increasing and strictly convex function of $\pi_n$. This property is reasonable, and can capture the fact that as the admitted fraction of SBSs for caching increases, the cost of maintaining leased SBSs increase as well as the revenue of backhaul transmission saving, however, the increasing speed of former is much faster than that of the latter. The dotted line in Fig. 11.2 illustrates the cost function of MNO when it leases different fractions of SBSs for VSP $n$ to providing caching service.

## 11.5  Double Auction Mechanism Design for Small-Cell Based Caching System

### 11.5.1  Social Welfare Maximization Problem

Based on the model and its analysis above, the objectives of the MNO and VSPs are opposite to each other. Specifically, VSPs are willing to get as many as fraction of SBSs for caching to achieve a maximum utility. On the other hand, the MNO tends to lease less fraction of SBSs to save its maintaining cost. Therefore, it is difficult for them to reach an agreement. Concerning this problem, a market broker (system controller) is necessary to operate the caching service market effectively and efficiently. In the market, the broker is paid according to the volume of transactions it facilitates. So we consider that the broker is honest and has no incentive to distort the transaction efficiency [35].

We consider a caching system with $N$ VSPs and an MNO having their caching requests and caching admission, respectively. Denote the request of SBS fraction for caching from VSP $n$ by $x_n$, then the caching request vector is written as $\mathbf{x} = (x_1, x_2, \cdots, x_N)$.

Then according to (11.6), the VSP $n$'s utility obtained from caching is given by

$$u_n^{VSP}(x_n) = \sum_{f=1}^{Q} K \psi p_f q_n \Pr\left\{I_{n,f}(x_n)\right\} r_{ld}. \tag{11.10}$$

Denote the admitted fraction of SBSs for caching service to VSP $n$ by $y_n$. Then the MNO's admitted fraction of SBSs for all VSPs can be given by $\mathbf{y} = (y_1, y_2, \cdots, y_N)$. According to (11.8) and (11.9), we can derive the cost of the MNO when admitting fraction $y_n$ of SBSs to VSP $n$ as

$$v_n^{MNO}(y_n) = c\varphi y_n - \sum_{f=1}^{Q} K \psi p_f q_n \Pr\left\{I_{n,f}(y_n)\right\} r_{bh}. \tag{11.11}$$

As the central controller of the entire caching system, the broker is willing to maximize the global social welfare by optimizing vectors $\mathbf{x}$ and $\mathbf{y}$. Thus the social welfare optimization problem for the broker can be formulated as

$$\max_{\mathbf{x},\mathbf{y}} \ U_1 (\mathbf{x}, \mathbf{y}) \triangleq \sum_{n=1}^{N} u_n^{\mathrm{VSP}} (x_n) - \sum_{n=1}^{N} v_n^{\mathrm{MNO}} (y_n), \tag{11.12a}$$

$$\text{s.t.} \ \sum_{n=1}^{N} y_n \leq 1, \tag{11.12b}$$

$$x_n \leq y_n, \quad \forall n \in \mathcal{N}, \tag{11.12c}$$

$$x_n \geq 0, \ y_n \geq 0, \quad \forall n \in \mathcal{N}. \tag{11.12d}$$

In the maximization problem above, constraint (11.12b) indicates that the total admitted SBSs to all VSPs cannot exceed the total SBSs operated by the MNO. Constraints (11.12c) indicate that the fraction of SBSs admitted to be assigned to VSP $n$ by the MNO must satisfy those requested by the respective VSP. In addition, we can notice that it will hold $x_n = y_n$ at the equilibrium, $\forall n \in \mathcal{N}$. Based on the analysis in Sect. 11.4.2, the objective function $U_1 (\mathbf{x}, \mathbf{y})$ in (11.12a) is strictly concave with respect to $\mathbf{x}$ and $\mathbf{y}$. Combining the compact and convex constraints in (11.12b), (11.12c) and (11.12d), the optimization problem formulated in (11.12) admits a unique optimal solution and can be characterized by applying necessary and sufficient Karush-Kuhn-Tucker (KKT) conditions. First, we define the Lagrangian of problem (11.12) by relaxing constraints (11.12b) and (11.12c):

$$\begin{aligned} \mathscr{L}_1 (\mathbf{x}, \mathbf{y}, \lambda, \boldsymbol{\mu}) = &\sum_{n=1}^{N} u_n^{\mathrm{VSP}} (x_n) - \sum_{n=1}^{N} v_n^{\mathrm{MNO}} (y_n) \\ &- \lambda \left( \sum_{n=1}^{N} y_n - 1 \right) - \sum_{n=1}^{N} \mu_n (x_n - y_n), \end{aligned} \tag{11.13}$$

where $\boldsymbol{\mu} \triangleq (\mu_1, \mu_2, \cdots, \mu_N)$ $(\mu_n \geq 0, \forall n \in \mathcal{N})$ and $\lambda \geq 0$ are Lagrange multiplies corresponding to constraints (11.12b) and (11.12c), respectively. Then the KKT conditions that yield optimal primal variables $\mathbf{x}^{\dagger}$ and $\mathbf{y}^{\dagger}$ and optimal dual variables $\lambda^{\dagger}$ and $\boldsymbol{\mu}^{\dagger}$ can be given by the following equations:

$$\frac{\partial u_n^{\mathrm{VSP}} \left( x_n^{\dagger} \right)}{\partial x_n} = \mu_n^{\dagger}, \ \frac{\partial v_n^{\mathrm{MNO}} \left( y_n^{\dagger} \right)}{\partial y_n} = \mu_n^{\dagger} - \lambda^{\dagger}, \ \forall n \in \mathcal{N}; \tag{11.14a}$$

$$\lambda^{\dagger} \left( \sum_{n=1}^{N} y_n^{\dagger} - 1 \right) = 0, \ \mu_n^{\dagger} \left( x_n^{\dagger} - y_n^{\dagger} \right) = 0, \ \forall n \in \mathcal{N}; \tag{11.14b}$$

$$x_n^{\dagger}, y_n^{\dagger}, \lambda^{\dagger}, \mu_n^{\dagger} \geq 0, \ \forall n \in \mathcal{N}. \tag{11.14c}$$

250 QoS-Aware Caching Resource Allocation

However, it is infeasible for the broker to derive the optimal solution of problem (11.12) by solving (11.14) directly, resulting from the insufficient information can be obtained by the broker. Specifically, VSP utility function set $\mathbf{u}^{\text{VSP}}(\mathbf{x}) \triangleq \{u_n^{\text{VSP}}(x_n), \forall n \in \mathcal{N}\}$ and MNO cost function set $\mathbf{v}^{\text{MNO}}(\mathbf{y}) \triangleq \{v_n^{\text{MNO}}(y_n), \forall n \in \mathcal{N}\}$ are local information for VSPs and the MNO, respectively, and the broker cannot be aware of such information. To eliminate this lack of information, it is necessary to design an incentive mechanism for the broker to encourage the MNO and VSPs to report their truthful admission and requirements of SBS storage for the video caching, respectively.

## 11.5.2  Iterative Double Auction Mechanism Design

A suitable scheme to deal with asymmetric information, i.e., the broker is unaware of the real needs and capability of VSPs and the MNO, respectively, is the DA mechanism. As summarized in Sect. 11.1, the DA mechanism should have the four economic properties: EE, IR, IC and BB. However, there is no such DA mechanism can satisfy the four properties at the same time [11, 44, 45]. To overcome this difficulty, we will propose an iterative DA (I-DA) mechanism, which considers the bidders (VSPs) as price-takers.[4] On this precondition, the four economic properties can be realized at the same time. In addition, the caching problem in small-cell based system with asymmetric information can also be solved.

The basic idea of the I-DA mechanism is that the broker solves a different optimization problem other than problem (11.12) to determine optimal vectors $\mathbf{x}$ and $\mathbf{y}$, according to which the pricing mechanism of the MNO and payments from VSPs can also be determined. Moreover, through applying the proposed I-DA mechanism to the caching system, the maximum social welfare formulated in (11.12a) can also be achieved.

In the I-DA mechanism, the broker facilitates the MNO and $N$ VSPs to interact iteratively and adjust their bids until the market reaches an optimal and feasible point. The detailed SBS storage resource allocation and pricing rules are introduced in the following sections.

### 11.5.2.1  I-DA Based Resource Allocation

There are two stages to implement the I-DA mechanism. In the first stage, each VSP submits a bid $\omega_n$ to the broker, and bid $\omega_n$ represents VSP $n$'s required fraction of SBSs for caching. On the other hand, the MNO submits a bid vector denoted by $\mathbf{g}$,

---

[4] The assumption of price-taking bidders is reasonable for bidders under the situation of asymmetric information, and the situation where there are a large number of participants having infinitesimal effect on market prices. Moreover, this assumption applies to the perfect competition market.

whose element $g_n$ represents the cost of the MNO when assigning SBSs to VSP $n$. Then we have bid vectors of VSPs and the MNO as

$$\boldsymbol{\omega} = (\omega_1, \omega_2, \cdots, \omega_N), \tag{11.15a}$$

$$\mathbf{g} = (g_1, g_2, \cdots, g_N), \tag{11.15b}$$

respectively.

In the second stage, after receiving the two bid vectors above, the broker determines the SBS storage allocation, i.e., how many fractions of SBS storage is assigned by the MNO to each VSP to place its videos, by solving the following optimization problem.

$$\max_{\mathbf{x}, \mathbf{y}} \ U_2(\mathbf{x}, \mathbf{y}) \triangleq \sum_{n=1}^{N} \omega_n \ln x_n - \sum_{n=1}^{N} \frac{g_n}{2} y_n^2, \tag{11.16a}$$

$$\text{s.t.} \ \sum_{n=1}^{N} y_n \leq 1, \tag{11.16b}$$

$$x_n \leq y_n, \quad \forall n \in \mathcal{N}, \tag{11.16c}$$

$$x_n \geq 0, \ y_n \geq 0, \quad \forall n \in \mathcal{N}, \tag{11.16d}$$

where the objective function (11.16a) is established based on the allocation rule in [35, 46]. In addition, the designed functions $\tilde{u}^{\text{VSP}}(x) \triangleq \ln x$ and $\tilde{v}^{\text{MNO}}(y) \triangleq y^2/2$ capture the concave and convex increasing properties of the utility function of VSP and the cost function of MNO, respectively. Therefore, the objective function of the optimization problem formulated in (11.16) is strictly concave. Considering the same constraints as problem (11.12), optimization problem (11.16) admits a unique optimal solution. Define the Lagrangian of problem (11.16) by relaxing constraints (11.16b) and (11.16c):

$$\mathcal{L}_2(\mathbf{x}, \mathbf{y}, \lambda, \boldsymbol{\mu}) = \sum_{n=1}^{N} \omega_n \ln x_n - \sum_{n=1}^{N} \frac{g_n}{2} y_n^2$$
$$- \lambda \left( \sum_{n=1}^{N} y_n - 1 \right) - \sum_{n=1}^{N} \mu_n (x_n - y_n). \tag{11.17}$$

Then the KKT conditions that yield optimal primal variables $\mathbf{x}^*$ and $\mathbf{y}^*$, and optimal dual variables $\lambda^*$ and $\boldsymbol{\mu}^*$ can be given by the following equations, $\forall n \in \mathcal{N}$.

$$\frac{\omega_n}{x_n^*} = \mu_n^*, \ g_n y_n^* = \mu_n^* - \lambda^*, \ \forall n \in \mathcal{N}; \tag{11.18a}$$

$$\lambda^* \left( \sum_{n=1}^{N} y_n^* - 1 \right) = 0, \ \mu_n^* \left( x_n^* - y_n^* \right) = 0, \ \forall n \in \mathcal{N}; \tag{11.18b}$$

$$x_n^*, y_n^*, \lambda^*, \mu_n^* \geq 0, \ \forall n \in \mathcal{N}. \tag{11.18c}$$

According to conditions in (11.18a), we have the SBS storage allocation rules as

$$x_n^* = \frac{\omega_n}{\mu_n^*}, \tag{11.19a}$$

$$y_n^* = \frac{\mu_n^* - \lambda^*}{g_n}. \tag{11.19b}$$

### 11.5.2.2   I-DA Based Pricing

Comparing KKT conditions (11.14) with (11.18), we can notice that when the submitted bids from VSPs and the MNO satisfy the following equations:

$$\omega_n = x_n^* \cdot \frac{\partial u_n^{\text{VSP}} \left( x_n^* \right)}{\partial x_n}, \tag{11.20a}$$

$$g_n = \frac{1}{y_n^*} \cdot \frac{\partial v_n^{\text{MNO}} \left( y_n^* \right)}{\partial y_n}, \tag{11.20b}$$

the optimization problems formulated in (11.12) and (11.16) have the same optimal solution, i.e., $\mathbf{x}^\dagger = \mathbf{x}^*$ and $\mathbf{y}^\dagger = \mathbf{y}^*$. An incentive pricing and payment rules for the MNO and VSPs, which are designed by the broker, should be capable of encouraging them to provide bids as (11.20). Next, we will design such payment rules required from the broker to VSPs and payments rules implemented from the MNO to the broker.

**For VSP Bidders**   Denote $\rho_n^{\text{VSP}}(x_n)$ as the payment requested to be paid by VSP $n$ to the broker when this VSP requests the fraction of SBSs as $x_n$. It should be noted that $x_n$ here is not requested directly from VSP $n$, but is determined by the VSP $n$'s bid $\omega_n$ and applying the allocation scheme ruled by (11.19a). In addition, this pricing rule is implemented by the broker, and each VSP just needs to optimize their bid $\omega_n$ ($\forall n \in \mathcal{N}$) to get a maximum utility after the payment. Then the optimization problem for VSP $n$ ($\forall n \in \mathcal{N}$) is given by

$$\max_{\omega_n} \ u_n^{\text{VSP}}(x_n) - \rho_n^{\text{VSP}}(x_n), \tag{11.21a}$$

$$\text{s.t.} \ \omega_n \geq 0. \tag{11.21b}$$

According to (11.6), one can notice that optimization problem (11.21) can be solved locally by each VSP. Then take the first derivative of objective function (11.21a) with respect to $\omega_n$, and according to allocation rule (11.19a), we derive that the unique optimal solution of problem (11.21) satisfies

$$\frac{\partial \rho_n^{\text{VSP}}(x_n)}{\partial \omega_n} = \frac{\partial u_n^{\text{VSP}}(x_n)}{\partial x_n} \cdot \frac{\partial x_n}{\partial \omega_n} = \frac{1}{\mu_n} \cdot \frac{\partial u_n^{\text{VSP}}(x_n)}{\partial x_n}. \tag{11.22}$$

Considering that the maximum social welfare can be achieved when VSP $n$ submits bid according to (11.20a), we have

$$\frac{\partial \rho_n^{\text{VSP}}(x_n)}{\partial \omega_n} = \frac{1}{\mu_n} \cdot \frac{\omega_n}{x_n} = 1. \tag{11.23}$$

Then the pricing rule requiring VSP $n$ to pay the broker is given

$$\rho_n^{\text{VSP}}(\omega_n) = \omega_n. \tag{11.24}$$

**For the MNO Bidder** The MNO needs to submit a bid vector denoted by $\mathbf{g} = (g_1, g_2, \cdots, g_N)$ to the broker. Denote $\rho_n^{\text{MNO}}(y_n)$ as the payment given by the broker to the MNO when the MNO admits fraction of SBSs $y_n$ to VSP $n$ for the caching service. Similarly, $y_n$ is determined by the broker by considering MNO's bid $g_n$ and applying the allocation rule in (11.19b), $\forall n \in \mathcal{N}$. Then the optimization problem for the MNO when submitting bid $\mathbf{g}$ can be given by

$$\max_{\mathbf{g}} \ \sum_{n=1}^{N} \rho_n^{\text{MNO}}(y_n) - v_n^{\text{MNO}}(y_n), \tag{11.25a}$$

$$\text{s.t.} \ g_n \geq 0, \ \forall n \in \mathcal{N}. \tag{11.25b}$$

Take the first derivative of objective function (11.25a) with respect to $g_n$, and according to allocation rule (11.19b), we can compute that the unique optimal solution of problem (11.25) satisfies

$$\frac{\partial \rho_n^{\text{MNO}}(y_n)}{\partial g_n} = \frac{\partial u_n^{\text{MNO}}(y_n)}{\partial y_n} \cdot \frac{\partial y_n}{\partial g_n} = \frac{\lambda - \mu_n}{g_n^2} \cdot \frac{\partial u_n^{\text{MNO}}(y_n)}{\partial y_n}. \tag{11.26}$$

Considering that the maximum social welfare can be achieved when the MNO submits bid according to (11.20a), we have

$$\frac{\partial \rho_n^{\text{MNO}}(y_n)}{\partial g_n} = \frac{\lambda - \mu_n}{g_n^2} \cdot g_n y_n = -\frac{\left(\mu_n^* - \lambda^*\right)^2}{g_n^2}. \tag{11.27}$$

Then the pricing rule that requires the broker to pay the MNO is given by

$$\rho_n^{\text{MNO}}(g_n) = \frac{\left(\mu_n^* - \lambda^*\right)^2}{g_n} = y_n \left(\mu_n^* - \lambda^*\right). \tag{11.28}$$

Summarily, the SBS storage allocation mechanisms as (11.19) and the pricing rules defined in (11.24) and (11.28) can ensure the maximum social welfare to be achieved.

## 11.6   Implementation of I-DA Mechanism

In this section, we will first design the I-DA algorithm to implement the I-DA based caching mechanism proposed in Sect. 11.5. Then we will demonstrate that the designed iterative algorithm is convergent and the proposed caching mechanism can satisfy the four economic properties, i.e., EE, IR, IC and BB, simultaneously.

### 11.6.1   I-DA Mechanism Based Algorithm

First, we explain how the designed caching mechanism in the previous section works to achieve the maximum social welfare. In order to elicit the hidden information among participants of the caching system, an iteration based algorithm is needed to gradually adjust that whether the submitted bids can lead to a desirable resource allocation and pricing rules. To this ends, a primal-dual Lagrange decomposition approach [46, 47] is introduced, and main operations of the algorithm are summarized as follows. In each iteration, the broker announces current Lagrange multiplies $\lambda$ and $\mu$. Then each VSP and the MNO compute their optimal bids $x_n$ ($\forall n \in \mathcal{N}$) and $\mathbf{y}$ by solving problem (11.21) and (11.25), respectively, and then submit them to the broker. After receiving VSPs' and MNO's bids, the broker decides SBS storage allocation vectors $\mathbf{x}$ and $\mathbf{y}$ according to scheme (11.19). Meanwhile, the broker updates $\lambda$ and $\boldsymbol{\mu}$ according to

$$
\begin{aligned}
\lambda^{(t+1)} &= \max \left\{ \lambda^{(t)} - \varepsilon \frac{\partial \mathcal{L}_2 (\mathbf{x}, \mathbf{y}, \lambda, \boldsymbol{\mu})}{\partial \lambda}, 0 \right\} \\
&= \max \left\{ \lambda^{(t)} + \varepsilon \left( \sum_{n=1}^{N} y_n - 1 \right), 0 \right\}, \\
\mu_n^{(t+1)} &= \max \left\{ \mu_n^{(t)} - \varepsilon \frac{\partial \mathcal{L}_2 (\mathbf{x}, \mathbf{y}, \lambda, \boldsymbol{\mu})}{\partial \mu_n}, 0 \right\} \\
&= \max \left\{ \mu_n^{(t)} + \varepsilon (x_n - y_n), 0 \right\}, \forall n \in \mathcal{N},
\end{aligned}
\tag{11.29}
$$

respectively, where $\varepsilon > 0$ is the step size, and $t$ is the index of iteration. Then the broker needs to adjust that whether the current bids reach stability, and if they do not, the procedure above is repeated until the system reaches equilibrium.

Next, in Algorithm 3, we summarize the implementation of the I-DA based mechanism, which can provide the SBS storage allocation scheme and caching service pricing rules in the small-cell based caching system. In Step 10, $0 < o_1, o_2 \ll 1$ are convergence indexes. As implemented in Algorithm 3, the MNO and VSPs find their optimal bids by solving (11.25) and (11.21), respectively, in a distributed way. On the other hand, the broker, who performs as a centralized controller, decides the optimal storage allocation strategies and pricing rules for the MNO and VSPs.

## 11.6.2   Convergence of I-DA Algorithm

Next, we provide the convergence behavior of Algorithm 3 in Theorem 11.1.

**Theorem 11.1** *Algorithm 3 designed for the small-cell based caching system converges to the unique optimal solution of the social welfare maximization problem formulated in (11.12).*

**Proof** According to (11.29), dynamics of $\lambda$ and $\boldsymbol{\mu}$ is given by

$$
\dot{\lambda} (t) =
\begin{cases}
\sum_{n=1}^{N} y_n - 1, & \lambda > 0, \\
0, & \lambda = 0;
\end{cases}
\tag{11.30a}
$$

$$
\dot{\mu}_n (t) =
\begin{cases}
x_n - y_n, & \mu_n > 0, \\
0, & \mu_n = 0,
\end{cases}
\quad \forall n \in \mathcal{N}.
\tag{11.30b}
$$

---

**Algorithm 3** Iterative Double Auction (I-DA) Algorithm

---

**Initialization:**
1: Create and initialize $\mathbf{x}^{(0)}$, $\mathbf{y}^{(0)}$, $\lambda^{(0)}$ and $\boldsymbol{\mu}^{(0)}$;
2: Initialize index: $t = 0$, conv $= 0$, $\varepsilon$, $o_1$, $o_2 > 0$.
3: **while** conv $= 0$ **do**
4:     The broker announces $\lambda^{(t)}$ and $\boldsymbol{\mu}^{(t)}$;
5:     VSP $n$ ($\forall n$): derives optimal bid $\omega_n^{(t)}$ by solving problem (11.21), and submits it to the broker;
6:     MNO: derives optimal bid vector $\mathbf{g}^{(t)}$ by solving problem (11.25), and submits it to the broker;
7:     Broker: decides the updated allocation schemes $\mathbf{x}^{(t)}$ and $\mathbf{y}^{(t)}$ according to

$$x_n^{(t)} = \frac{\omega_n^{(t)}}{\mu_n^{(t)}}, \quad y_n^{(t)} = \frac{\mu_n^{(t)} - \lambda^{(t)}}{g_n^{(t)}};$$

8:     Broker: computes the updated $\lambda^{(t)}$ and $\boldsymbol{\mu}^{(t)}$ through

$$\lambda^{(t+1)} = \max\left\{\lambda^{(t)} + \varepsilon\left(\sum_{n=1}^{N} y_n^{(t)} - 1\right), 0\right\},$$

$$\mu_n^{(t+1)} = \max\left\{\mu_n^{(t)} + \varepsilon\left(x_n^{(t)} - y_n^{(t)}\right), 0\right\}, \forall n \in \mathcal{N}.$$

9:     Broker: checks convergence
10:    **if** $\left|\frac{\omega_n^{(t)} - \omega_n^{(t-1)}}{\omega_n^{(t-1)}}\right| < o_1$ and $\left|\frac{g_n^{(t)} - g_n^{(t-1)}}{g_n^{(t-1)}}\right| < o_2$ **then**
11:        conv $= 1$;
12:        decides pricing rules $\rho_n^{\text{VSP}}(x_n)$ and $\rho_n^{\text{MNO}}(y_n)$ by

$$\rho_n^{\text{VSP}}\left(\omega_n^{(t)}\right) = \omega_n^{(t)}, \rho_n^{\text{MNO}}\left(g_n^{(t)}\right) = \frac{\left(\mu_n^{(t)} - \lambda^{(t)}\right)^2}{g_n^{(t)}}.$$

13:    **else**
14:        $t = t + 1$.
15:    **end if**
16: **end while**
**Output:**
17: Optimal storage allocation: $\mathbf{x}^*$, $\mathbf{y}^*$;
18: Optimal pricing: $\rho_n^{\text{VSP}}(\omega_n^*)$, $\rho_n^{\text{MNO}}(g_n^*)$, $\forall n \in \mathcal{N}$.

---

Define the Lyapunov function as

$$V(\lambda, \boldsymbol{\mu}) = \frac{1}{2}(\lambda - \lambda^*)^2 + \frac{1}{2}\sum_{n=1}^{N}(\mu_n - \mu_n^*)^2. \tag{11.31}$$

Taking the first derivative of Lyapunov function $V(\lambda, \mu)$ with respect to $t$, we have

$$\frac{dV(\lambda, \mu)}{dt} = (\lambda - \lambda^*)\dot{\lambda}(t) + \sum_{n=1}^{N}(\mu_n - \mu_n^*)\dot{\mu}_n(t)$$

$$\leq (\lambda - \lambda^*)\left(\sum_{n=1}^{N}y_n - 1\right) + \sum_{n=1}^{N}(\mu_n - \mu_n^*)(x_n - y_n)$$

$$= (\lambda - \lambda^*)\left(\sum_{n=1}^{N}y_n - \sum_{n=1}^{N}y_n^*\right) + (\lambda - \lambda^*)\left(\sum_{n=1}^{N}y_n^* - 1\right)$$

$$+ \sum_{n=1}^{N}(\mu_n - \mu_n^*)\left[x_n - x_n^* - (y_n - y_n^*)\right] + \sum_{n=1}^{N}(\mu_n - \mu_n^*)(x_n^* - y_n^*)$$

$$= \sum_{n=1}^{N}(\mu_n - \mu_n^*)(x_n - x_n^*) + \sum_{n=1}^{N}(\lambda - \lambda^* - \mu_n + \mu_n^*)(y_n - y_n^*)$$

$$+ \lambda\left(\sum_{n=1}^{N}y_n^* - 1\right) + \sum_{n=1}^{N}\mu_n(x_n^* - y_n^*)$$

$$\leq \sum_{n=1}^{N}\left(\frac{\partial u_n^{VSP}(x_n)}{\partial x_n} - \frac{\partial u_n^{VSP}(x_n^*)}{\partial x_n}\right)(x_n - x_n^*)$$

$$+ \sum_{n=1}^{N}\left(-\frac{\partial v_n^{MNO}(y_n)}{\partial y_n} + \frac{\partial v_n^{MNO}(y_n^*)}{\partial y_n}\right)(y_n - y_n^*).$$

Considering the strictly concave property of VSP utility function $u_n^{VSP}(x_n)$ and the strictly convex property of MNO cost function $v_n^{MNO}(y_n)$, we have $\forall n \in \mathcal{N}$,

$$\sum_{n=1}^{N}\left(\frac{\partial u_n^{VSP}(x_n)}{\partial x_n} - \frac{\partial u_n^{VSP}(x_n^*)}{\partial x_n}\right)(x_n - x_n^*) < 0, \tag{11.32a}$$

$$\sum_{n=1}^{N}\left(-\frac{\partial v_n^{MNO}(y_n)}{\partial y_n} + \frac{\partial v_n^{MNO}(y_n^*)}{\partial y_n}\right)(y_n - y_n^*) < 0. \tag{11.32b}$$

Thus we conclude that

$$\frac{dV(\lambda, \mu)}{dt} < 0. \tag{11.33}$$

This completes the proof of the algorithm convergence.

### 11.6.3   Economic Properties of I-DA Mechanism

In this part, we provide economic properties processed by the designed I-DA based caching mechanism in Theorem 11.2, and show that the four economic properties can be satisfied at the same time ideally.

**Theorem 11.2** *The I-DA mechanism designed for the small-cell based caching system satisfies the four economic properties, i.e., Economic Efficiency, Individual Rationally, Incentive Compatibility and Budge Balance, at the same time.*

*Proof*

(1) **Economic Efficiency:** According to Theorem 11.1, we can conclude that Algorithm 3 converges to an optimal solution satisfying KKT conditions (11.18a). In addition, applying pricing rules regulated as (11.24) and (11.28) when the broker asks payment from VSPs and pays the MNO, respectively, the optimal bids obtained by Algorithm 3 are used to deduce optimal allocation mechanisms $\mathbf{x}^*$ and $\mathbf{y}^*$. These allocation mechanisms are equal to the optimal solutions of problem (11.12), under the situation that each VSP and the MNO submit bids $\omega_n$ and $\mathbf{g}$ according to (11.20). This completes the proof of efficiency of the I-DA mechanism.

(2) **Individual Rationality:** Since $u_n^{\mathrm{VSP}}(x_n)$ is a strictly concave function of $x_n$ and $u_n^{\mathrm{VSP}}(0) = 0$, then we have

$$u_n^{\mathrm{VSP}}\left(x_n^*\right) > u_n^{\mathrm{VSP}}(0) + x_n^* \frac{\partial u_n^{\mathrm{VSP}}\left(x_n^*\right)}{\partial x_n} = x_n^* \frac{\partial u_n^{\mathrm{VSP}}\left(x_n^*\right)}{\partial x_n}. \qquad (11.34)$$

Considering the optimal bid of VSP satisfying (11.20a) and the pricing rule in (11.24), we derive that

$$x_n^* \frac{\partial u_n^{\mathrm{VSP}}\left(x_n^*\right)}{\partial x_n} = x_n^* \cdot \frac{\omega_n^*}{x_n^*} = \omega_n^* = \rho_n^{\mathrm{VSP}}\left(\omega_n^*\right), \qquad (11.35)$$

which implies that

$$u_n^{\mathrm{VSP}}\left(x_n^*\right) > \rho_n^{\mathrm{VSP}}\left(\omega_n^*\right), \quad \forall n \in \mathcal{N} \qquad (11.36)$$

always holds. Similarly, considering the strictly convex property of MNO's cost function $v_n^{\mathrm{MNO}}(y_n)$, optimal bid of the MNO $g_n^*$ satisfying (11.20b) and the pricing rule as (11.28) for VSP $n$ ($\forall n \in \mathcal{N}$), we can deduce that $v_n^{\mathrm{MNO}}\left(x_n^*\right) < \rho_n^{\mathrm{MNO}}\left(g_n^*\right)$, and then for the MNO, the total net utility

$$\sum_{n=1}^{N}\left[\rho_n^{\mathrm{MNO}}\left(g_n^*\right) - v_n^{\mathrm{MNO}}\left(y_n^*\right)\right] > 0 \qquad (11.37)$$

always holds. So in conclusion, each VSP and the MNO will always obtain a positive net utility when they honestly participate the I-DA based caching mechanism. Moreover, each participant gets a zero utility if it does not participate the auction. Therefore, the designed I-DA mechanism always satisfies IR conditions.

(3) **Incentive Compatibility:** According to the derivation in Sects. 11.5 and 11.6, we can observe that, VSPs and the MNO do not have to report their local-known and real caching storage requests and admission, but submit their current optimal bids to the broker by locally solving optimization problems (11.21) and (11.25), respectively. In addition, these iteratively updated optimal bids can gradually reveal VSPs' hidden utility and MNO's hidden cost. In other words, although VSPs and the MNO do not share their information with each other and the broker, their optimal bids to the broker can elicit the asymmetric information, and the maximum social welfare can be reached by proper pricing rules defined by (11.24) and (11.28). Therefore, we can conclude that the designed I-DA based caching mechanism processes the characteristics of IC.

(4) **Budge Balance:** According to local optimization problems of each VSP and MNO formulated respectively by (11.21) and (11.25), the budget of the broker can be given by

$$
\Gamma_b(\boldsymbol{\omega}, \mathbf{g}) = \sum_{n=1}^{N} \left[ \rho_n^{VSP}(\omega_n) - \rho_n^{MNO}(g_n) \right]
$$

$$
= \sum_{n=1}^{N} \left[ \omega_n - \frac{1}{g_n}(\mu_n - \lambda)^2 \right] = \sum_{n=1}^{N} [\lambda y_n - \mu_n (y_n - x_n)].
\tag{11.38}
$$

When optimal bids from the both sides of participants in DA are submitted, we have

$$
\Gamma_b(\boldsymbol{\omega}^*, \mathbf{g}^*) = \sum_{n=1}^{N} \left[ y_n^* \lambda^* - (y_n^* - x_n^*) \mu_n^* \right]
$$

$$
= \lambda^* \left( \sum_{n=1}^{N} y_n^* - 1 \right) - \sum_{n=1}^{N} (y_n^* - x_n^*) \mu_n^* + \lambda^* = \lambda^* \geq 0,
\tag{11.39}
$$

which indicates that the broker will always get a nonnegative budget when implementing pricing rules defined in (11.24) and (11.28).

Summarizing the proof of the four economic properties above, we can conclude that the designed I-DA based caching mechanism is capable of satisfying *Economic Efficiency*, *Individual Rationality*, *Incentive Capability* and *Budget Balance* at the same time. This completes the proof of Theorem 11.2.

*Remark 11.3* As summarized in Theorems 11.1 and 11.2, the maximum social welfare will always be achieved by the I-DA based caching mechanism proposed. Moreover, resulting from the IR and IC properties, both the MNO and VSPs are encourage to participate the caching resource auction actively, and reflect their caching costs and requests truthfully, respectively. Furthermore, the BB property will ensure the feasibility of designed caching mechanism for the third-party broker, which means that the broker does not have to invest additional "money" to make the mechanism go round.

## 11.7   Evaluation Results

This part provides numerical results to demonstrate and test the validity and effectiveness of the designed I-DA algorithm for the small-cell based video caching system. In addition, the convergence and economic properties of the proposed I-DA based caching mechanism are also verified through the simulation.

First of all, we introduce the scenario setup for simulations. We consider a video caching system with an MNO and multiple VSPs. The unit cost of the MNO to maintain the assigned SBSs is set as $c = 0.005$, and detailed settings of other parameters are shown in Table 11.2.

To test the performance of the I-DA based caching mechanism and observe the related optimal behaviors of each SBS and the MNO, we first consider a small number of VSPs in the system, i.e., $N = 3$. Set the step size as $\varepsilon = 0.1$. According to the definition of VSP preference in (11.3), the three VSP preferences can be calculated, i.e., $q_1 = 0.5455$, $q_2 = 0.2727$ and $q_3 = 0.1818$. Then we apply

**Table 11.2** Detailed system parameters

| Parameter | Value |
| --- | --- |
| Transmission power of each SBS $P$ | 10 W [41] |
| Number of videos can be stored at each SBS $Q$ | 40 [41] |
| Number of video files $F$ | 100 [41] |
| Average number of video requests $K$ | 50/month [41] |
| Density of SBS $\varphi$ | 20/km$^2$ [41] |
| Density of MU $\psi$ | 80/km$^2$ [41] |
| Exponent of video popularity $\alpha$ | 0.2 [41] |
| Exponent of VSP preference $\beta$ | 1 [41] |
| Path loss exponent $\gamma$ | 4 [39] |
| Variance of the AWGN $\sigma^2$ | $10^{-10}$ Watt [39] |
| SINR threshold $\delta$ | 0.02 [41] |
| LDS $r_{ld}$ | $10^{-3}$ |
| ABC $r_{bh}$ | $5 \times 10^{-5}$ |
| Convergence indexes $o_1, o_2$ | $10^{-5}$ |

**Table 11.3** Simulation results of allocation $(\mathbf{x}, \mathbf{y})$, bids $(\boldsymbol{\omega}, \mathbf{g})$, pricing $(\boldsymbol{\rho}^{\text{VSP}}, \boldsymbol{\rho}^{\text{MNO}})$ and lagrange multiplies $(\boldsymbol{\lambda}, \boldsymbol{\mu})$

| Parameter | | $\text{VSP}_1$ | $\text{VSP}_2$ | $\text{VSP}_1$ |
|---|---|---|---|---|
| Allocation | $\mathbf{x}^*$ | 0.4154 | 0.2994 | 0.2441 |
| | $\mathbf{y}^*$ | 0.4154 | 0.2994 | 0.2441 |
| Bids | $\boldsymbol{\omega}^*$ | 0.6571 | 0.3414 | 0.2269 |
| | $\mathbf{g}^*$ | 3.8095 | 3.8095 | 3.8096 |
| Pricing | $\left(\boldsymbol{\rho}^{\text{VSP}}\right)^*$ | 0.6571 | 0.3414 | 0.2269 |
| | $\left(\boldsymbol{\rho}^{\text{MNO}}\right)^*$ | 0.6571 | 0.3414 | 0.2269 |
| VSP utility | $\left(\mathbf{u}^{\text{VSP}}\right)^*$ | 1.6374 | 0.7085 | 0.4258 |
| MNO cost | $\left(\mathbf{v}^{\text{MNO}}\right)^*$ | 0.2631 | 0.1438 | 0.0978 |
| Lagrange multiplies | $\lambda^*$ | 0 | | |
| | $\boldsymbol{\mu}^*$ | 1.5821 | 1.1405 | 0.9298 |

Algorithm 3 to the caching system, and optimal allocation (request of fraction of SBS storage) $\mathbf{x}^*$, submitted bid $\boldsymbol{\omega}^*$, payment to the broker $\left(\boldsymbol{\rho}^{\text{VSP}}\right)^*$ and maximized utility $\left(\mathbf{u}^{\text{VSP}}\right)^*$ of each VSP, as well as optimal allocation (acceptance of fraction of SBS storage) $\mathbf{y}^*$, submitted bid $\mathbf{g}^*$, payment obtained from the broker $\left(\boldsymbol{\rho}^{\text{MNO}}\right)^*$ and minimized cost $\left(\mathbf{v}^{\text{MNO}}\right)^*$ of the MNO, are obtained, and results are shown in Table 11.3. In addition, we record final values of Lagrange multiplies $\lambda^*$ and $\boldsymbol{\mu}^*$ when the iteration stops at the equilibrium, and also provide them in Table 11.3.

Observing the optimal results in Table 11.3, we notice that $\mathbf{x}^* = \mathbf{y}^*$ and $\left(\boldsymbol{\rho}^{\text{VSP}}\right)^* = \left(\boldsymbol{\rho}^{\text{MNO}}\right)^*$. The former equilibrium indicates that, by applying the designed allocation schemes in (11.19), the MNO and all VSPs can finally agree on the fraction of SBSs that should be assigned to each VSP for caching service, i.e., $x_n = y_n$, $\forall n \in \mathcal{N}$. The later equilibrium demonstrates that the broker makes ends meet when the iteration stops. In addition, equilibrium $\left(\boldsymbol{\rho}^{\text{VSP}}\right)^* = \left(\boldsymbol{\rho}^{\text{MNO}}\right)^*$ also validates the optimal budget of the broker that is derived in (11.39), i.e., $\Gamma_b (\boldsymbol{\omega}^*, \mathbf{g}^*) = \lambda^*$, since $\lambda^* = 0$ shown in Table 11.3. Furthermore, this balance of payments at the broker also verifies that the property of **BB** can be satisfied by applying the designed I-DA algorithm.

In addition, results in Table 11.3 show that $u_n^{\text{VSP}} > \rho_n^{\text{VSP}}$ and $v_n^{\text{MNO}} < \rho_n^{\text{MNO}}$, $\forall n \in \mathcal{N}$. These results imply that the utility obtained by every VSP exceeds its payment to the broker, and the payment received by the MNO from the broker can cover its cost caused by providing caching service. Thus, the property of **IR** can be verified by these results.

Next, we test the convergence and other characteristics of the designed I-DA based caching mechanism. We record values of parameters, which are listed in Table 11.3, updated in every iteration before convergence conditions being satisfied. The evolution of these parameters are shown in Figs. 11.3, 11.4, 11.5, and 11.6.

In Fig. 11.3a, we notice that bids submitted by VSP 1 and VSP 2 are always larger than their allocated fractions of SBSs decided by the broker, i.e., $\omega_n^{(t)} > x_n^{(t)}$, $\forall t$ and $n = 1, 2$. Conversely, for VSP 3, $\omega_3^{(t)} < x_3^{(t)}$ always holds $\forall t$. This phenomenon results from the decreasing preference of the three VSPs, i.e., $q_1 > q_2 > q_3$.

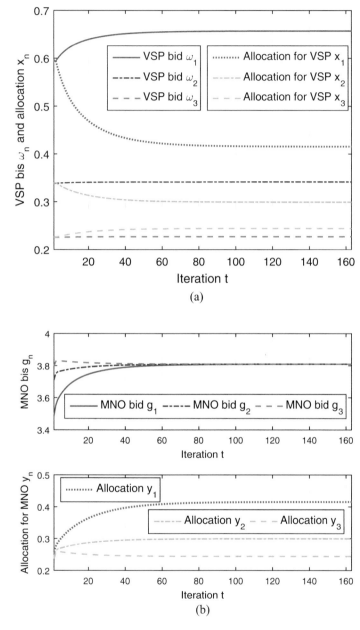

**Fig. 11.3** Evolution of participants' (VSPs and the MNO) bids and SBSs' storage allocation (storage requests and storage admission) by the I-DA algorithm based caching mechanism ($\varepsilon = 0.1$ and $N = 3$). (**a**) Evolution of VSPs' bids and SBSs allocation for them. (**b**) Evolution of MNO's bids and SBSs allocation for it

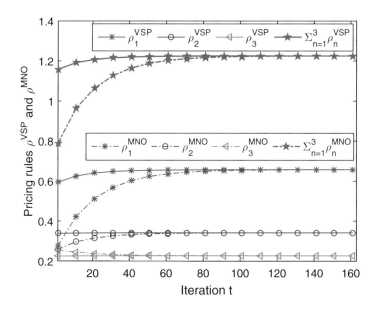

**Fig. 11.4** Budge balance property of I-DA algorithm and evolution of payment (pricing rules) regulated from VSPs to the broker $\rho^{VSP}$ and from the broker to the MNO $\rho^{MNO}$ ($\varepsilon = 0.1$ and $N = 3$)

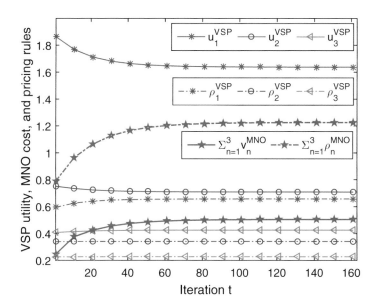

**Fig. 11.5** IR property of I-DA algorithm and evolution of VSPs' utility $\mathbf{u}^{VSP}$, MNO's cost $\mathbf{v}^{MNO}$ and payment (pricing rules) regulated from VSPs to the broker $\rho^{VSP}$ and from the broker to the MNO $\rho^{MNO}$ ($\varepsilon = 0.1$ and $N = 3$)

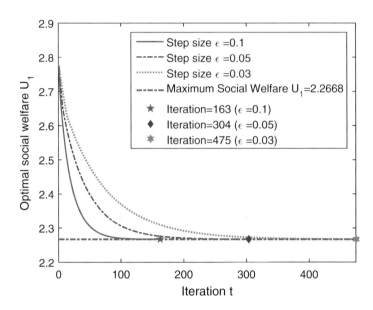

**Fig. 11.6** Impact of step size $\varepsilon$ on the evolution of social welfare reached by the I-DA based caching mechanism ($N = 3$)

Specifically, VSP's bids are calculated locally and separately at each VSP by solving optimization problem (11.21). Resulting from the hidden information among VSPs, each VSP is unaware of other VSPs' bids, and VSP 1 and VSP 2 tend to submit high bids due to their high preferences. However, the broker, as a central controller, cannot allocate the fraction of SBSs as high as bids from VSP 1 and VSP 2. The broker also has to ensure its allocation for VSPs to satisfy $\sum_{n=1}^{N} x_n \leq \sum_{n=1}^{N} y_n \leq 1$. Therefore, $x_1$ and $x_2$ decrease with the increasing of iteration times. In addition, to optimize the social welfare, the broker tends to allocate more SBS storage to these two VSPs, which have a potential ability of saving more backhaul cost, than to VSP 3. Furthermore, we can notice that VSP 1 submits a sequence of increasing bids as the iteration progress. This result verifies the **IC** property of the designed I-DA algorithm, which is capable of prompting VSPs with higher preferences to submit higher bids to "win" more SBS storage.

Results in Fig. 11.3b show that bids from the MNO to the broker for VSP 1 and VSP 2 increase with increasing iteration times, which results from the larger preferences of the two VSPs than that of VSP 3. So the MNO is willing to assign more SBS storage to VSP 1 and VSP 2 to save more backhaul cost with a high probability. Because of the small preference of VSP 3, $g_3$ decreases as the iterations progress, although it increases at the beginning. Meanwhile, due to the same reason discussed above, the allocations of caching admission **y** for the MNO, decided by the broker, present a similar tendency to **g**, as shown in the second figure in Fig. 11.3b. In addition, the **IC** property of the I-DA algorithm can also be verified since that results show an incentive capability for the MNO, who is promoted to submit higher bids to ensure VSPs with higher preferences can be assigned more SBS storage.

The evolution of the payment regulated from each VSP to the broker $\rho^{\text{VSP}}$ and from the broker to the MNO $\rho^{\text{MNO}}$ is shown in Fig. 11.4. we notice that $\rho^{\text{VSP}}$ and $\rho^{\text{MNO}}$ present similar tendencies to $\mathbf{x}$ and $\mathbf{y}$, respectively. This phenomenon results from pricing rules defined in (11.24) and (11.28). In addition, $\rho^{\text{VSP}}$ and $\rho^{\text{MNO}}$ ($\forall n \in \mathcal{N}$) converge to same values with increasing iteration times, and $\rho^{\text{VSP}} \leq \rho^{\text{MNO}}$ ($\forall n \in \mathcal{N}$) before the convergence points. Such results validate the **BB** property of the I-DA algorithm.

Figure 11.5 shows the evolution of VSP utilities, MNO costs and their payments. As shown in Fig. 11.5, $u_1^{\text{VSP}}$ and $u_2^{\text{VSP}}$ decrease with the increasing of iterations before the equilibrium, which results from that the allocated SBS storage for VSP 1 and VSP2 decrease as shown in Fig. 11.3a. In addition, results in Fig. 11.5 show that $\left(u_n^{\text{VSP}}\right)^{(t)} > \left(\rho_n^{\text{VSP}}\right)^{(t)}$, $\forall n \in \mathcal{N}$, and $\left(\sum_{n=1}^{3} v_n^{\text{MNO}}\right)^{(t)} < \left(\sum_{n=1}^{3} \rho_n^{\text{MNO}}\right)^{(t)}$, $\forall t$, which verify the **IR** property of the I-DA algorithm.

Then we test the social welfare can be reached during iterations. When iteration step size is set as $\epsilon = 0.1$, the evolution of social welfare $U_1$ is shown as the solid curve in Fig. 11.6. The straight pecked line in Fig. 11.6 indicates the value of $U_1^{\dagger}$, and is calculated by solving problem (11.12) when assuming that all necessary information is known. Results of these two lines indicate that by applying the designed I-DA based caching mechanism, the allocated $\mathbf{x}$ and $\mathbf{y}$ can lead the social welfare of the caching system to converge to the maximum social welfare. This result verifies the **EE** property of the designed algorithm.

In addition, the evolution of allocation $(\mathbf{x}, \mathbf{y})$, bids $(\boldsymbol{\omega}, \mathbf{g})$, pricing $\left(\rho^{\text{VSP}}, \rho^{\text{MNO}}\right)$ and the social welfare, shown in Figs. 11.3, 11.4, 11.5, and 11.6, demonstrate that the designed iterative algorithm for the caching optimization problem is of fast **convergence** property. Furthermore, obtained optimal allocation solutions $(\mathbf{x}^*, \mathbf{y}^*)$ by applying Algorithm 3 are equal to $(\mathbf{x}^{\dagger}, \mathbf{y}^{\dagger})$, and $\mathbf{x}^* = \mathbf{y}^* = \mathbf{x}^{\dagger} = \mathbf{y}^{\dagger}$ is the optimal solution of social welfare optimization problem (11.12), and then makes sure that the maximum social welfare can be reached.

With the same parameter settings of the small-cell based caching system above, we explore that how step size $\varepsilon$ affects the convergence speed of Algorithm 3. We consider another two smaller step sizes selecting values in $\{0.05, 0.03\}$, which are used for updating Lagrange multiplies (Step 8 of Algorithm 3). For this experiment, the convergence indexes are still set as $o_1 = o_2 = 10^{-5}$. Figure 11.6 shows the evolution of social welfare for three values of the three step size. Results in Fig. 11.6 indicate that the iteration time that the algorithm needs in order to converge to the optimal solution increases with decreasing $\varepsilon$. Specifically, when $\varepsilon = 0.1$, as set in the previous experiment, the I-DA algorithm converges after 163 iterations, while for $\varepsilon = 0.05$ and $\varepsilon = 0.03$, it needs 304 and 475 iterations in order to converge to the optimal solution, respectively. However, all these three small step sizes can ensure the I-DA algorithm to be convergent, as shown in Fig. 11.6. Nevertheless, when we increase $\varepsilon$ to the values larger than 2, we will see that the algorithm is not convergent any more.

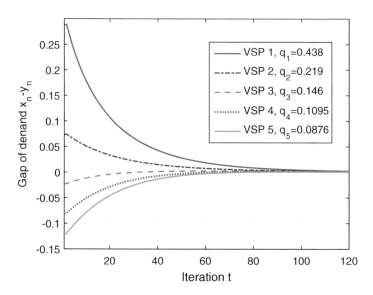

**Fig. 11.7** Evolution of the demand gap between caching request **x** and caching admission **y** by applying the SBSs allocation mechanism ruled in (11.19) (Iteration time before convergency is 324 when $\varepsilon = 0.1$ and $N = 5$)

Next, we test the convergence of the gap between **x** and **y**. To better reveal the relationship between **x** and **y**, we consider there are $N = 5$ VSPs in the system. Set $\varepsilon = 0.1$, and other parameters remain the same. In this simulation, the I-DA algorithm needs 324 iterations to reach the optimal solution. While after $t = 120$, the gap between **x** and **y** is not change observably. So in Fig. 11.7, we only show how $x_n - y_n$ ($\forall n \in \mathcal{N}$) changes in the first 120 iterations. Results in Fig. 11.7 indicate that for VSP 1 and VSP 2, who are with higher VSP preference, the value of $x_n - y_n$ ($n = 1,\ 2$) is positive at the beginning of the evolution, and decreases with the iterations progress. Conversely, for VSP $n = 3,\ 4,\ 5$, who are with smaller VSP preference, the value of $x_n - y_n$ is negative at the beginning of the evolution, and increases with the iterations progress. However, for all the five VSPs, the value of $x_n - y_n$ ($\forall n$) converges to zero with increasing iteration times. This means that VSPs and the MNO agree on the fraction of SBS storage that should be assigned to each VSP. In addition, results in Figs. 11.6 and 11.7 imply that the designed I-DA based caching mechanism elicits the true hidden information, i.e., the VSP utility function and MNO cost function.

Figure 11.8, we present the evolution of optimal social welfare when increasing the number of VSPs in the caching system from $N = 3$ to $N = 10$ and $N = 15$. We see that for the three cases, their social welfare gradually converges to different values, which are decreasing with increasing $N$. This phenomenon results from the increasing and concave properties of the VSP utility function and the increasing convex properties of the MNO cost function. Specifically, when $N$ increases, the allocated fraction of SBS storage to each VSP decreases due

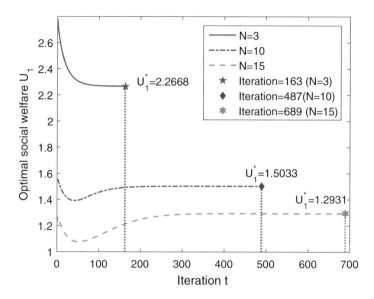

**Fig. 11.8** Impact of number of VSPs $N$ on the evolution of social welfare reached and iteration stop points by applying the I-DA algorithm based caching mechanism ($\varepsilon = 0.1$)

to $\sum_{n=1}^{N} x_n \leq \sum_{n=1}^{N} y_n \leq 1$. When $x_n$ decreases, $u_n^{\text{VSP}}$ decreases sharply; and when $y_n$ decreases, $v_n^{\text{MNO}}$ decreases gently, $\forall n \in \mathcal{N}$. Therefore, increasing $N$ will reduce the social welfare of the caching system. In addition, the designed algorithm needs more iterations in order to converge to the optimal solution when $N$ increases because of the increasing computation complexity.

## 11.8  Conclusion

In this part, we have considered a small-cell base video caching system with one MNO who leases its SBS storage resource and multiple VSPs with caching requirements. The social welfare optimization problem was established and analyzed for the caching system. To achieve the maximum social welfare when there exists hidden information, i.e., the VSP utility function and the MNO cost function cannot be known among the MNO, VSPs and the broker, a double auction mechanism was designed to elicit the hidden information, and an alternative optimization problem for the broker was formulated. To solve this problem, an iteration based algorithm was introduce to make sure that the maximum social welfare can be achieved, with assistance of properly designed resource allocation scheme and pricing rule. In addition, the designed allocation scheme and pricing rule for VSPs and the MNO reflect the VSPs' caching requests and MNO's caching cost, respectively. Furthermore, the convergence, EE, IR, IC and BB properties of the designed auction based caching mechanism have been demonstrated in this work.

Simulation results revealed the effect of some parameters, such as the step size of iterative algorithm, the number of VSPs, on the system performance and algorithm performance. Moreover, the convergence, EE, IR, IC and BB properties of the I-DA based caching mechanism were also validated by the simulation results.

# References

1. P. Juluri, V. Tamarapalli, and D. Medhi, "Measurement of quality of experience of video-on-demand services: A survey," *IEEE Communications Surveys & Tutorials*, vol. 18, no. 1, pp. 401–418, Jan. 2016.
2. C. Jiang, H. Zhang, Y. Ren, Z. Han, K.-C. Chen, and L. Hanzo, "Machine learning paradigms for next-generation wireless networks," *IEEE Wireless Communications.*, vol. 24, no. 2, pp. 98–105, Apr. 2017.
3. J. Du, E. Gelenbe, C. Jiang, H. Zhang, and Y. Ren, "Contract design for traffic offloading and resource allocation in heterogeneous ultra-dense networks," *IEEE J. Sel. Areas Commun.*, vol. 35, no. 11, pp. 2457–2467, Nov. 2017.
4. M. Ji, A. M. Tulino, J. Llorca, and G. Caire, "Order-optimal rate of caching and coded multicasting with random demands," *IEEE Trans. Inform. Theory*, vol. 63, no. 6, pp. 3923–2349, Jun. 2017.
5. M. Ji, G. Caire, and A. F. Molisch, "Fundamental limits of caching in wireless D2D networks," *IEEE Trans. Inform. Theory*, vol. 62, no. 2, pp. 849–869, Feb. 2016.
6. J. Du, E. Gelenbe, C. Jiang, H. Zhang, Z. Han, and Y. Ren, "Data transaction modeling in mobile networks: Contract mechanism and performance analysis," in *IEEE Global Commun. Conf. (GLOBECOM'17)*.　Singapore, 4-8 Dec. 2017.
7. Z. Zhao, M. Peng, Z. Ding, W. Wang, and H. V. Poor, "Cluster content caching: An energy-efficient approach to improve quality of service in cloud radio access networks," *IEEE J. Sel. Areas Commun.*, vol. 34, no. 5, pp. 1207–1221, May 2016.
8. K. Poularakis, G. Iosifidis, and L. Tassiulas, "Approximation algorithms for mobile data caching in small cell networks," *IEEE Trans. Wireless Commun.*, vol. 62, no. 10, pp. 3665–3677, Oct. 2014.
9. K. Poularakis, G. Iosifidis, A. Argyriou, I. Koutsopoulos, and L. Tassiulas, "Caching and operator cooperation policies for layered video content delivery," in *The 35th Annual IEEE Int. Confe. Comput. Commun. (INFOCOM)*.　San Francisco, CA, USA, 10–14 Apr. 2016.
10. H. Liu, Z. Chen, X. Tian, X. Wang, and M. Tao, "On content-centric wireless delivery networks," *IEEE Wireless Commun.*, vol. 21, no. 6, pp. 118–125, Dec. 2014.
11. R. B. Myerson and M. A. Satterthwaite, "Efficient mechanisms for bilateral trading," *J. Economic Theory*, vol. 29, no. 2, pp. 265–281, Apr. 1983.
12. W. Vickrey, "Counterspeculation, auctions, and competitive sealed tenders," *J. Tinance*, vol. 16, no. 1, pp. 8–37, Mar. 1961.
13. S. Paris, F. Martignon, I. Filippini, and L. Chen, "An efficient auction-based mechanism for mobile data offloading," *IEEE Trans. Mobile Computing*, vol. 14, no. 8, pp. 1573–1586, Aug. 2015.
14. M. Dehghan, B. Jiang, A. Seetharam, T. He, T. Salonidis, J. Kurose, D. Towsley, and R. Sitaraman, "On the complexity of optimal request routing and content caching in heterogeneous cache networks," *IEEE/ACM Trans. Networking*, vol. 25, no. 3, pp. 1635–1648, Jun. 2017.
15. B. Bharath, K. Nagananda, and H. V. Poor, "A learning-based approach to caching in heterogenous small cell networks," *IEEE Trans. Commun.*, vol. 64, no. 4, pp. 1674–1686, Apr. 2016.
16. C. Yang, Y. Yao, Z. Chen, and B. Xia, "Analysis on cache-enabled wireless heterogeneous networks," *IEEE Trans. Wireless Commun.*, vol. 15, no. 1, pp. 131–145, Jan. 2016.

17. M. Gerami, M. Xiao, and M. Skoglund, "Partial repair for wireless caching networks with broadcast channels," *IEEE Wireless Commun. Lett.*, vol. 4, no. 2, pp. 145–148, Jun. 2015.

18. N. Golrezaei, P. Mansourifard, A. F. Molisch, and A. G. Dimakis, "Base-state assisted device-to-device communications for high-throughput wireless video networks," *IEEE Trans. Wireless Commun.*, vol. 13, no. 7, pp. 3665–3676, Jul. 2014.

19. S. Krishnan and H. S. Dhillon, "Effect of user mobility on the performance of device-to-device networks with distributed caching," *IEEE Wireless Commun. Lett.*, vol. 6, no. 2, pp. 194–197, Apr. 2017.

20. S.-W. Jeon, S.-N. Hong, J. Mingyue, G. Caire, and A. F. Molisch, "Wireless multihop device-to-device caching networks," *IEEE Trans. Inform. Theory*, vol. 63, no. 3, pp. 1662–1676, Jan. 2016.

21. L. Zhang, M. Xiao, G. Wu, and S. Li, "Efficient scheduling and power allocation for D2D-assisted wireless caching networks," *IEEE Trans. Commun.*, vol. 64, no. 6, pp. 2438–2452, Jun. 2016.

22. A. Ferragut, I. Rodriguez, and F. Paganini, "Optimal timer-based caching policies for general arrival processes," *Queueing Syst.*, pp. 1–35, Jul. 2017.

23. E. Gelenbe, "A unified approach to the evaluation of a class of replacement algorithms," *IEEE Trans. Comput.*, vol. 100, no. 6, pp. 611–618, Jun. 1973.

24. D. Huang, X. Tao, C. Jiang, Y. Li, and J. Lu, "Latency-efficient video streaming in metropolis: A caching framework," in *IEEE Global Commun Conf. (GLOBECOM)*. Singapore, 4-8 Dec. 2017.

25. D. S. Berger, R. K. Sitaraman, and M. Harchol-Balter, "Adaptsize: Orchestrating the hot object memory cache in a content delivery network." in *ACM Int. Conf. Measurement and Modeling of Comput. Syst. (SIGMETRICS)*. Urbana-Champaign, Illinois, USA, 5–9 Jun. 2017, pp. 483–498.

26. S. Basu, A. Sundarrajan, J. Ghaderi, S. Shakkottai, and R. Sitaraman, "Adaptive TTL-based caching for content delivery," vol. 26, no. 3, pp. 1063–1077, Jun. 2018.

27. J. Luo, Z. Guan, Y. Cui, and Y. Lei, "Performance optimization of self-represented systems supporting run-time meta-modeling," *DEStech Trans. Comput. Sci. and Eng.*, no. cece, 2017.

28. J. Li, W. Chen, M. Xiao, F. Shu, and X. Liu, "Efficient video pricing and caching in heterogeneous networks." *IEEE Trans. Veh. Technol.*, vol. 65, no. 10, pp. 8744–8751, Oct. 2016.

29. E. Gelenbe, "Analysis of single and networked auctions," *ACM Trans. Internet Technol.*, vol. 9, no. 2, p. 8, 2009.

30. J. Du, C. Jiang, Z. Han, H. Zhang, S. Mumtaz, and Y. Ren, "Contract mechanism and performance analysis for data transaction in mobile social networks," *IEEE Trans. Network Sci. Eng.*, vol. 6, no. 2, pp. 103–115, Apr. –Jun. 2019.

31. J. Du, E. Gelenbe, C. Jiang, Z. Han, Y. Ren, and M. Guizani, "Cognitive data allocation for auction-based data transaction in mobile networks," in *IEEE Int. Wireless Commun. Mobile Comput. Conf. (IWCMC)*. Limassol, Cyprus, 25-29 Jun. 2018.

32. J. Du, C. Jiang, E. Gelenbe, Z. Han, Y. Ren, and M. Guizani, "Networked data transaction in mobile networks: A prediction-based approach using auction," in *IEEE Int. Wireless Commun. Mobile Comput. Conf. (IWCMC)*. Limassol, Cyprus, 25-29 Jun. 2018.

33. Z. Hu, Z. Zheng, T. Wang, L. Song, and X. Li, "Caching as a service: Small-cell caching mechanism design for service providers," *IEEE Trans. Wireless Commun.*, vol. 15, no. 10, pp. 6992–7004, Oct. 2016.

34. ——, "Game theoretic approaches for wireless proactive caching," *IEEE Commun. Mag.*, vol. 54, no. 8, pp. 37–43, Sept. 2016.

35. G. Iosifidis, L. Gao, J. Huang, and L. Tassiulas, "A double-auction mechanism for mobile data-offloading markets," *IEEE/ACM Trans. Networking (TON)*, vol. 23, no. 5, pp. 1634–1647, 2015.

36. F. P. Kelly, A. Maulloo, and D. Tan, "Rate control for communication networks: Shadow prices, proportional fairness and stability," *J. Oper. Res. Soc.*, vol. 29, no. 3, pp. 237–252, Mar. 1998.

37. C. Jiang, Y. Chen, Y. Gao, and K. R. Liu, "Joint spectrum sensing and access evolutionary game in cognitive radio networks," *IEEE Trans. Wireless Commun.*, vol. 12, no. 5, pp. 2470–2483, May 2013.

38. C. Jiang, Y. Chen, K. R. Liu, and Y. Ren, "Renewal-theoretical dynamic spectrum access in cognitive radio network with unknown primary behavior," *IEEE J. Sel. Areas Commun.*, vol. 31, no. 3, pp. 406–416, Mar. 2013.

39. J. Li, H. Chen, Y. Chen, Z. Lin, B. Vucetic, and L. Hanzo, "Pricing and resource allocation via game theory for a small-cell video caching system," *IEEE J. Sel. Areas Commun.*, vol. 34, no. 8, pp. 2115–2129, Aug. 2016.

40. J. Du, C. Jiang, H. Zhang, X. Wang, Y. Ren, and M. Debbah, "Secure satellite-terrestrial transmission over incumbent terrestrial networks via cooperative beamforming," *IEEE J. Sel. Areas Commun.*, vol. 36, no. 7, pp. 1367–1382, Jul. 2018.

41. T. Liu, J. Li, F. Shu, M. Tao, W. Chen, and Z. Han, "Design of contract-based trading mechanism for a small-cell caching system," *IEEE Trans. Wireless Commun.*, vol. 16, no. 10, pp. 6602–6617, Oct. 2017.

42. F. You, J. Li, J. Lu, and F. Shu, "On the auction-based resource trading for a small-cell caching system," *IEEE Commun. Lett.*, vol. 21, no. 7, pp. 1473–1476, Jul. 2017.

43. M. Cha, H. Kwak, P. Rodriguez, Y.-Y. Ahn, and S. Moon, "I tube, you tube, everybody tubes: analyzing the world's largest user generated content video system," in *Proc. 7th ACM SIGCOMM Conf. Internet measurement.* ACM, San Diego, California, USA, Oct. 24–26, 2007, pp. 1–14.

44. M. O. Jackson, "Mechanism theory," 2014.

45. T. Matsuo, "On incentive compatible, individually rational, and ex post efficient mechanisms for bilateral trading," *J. Econ. theory*, vol. 49, no. 1, pp. 189–194, Oct. 1989.

46. D. P. Palomar and M. Chiang, "A tutorial on decomposition methods for network utility maximization," *IEEE J. Sel. Areas Commun.*, vol. 24, no. 8, pp. 1439–1451, Aug. 2006.

47. J.-H. Hours and C. N. Jones, "A parametric nonconvex decomposition algorithm for real-time and distributed NMPC," *IEEE Trans. Automat. Control*, vol. 61, no. 2, pp. 287–302, Feb. 2016.

# Chapter 12
# Priority-Aware Computational Resource Allocation

**Abstract** Vehicular fog computing (VFC) has been expected as a promising scheme that can increase the computational capability of vehicles without relying on servers. Comparing with accessing the remote cloud, VFC is suitable for delay-sensitive tasks because of its low-latency vehicle-to-vehicle (V2V) transmission. However, due to the dynamic vehicular environment, how to motivate vehicles to share their idle computing resource while simultaneously evaluating the service availability of vehicles in terms of vehicle mobility and vehicular computational capability in heterogeneous vehicular networks is a main challenge. Meanwhile, tasks with different priorities of a vehicle should be processed with different efficiencies. In this work, we propose a task offloading scheme in the context of VFC, where vehicles are incentivized to share their idle computing resource by dynamic pricing, which comprehensively considers the mobility of vehicles, the task priority, and the service availability of vehicles. Given that the policy of task offloading depends on the state of the dynamic vehicular environment, we formulate the task offloading problem as a Markov decision process (MDP) aiming at maximizing the mean latency-aware utility of tasks in a period. To solve this problem, we develop a soft actor-critic (SAC) based deep reinforcement learning (DRL) algorithm for the sake of maximizing both the expected reward and the entropy of policy. Finally, extensive simulation results validate the effectiveness and superiority of our proposed scheme benchmarked with traditional algorithms.

**Keywords** Vehicular Fog Computing (VFC) · Task Offloading · Task Priority · Dynamic Pricing · Soft Actor-critic (SAC)

## 12.1 Introduction

The development of Internet of Vehicles (IoV) and artificial intelligence (AI) technologies facilitates a range of compelling vehicular computation-intensive applications [1], such as autonomous driving, crowd-sensing, virtual reality [2], etc. To meet such increasing computational demands, on-board computers of vehicles will face great challenges of providing high-quality services, resulting from their

© The Author(s), under exclusive license to Springer Nature Singapore Pte Ltd. 2023    271
J. Du, C. Jiang, *Cooperation and Integration in 6G Heterogeneous Networks*,
Wireless Networks, https://doi.org/10.1007/978-981-19-7648-3_12

limit computational power and high cost of upgrading. As a promising solution to these challenges, cloud offloading has been delivered to process computational tasks in part at remote and centralized cloud to obtain a plenty of computing resource. However, cloud offloading will lead to long transmission delay and high energy consumption of communication. Therefore, offloading to centralized cloud is not applicable for delay-sensitive applications [3–5].

To meet the requirement of low latency in task offloading, mobile edge computing (MEC) is introduced in vehicular networks, i.e., vehicular edge computing (VEC), where road side unit (RSU) or base station (BS) is equipped with a certain amount of computation and storage resources to provide computing services for vehicles [6]. However, with the increase of traffic density, it is difficult to guarantee the quality of service (QoS) of all vehicular applications for the limited computing resource in RSU or BS. Furthermore, RSU/BS servers are usually deployed sparsely, and their performance is constraint by the radio coverage [7]. Moreover, since BS and RSU are both stationary infrastructures, the relative speed between vehicle and RSU/BS is high, which leads to short link duration. Instead, the relative speed between two vehicles driving in the same direction is smaller, thus a longer link duration becomes possible [8]. Nowadays, many high-end vehicles are equipped with a fair amount of computational capacity, and with the rapid development of autonomous driving and the fifth-generation (5G) communication technology, there will be a growing number of vehicles equipped with plenty of computational capacity in the near future.

Vehicular fog computing (VFC) is assumed to be a promising scheme where vehicles can share their idle computing resource among each other. There exist some works that investigated computational resource allocation in VFC. Some works focused on the task allocation in VFC with the aim of minimizing the task offloading delay, and proposed various methods to solve the problem, such as Particle Swarm Optimization (PSO) [9, 10], modified genetic algorithm [6], Lyapunov optimization [11], etc. Besides, some works aimed at minimizing the energy consumption [12] or computation cost [13] in VFC task offloading. Moreover, in order to incentivize vehicles to share their computing resource, some works employed contract theory [14–16] and auction mechanism [17] in VFC. Considering the high dynamic vehicular environment, the policy of computational resource allocation should vary in response to the vehicular environment in real time, and it is hard to obtain the complete system model and dynamics of the environment. Therefore, some works [18–20] utilized model-free DRL to obtain the optimal resource allocation policy by observing the system states of VFC. Additionally, some works also investigated the task allocation in heterogeneous vehicular networks, where vehicles and fixed road infrastructures were integrated to provide computing service cooperatively, and provided various methods to optimize the efficiency and reliability in task offloading, such as proximal policy optimization (PPO) [21], fault-tolerant PSO [22], DRL-based adaptive algorithm [23], etc. However, few of existing works considered the priority of tasks and the service availability of neighboring vehicles. As a result, all the tasks have the same probability to be offloaded to servers and obtain corresponding computing resource, some tasks with strict latency may not be completed within the maximum latency. Meanwhile, offloading failure may occur due to the short V2V link duration or insufficient vehicular computing resource. In addition, considering the energy consumption and vehicle safety, it can be expected that vehicles on the road may not be willing to contribute their idle computing resource without any incentive.

To solve the problems mentioned above in VFC, we design a distributed vehicle-to-vehicle (V2V) task offloading scheme that mainly concerns with the following aspects: 1) The priority of computational tasks is considered, and vehicular task with high-priority is ensured to be executed preferentially. 2) The service availability of vehicles is evaluated in the process of task offloading. 3) The mobility of vehicles is taken into consideration as well, and the effect of vehicle mobility in task offloading is evaluated in the calculation of transmission rate and service availability of vehicles. The problem of task offloading is formulated as a Markov decision process (MDP) with the objective of maximizing the mean latency-aware utility of offloading tasks in a period. To solve the problem, we propose a model-free reinforcement learning algorithm, which utilizes the actor-critic framework to evaluate and improve the policy of task offloading. Meanwhile, both the expected reward and the entropy of policy are maximized to improve the robustness and sample-efficiency by applying the proposed algorithm.

The main contributions of this part are summarized as follows:

1. **Model:** We provide a VFC framework where vehicles with limited computational capability can offload part of their computational tasks to neighboring vehicles with idle computing resource, and the computational resource allocation is in the charge of BSs. In the proposed framework, vehicular tasks are classified by different priorities. In order to ensure high-priority tasks executed preferentially, we conceive two utility functions for high-priority task and low-priority task respectively. Moreover, with the consideration of vehicle mobility and limited vehicular computing resource, we model vehicular service availability which is regarded as a basis in the choice of service vehicles according to the link duration and the vehicle state. Additionally, with the aim of incentivizing vehicles to share their idle computing resource and improving the efficiency of resource utilization, a dynamic pricing scheme is introduced, where the resource allocation in a service vehicle is based on both the idle computing resource and the service price paid for task offloading.

2. **Algorithm:** Considering the stochastic vehicular environment and uncertain communication conditions, the problem is formulated as an MDP with the objective of maximizing the mean latency-aware utility of offloading tasks in a period, where the selection of service vehicle and the service price of task offloading are determined at the same time. We develop a model-free deep reinforcement learning algorithm based on soft actor-critic (SAC), in which the expected utility and the entropy of policy are maximized at the same time.

3. **Simulation:** The performance of our proposed algorithm is evaluated by extensive simulations. Simulation results validate that our algorithm achieves better performance comparing to the regular algorithms under both low traffic density and high traffic density.

The rest of the part is organized as follows. Section 12.2 analyzes the research work of resource allocation problem in VFC. In Sect. 12.3, the system architecture and models are detailed. The optimization problem is formulated in Sect. 12.4, and Sect. 12.5 presents the algorithm of task offloading. The performance of our proposed scheme is evaluated in Sect. 12.6. Finally, we conclude this part in Sect. 12.7.

## 12.2  Related Work

### 12.2.1  Computation Offloading Optimization In VEC

Some literatures have investigated the computational resource allocation in VEC. In [24], a continuous alternating direction method of multipliers (ADMM) based optimization algorithm was proposed with the consideration of the overall energy consumption and latency of task offloading in VEC. Furthermore, Dai et al. [25] investigated a multi-user multi-server VEC system where the load balance of servers and task offloading were jointly optimized. In [26], a hybrid optimization algorithm that combined partheno genetic algorithm and heuristic rules was presented in a VEC scenario, where multiple adjacent VEC servers at the roadside provided computing services for passing vehicles. Moreover, the system costs and offloading latency of VEC were investigated in [27], in which a mobility-aware task offloading scheme was proposed in independent VEC servers while a location-based offloading scheme was proposed in cooperative VEC servers. Specifically, Zhang et al. [28] presented an MEC-enabled LTE-V network and adopted a deep Q-learning optimization method in task offloading where the selection of target server and data transmission mode were both considered. In addition, based on the historical association experiences in the dynamic scenario of VEC, each BS in [29] was regarded as an agent that executed online reinforcement learning and decided which vehicles were associated with the BS. In [30], vehicles were able to offload computational tasks to nearby RSUs or BSs, and a two-side matching approach and a double deep Q-network (DDQN) algorithm were proposed for offloading scheduling and resource allocation respectively.

However, all of the works mentioned above assumed that the infrastructures deployed by the road are equipped with MEC servers that have enough computing resource for task offloading, which are not well-suited for some scenarios with no MEC servers at the roadside.

### 12.2.2  Computation Offloading Optimization in VFC

VFC is deemed to be a promising solution that can increase computational capability of vehicles in a distributed manner. In [8], Feng et al. proposed a VFC aided computational resource allocation framework solved by a modified ant colony optimization algorithm for the sake of supporting task offloading in autonomous vehicular networks. Considering the dwell time of vehicles and heterogeneous vehicular computational capabilities in VFC, Sun et al. [6] presented a vehicular cooperative computation offloading scheme solved by a low-complexity modified genetic algorithm. Moreover, in consideration of the opportunistic V2V communication, a VFC framework based on graph jobs was proposed in [31], where a computation-intensive task was represented as a graph whose components represented the interdependent subtasks. Specifically, Pu et al. [11] presented a hybrid VFC framework where vehicles can obtain computing resource from cooperative vehicles and the virtual machine (VM) pool, and an online task scheduling algorithm based on Lyapunov optimization was developed. Additionally, Zhu et al. [10] proposed a dynamic task allocation framework that optimized service latency and quality loss of results in VFC, and the bi-optimization problem was solved by linear-programming and binary particle swarm. In VFC, a vehicle can act as either service provider or requester. Therefore, a Stackelberg game based opportunistic V2V offloading scheme was conceived in [32] to decide the selection of servers and the price of the service under situations involving complete and incomplete information. Moreover, considering that vehicles may not be willing to share their computing resource voluntarily, some works also investigated how to incentivize vehicles to contribute their idle computing resource to nearby computation-intensive vehicles in VFC. In [17], a Vickrey-Clarke-Groves based reverse auction mechanism was proposed, while in [33], a market mechanism was developed with the consideration of both task with time to live (TTL) and task without TTL. Furthermore, a contract-matching method was developed in [14], and the problem of task assignment was solved by a pricing-based stable matching algorithm. In [15], Zhao et al. employed contract theory and developed a distributed deep reinforcement learning (DRL) algorithm to reduce the complexity of system implementation. However, most of the studies mentioned above assumed that the dynamics of vehicular environment can be accurately modeled, which is usually impractical in VFC.

### 12.2.3   DRL-Based Computation Offloading Optimization in VFC

With the advancement of machine learning, reinforcement learning has been applied in some works to solve the problems of computational resource allocation in VFC. Sun et al. [34] designed a multi-armed bandit theory based V2V task offloading algorithm aiming at minimizing the average offloading delay in VFC. In addition, Wang et al. [35] proposed an online learning method based on Combinatorial Multi-Armed Bandits (CMAB) in dynamic VFC to jointly optimized task allocation decision and spectrum scheduling. With the aim of maximizing the long-term reward in terms of the allocation of heterogeneous vehicular computing resource in VFC, the task offloading problem in [36] was formulated as a semi-Markov decision process and solved by an iterative algorithm based on Bellman equation. Moreover, in the VFC scenario of [3], users were able to offload a group of tasks to nearby vehicles, and a constrained Markov decision process was proposed to maximize the utility of users with the consideration of QoS requirement. Furthermore, to reduce the complexity of resource allocation in VFC, a multi-timescale framework based on Q-learning was presented in [18], where mobility-aware reward was estimated in large timescale while exact immediate reward was calculated in small timescale. With the consideration of stochastic traffic and uncertain communication conditions in VFC, the problem of computation offloading in [19] was conceived as a semi-Markov process which was solved by both Q-learning and DRL based method. In [37], both parking vehicles and moving vehicles can act as fog servers, and a DRL algorithm based on queuing theory was developed to schedule offloading task flows so that the overall energy consumption is minimized.

In contrast to the existing works, this work considers the integrated impacts of vehicle mobility, priority of computational tasks, service availability of vehicles, and incentive of resource contribution. In addition, we propose a model-free DRL algorithm based on SAC to learn the task offloading policy that maximizes both the mean utility of offloading tasks and the entropy of the policy.

## 12.3  System Model

### 12.3.1  System Architecture

Consider a distributed V2V communication system as illustrated in Fig. 12.1. There is a one-way road which is covered by several BSs, and neighboring BSs can communicate with each other. We divide the system time into several periods, and discretize each period into a number of slots. In each slot, the system status can be regarded as constant, but changes over slots [38]. In the system, BSs take charge of the computational resource allocation all the time, and the communication range of BS is much longer than vehicular communication range. We assume that if a vehicle enters the coverage area of a BS and is willing to participate in VFC, it will send a message that contains its position, velocity and computational capability to the BS. Then, the traffic of the road within the coverage of a BS can be known by the agent deployed in the BS. In a certain period $T$, we assume that there are several vehicles that do not have enough computational capability to execute some of the on-board applications, and we focus on one of these vehicles $V_t$ and call this vehicle as task vehicle.

We assume that during period $T$, there are $K$ vehicles within the communication range of $V_t$, which are denoted as $\mathscr{K} = \{V_1, V_2, \ldots, V_K\}$, and the total computational capability of each vehicle is represented as $\{F_1, F_2, \ldots, F_K\}$, respectively. All of these vehicles have the potential to be selected as the service vehicle that provides idle computing resource for task vehicle. In each slot, if a task vehicle

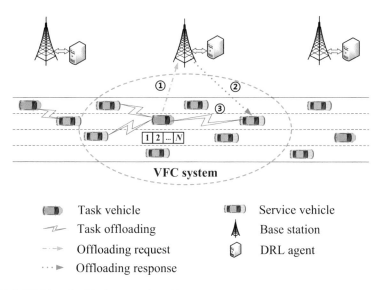

**Fig. 12.1** VFC based vehicular network architecture

has a computational task to offload, it first sends an offloading request to the BS, and then the BS decides which service vehicle the task should be offloaded to and determines the corresponding price that the task vehicle pays to the service vehicle, and transmits the message of service assignment back to the task vehicle and selected service vehicle. Considering that the message of offloading request and service assignment only contains several bits, we ignore the transmission delay of offloading request and service assignment in our system. Furthermore, we assume that there are $N$ tasks generated by the task vehicle in period $T$, the task set is denoted as $\mathcal{N} = \{\phi_1, \phi_2, \ldots, \phi_N\}$, and the profile of task $\phi_n$ is represented as $\{D_n, C_n, \tau_n, \kappa_n\}$, where $D_n$ denotes the input data size, $C_n$ is the computation size which represents CPU cycles required to complete the task, $\tau_n$ denotes the delay constraint, and $\kappa_n$ represents the priority of the task.

## 12.3.2 Mobility Model

In this work, we focus on a task vehicle and multiple service vehicles within the communication range of the task vehicle on a road, all of which are in the coverage of a BS during the period of consideration, and the traffic density and the average vehicle velocity both affect the performance of task offloading. Similar to [39, 40], we consider a free flow traffic model, where all the vehicles within the coverage of BS drive at a constant speed, and the velocity of vehicles forms a Gaussian distribution. The mean and variance of the velocity are denoted as $\bar{v}$ and $\sigma_v$, respectively. The relationship between the traffic density and the average velocity of vehicles follows

$$\bar{v} = v_{max} \left( 1 - \frac{\rho}{\rho_{max}} \right),$$ (12.1)

where $\rho$ represents the traffic density in the coverage of the BS, $v_{max}$ and $\rho_{max}$ are the maximum velocity and density of vehicles, respectively. Considering that when the traffic density becomes higher, the average distance between vehicles will be shorter, and the variance of velocity should be lower, otherwise the high relative speed between vehicles may cause vehicle collision. Therefore, similar to [33] and according to (12.1), we assume that the variance of the velocity is proportional to the average velocity, which is given as

$$\sigma_v = \alpha_v \bar{v},$$ (12.2)

where $\alpha_v$ is a constant.

### 12.3.3  Communication Model

We assume that the wireless channel state between task vehicle and service vehicle remains static during the data transmission of each computational task. The transmission rate between task vehicle and service vehicle is calculated by

$$r_{t,k} = B_{t,k} \log_2 \left( 1 + \gamma_{t,k} \right),$$ (12.3)

where $B_{t,k}$ is the allocated bandwidth of the V2V channel between task vehicle $V_t$ and service vehicle $V_k$, $\gamma_{t,k}$ represents the signal to noise ratio (SNR) of the wireless channel, which is calculated by

$$\gamma_{t,k} = \frac{P_t d_{t,k}^{-\alpha} h_{t,k}^2}{N_0 + \sum_{j \in \mathcal{K}, j \neq k} P_j d_{j,k}^{-\alpha} h_{j,k}^2},$$ (12.4)

where $P_t$ represents the power of transmitter, $d_{t,k}$ is the distance between task vehicle $V_t$ and service vehicle $V_k$, $\alpha$ denotes the path loss exponent related to the distance, $h_{t,k}$ represents the channel gain, $N_0$ is the power of Gaussian White Noise, and $\sum_{j \in \mathcal{K}, j \neq k} P_j d_{j,k}^{-\alpha} h_{t,k}^2$ indicates the interference introduced by other V2V transmissions.

### 12.3.4  Computation Model

There are two ways to execute a computational task in a vehicle, executing in local processor or offloading to another neighboring vehicle. If a computational task $\phi_n$ is executed locally, then the delay of the task only depends on the allocated frequency in the local processor, which can be given as

$$t_n^{loc} = \frac{C_n}{F^n}, \tag{12.5}$$

where $F^n$ is the frequency allocated for task $\phi_n$. If task $\phi_n$ is offloaded to a service vehicle $V_k$, then the delay of the task is calculated by

$$t_n^o = t_n^{up} + t_n^{comp} + t_n^{down} = \frac{D_n}{r_{t,k}} + \frac{C_n}{F_k^n} + \frac{\delta D_n}{r_{k,t}}, \tag{12.6}$$

where $t_n^{up}$, $t_n^{comp}$ and $t_n^{down}$ are the time for uploading the task, computing the task and downloading the computation result, respectively. $F_k^n$ denotes the allocated frequency in $V_k$. According to [41], for many applications, the data size of computation result in general is much smaller than the size of input data, i.e., $\delta \ll 1$. In our work, we ignore the transmission time of computation result in the calculation of offloading delay. Besides, if a computational task generated by a task vehicle is too large, the task vehicle can divide the computational task into several subtasks, which can be executed parallelly or sequentially, and gives each subtask a delay constraint. Then the BS schedules the subtask according to the task profile that contains the data size, computation size and delay constraint, which is the same as other tasks.

### 12.3.5  Task Model

In general, vehicle applications can be classified into three classes: crucial tasks, high-priority tasks and low-priority tasks [8]. Crucial tasks contain the core tasks and safety-related tasks of vehicle system, which must be executed locally without relying on vehicular environment. Therefore, vehicles must reserve some computing resource for local crucial tasks. High-priority tasks represent a class of tasks that have strict delay constraint, such as navigation, road-sensing, etc. If a high-priority task cannot be finished within the maximum tolerable delay, the task will be failed and may bring some loss to vehicle. Compared to low-priority tasks, vehicles should ensure high-priority tasks executed first. Low-priority tasks are a class of delay-tolerant tasks, such as vehicular entertainment applications, value-added services, etc. If the completion time of a low-priority task exceeds the reference delay, the result will still be of use, but the availability of the result will decline with the execution time increasing.

In our work, tasks with high priority should meet their deadline without any negotiation. If a high-priority task is completed before the deadline, the utility of the task is non-negative and depends on the completion time. On the other hand, if a high-priority task fails because of the completion time exceeding the deadline, the utility will be negative as a penalty. Moreover, since the relay of execution result may introduce queuing time in relay node and extra transmission time, which degrades the efficiency of task offloading, in our proposed scheme, we do not consider the relay of execution result when the communication link interruption occurs in the task offloading, and the task offloading will be regarded as a failure and the utility will be the same as in the condition of completion time exceeding deadline. Similar to [42], a logarithm function is employed to represent the utility of executing a task, which is shown as

$$
\mathscr{U}_n^H = \begin{cases} \log\left(1 + \tau_n - t_n\right), & t_n \leq \tau_n, \\ -\Gamma^H, & t_n > \tau_n. \end{cases} \tag{12.7}
$$

where $t_n$ is the completion time of task $\phi_n$, $-\Gamma^H$ is a negative constant, which represents the penalty of the offloading failure of high-priority task.

For low-priority task, if the completion time is less than the reference delay, the utility will be a positive constant. Instead, if the completion time exceeds the reference delay, the result of the task is still regarded as available, but the utility declines exponentially with the increase of the time beyond the reference delay. We define the utility function of low-priority task as

$$
\mathscr{U}_n^L = \begin{cases} \Gamma^L, & t_n \leq \tau_n, \\ \Gamma^L e^{-c(t_n - \tau_n)}, & t_n > \tau_n, \end{cases} \tag{12.8}
$$

where $\Gamma^L$ is a positive constant, which represents the reward of completing a low-priority task within the reference delay, $c > 0$ is a constant. Specifically, if a low-priority task fails due to the communication link interruption, the completion time will be infinite, i.e., $t_n = \infty$, then the utility will be zero.

We denote $p_n$ as the unit price paid by task vehicle for offloading task $\phi_n$, $f_n$ is the frequency allocated in service vehicle for task $\phi_n$, and $t_n'$ represents the computing time of task $\phi_n$, then we have $C_n = f_n t_n'$. We assume that the price paid by task vehicle is proportional to the energy consumed in service vehicle. In a service vehicle, given a certain $f_n$, the energy consumption is proportional to the completion time of task $\phi_n$, while given a certain $t_n'$, the energy consumption is proportional to the allocated frequency. As a result, the price paid for task $\phi_n$ should be proportional to the computation size $C_n$. Finally, the utility of task vehicle for offloading a task can be given as

$$
U_n = \mathbf{1}\left(\kappa_n = \kappa_H\right) \mathscr{U}_n^H + \mathbf{1}\left(\kappa_n = \kappa_L\right) \mathscr{U}_n^L - p_n C_n, \tag{12.9}
$$

where $\mathbf{1}(\cdot)$ is the indicator function, $\kappa_H$ and $\kappa_L$ represent the priority level of high-priority task and low-priority task respectively, and both of which are constants.

### 12.3.6 Service Availability

In the VFC system, not all of the vehicles within the communication range of task vehicle are suitable for task offloading, because some vehicles may have heavy workload or the total delay of task offloading is beyond the contact time of task vehicle and service vehicle. We use $\eta_n^k$ to represent the service availability of service vehicle $V_k$ for offloading task $\phi_n$. Then, we assume that the service availability depends on both the service probability of the service vehicle and the V2V link duration between the task vehicle and the service vehicle, which is given as

$$\eta_n^k = \varepsilon_k \left( \frac{T_k - \tau_n}{T_k} \right)^+, \tag{12.10}$$

where $\varepsilon_k$ is the probability that an offloading task is accepted by service vehicle $V_k$, $T_k$ denotes the duration of the V2V transmission link between task vehicle $V_t$ and service vehicle $V_k$, $\tau_n$ represents the maximum tolerable delay of high-priority task or the reference delay of low-priority task. $\frac{T_k - \tau_n}{T_k}$ indicates that vehicles with longer dwell time within the communication range of task vehicle are more likely to be chosen as the service vehicle. Function $(x)^+ = \max(x, 0)$ ensures that $\eta_k$ is non-negative.

In order to evaluate the value of $T_k$, we consider a one-dimensional road, on which all the vehicles drive in the same direction, then $T_k$ can be evaluated by

$$T_k = \frac{R}{|v_k - v_t|} - \frac{x_k - x_t}{v_k - v_t}, \tag{12.11}$$

where $x_t$ and $v_t$ denote the current position and velocity of $V_t$ respectively, $x_k$ and $v_k$ represent the current position and velocity of $V_k$ respectively, $R$ is the effective V2V communication range.

To evaluate the value of $\varepsilon_k$, we first introduce a concept named service ratio (SR) for service vehicle similar to [43]. Since a service vehicle may receive task offloading requests from multiple task vehicles in a slot, considering the limited computing resource and instability of V2V links, service vehicles have to reject some of the requests which are low-priority or cannot be completed within the maximum tolerable delay. We denote the SR of service vehicle $V_k$ as $\beta_k$, which can be obtained by

$$\beta_k = \frac{\kappa r^{-c}}{\kappa_0 \widetilde{F}_k + \sum_{j \in \mathcal{Q}} \kappa_j r_j^{-c}}, \tag{12.12}$$

where $\kappa$ denotes the priority of the offloading task from $V_t$, $r$ is the distance between $V_t$ and $V_k$, $\kappa_0$ represents the weight factor and can be assumed as a constant, $\mathcal{Q}$ denotes a set of service requests from other task vehicles within the communication range of $V_k$ in a slot. Let $F_k^{min}$ represent the minimum computing resource reserved for local tasks in $V_k$, then $\widetilde{F}_k = F_k^{min}/F_k$ denotes the fraction of the minimum computing resource occupied by local tasks in $V_k$. From (12.12), one can observe that the SR of a service vehicle not only depends on the number of task requests from different task vehicles in a slot, but also depends on the fraction of the computing resource occupied by local tasks in the service vehicle. Moreover, we assume that each service vehicle has the same predetermined threshold of SR $\hat{\beta}$, then the probability that a computational task is accepted by $V_k$ can be defined as

$$\varepsilon_k = P\left(\beta_k > \hat{\beta}\right). \tag{12.13}$$

In order to estimate the admission probability $\varepsilon_k$, a virtual zone centered at service vehicle $V_k$ with radius $R_{v,k}$ is defined. Once other task vehicles appear in the virtual zone and send $V_k$ the requests of computing services, the SR of $V_k$ for offloading a task from $V_t$ drops below $\hat{\beta}$, then the offloading request from $V_t$ is rejected by $V_k$.

**Theorem 12.1** *Given the threshold of SR $\hat{\beta}$, the radius of the virtual zone centered at $V_k$ is obtained by*

$$R_{v,k} = \kappa_H^{\frac{1}{c}}\left(\frac{\kappa r^{-c}}{\hat{\beta}} - \kappa_0\widetilde{F}_k\right)^{-\frac{1}{c}}. \tag{12.14}$$

**Proof** According to the definition of $R_{v,k}$, once any task vehicle besides $V_t$ appears in the virtual zone, the SR of $V_k$ for $V_t$ will drop below $\hat{\beta}$. We assume that there is a task vehicle $V_j$ appearing in the virtual zone, from (12.12), we have

$$\frac{\kappa r^{-c}}{\kappa_0\widetilde{F}_k + \kappa_j r_j^{-c}} \leq \hat{\beta}, \tag{12.15}$$

then, the range of $r_j$ is obtained by transforming (12.15), i.e.,

$$r_j \leq \kappa_i^{\frac{1}{c}}\left(\frac{\kappa r^{-c}}{\hat{\beta}} - \kappa_0\widetilde{F}_k\right)^{-\frac{1}{c}}, \tag{12.16}$$

Notice that $\kappa_i$ can be either $\kappa_L$ or $\kappa_H$, and $\kappa_L < \kappa_H$. Since $R_{v,k}$ is the upper bound of $r_j$, then we have

$$R_{v,k} = \kappa_H^{\frac{1}{c}} \left( \frac{\kappa r^{-c}}{\hat{\beta}} - \kappa_0 \widetilde{F}_k \right)^{-\frac{1}{c}}. \tag{12.17}$$

This completes the proof of Theorem 12.1.

Furthermore, according to [44], the probability of an offloading task being rejected by $V_k$ can be given as follows:

$$1 - \varepsilon_k = 1 - e^{-2R_{v,k}\rho_t}, \tag{12.18}$$

where $\rho_t$ is the density of task vehicles in the communication range of service vehicle. We assume that $\rho_t$ is proportional to the traffic density $\rho$, i.e., $\rho_t = c_t\rho$, where $c_t$ is assumed as a constant during the period of consideration. From (12.18), we obtain $\varepsilon_k = e^{-2R_{v,k}c_t\rho}$. Finally, service availability $\eta_k$ is calculated by (12.10).

### 12.3.7 Pricing Model

In a service vehicle, there may be local computational tasks and offloading tasks from task vehicles to be executed at the same time. Considering the limited on-board computational capability, the service vehicle may not ensure that all the computational tasks obtain enough computing resource. Therefore, the computational resource allocation should first guarantee the local high-priority tasks completed before deadline. Consider that there are $L_k$ local high-priority tasks in $V_k$, the computation size and the deadline of each local task are denoted as $C_l$ and $\tau_l$, respectively. Then, the least required frequency of local tasks can be given as $F_k^{min} = \sum_{l=1}^{L} \frac{C_l}{\tau_l}$. If a vehicle rejects all the offloading tasks and allocates all of its computing resource to local tasks, then we have $F_k = \sum_{l=1}^{L} \frac{C_l}{\theta_k \tau_l}$, where $\theta_k = F_k^{min} / F_k$. It can be seen that the domain of frequency reserved for local tasks is $\left[ F_k^{min}, F_k \right]$. We use $\theta$ to denote the ratio of $F_k^{min}$ to the computing resource reserved for local tasks, the domain of $\theta$ can be given as $[\theta_k, 1]$. According to (12.7), the total utility of local tasks is represented as

$$U_k^{local}(\theta) = \sum_{l=1}^{L_k} \log(1 + \tau_l - \theta\tau_l), \quad \theta \in [\theta_k, 1]. \tag{12.19}$$

Once a vehicle decides to accept an offloading request from other vehicle and allocates frequency $f_n$ to an offloading task $\phi_n$, the utility of local tasks changes from $U_k^{local}(\theta_k)$ to $U_k^{local}(\theta_k')$, and the price $p_n C_n$ paid for task $\phi_n$ should satisfy

$$p_n C_n = U_k^{local}(\theta_k) - U_k^{local}(\theta_k'),  \tag{12.20}$$

i.e., the service price of an offloading task should compensate for the loss of the utility of local tasks. After given the price $p_n C_n$, $\theta_k'$ can be determined by (12.20), and the frequency allocated to task $\phi_n$ is calculated by $f_n = F_k - F_k^{min}/\theta_k'$. The domain of the unit price of task $\phi_n$ is $\left(0, U_k^{local}(\theta_k)/C_n\right]$. As a result, if an offloading task requires more computing resource in $V_k$, the task vehicle should pay $V_k$ a higher price.

In addition, during the process of executing task $\phi_n$, if there is another task $\phi_{n'}$ offloaded to $V_k$, then the utility of local tasks changes from $U_k^{local}(\theta_k')$ to $U_k^{local}(\theta_k'')$, and $\theta_k''$ is determined by the service price of task $\phi_{n'}$, and so on.

## 12.4  Formulation of Optimization Problem for Task Offloading

In the proposed VFC system, we consider a certain period of time, there are some vehicles in the coverage of the BS. We focus on a task vehicle which has to offload a group of computational tasks due to the limited on-board computational capability. For each offloading task in the task vehicle, the BS chooses a service vehicle surrounding the task vehicle and determines the unit price that the task vehicle should pay to the selected service vehicle. The utility of a computational task depends on the completion time of the task. Our aim is to maximize the mean utility of offloading tasks in the period. The optimization problem is formulated as follows:

$$\max \quad \frac{1}{N} \sum_{n=1}^{N} \sum_{k=1}^{K} c_n^k \left( \mathbf{1}(\kappa_n = \kappa_H) \mathcal{U}_n^H + \mathbf{1}(\kappa_n = \kappa_L) \mathcal{U}_n^L - p_n C_n \right),$$
$$\tag{12.21a}$$

$$\text{s.t.} \quad 0 < p_n^k \le \frac{U_k^{local}(\tilde{\theta}_k)}{C_n}, \quad \forall n \in \mathcal{N}, k \in \mathcal{K},  \tag{12.21b}$$

$$c_n^k \in \{0, 1\}, \quad \forall n \in \mathcal{N}, k \in \mathcal{K},  \tag{12.21c}$$

$$\sum_{k=1}^{K} c_n^k = 1, \quad \forall n \in \mathcal{N},  \tag{12.21d}$$

$$\mathbf{1}\left(c_n^k = 1\right) \eta_n^k > 0, \quad \forall n \in \mathcal{N}, k \in \mathcal{K}.  \tag{12.21e}$$

In the above problem, constraint (12.21b) guarantees that the price is positive and not exceeds the maximum value, and $\tilde{\theta}_k$ represents the current value of $\theta$ in $V_k$ in the step of offloading task $\phi_n$. In constraint (12.21c), $c_n^k = 1$ means that task $\phi_n$ is offloaded to $V_k$, while $c_n^k = 0$ means not. Constraint (12.21d) ensures that a task is executed only in one vehicle. Constraint (12.21e) indicates that the service vehicle selected for task $\phi_n$ must be available. The objective of the problem is to obtain the maximum mean latency-aware utility of offloading tasks in the period. Notice that the utility of computational task is non-linear, and the remaining computing resource of a service vehicle varies with the decision of task offloading. Meanwhile, due to the mobility of vehicles, the relative distance between vehicles and V2V link duration are both time-variant, which makes it hard to accurately give a complete model for the V2V channel state and the service availability of vehicles. Therefore, regular optimization methods are not appropriate for solving the problem. In the next section, we transform the optimization problem as a Markov decision process which can be solved by model-free reinforcement learning method.

## 12.5  SAC Based DRL Algorithm for Task Offloading

In the VFC system, we focus on a task vehicle that has a group of computational tasks to offload in a period, and in each slot, the agent in BS determines the service vehicle and corresponding unit service price for an offloading task. Thus, the task offloading problem can be regarded as an optimal sequential decision-making problem under the dynamic vehicular environment. Due to the mobility of vehicles, the contact time and the wireless channel state between vehicles are both time-variant, and the available computing resource of vehicles varies over time as well. Since the current state of the system changes over time slots, we cannot simply make decision according to the current observed state. Therefore, we formulate the problem as the Markov decision process whose dynamics are unknown to the agent in BS. To solve the problem, we propose a model-free reinforcement learning algorithm based on SAC [45].

SAC is a maximum entropy reinforcement learning algorithm based on off-policy actor-critic model. Comparing to the value-based DRL algorithms, such as DQN and double DQN, SAC is more efficient in solving the problems with high-dimensional action space, and therefore is more suitable for the vehicular task allocation under different traffic densities. Moreover, by incorporating the entropy measure of the policy into the reward, SAC is able to explore more feasible strategies. If there exist multiple optimal options, the policy in SAC will choose each option with equal probability. Therefore, comparing to some other policy-based DRL algorithms, such as Deep Deterministic Policy Gradients (DDPG), Asynchronous Advantage Actor Critic (A3C), SAC is more robust and generalized, and thus makes it easier to make adjustment in stochastic vehicular environment [46, 47].

In this section, we first demonstrate the state space, the action space, and the reward function of the formulated MDP. Then, to illustrate how SAC works, the policy and value function are presented. Furthermore, we elaborate the critic part and the actor part of SAC. Finally, the detailed implementation of our proposed algorithm is presented.

### 12.5.1  State Space

In each slot, the DRL agent in BS observes the vehicular environment in the coverage of the BS and collects the following parameters:

1. $\gamma_k(t)$: The SNR of the wireless link between task vehicle $V_t$ and service vehicle $V_k$ at time $t$.
2. $F_k^r(t)$: The remaining computing resource of service vehicle $V_k$ at time $t$. For the sake of simplification, we assume that during a period, the local tasks to be executed in $V_k$ remain unchanged, and the variation of the remaining computing resource only depends on the amount of the accepted offloading tasks in $V_k$.
3. $\eta_k(t)$: The service availability of $V_k$ for offloading a computational task of $V_t$. According to the definition of service availability, $\eta_k(t)$ can be evaluated by the fraction of the minimum computing resource occupied by local tasks in $V_k$, the priority of offloading task, the relative distance and velocity between $V_t$ and $V_k$, and the traffic density at time $t$.
4. $u_k(t)$: The total utility of local tasks in $V_k$ at time $t$, which equals to $U_k^{local}(\theta)$ defined in (12.19).
5. $D(t), C(t), \tau(t), \kappa(t)$: The data size, the computation size, the deadline for high-priority task or the reference delay for low-priority task, and the priority of the computational task at time $t$, respectively.

Let $\mathscr{S}$ denote the state space, and the state vector at time $t$ is represented as follows:

$$\mathbf{s}_t = \left[\gamma_1(t), \gamma_2(t), \cdots, \gamma_K(t), F_1^r(t), F_2^r(t), \cdots, F_K^r(t),\right.$$
$$\eta_1(t), \eta_2(t), \cdots, \eta_K(t), u_1(t), u_2(t), \cdots, u_K(t), \qquad (12.22)$$
$$\left. D(t), C(t), \tau(t), \kappa(t)\right].$$

### 12.5.2  Action Space

By observing the state at time $t$, the agent conducts an action to determine the service vehicle and the unit price that the task vehicle should pay to the service vehicle. We denote the action space as $\mathscr{A}$, and represent the action vector conducted by the agent at time $t$ as

$$\mathbf{a}_t = [c_1(t), c_2(t), \cdots, c_K(t), p_1(t), p_2(t), \cdots, p_K(t)]. \qquad (12.23)$$

In the action vector, $c_k(t)$ is a binary variable, $c_k(t) = 1$ indicates that the computational task is offloaded to $V_k$. $p_k(t)$ denotes the unit price that $V_t$ pays to $V_k$ per CPU cycle, and the domain of $p_k(t)$ is $\left(0, U_k^{local}\left(\tilde{\theta}_k\right)/C(t)\right]$, which is related to the fraction of the available computing resource in $V_k$.

### 12.5.3  Reward Function

In each step, the agent conducts action $\mathbf{a}_t$ by observing state $\mathbf{s}_t$, and then gets an immediate reward, which can be represented as $R(\mathbf{s}_t, \mathbf{a}_t) = \sum_{k=1}^{K} c_t^k U_t^k$, where $U_t^k$ represents the utility of a task in (12.9). Our goal is to maximize the mean utility of $V_t$ in a period. Assume that there are $T$ time slots in a period, and in each slot, a computational task of the task vehicle is offloaded to a service vehicle, then the mean reward can be given as

$$R = \frac{1}{T}\sum_{t=0}^{T-1}\sum_{k=1}^{K} c_t^k U_t^k. \tag{12.24}$$

### 12.5.4  Policy and Value Function

In reinforcement learning, policy is an action-selection strategy that determines the long-term expected return, it can be either deterministic or stochastic [48, 49]. In the algorithm, the policy is stochastic and can be denoted as $\pi(\mathbf{a}|\mathbf{s})$, which means a probability distribution over actions given a certain observed state. The goal of the agent is to learn an optimal policy $\pi^*(\mathbf{a}|\mathbf{s})$ that maximizes the expected return corresponding to the reward defined in (12.24), and the policy evaluation and improvement are presented in the following subsections.

We then define two functions, the action-value function and the state-value function. Since the current action can affect future returns, the action-value function $Q^\pi(\mathbf{s}, \mathbf{a})$ is defined as the expected discounted return of a trajectory starting at time 0 with state $\mathbf{s}$ and selecting action $\mathbf{a}$, which is shown as follows:

$$Q^\pi(\mathbf{s}, \mathbf{a}) = \mathop{\mathbb{E}}_{\mu \sim \pi}\left[\sum_{t=0}^{T-1}\gamma^t R_t | \mathbf{s}_0 = \mathbf{s}, \mathbf{a}_0 = \mathbf{a}\right], \tag{12.25}$$

where $R_t$ is the short name for $R\,(\mathbf{s}_t, \mathbf{a}_t)$, $\gamma \in (0, 1)$ is discount factor, $\mu$ denotes the state-action trajectory $\{\mathbf{s}_0, \mathbf{a}_0, \mathbf{s}_1, \mathbf{a}_1, \ldots, \mathbf{s}_{T-1}, \mathbf{a}_{T-1}, \mathbf{s}_T\}$. Similarly, we define $V^\pi\,(\mathbf{s})$ as the state-value function, which means the expected discounted return starting from state $\mathbf{s}$ and taking actions following policy $\pi$, which is presented as follows:

$$V^\pi\,(\mathbf{s}) = \underset{\mu \sim \pi}{\mathbb{E}} \left[ \sum_{t=0}^{T-1} \gamma^t R_t | \mathbf{s}_0 = \mathbf{s} \right]. \tag{12.26}$$

In SAC, the optimization objective not only maximizes the expected reward, but also maximizes the entropy of the policy at the same time. Therefore, the action-value function and the state-value function should be modified in accordance with the optimization objective of the algorithm. Similar to the regularized value function defined in [50] and according to (12.25), the soft action-value function can be given as

$$Q_h^\pi\,(\mathbf{s}, \mathbf{a}) = \underset{\mu \sim \pi}{\mathbb{E}} \left[ \sum_{t=0}^{T-1} \gamma^t R_t + \alpha \sum_{t=1}^{T-1} \gamma^t H\,(\pi\,(\cdot|\mathbf{s}_t)) | \mathbf{s}_0 = \mathbf{s}, \mathbf{a}_0 = \mathbf{a} \right], \tag{12.27}$$

where $\alpha$ is the temperature factor that determines the importance of the policy entropy in the optimization objective. In the same way, according to (12.26), the soft state-value function is defined as follows:

$$V_h^\pi\,(\mathbf{s}) = \underset{\mu \sim \pi}{\mathbb{E}} \left[ \sum_{t=0}^{T-1} \gamma^t\,(R_t + \alpha H\,(\pi\,(\cdot|\mathbf{s}_t))) | \mathbf{s}_0 = \mathbf{s} \right]. \tag{12.28}$$

Since the dimensions of state space and actor space can be extremely high, and the process of value iteration until convergence is computationally too expensive, it is necessary to employ function approximators for the soft action-value function and the policy. In the algorithm, deep neural network (DNN) is employed to represent the soft action-value function and the policy. Soft action-value function $Q_h^\pi\,(\mathbf{s}, \mathbf{a})$ can be parameterized as $Q_\theta\,(\mathbf{s}, \mathbf{a})$ by utilizing a fully connected DNN which contains multiple hidden layers, and $\theta$ represents the parameters of the network. In the same way, policy $\pi\,(\mathbf{a}|\mathbf{s})$ is parameterized as $\pi_\phi\,(\mathbf{a}|\mathbf{s})$ with a fully connected DNN, and $\phi$ represents the parameters of the network.

### 12.5.5  Policy Evaluation

In the algorithm, the soft value functions can be computed iteratively with a given policy $\pi$. According to the Bellman equation, the relationship between the soft action-value function and the soft state-value function is shown as follows [45]:

$$Q_h^\pi (\mathbf{s}_t, \mathbf{a}_t) = R_t + \gamma \mathbb{E}_{\mathbf{s}_{t+1} \sim \rho_\pi} \left[ V_h^\pi (\mathbf{s}_{t+1}) \right], \tag{12.29}$$

$$V_h^\pi (\mathbf{s}_t) = \mathbb{E}_{\mathbf{a}_t \sim \pi} \left[ Q_h^\pi (\mathbf{s}_t, \mathbf{a}_t) - \alpha \log \pi (\mathbf{a}_t | \mathbf{s}_t) \right], \tag{12.30}$$

where $\rho_\pi$ is the state marginals of the trajectory distribution induced by policy $\pi$.

In order to stabilize the training in the iteration of the soft action-value function, we employ a target soft action-value function with parameters $\bar{\theta}$ which can be obtained as an exponentially moving average of $\theta$. We define the target soft action-value function as

$$\hat{Q}_{\bar{\theta}} (\mathbf{s}_t, \mathbf{a}_t) = R_t + \gamma \mathbb{E}_{\mathbf{s}_{t+1} \sim \rho_\pi} \left[ V_{\bar{\theta}} (\mathbf{s}_{t+1}) \right]. \tag{12.31}$$

To break up the temporal correlations during the training process, we set an experience replay buffer $\mathcal{M}$ with a fixed size. In each time slot, the transition of the state of the vehicular environment, the conducted action and the immediate reward form a tuple $(\mathbf{s}_t, \mathbf{a}_t, R_t, \mathbf{s}_{t+1})$, which is then stored in $\mathcal{M}$. By sampling a mini-batch of tuples from the buffer, the actor and the critic are updated. Then the loss function of the critic becomes

$$J_Q (\theta) = \mathbb{E}_{(\mathbf{s}_t, \mathbf{a}_t) \sim \mathcal{M}} \left[ \frac{1}{2} \left( Q_\theta (\mathbf{s}_t, \mathbf{a}_t) - \hat{Q}_{\bar{\theta}}(\mathbf{s}_t, \mathbf{a}_t) \right)^2 \right], \tag{12.32}$$

and the parameters of the soft action-value function is trained by minimizing loss function $J_Q (\theta)$.

### 12.5.6  Policy Improvement

In the algorithm, if a policy of task offloading is optimal, then all the offloading tasks in a period will be completed with the least time and price, and the total utility will be maximum as well. On the other hand, if some of the offloading tasks are completed with much long time or cannot be completed before deadline, the utility will be a smaller value, which reflects that the current policy is poor. As a result, the policy should be improved. In SAC, the policy parameters can be learned by minimizing the expected KL-divergence, which is shown as follows [45]:

$$\pi_t = \arg \min_{\pi' \in \Pi} D_{\text{KL}} \left( \pi' (\cdot | \mathbf{s}_t) \, \middle\| \, \frac{\exp \left( \frac{1}{\alpha} Q^\pi (\mathbf{s}_t, \cdot) \right)}{Z^\pi (\mathbf{s}_t)} \right), \tag{12.33}$$

where $\Pi$ represents a set of policies which correspond to a parameterized family of distributions such as Gaussians. $Z^{\pi}(\mathbf{s}_t)$ is a function that normalizes the distribution and does not have effect on the gradient with respect to the new policy. We further transform the KL-divergence in (12.33) by multiplying $\alpha$ and ignore the constant normalization term $\mathbb{E}_{\mathbf{a}_t \sim \pi_{\phi}}\left[\alpha \log Z^{\pi}(\mathbf{s}_t)\right]$. Afterwards, the policy parameters can be trained by minimizing the following function:

$$J_{\pi}(\phi) = \mathbb{E}_{\mathbf{s}_t \sim \mathcal{M}}\left[\mathbb{E}_{\mathbf{a}_t \sim \pi_{\phi}}\left[\alpha \log\left(\pi_{\phi}(\mathbf{a}_t|\mathbf{s}_t)\right) - Q_{\theta}(\mathbf{s}_t, \mathbf{a}_t)\right]\right]. \tag{12.34}$$

In the process of policy iteration, the soft policy evaluation and the soft policy improvement alternate until the iteration converges to an optimal policy with maximum entropy among the policies in $\Pi$.

### 12.5.7  Algorithm Design Based on SAC

In the process of policy evaluation and policy improvement illustrated above, the temperature parameter is treated as a constant. Here, we will demonstrate how to choose the optimal temperature automatically in SAC. Because the entropy can vary unpredictably both across different training tasks and during training process, it is difficult to adjust the temperature. To solve the problem, a constrained optimization problem is formulated. While maximizing the expected return, the entropy of policy should satisfy a minimum constraint, which is shown as follows [51]:

$$\max \quad \mathbb{E}_{\rho_{\pi}}\left[\sum_{t=0}^{T-1} R(\mathbf{s}_t, \mathbf{a}_t)\right],$$

$$\text{s.t.} \quad \mathbb{E}_{(\mathbf{s}_t, \mathbf{a}_t) \sim \rho_{\pi}}\left[-\log\left(\pi_t(\mathbf{a}_t|\mathbf{s_t})\right)\right] \geq \mathcal{H}_0, \quad \forall t, \tag{12.35}$$

where $\mathcal{H}_0$ is a predefined minimum policy entropy threshold. The temperature parameter can be learned in every time slot by minimizing the following objective function [51]:

$$J(\alpha) = \mathbb{E}_{\mathbf{a}_t \sim \pi_t}\left[-\alpha \log \pi_t(\mathbf{a}_t|\mathbf{s}_t) - \alpha \mathcal{H}_0\right]. \tag{12.36}$$

In addition, in order to mitigate the positive bias in the policy improvement, two soft action-value functions are employed in the algorithm. As is proposed in [52], we parameterize two soft action-value functions and train them independently. Then we use the minimum of the soft action-value functions to compute the stochastic gradient of $J_Q(\theta)$ and the policy gradient of $J_{\pi}(\phi)$.

---

**Algorithm 4** Task Offloading Algorithm based on SAC

---

1: **Initialize:**
2: Initialize main soft Q-networks $Q_{\theta_1}$ (**s**, **a**) and $Q_{\theta_2}$ (**s**, **a**) with weights $\theta_1$ and $\theta_2$.
3:
4: Initialize target soft Q-networks $Q_{\bar{\theta}_1}$ (**s**, **a**) and $Q_{\bar{\theta}_2}$ (**s**, **a**) with weights $\bar{\theta}_1 = \theta_1$ and $\bar{\theta}_2 = \theta_2$.
5:
6: Initialize policy $\pi_\phi$ (**a**|**s**) with weights $\phi$.
7: Initialize replay memory $\mathscr{M} = \varnothing$.
8:
9: **for** each period **do**
10:     Collect initial observation state $\mathbf{s}_0$.
11:
12:     **for** time slot $t = 0, 1, \ldots, T - 1$ **do**
13:         Receive offloading request from task vehicle and collect information from vehicular environment, estimate state $\mathbf{s}_t$.
14:         Generate action $\mathbf{a}_t$ that determines service vehicle $V_k$ and unit price $p_k$, and send it to $V_k$ and the task vehicle.
15:         Compute immediate reward $R_t$ according to the resource allocation of $V_k$, and estimate the next state $\mathbf{s}_{t+1}$.
16:         Store tuple $(\mathbf{s}_t, \mathbf{a}_t, R_t, \mathbf{s}_{t+1})$ in $\mathscr{M}$.
17:         Sample a batch of tuples $\mathscr{B}$ from $\mathscr{M}$.
18:         Update $\theta_i$ by computing the gradient of $J_Q$ $(\theta_i)$ defined in (12.32),

$$\theta_i = \theta_i - \delta_Q \nabla_{\theta_i} \frac{1}{|\mathscr{B}|} \sum_{\mathscr{B}} J_Q\ (\theta_i)\,, \quad \forall i = 1, 2.$$

19:         Update policy parameters $\phi$ by calculating the gradient of $J_\pi$ $(\phi)$ defined in (12.34),

$$\phi = \phi - \delta_\pi \nabla_\phi \frac{1}{|\mathscr{B}|} \sum_{\mathscr{B}} J_\pi\ (\phi)\,.$$

20:         Update temperature parameter $\alpha$ by computing the gradient of $J$ $(\alpha)$ defined in (12.36),

$$\alpha = \alpha - \delta_\alpha \nabla_\alpha \frac{1}{|\mathscr{B}|} \sum_{\mathscr{B}} J\ (\alpha)\,.$$

21:         Update parameters of target soft action-value functions $\bar{\theta}_i$ by

$$\bar{\theta}_i = \omega\theta_i + (1 - \omega)\,\bar{\theta}_i\,, \quad \forall i = 1, 2.$$

22:     **end for**
23: **end for**

---

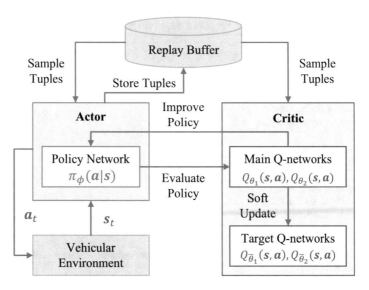

**Fig. 12.2** The structure of SAC-based task offloading algorithm

Finally, the structure of our proposed SAC-based algorithm is presented in Fig. 12.2, which mainly contains two parts, the actor and the critic. The actor contains the policy network, which takes charge of giving offloading decision by observing the system state and meanwhile improving the policy. The critic contains the main soft Q-networks and the target soft Q-networks, which are responsible for policy evaluation. In addition, the replay buffer is employed to store the task offloading experiences, which can be used to train the networks in the actor and the critic. We present the SAC-based task offloading algorithm in Algorithm 4, and illustrate the main steps of the algorithm in the following:

1. Initialize parameters of two action-value functions $Q_{\theta_1}(\mathbf{s}, \mathbf{a})$ and $Q_{\theta_2}(\mathbf{s}, \mathbf{a})$ with weights $\theta_1$ and $\theta_2$ that form some distributions, respectively. The parameters of target soft action-value networks $\bar{\theta}_1$ and $\bar{\theta}_2$ are assigned with the weights $\theta_1$ and $\theta_2$, respectively. Initialize parameters of the stochastic policy $\pi_\phi(\mathbf{a}|\mathbf{s})$ with random weights $\phi$ and set the experience replay buffer $\mathcal{M}$ as $\varnothing$.
2. In each period, the agent in BS first collects the initial state $\mathbf{s}_0$ by observing the vehicular environment.
3. In each time slot, the agent receives a task offloading request from task vehicle, and by collecting state information from the vehicular environment, state $\mathbf{s}_t$ can be estimated. Then the actor generates action $\mathbf{a}_t$ according to state $\mathbf{s}_t$ and policy $\pi_\phi(\mathbf{a}_t|\mathbf{s}_t)$.
4. Given action $\mathbf{a}_t$, the agent sends service assignment which contains the selected service vehicle $V_k$ and unit price $p_k$ to the task vehicle and $V_k$, and $V_k$ allocates computing resource for the offloading task according to $p_k$. Then $V_k$ sends the

allocation information back to the agent, and the agent computes the immediate reward $R_t$. At last, the task vehicle transmits the task data to $V_k$.

5. The agent observes the vehicular environment and estimates the next state $\mathbf{s}_{t+1}$, and stores tuple $(\mathbf{s}_t, \mathbf{a}_t, R_t, \mathbf{s}_{t+1})$ into the experience replay buffer $\mathcal{M}$.

6. The agent samples a batch of tuples $\mathcal{B}$ from $\mathcal{M}$ randomly. The parameters $\theta_1$ and $\theta_2$ of soft action-value functions are updated by calculating the gradients of $J_Q(\theta_1)$ and $J_Q(\theta_2)$ defined in (12.32), the parameters $\phi$ of policy is updated by computing the gradient of $J_\pi(\phi)$ defined in (12.34), and the temperature parameter is updated according to the gradient of $J(\alpha)$ defined in (12.36).

7. The agent updates the parameters of target soft action-value functions, $\bar{\theta}_1$ and $\bar{\theta}_2$, with the exponentially moving average of $\theta_1$ and $\theta_2$, respectively.

## 12.5.8  Complexity Analysis

The complexity of our proposed algorithm mainly contains two parts, one is the complexity of action generation, and the other is the complexity of training actor and critic networks. We assume that there are $K$ service vehicles surrounding a task vehicle in a period. For each offloading task, the agent generates an action in terms of task offloading according to the system state. In our proposed algorithm, the actor and critic are both fully-connected networks that contain two hidden layers. We denote the hidden layers of actor as $\left(L_1^a, L_2^a\right)$. From (12.22) and (12.23), the dimension of state space is $4K + 4$, and the dimension of action space is $2K$. Then, the complexity of action generation in actor is $O\left((4K + 4) L_1^a + L_1^a L_2^a + 2L_2^a K\right)$. Since the size of hidden layers of actor and critic in our system is fixed, the complexity of action generation can be given as $O(K)$. In the training process of actor and critic networks, we assume there are $T$ time slots in a period, then the actor and critic networks will be trained in $T$ steps. In each step, similar to the action generation in actor, the complexity of training actor and critic is $O(K)$. Finally, in a period, the complexity of our proposed algorithm is $O(KT)$.

## 12.6  Performance Evaluation

In this section, we conduct a number of simulations to evaluate the performance of our proposed scheme. We first present the simulation parameters of the system. Then, we evaluate the mean utility of the proposed scheme under different traffic densities and task arrival rates. Finally, we analyse the completion ratio and the mean delay of high-priority tasks and low-priority tasks under different traffic densities.

## 12.6.1   Simulation Setup

In the simulation, we consider an one-way road, where a BS equipped with a DRL agent is located along the road, and there are several vehicles passing the coverage of the BS all the time, the traffic density is set as $5 \sim 50$ vehicles/km. Considering that the normal velocity of a vehicle is $60 \sim 110$ km/h, we set the relative velocity between task vehicle and service vehicle distributed in $[-50, 50]$ km/h, and generate the traffic data by the free flow traffic model presented in Sect. 12.3. We assume that different vehicles have different computational capabilities, the computational capability of a vehicle is randomly chosen from the set $\{5, 6, \cdots, 10\}$ GHz. Since the data transmission between the BS and vehicles in each slot only contains a few bits that describe offloading request and response, the transmission time between the BS and vehicles can be ignored, we only consider the transmission time of V2V link. According to the bandwidth of subchannels in dedicated short-range communications (DSRC), the bandwidth of a V2V channel is set as 10 MHz, the transmission power of each vehicle is assumed to be identical, and then the V2V transmission rate is estimated by the channel capacity in (12.3). In the simulation, we mainly consider the computation-intensive tasks, and due to the unstable V2V link and short link duration, we set the data size of task uniformly distributed in $[0.02, 0.2]$ MB, the computation size of task uniformly distributed in $[0.2, 3.2] \times 10^9$ CPU cycles, the low priority of task $\kappa_L = 0.5$ and the high priority of task $\kappa_H = 1$, and the maximum tolerable delay of high-priority task or the reference delay of low-priority task is set as $\{0.5, 1, 2, 4\}$ s. For a computational task with large data size and computation size, the task vehicle can divide the task into several subtasks to offload. In addition, according to DSRC, we set the maximum communication range of a vehicle as 500 m. The maximum communication range of a BS is set as 3 km. Finally, the parameters in our simulation are summarized in Table 12.1.

The simulations are conducted on a desktop which has two NVIDIA TITAN Xp GPUs, a 128G RAM and an Intel Xeon CPU. The simulation platform is based on Pytorch with Python 3.7 on Ubuntu 16.04 LTS. In our proposed algorithm, the soft action-value function and the policy function are approximated by DNNs. We design an actor network with two hidden layers, and the size of hidden layers is (1200, 1200), the learning rate of the actor is set as $\delta_\pi = 8 \times 10^{-4}$. Similarly, we design a critic network with two hidden layers with the size (1200, 1200), the learning rate of the critic is set as $\delta_Q = 8 \times 10^{-4}$. We set the learning rate of temperature parameter $\delta_\alpha = 8 \times 10^{-4}$, the batch size $|\mathscr{B}| = 256$, and the delay factor $\omega = 0.005$.

In order to verify our proposed algorithm, we introduce the following regular algorithms for comparison.

1. *Random Based Algorithm (RBA)*: The RBA selects a service vehicle surrounding the task vehicle randomly, and gives a random price for the offloading task. In the simulation, the agent runs RBA 2000 times in each slot and selects the action corresponding to the maximum utility.

**Table 12.1** Simulation parameters

| Parameter | Value |
|---|---|
| Traffic density $\rho$ (vehicles/km) | $5 \sim 50$ |
| Relative velocity $v_k - v_t$ (km/h) | $[-50, 50]$ |
| Vehicular computational capability $F_k$ (GHz) | $\{5, 6, 7, 8, 9, 10\}$ |
| Data size of task $D_n$ (MB) | $[0.02, 0.2]$ |
| Computation size of task $C_n$ ($10^9$ cycles) | $[0.2, 3.2]$ |
| Maximum delay of task $\tau_n$ (s) | $\{0.5, 1, 2, 4\}$ |
| High-priority $\kappa_H$ | 1 |
| Low-priority $\kappa_L$ | 0.5 |
| V2V bandwidth $B_{t,k}$ (MHz) | 10 |
| Maximum V2V transmission range $R$ (m) | 500 |
| Maximum BS transmission range $R_B$ (km) | 3 |

2. *Greedy Based Algorithm (GBA)*: The GBA selects the service vehicle with maximum remaining computing resource and determines the service price that maximizes the utility of the offloading task in each slot.
3. *Double Deep Q-Network (DDQN)*: DDQN [53] is a Q-learning-based DRL algorithm. In the simulation, the state space, the action space and the reward function of DDQN are the same as SAC.

## *12.6.2 Average Utility*

We first present the mean utility of offloading tasks in our proposed algorithm with different learning rates. As shown in Fig. 12.3, when $\delta_Q \leq 2 \times 10^{-3}$ and $\delta_\pi \leq 2 \times 10^{-3}$, the mean utility of offloading tasks in our proposed algorithm reaches convergence around 2000 training episodes, and the difference among the mean utilities w.r.t. different learning rates is small, which demonstrates that our proposed algorithm is not very sensitive to small learning rates. On the other hand, Fig. 12.3 shows that when the learning rates of the actor and the critic become large enough (e.g., $\delta_Q = 8 \times 10^{-3}, \delta_\pi = 8 \times 10^{-3}$), the proposed algorithm falls into a local optimum and cannot reach convergence.

Then, we compare the mean utility of offloading tasks in a period by utilizing RBA, GBA, DDQN, and our proposed algorithm. To evaluate the performance of the proposed algorithm both in low traffic density and high traffic density, we conduct several simulations with the vehicle density varies from 5 vehicles/km to 50 vehicles/km. As shown in Fig. 12.4, the mean utility of our proposed algorithm is higher than the other algorithms under various traffic densities. It is because that the objective of our proposed algorithm is to maximize the mean utility of all offloading tasks in a period, and in each time slot, the policy of task offloading aims to maximize the expected discounted return starting from the current time, while in

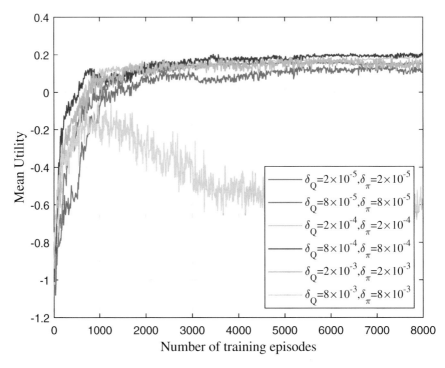

**Fig. 12.3** Mean utility of offloading tasks in SAC with different learning rates

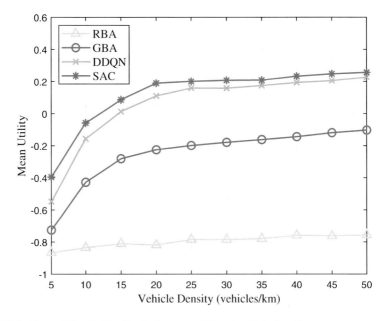

**Fig. 12.4** Mean utility of offloading tasks versus different vehicle densities

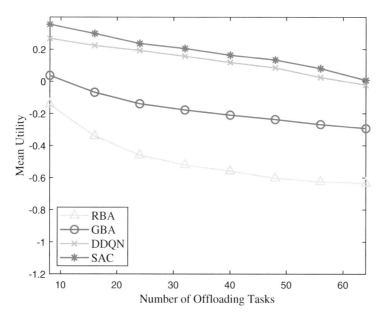

**Fig. 12.5** Mean utility of offloading tasks versus different tasks arrival rates when the vehicle density is 30 vehicles/km

GBA, the policy aims to maximize only the current reward in each time slot, and thus may not obtain the maximum mean utility of all offloading tasks in a period. In addition, although the optimization objective of DDQN is maximizing the expected long-term return, the mean utility of DDQN is slightly less than the mean utility of SAC, this is because SAC employs actor-critic framework while DDQN utilizes the framework that includes main Q-network and target Q-network. Besides, SAC includes the entropy of policy in the optimization objective, which can explore more suitable actions in dynamic vehicular environment.

We also simulate the performance of the four algorithms with different task arrive rates when the vehicle density is 30 vehicles/km. In the simulation, the value of task arrive rate means the number of offloading tasks in a period. In Fig. 12.5, we can see that the mean utilities of all the algorithms decline with the increase of task arrive rate. With the number of offloading tasks increasing, the average allocated vehicular computing resource for each offloading task reduces, thus makes the average completion time longer, and further makes the mean delay-aware utility decline. In addition, due to the optimization of the long-term expected return in DRL, the mean utilities of DDQN and our proposed algorithm are higher than GBA and RBA versus various task arrive rates.

### 12.6.3   Completion Ratio

Figure 12.6a shows the completion ratio of high-priority tasks by utilizing the four algorithms under different vehicle densities. It can be seen that the completion ratios of DDQN and our proposed algorithm are higher than GBA and RBA when vehicle density is less than 15 vehicles/km, it is because in DDQN and our proposed algorithm, once a high-priority task cannot be completed before the deadline, the loss of utility will be much higher than low-priority task. Since the aim of DDQN and our proposed algorithm is to maximize the mean utility of offloading tasks, it must ensure that high-priority tasks completed with the maximum probability. In addition, when vehicle density is higher than 15 vehicles/km, the completion ratio of GBA, DDQN, and our proposed algorithm are all close to 100%, which means that the computing resource of service vehicles are enough to execute all of the high-priority tasks.

In Fig. 12.6b, the completion ratio of low-priority tasks by employing different algorithms is presented. Notice that the completion ratio of DDQN and our proposed algorithm is lower than GBA when the traffic density is $5 \sim 10$ vehicles/km, it is because the service vehicles are very few, the on-board computing resource of service vehicles is not enough to execute all the offloading tasks of task vehicle, DDQN and our proposed algorithm ensure the high-priority tasks completed first, and then executes the low-priority tasks, while GBA ensures all the offloading tasks completed with the maximum utility in each step, which makes the completion ratio of low-priority tasks higher but the completion ratio of high-priority tasks lower than our proposed algorithm. Moreover, from Fig. 12.6a and b, the completion ratio of high-priority tasks in DDQN is lower than that in our proposed algorithm, but the completion of low-priority tasks in DDQN is higher than that in our proposed algorithm when vehicle density is less than 10 vehicles/km, which demonstrates that our algorithm can ensure more high-priority tasks completed first when there are fewer service vehicles.

### 12.6.4   Average Delay

We also simulate the completion time of offloading tasks under both low traffic density and high traffic density. Figure 12.7a shows the average delays of successfully completed high-priority tasks with different deadlines under low traffic density. Although the average delay of high-priority tasks by applying RBA is lower than the other algorithms, it suffers from the low completion ratio as shown in Fig. 12.6a. From Fig. 12.7a, we can see that the average delays of high-priority tasks with different deadlines by applying our proposed algorithm are all lower

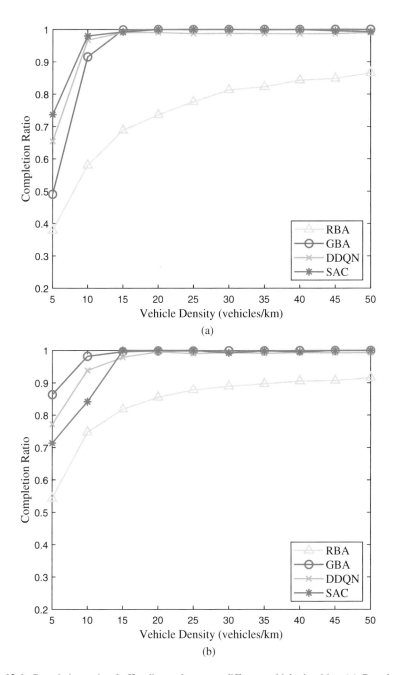

**Fig. 12.6** Completion ratio of offloading tasks versus different vehicle densities. (**a**) Completion ratio of high-priority tasks. (**b**) Completion ratio of low-priority tasks

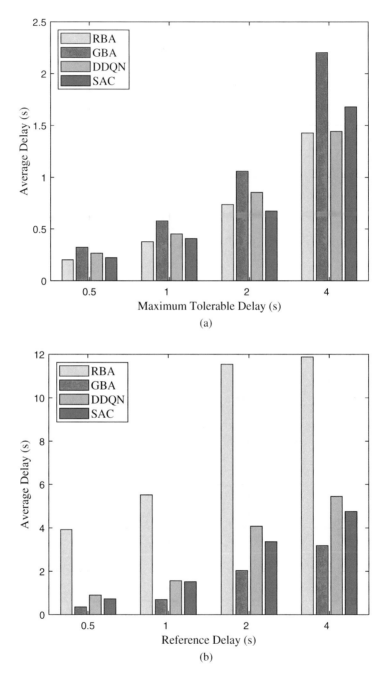

**Fig. 12.7** Average delay of offloading tasks with different maximum tolerable / reference delays when the vehicle density is 15 vehicles/km. (**a**) Average delay of high-priority tasks with different maximum tolerable delays. (**b**) Average delay of low-priority tasks with different reference delays

than GBA, while the average delays of low-priority tasks with various reference delays by applying our proposed algorithm are all higher than GBA, as shown in Fig. 12.7b. Compared to RBA and GBA, our proposed algorithm ensures all the high-priority tasks completed with the least time, and meanwhile makes low-priority tasks completed with a relative appropriate delay comparing with the reference delay under low traffic density. Moreover, from the overall perspective, the average delays of high-priority tasks and low-priority tasks in our proposed algorithm are less than that in DDQN. Therefore, the proposed algorithm has better performance in executing tasks with different priorities than other algorithms.

Figure 12.8a and b show the average delays of high-priority tasks and low-priority tasks with different maximum tolerable or reference delays under high traffic density, respectively. Comparing with the average delays under low vehicle density, the average delays of high-priority tasks and low-priority tasks under high vehicle density are both lower due to more service vehicles surrounding the task vehicle.

## 12.7 Conclusion

This part targets the problem of task allocation in dynamic VFC environment. Our work goes beyond existing approaches by jointly considering task priority, service availability of vehicles, and incentive of computing resource sharing to obtain the optimal offloading policy that maximize the utility of offloading tasks. Moreover, in order to make the offloading policy adapt to changes in dynamic vehicular environment, we formulate the task allocation problem as an MDP, and propose a DRL method based on SAC to solve the problem. By incorporating the entropy measure of the policy into the reward, our proposed algorithm becomes more robust and sample-efficient. Simulation results validate that our proposed algorithm can effectively ensure high-priority tasks completed preferentially, and meanwhile has better performance in task completion ratio and offloading delay comparing with the conventional approaches.

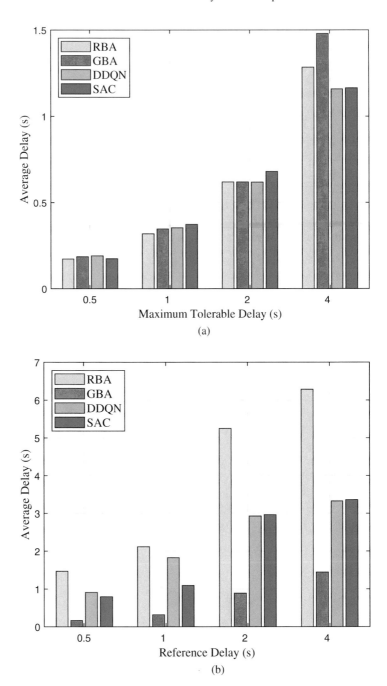

**Fig. 12.8** Average delay of low-priority tasks with different maximum tolerable / reference delays when the vehicle density is 35 vehicles/km. (**a**) Average delay of high-priority tasks with different maximum tolerable delays. (**b**) Average delay of low-priority tasks with different reference delays

# References

1. J. Wang, C. Jiang, Z. Han, Y. Ren, and L. Hanzo, "Internet of vehicles: Sensing-aided transportation information collection and diffusion," *IEEE Trans. Veh. Technol.*, vol. 67, no. 5, pp. 3813–3825, May 2018.
2. J. Du, C. Jiang, J. Wang, Y. Ren, and M. Debbah, "Machine learning for 6G wireless networks: Carry-forward-enhanced bandwidth, massive access, and ultrareliable/low latency," *IEEE Veh. Technol. Mag*, vol. 15, no. 4, pp. 123–134, Dec. 2020.
3. D. Van Le and C. Tham, "Quality of service aware computation offloading in an ad-hoc mobile cloud," *IEEE Trans. Veh. Technol.*, vol. 67, no. 9, pp. 8890–8904, Sep. 2018.
4. J. Du, C. Jiang, A. Benslimane, S. Guo, and Y. Ren, "Stackelberg differential game based resource sharing in hierarchical fog-cloud computing," in *Proc. IEEE Global Commun. Conf. (GLOBECOM'19)*. Waikoloa, Hawaii, USA, 9-13 Dec. 2019.
5. J. Zhang, J. Du, Y. Shen, and J. Wang, "Dynamic computation offloading with energy harvesting devices: A hybrid-decision-based deep reinforcement learning approach," *IEEE Internet Things J.*, vol. 7, no. 10, pp. 9303–9317, Oct. 2020.
6. F. Sun, F. Hou, N. Cheng, M. Wang, H. Zhou, L. Gui, and X. Shen, "Cooperative task scheduling for computation offloading in vehicular cloud," *IEEE Trans. Veh. Technol.*, vol. 67, no. 11, pp. 11 049–11 061, Nov. 2018.
7. J. Zhang and K. B. Letaief, "Mobile edge intelligence and computing for the internet of vehicles," *Proc. IEEE*, vol. 108, no. 2, pp. 246–261, Feb. 2020.
8. J. Feng, Z. Liu, C. Wu, and Y. Ji, "AVE: Autonomous vehicular edge computing framework with ACO-based scheduling," *IEEE Trans. Veh. Technol.*, vol. 66, no. 12, pp. 10 660–10 675, Dec. 2017.
9. C. Chen, L. Chen, L. Liu, S. He, X. Yuan, D. Lan, and Z. Chen, "Delay-optimized V2V-based computation offloading in urban vehicular edge computing and networks," *IEEE Access*, vol. 8, pp. 18 863–18 873, 2020.
10. C. Zhu, J. Tao, G. Pastor, Y. Xiao, Y. Ji, Q. Zhou, Y. Li, and A. Yl?-J??ski, "Folo: Latency and quality optimized task allocation in vehicular fog computing," *IEEE Internet Things J.*, vol. 6, no. 3, pp. 4150–4161, Jun. 2019.
11. L. Pu, X. Chen, G. Mao, Q. Xie, and J. Xu, "Chimera: An energy-efficient and deadline-aware hybrid edge computing framework for vehicular crowdsensing applications," *IEEE Internet Things J.*, vol. 6, no. 1, pp. 84–99, Feb. 2019.
12. T. Liu, J. Li, F. Shu, and Z. Han, "Optimal task allocation in vehicular fog networks requiring URLLC: An energy-aware perspective," *IEEE Trans. Netw. Sci. Eng.*, vol. 7, no. 3, pp. 1879–1890, Jul.-Sep. 2020.
13. H. Li, X. Li, M. Zhang, and B. Ulziinyam, "Multicast-oriented task offloading for vehicle edge computing," *IEEE Access*, vol. 8, pp. 187 373–187 383, 2020.
14. Z. Zhou, P. Liu, J. Feng, Y. Zhang, S. Mumtaz, and J. Rodriguez, "Computation resource allocation and task assignment optimization in vehicular fog computing: A contract-matching approach," *IEEE Trans. Veh. Technol.*, vol. 68, no. 4, pp. 3113–3125, Apr. 2019.
15. J. Zhao, M. Kong, Q. Li, and X. Sun, "Contract-based computing resource management via deep reinforcement learning in vehicular fog computing," *IEEE Access*, vol. 8, pp. 3319–3329, 2020.
16. Z. Zhou, H. Liao, X. Zhao, B. Ai, and M. Guizani, "Reliable task offloading for vehicular fog computing under information asymmetry and information uncertainty," *IEEE Trans. Veh. Technol.*, vol. 68, no. 9, pp. 8322–8335, Sep. 2019.
17. M. Liwang, S. Dai, Z. Gao, Y. Tang, and H. Dai, "A truthful reverse-auction mechanism for computation offloading in cloud-enabled vehicular network," *IEEE Internet Things J.*, vol. 6, no. 3, pp. 4214–4227, Jun. 2019.
18. L. T. Tan and R. Q. Hu, "Mobility-aware edge caching and computing in vehicle networks: A deep reinforcement learning," *IEEE Trans. Veh. Technol.*, vol. 67, no. 11, pp. 10 190–10 203, Nov. 2018.

19. Y. Liu, H. Yu, S. Xie, and Y. Zhang, "Deep reinforcement learning for offloading and resource allocation in vehicle edge computing and networks," *IEEE Trans. Veh. Technol.*, vol. 68, no. 11, pp. 11 158–11 168, Nov. 2019.
20. Q. Luo, C. Li, T. H. Luan, and W. Shi, "Collaborative data scheduling for vehicular edge computing via deep reinforcement learning," *IEEE Internet Things J.*, vol. 7, no. 10, pp. 9637–9650, Oct. 2020.
21. S. Lee and S. Lee, "Resource allocation for vehicular fog computing using reinforcement learning combined with heuristic information," *IEEE Internet Things J.*, vol. 7, no. 10, pp. 10 450–10 464, Oct. 2020.
22. X. Hou, Z. Ren, J. Wang, W. Cheng, Y. Ren, K. C. Chen, and H. Zhang, "Reliable computation offloading for edge-computing-enabled software-defined IoV," *IEEE Internet Things J.*, vol. 7, no. 8, pp. 7097–7111, Aug. 2020.
23. H. Ke, J. Wang, L. Deng, Y. Ge, and H. Wang, "Deep reinforcement learning-based adaptive computation offloading for MEC in heterogeneous vehicular networks," *IEEE Trans. Veh. Technol.*, vol. 69, no. 7, pp. 7916–7929, Jul. 2020.
24. Z. Zhou, J. Feng, Z. Chang, and X. Shen, "Energy-efficient edge computing service provisioning for vehicular networks: A consensus ADMM approach," *IEEE Trans. Veh. Technol.*, vol. 68, no. 5, pp. 5087–5099, May 2019.
25. Y. Dai, D. Xu, S. Maharjan, and Y. Zhang, "Joint load balancing and offloading in vehicular edge computing and networks," *IEEE Internet Things J.*, vol. 6, no. 3, pp. 4377–4387, Jun. 2019.
26. J. Sun, Q. Gu, T. Zheng, P. Dong, A. Valera, and Y. Qin, "Joint optimization of computation offloading and task scheduling in vehicular edge computing networks," *IEEE Access*, vol. 8, pp. 10 466–10 477, 2020.
27. C. Yang, Y. Liu, X. Chen, W. Zhong, and S. Xie, "Efficient mobility-aware task offloading for vehicular edge computing networks," *IEEE Access*, vol. 7, pp. 26 652–26 664, 2019.
28. K. Zhang, Y. Zhu, S. Leng, Y. He, S. Maharjan, and Y. Zhang, "Deep learning empowered task offloading for mobile edge computing in urban informatics," *IEEE Internet Things J.*, vol. 6, no. 5, pp. 7635–7647, Oct. 2019.
29. Z. Li, C. Wang, and C. Jiang, "User association for load balancing in vehicular networks: An online reinforcement learning approach," *IEEE Trans. Intell. Transp. Syst.*, vol. 18, no. 8, pp. 2217–2228, Aug. 2017.
30. Z. Ning, P. Dong, X. Wang, J. J. P. C. Rodrigues, and F. Xia, "Deep reinforcement learning for vehicular edge computing: An intelligent offloading system," *ACM Trans. Intell. Syst. Technol.*, vol. 10, no. 6, Oct. 2019.
31. M. LiWang, S. Hosseinalipour, Z. Gao, Y. Tang, L. Huang, and H. Dai, "Allocation of computation-intensive graph jobs over vehicular clouds in IoV," *IEEE Internet Things J.*, vol. 7, no. 1, pp. 311–324, Jan. 2020.
32. M. Liwang, J. Wang, Z. Gao, X. Du, and M. Guizani, "Game theory based opportunistic computation offloading in cloud-enabled IoV," *IEEE Access*, vol. 7, pp. 32 551–32 561, 2019.
33. Z. Su, Y. Hui, and T. H. Luan, "Distributed task allocation to enable collaborative autonomous driving with network softwarization," *IEEE J. Sel. Areas Commun.*, vol. 36, no. 10, pp. 2175–2189, Oct. 2018.
34. Y. Sun, X. Guo, J. Song, S. Zhou, Z. Jiang, X. Liu, and Z. Niu, "Adaptive learning-based task offloading for vehicular edge computing systems," *IEEE Trans. Veh. Technol.*, vol. 68, no. 4, pp. 3061–3074, Apr. 2019.
35. K. Wang, Y. Tan, Z. Shao, S. Ci, and Y. Yang, "Learning-based task offloading for delay-sensitive applications in dynamic fog networks," *IEEE Trans. Veh. Technol.*, vol. 68, no. 11, pp. 11 399–11 403, Nov. 2019.
36. Q. Wu, H. Liu, R. Wang, P. Fan, Q. Fan, and Z. Li, "Delay-sensitive task offloading in the 802.11p-based vehicular fog computing systems," *IEEE Internet Things J.*, vol. 7, no. 1, pp. 773–785, Jan. 2020.

37. Z. Ning, P. Dong, X. Wang, L. Guo, J. J. P. C. Rodrigues, X. Kong, J. Huang, and R. Y. K. Kwok, "Deep reinforcement learning for intelligent internet of vehicles: An energy-efficient computational offloading scheme," *IEEE Trans. Cogn. Commun. Netw.*, vol. 5, no. 4, pp. 1060–1072, Dec. 2019.
38. J. Wang, C. Jiang, K. Zhang, T. Q. S. Quek, Y. Ren, and L. Hanzo, "Vehicular sensing networks in a smart city: Principles, technologies and applications," *IEEE Wireless Commun.*, vol. 25, no. 1, pp. 122–132, Feb. 2018.
39. W. L. Tan, W. C. Lau, O. Yue, and T. H. Hui, "Analytical models and performance evaluation of drive-thru internet systems," *IEEE J. Sel. Areas Commun.*, vol. 29, no. 1, pp. 207–222, Jan. 2011.
40. Y. Hui, Z. Su, T. H. Luan, and J. Cai, "Content in motion: An edge computing based relay scheme for content dissemination in urban vehicular networks," *IEEE Trans. Intell. Transp. Syst.*, vol. 20, no. 8, pp. 3115–3128, Aug. 2019.
41. X. Chen, L. Jiao, W. Li, and X. Fu, "Efficient multi-user computation offloading for mobile-edge cloud computing," *IEEE/ACM Trans. Netw.*, vol. 24, no. 5, pp. 2795–2808, Oct. 2016.
42. J. Zhao, Q. Li, Y. Gong, and K. Zhang, "Computation offloading and resource allocation for cloud assisted mobile edge computing in vehicular networks," *IEEE Trans. Veh. Technol.*, vol. 68, no. 8, pp. 7944–7956, Aug. 2019.
43. Y. Zhang, D. Niyato, and P. Wang, "Offloading in mobile cloudlet systems with intermittent connectivity," *IEEE Trans. Mobile Comput.*, vol. 14, no. 12, pp. 2516–2529, Dec. 2015.
44. M. Kaynia and N. Jindal, "Performance of ALOHA and CSMA in spatially distributed wireless networks," in *Proc. IEEE Int. Conf. Commun. (ICC)*.    Beijing, China, May 2008, pp. 1108–1112.
45. T. Haarnoja, A. Zhou, P. Abbeel, and S. Levine, "Soft actor-critic: Off-policy maximum entropy deep reinforcement learning with a stochastic actor," in *Proc. 35th Int. Conf. Mach. Learn. (ICML)*, vol. 80.    Stockholm, Sweden, Jul. 2018, pp. 1861–1870.
46. Z. Xia, J. Du, J. Wang, C. Jiang, Y. Ren, G. Li, and Z. Han, "Multi-agent reinforcement learning aided intelligent UAV swarm for target tracking," *IEEE Trans. Veh. Tech.*, vol. 71, no. 1, pp. 931–945, Jan. 2022.
47. Z. Xia, J. Du, C. Jiang, J. Wang, Y. Ren, and G. Li, "Multi-UAV cooperative target tracking based on swarm intelligence," in *IEEE Int. Conf. Commun. (ICC'21)*.    Montreal, QC, Canada, 14-23 Jun. 2021.
48. J. Shi, J. Du, J. Wang, J. Wang, and J. Yuan, "Priority-aware task offloading in vehicular fog computing based on deep reinforcement learning," *IEEE Trans. Veh. Tech.*, pp. 1–1, Early Access, Dec. 2020.
49. J. Shi, J. Du, J. Wang, and J. Yuan, "Distributed V2V computation offloading based on dynamic pricing using deep reinforcement learning," in *IEEE Wireless Commun. Networking Conf. (WCNC'20)*.    Seoul, Korea (South), 25-28 May 2020.
50. W. Wang, N. Yu, Y. Gao, and J. Shi, "Safe off-policy deep reinforcement learning algorithm for volt-var control in power distribution systems," *IEEE Trans. Smart Grid*, vol. 11, no. 4, pp. 3008–3018, Jul. 2020.
51. T. Haarnoja, A. Zhou, K. Hartikainen, G. Tucker, S. Ha, J. Tan, V. Kumar, H. Zhu, A. Gupta, P. Abbeel, and S. Levine, "Soft actor-critic algorithms and applications," *arXiv:1812.05905*, 2018.
52. S. Fujimoto, H. Van Hoof, and D. Meger, "Addressing function approximation error in actor-critic methods," *arXiv:1802.09477*, 2018.
53. H. Van Hasselt, A. Guez, and D. Silver, "Deep reinforcement learning with double Q-learning," in *Thirtieth AAAI Conf. on Artificial Intell.*    Phoenix, Arizona, USA, 12-17 Feb. 2016.

# Chapter 13
# Energy-Aware Computational Resource Allocation

**Abstract**  Mobile edge computing (MEC) with energy harvesting (EH) is becoming an emerging paradigm to improve the computation experience for the Internet of Things (IoT) devices. For a multi-device multi-server MEC system, the frequently varied harvested energy, along with changeable computation task loads and time-varying computation capacities of servers, increase the system's dynamic. Therefore, each device should learn to make coordinated actions, such as the offloading ratio, local computation capacity and server selection, to achieve a satisfactory computation quality. Thus, the MEC system with EH devices are highly dynamic and face two challenges: continuous-discrete hybrid action spaces, and coordination among devices. To deal with such problem, we propose two deep reinforcement learning (DRL) based algorithms: hybrid decision based actor-critic learning (Hybrid-AC), and multi-device hybrid decision based actor-critic learning (MD-Hybrid-AC) for dynamic computation offloading. Hybrid-AC solves the hybrid action space with an improvement of *actor-critic* architecture, where the *actor* outputs continuous actions (offloading ratio and local computation capacity) corresponding to every server, the *critic* evaluates the continuous actions and outputs the discrete action of server selection. MD-Hybrid-AC adopts the framework of centralized training with decentralized execution. It learns coordinated decisions by constructing a centralized *critic* to output server selections, which considers the continuous action policies of all devices. Simulation results show that the proposed algorithms achieve a good balance between consumed time and energy, and have a significant performance improvement compared with baseline offloading policies.

**Keywords**  Computation Offloading · Mobile Edge Computing (MEC) · Energy Harvesting · Internet of Things (IoT) · Continuous-discrete Hybrid Decision · Deep Reinforcement Learning

## 13.1  Introduction

With the rapid growth and application of the Internet of Thing (IoT) devices, such as smartphones, sensors, and wearable devices, new advanced applications with computation-intensive tasks are emerging [2, 3]. However, IoT devices usually

J. Du, C. Jiang, *Cooperation and Integration in 6G Heterogeneous Networks*,
Wireless Networks, https://doi.org/10.1007/978-981-19-7648-3_13

have limited computation capacity, energy, and memory restrictions [4]. To address the conflict between computation-intensive applications and resource-limited IoT devices, some computation tasks have to be offloaded to the servers with sufficient computation capability. Thus, cloud computing has been proposed to provide a strong ability for computation and storage [5]. However, the cloud server is physically or logically far from devices and may incur long latency which cannot satisfy the ultra-low latency requirements especially for time-sensitive applications or services [6].

This challenge can be relieved by mobile edge computing (MEC) [7, 8], which provides computing services at the edge of networks. In MEC systems, the MEC server is much closer to IoT devices compared to the traditional cloud server. Besides, the transmission latency is significantly reduced as the data transmission would not be congested due to the distributed structure of MEC servers. As a result, MEC is a promising paradigm to support latency-critical services and a variety of IoT applications compared with cloud computing. Offloading computation tasks to relatively resource-rich edge servers can not only improve the computation quality of service (QoS) but also augment the capabilities of end devices for resource-demanding applications.

Compared to traditional cloud servers, MEC servers may be less resourceful and more dynamic, and devices need to compete for the finite computation resources of servers. Therefore, resource allocation and schedule, such as server selection, allocation of offloading ratios and local computation capacity, are quite important for such resource-constrained systems. To achieve efficient utilization of computing resources and satisfy the computation requirements of devices, an intelligent computation offloading strategy is needed. Hence, computation offloading attracts more researchers' attention [9].

In reality, as the IoT has a crucial need for long-term operations to support various applications [10–12], the energy limitation of IoT devices is a critical problem to the development of MEC systems. IoT devices are usually battery-powered, the computation performance may be compromised due to insufficient battery energy for offloading. As a promising solution, energy harvesting (EH) is an emerging technique to significantly increase the device's lifetime by capturing ambient energies [13, 14]. An EH enabled IoT device can collect energy from external sources, such as solar, wind, radiation and radio-frequency (RF) signals [15]. By integrating EH into MEC sustained satisfactory computation performance can be achieved. However, it also brings some new challenges. For example, as the harvested energy is unpredictable, the computation offloading strategies of EH enabled devices are not easily obtained, and the policy without considering energy consumption cannot be directly adopted.

In this work, we will investigate hybrid decision based dynamic computation offloading problem with EH enabled devices. Specifically, the problem is to optimize the server selection, the continuous offloading ratio, and local computation capacity by minimizing the execution time and consumed energy. Besides, in this hybrid decision based dynamic system, the decision of each device is interdependent and will affect decisions of the following states. To address the difficulties brought by hybrid decision and collaboration among different devices, we solve the problem in two steps and propose two DRL approaches. The first approach is a **hybrid** decision based *actor-critic* learning method for dynamic computation offloading (hybrid-AC), which is an improvement of *actor-critic* methods and can output both continuous and discrete actions. It combines the structure of deep deterministic policy gradient (DDPG) [16] and deep Q network (DQN) [17]. *actor* outputs continuous actions (offloading ratio and local computation capacity) corresponding to every server, *critic* evaluates the $Q$ values for continuous actions of each server and also outputs the selected server which has the maximum $Q$ value. To solve the coordination between devices, we propose the second approach, **multidevice hybrid** decision based *actor-critic* learning method for dynamic computation offloading (MD-Hybrid-AC). Built on Hybrid-AC, MD-Hybrid-AC adopts the widely used framework of centralized training with decentralized execution. It considers action policies of all devices and can learn coordinated server selection policies.

We highlight the main contributions and our main ideas as follows:

1. We establish a dynamic computation offloading framework for multiple EH enabled devices with multiple servers, where the dynamics of both devices and servers are considered. Our framework accounts for continuous and discrete decisions, as well as independent and coordinated decisions.
2. We propose a hybrid decision-based DRL algorithm to solve the dynamic computation offloading problem with continuous-discrete hybrid action space. The discrete action (server selection) and continuous actions (offloading ratio and local computation capacity) can be jointly obtained in an end-to-end fashion, without discretizing or relaxing the action space.
3. We propose a multi-device hybrid decision-based DRL algorithm to coordinate among different devices. With a centralized *critic* considering the continuous actions of all devices, devices learn to coordinate in server selection.
4. We derive the optimal offloading policies for different environment conditions, and demonstrate the effectiveness of the proposed algorithms by comparing with three benchmark policies via simulation. It is indicated that Hybrid-AC performs better than discrete action based DRL method and MD-Hybrid-AC achieves better coordination than Hybrid-AC in the multi-device scenario.

The rest of this part is organized as follows. In Sect. 13.2, we discuss some related works associated with computation offloading mechanisms. The system model is defined in Sect. 13.3. The MDP modeling and the hybrid decision based DRL algorithm for dynamic computation offloading are described in Sect. 13.4, and multi-device hybrid decision based DRL algorithm is proposed in Sect. 13.5. Section 13.6 analyses the performance of proposed optimal offloading policy. Simulations are shown in Sect. 13.7, and conclusions are drawn in Sect. 13.8.

## 13.2  Related Works

Researchers have proposed many methods to the design of computation offloading policies. Most of the previous works have adopted the optimization or game based methods to solve the computation offloading problem [1, 18–20]. A Lyapunov optimization based dynamic computation offloading policy is developed in [19] for a MEC system with wireless EH enabled mobile devices. In [20], an online multi-tier operations scheduling scheme based on Lyapunov optimization in fog networks is considered. The network throughput, service delay, and fairness are jointly considered in this scheme. Although the Lyapunov optimization method is widely used to solve the long-term optimization problems, only an approximate optimal solution can be obtained. Besides, the prior information of environment statistics is needed, which may not be practically available in dynamic MEC systems. To tackle these problems, researchers have been turning to model the computation offloading problem as a Markov Decision Process (MDP) and solve it with reinforcement learning (RL) or Deep reinforcement learning (DRL) [21] methods.

DRL has recently made great progress [17] and various algorithms [16, 22] have been proposed. It has been applied to various applications, such as robotics [23], computer vision [24] and uav navigation [25, 26]. With the great success of DRL, there are also a few works that apply RL or DRL to the computation offloading problem [27–31]. In [27], a Q-learning based method is proposed to solve the task offloading problem considering both the task execution time and energy consumption. A deep Q-learning method is developed to offload computation tasks in [28], where edge device selection and offloading rate in discrete action spaces are selected without knowledge of the MEC model. The authors of [29] developed a double deep Q-network and a linear Q-function decomposition based SARSA method to deal with the state space explosion. A Q-learning method with after-state is proposed in [30] to select computation mode, which consists of three modes: dropping data, locally processing data and totally offloading to the server. To manage continuous energy consumption in different EH networks, deep deterministic policy gradient is utilized [31]. A DNN based actor-critic algorithm with the Natural policy gradient is utilized in [32] to solve the joint optimization of caching, computing and radio resource. A DRL-based online offloading framework is proposed in [33]. It uses DRL to generate continuous offloading action and quantizes the action into a binary one to avoid iteratively searching. The above RL or DRL based methods achieved a good performance without requiring the prior knowledge of environment statistics. However, they were modeled in either a discrete action space or a continuous action space, which restricts the optimization of offloading decisions in limited action space. The action space of offloading problem in reality is often continuous-discrete hybrid, each device needs to jointly decide continuous and discrete actions to accomplish the offloading process. For example, the device should not only decide whether to offload task or which server to select, but also select the offloading ratio, or the local computation capacity to balance the time and energy consumption. Therefore, these methods may not perform well as a fine discretization of continuous action will be problematic when the action space becomes large, and relaxing discrete action into a continuous set might significantly increase the complexity of the action space.

## 13.3 System Model

Considering a MEC system consisting of $M$ edge servers to provide edge computing services to $N$ devices as shown in Fig. 13.1. In this MEC system, each device $n \in \mathcal{N} = \{1, \cdots, N\}$, with wireless charging, has a computation-intensive task to be processed timely during each time slot $t \in \mathcal{T} = \{1, 2, \cdots\}$. We assume that the data of computation task are fine-grained and can be partitioned into subsets of any size. Therefore, each computation task can either be totally executed at the device, or be partially offloaded to the one MEC server $m \in \mathcal{M} = \{1, \ldots, M\}$ for computing. The key notations of our system model are listed in Table 13.1 for ease of reference.

At time slot $t$, device $n$ needs to decide which server $m_n^t$ to offload the task, the offloading ratio $\alpha_n^t \in [0, 1]$, which can be consider as the percentage of the task's data size (in bit) to be offloaded to the server, and local computation capacity $f_n^t \in [0, f_{\max}]$, which can be viewed as the CPU-cycle frequency to process the task data locally. Specifically, device $n$ offloads $\alpha_n^t$ parts of the task data to server $m_n^t$ and locally processes the remaining $1 - \alpha_n^t$ parts with specified CPU frequency $f_n^t$.

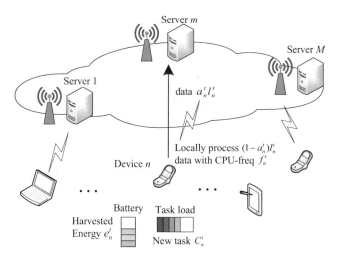

**Fig. 13.1** Computation offloading model for mobile edge computing

**Table 13.1** List of key notations

| Notation | Description |
|---|---|
| $M$ | Number of MEC servers |
| $N$ | Number of IoT devices |
| $\alpha_n^t$ | Percentage of the data size to be offloaded |
| $f_n^t$ | CPU-cycle frequency to process the task data |
| $l_n^t$ | Processed task load |
| $L_n^t$ | Size of computation data in the task queue |
| $X$ | Number of CPU cycles to process one-bit task |
| $Z_n$ | Important factor |
| $\varphi$ | Maximum tolerable delay |
| $C_n^t$ | Requested computation task data |
| $D_n^t / E_n^t$ | Total execution delay / consumed energy |
| $\kappa$ | The effective capacitance coefficient |
| $\Omega$ | System bandwidth |
| $p_n^t$ | Transmission power |
| $g_{n,m}^t$ | Normalized channel gain of device $n$ to server $m$ |
| $v_{n,m}^t$ | Transmission rate of device $n$ to server $m$ |
| $e_n^t$ | Harvested energy |
| $b_n^t$ | Battery energy level |

### 13.3.1   Task Model

The computation task may fail to finish as the limited computation capacity of devices, so the current task load is connected with the previous execution results, and a task queue is needed to represent this dynamic nature. We represent a computation task for device $n$ at time $t$ as $\left(L_n^t, l_n^t, Z_n, X, \varphi\right)$, where task load queue $L_n^t$ (in bits) denotes the size of computation data in the task queue, currently processed task load $l_n^t$ (in bits) denotes the actual processed data at time $t$, $X$ is the number of CPU cycles required to process one-bit task, and $\varphi$ is the maximum tolerable delay. We use an index $Z_n$ to denote the importance of different devices.[1] The device with higher priority has a larger value of $Z_n$.

At each time slot, the device could process task load $l_n^t$ which is no more than its maximum capacity. Mathematically, $l_n^t$ is defined as

$$l_n^t = \min\left\{l_{\max}, L_n^t\right\}, \tag{13.1}$$

---

[1] Devices may run various tasks with different concerns. For instance, autonomous navigation should have strict delay constraints, while healthcare data analytics will be more tolerant of delays. Therefore, we should pay more attention on the device with more strict constraints.

where $l_{\max} = \eta L_b$ denotes the maximum processed task load, with $L_b$ is the meta task load[2] and $\eta$ is a predefined coefficient.

The task queue $L_n^t$ is correlated with the remaining task from last time slot and the new computation task generated at this time slot, which can be formulated as

$$L_n^{t+1} = \min \left\{ L_n^t - l_n^t \mathbf{1}_{\mathrm{drop}_n^t = \mathrm{False}} + C_n^t, L_{\max} \right\}, \tag{13.2}$$

where $C_n^t = \mathrm{Pr}\,(\zeta)\,L_b$ is the computation task requested at time $t$ in device $n$, which can be modeled as an independent and identically distributed (i.i.d.) Poisson process with parameter $\zeta$. $\mathrm{drop}_n^t$ is a bool variable. When the task of device $n$ in time slot $t$ is processed successfully, $\mathrm{drop}_n^t = \mathrm{False}$ and will be defined mathematically in Sect. 13.3.4. In addition, $L_{\max}$ in (13.2) is the maximum number of computation task loads that can be queued at the device.

### 13.3.2 Local Computing

The device needs $(1 - \alpha_n^t)\,l_n^t X$ CPU cycles to locally execute $(1 - \alpha_n^t)\,l_n^t$ bits of task data. By applying dynamic voltage and frequency scaling techniques [34], the device can balance the execution time and energy consumption by adjusting the CPU frequency for each cycle. For simplicity, we assume the executing CPU frequency $f_n^t$ stays unchanged in one time slot. The execution time for locally processing the computation task of $(1 - \alpha_n^t)\,l_n^t$ data is given by:

$$D_{\mathrm{device}_n}^t = \frac{(1 - \alpha_n^t)\,l_n^t X}{f_n^t}, \tag{13.3}$$

and the corresponding energy consumption for locally computing $(1 - \alpha_n^t)\,l_n^t$ data is expressed as

$$E_{\mathrm{device}_n}^t = \kappa\,(1 - \alpha_n^t)\,l_n^t X (f_n^t)^2, \tag{13.4}$$

where $\kappa$ is the effective capacitance coefficient that depends on the chip architecture [35].

---

[2] Specifically, $L_b$ is the averaged data size of a typical computation task, e.g. 1000 bits in the experiments.

### 13.3.3   Offloading Computing

To utilize the rich computation resources of the MEC server, the device will offload $\alpha_n^t l_n^t$ data to an appropriate server, which processes the data and returns results to the device. We assume that the size of results obtained from the server is small [19], the time and energy consumption of feedback transmission is negligible in this work. The frequency-division multiple access (FDMA) is commonly used for the uplink transmission between devices and servers, which orthogonalizes different users' transmissions, and fully mitigates intra-cell interference [20, 36].

According to the Shannon formula, the achievable transmission rate for device $n$ to server $m$ is

$$v_{n,m}^t = \Omega \log_2 \left( 1 + g_{n,m}^t p_n^t \right), \tag{13.5}$$

where $\Omega$ and $p_n^t$ are the system bandwidth and transmission power. $g_{n,m}^t = h_{n,m}^t / \sigma$ is the normalized channel gain with the channel gain $h_{n,m}^t$ at time $t$ and the noise power $\sigma$.

The transmission time to offload $\alpha_n^t l_n^t$ data of device $n$ to the edge server $m$ in time slot $t$ is defined as

$$D_{\text{transmit}_n}^t = \frac{\alpha_n^t l_n^t}{v_{n,m}^t}, \tag{13.6}$$

and the corresponding energy consumption $E_{\text{transmit}_n}^t$ is

$$E_{\text{transmit}_n}^t = p_n^t D_{\text{transmit}_n}^t. \tag{13.7}$$

Note that the computation capacity of each server should be time-varying, because one server may also support the computation requests from other device groups. To model the server's dynamics, we assume its available computation capacity varies randomly between different time slots but remains unchanged in each time slot. Mathematically, we define the computation capacity of server $m$ is

$$f_{\text{server}_m}^t = f_{\text{server}}^{\max} - f_o^t, \tag{13.8}$$

where $f_o^t = \Pr(\lambda) f_{\text{server}}^{\text{unit}}$ is the occupied server resources, which is modeled as an i.i.d. Poisson process with parameter $\lambda$. $f_{\text{server}}^{\text{unit}}$ is the occupied computation resource for each unit.

After receiving the offloaded tasks, the server needs to process the offloaded data of all devices in each time slot and then return computation results. For server $m$, the execution time can be given by

$$D^t_{\text{server}_m} = \frac{\sum_{n=1}^{N} \mathbf{1}_{m^t_n=m} \alpha^t_n l^t_n X}{f^t_{\text{server}_m}}. \tag{13.9}$$

Based on (13.6), (13.9), the execution delay of offloading computing part is

$$D^t_{\text{remote}_n} = \max_{n \in \{\mathcal{N}|m^t_n=m\}} D^t_{\text{transmit}_n} + D^t_{\text{server}_m}. \tag{13.10}$$

Finally, according to (13.3), (13.10) and (13.4), (13.7), the total amount of time consumption $D^t_n$ and energy consumption $E^t_n$ of device $n$ at time slot $t$ are calculated as

$$D^t_n = \max\left\{ D^t_{\text{device}_n}, D^t_{\text{remote}_n} \right\}, \tag{13.11a}$$

$$E^t_n = E^t_{\text{device}_n} + E^t_{\text{transmit}_n}. \tag{13.11b}$$

### 13.3.4 Energy Harvesting

The EH process can be modeled as receiving sequential energy packets at the beginning of each time slot [19]. The harvested energy $e^t_n$ is stored in the battery, which is stochastic and intermittent and usually modeled as a random process [37]. We use an i.i.d. Uniform process with maximum value of $e_{\text{max}}$ to model the harvested energy in each time slot. In this part, we focus on the energy consumption of local computation and transmission and ignore the others for simplicity. Therefore, the battery status in the next time slot depends on both the energy consumption and harvesting, which evolves according to the following equation:

$$b^{t+1}_n = \min\left\{\max\left\{b^t_n - E^t_n + e^t_n, 0\right\}, b_{\text{max}}\right\}, \tag{13.12}$$

where $b_{\text{max}}$ is the maximum energy capacity of battery.

Besides $\text{drop}^t_n = \text{False}$ mentioned in (13.2), $\text{drop}^t_n = \text{True}$ when the battery energy of the device is insufficient ($b^{t+1}_n = 0$) or the execution time exceeds $\varphi$, which means the task will be dropped. Therefore, $\text{drop}^t_n$ is defined as

$$\text{drop}^t_n = \begin{cases} \text{True}, & b^{t+1}_n = 0 \text{ or } D^t_n > \varphi, \\ \text{False}, & \text{otherwise}. \end{cases} \tag{13.13}$$

## 13.4   Hybrid Decision Based DRL For Dynamic Computation Offloading

Based on the system model built in Sect. 13.3, the objective of the MEC system is to minimize the average weighted cost of the execution time and consumed energy in the long-term, which can be formulated as follows:

$$\min_{m_n^t, \alpha_n^t, f_n^t} \quad \frac{1}{T} \sum_{t=1}^{T} \sum_{n=1}^{N} \left( \omega_1 D_n^t + \omega_2 E_n^t \right), \tag{13.14a}$$

$$\text{s.t.} \quad (13.11\text{a}), (13.11\text{b}), \tag{13.14b}$$

$$\alpha_n^t \in [0, 1], n \in \mathcal{N}, t \in \mathcal{T}, \tag{13.14c}$$

$$f_n^t \in [0, f_{\max}], n \in \mathcal{N}, t \in \mathcal{T}, \tag{13.14d}$$

$$m_n^t \in \mathcal{M}, n \in \mathcal{N}, t \in \mathcal{T}, \tag{13.14e}$$

$$D_n^t < \varphi, n \in \mathcal{N}, t \in \mathcal{T}, \tag{13.14f}$$

$$E_n^t < b_n^t + e_n^t, n \in \mathcal{N}, t \in \mathcal{T}, \tag{13.14g}$$

where $\omega_1$ and $\omega_2$ are weighted parameters to get tradeoff between the consumed time and energy. The inequalities (13.14f) and (13.14g) are the time and energy constraints respectively, which guarantee the total execution time should not exceed the maximum tolerable delay and the battery energy should not be run out for each computing. The object function and constraints of problem (13.14) are non-convex and the difficulty of this dynamic computation offloading problem lies in many aspects: it is a decision process containing both continuous decisions and discrete decisions; the states of devices and servers are highly dynamic and dependent; the optimal offloading strategy should coordinate among devices. Therefore, it is difficult or impossible to find optimal solutions, which are time-dependent, using traditional optimization-based methods. Thus, RL algorithms are introduced to solve this dynamic offloading problem. However, the existing works deal with either a discrete action space or a continuous action space, which are not suitable for hybrid action space. To solve the problem, we will present our approach in two steps. To solve the problem of continuous-discrete hybrid decisions, Sect. 13.4.2 is intended to elaborate on the proposed hybrid decision based DRL algorithm. The proposed method is built on the single device scenario, which is convenient and effective to focus on the impact of hybrid action space. In the next Sect. 13.5, we will extend the concept of hybrid decision based method into the multi-device scenario with multi-agent settings.

### 13.4.1 MDP Modeling

To model the system as an MDP, three key elements of device $n$ are constructed as follows.

#### 13.4.1.1 States

$$s_n^t = \{L_n^t, b_n^t, e_n^t, Z_n, g_{n,1}^t, f_{\text{server}_1}^t, \ldots, g_{n,M}^t, f_{\text{server}_M}^t\} \in \mathscr{S}, \qquad (13.15)$$

which contains the states of device $n$ and $M$ servers.

#### 13.4.1.2 Action

$a_n^t = \{m_n^t, \alpha_n^t, f_n^t\} \in \mathscr{A}$ consists of the server selection $m_n^t$, the offloading ratio $\alpha_n^t$, and local computation capacity $f_n^t$. With the offloading decision of every device, the system executes computation tasks and steps into the next state.

#### 13.4.1.3 Reward

Reward function should be related to both the objective function and the constraints of the system. In the proposed algorithm, the reward is composed of the following four parts. The first two parts are the reward of normalized execution time and energy consumption, which straightly reflect the performance of computation offloading. The third part is a penalty when the task is dropped. The fourth part is a penalty when the task queue exceeds the maximum queue length $L_{max}$. The normalized time consumption is defined as

$$\bar{D}_n^t = \frac{Z_n}{\sum_{i=1}^{N} Z_i} \frac{D_{\text{Local}} - (L_b/l_n^t) D_n^t}{D_{\text{Local}}}, \qquad (13.16)$$

where $D_{\text{Local}} = L_b X / f_{\max}$ is the execution time for totally local computing with maximum computation capability $f_{\max}$.

Similarly, the normalized energy consumption is

$$\bar{E}_n^t = \frac{Z_n}{\sum_{i=1}^{N} Z_i} \frac{E_{\text{Local}} - (L_b/l_n^t) E_n^t}{E_{\text{Local}}}, \qquad (13.17)$$

where $E_{\text{Local}} = \kappa L_b X (f_{\max})^2$ is the energy consumption for totally local computing with maximum computation capability.

The penalty for exceeding the task queue is proportional to the amount of tasks exceeded:

$$\bar{P}_n^t = \max\left\{\left(\left(L_n^t - l_n^t \mathbf{1}_{\mathrm{drop}_n^t = \mathrm{False}} + C_n^t\right) - L_{\max}\right)/L_b, 0\right\}. \qquad (13.18)$$

Consequently, the reward function is defined as

$$r^t = \sum_{n=1}^{N} \omega_d \bar{D}_n^t + \omega_e \bar{E}_n^t - \mathbf{1}_{\mathrm{drop}_n^t = \mathrm{True}} - \bar{P}_n^t, \qquad (13.19)$$

where $\omega_d$ and $\omega_e$ are the weighted parameters of $\bar{D}_n^t$ and $\bar{E}_n^t$.

### 13.4.2  Hybrid Decision Based DRL Method

In reinforcement learning, the policy is a mapping from *states* to an *action* $\pi$ : $\mathscr{S} \to \mathscr{A}$. The target of the agent is to learn a policy which maximizes the expected future rewards by interacting with the environment. The future rewards are usually discounted by a factor and the total discounted future reward, also named return, from time slot $t$ is: $R^t = \sum_{i=t}^{\infty} \gamma^{i-t} r\left(s^i, a^i\right)$, where $\gamma \in [0, 1]$ is a discount factor. Note that the return depends on the state and action, therefore, action-value function (also named $Q$ function) should also be introduced [21]. It describes the expected return after taking an action $a$ in state $s$ and thereafter following policy $\pi$:[3]

$$Q(s, a) = \mathbb{E}(R|s, a; \pi). \qquad (13.20)$$

As defined in Sect. 13.4.1.2, the *action* $a = (m, \alpha, f)$ is the offloading decision, which contains both discrete and continuous action space, thereby it is computationally intractable to directly calculate the gradient for the policy as existing methods do [16, 17]. We divide the action space into two parts: $a = (m, u)$. The discrete part $m$ represents the server selection operation and the continuous part $u = (\alpha, f)$ represents the decisions of offloading ratio and local computation capacity. This separation has two advantages: firstly, the action space is decomposed into independently discrete action space and continuous action space, which makes the originally intractable action can be solved with two feasible sub actions; secondly, as we observe that the decision of server selection $m$ depends on the offloading ratio $\alpha$, we can find the optimal $(\alpha, f)$ for each server, then select the optimal server $m$ based on the continuous actions of all servers.

---

[3] In the following part, we omit the superscript of time $t$ for the ease of descriptions. In this Section, we omit the subscript of device $n$, as we only consider a single device scenario.

With the above action space separation, we then adopt the widely used *actor-critic* architecture [21] to design our hybrid decision based DRL algorithm for dynamic computation offloading. The *actor* can be used to generate continuous actions $\hat{u} = (u_1, \cdots u_M)$ concerning all servers, which is parameterized by a deterministic policy network $\mu(s; \phi_\mu)$. Meanwhile, the *critic* can output the discrete action $m$ and also evaluate the *actor*'s output, which is parameterized by an action-value network $Q(s, m, \hat{u}; \phi_Q)$. The action-value network $Q(s, m, \hat{u}; \phi_Q)$ takes states $s$ and continuous actions $\hat{u}$ corresponding to all servers as input, and outputs $Q$ values for continuous actions related to each server. Then, the server corresponding to the maximum $Q$ value is selected as the optimal server $m$. The final action can be obtained by:

$$m = \arg\max Q(s, m, \hat{u}; \phi_Q),\tag{13.21a}$$

$$(\alpha, f) = u_m = \hat{u}[m].\tag{13.21b}$$

The complete interaction process can be referred to Fig. 13.2.

### 13.4.2.1 Continuous Action Updating

The action-value function with continuous actions corresponding to all servers is $Q(s, m, \hat{u}) = Q(s, m, u_1, \cdots, u_M)$. Suppose server selection $m$ is known, the

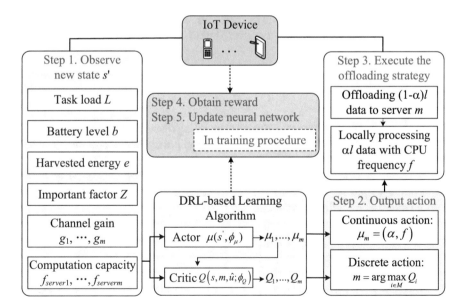

**Fig. 13.2** Illustration of hybrid decision controlled DRL based dynamic computation offloading scheme for IoT devices

original dynamic computation offloading problem degenerates to a sub problem which can be modeled as a continuously controlled MDP. The action-value function for the continuously controlled sub problem is:

$$Q^{\mu}\left(s, \hat{u}\right) = \max_{m \in \mathcal{M}} Q\left(s, m, \hat{u}\right). \tag{13.22}$$

The maximum $Q$ value represents the evaluation of the selected offloading ratio and local computation capacity. Because the final actions of offloading ratio and local computation capacity correspond only on the optimal server, it is reasonable to take the maximum $Q$ value as the target $Q$ value for the policy, instead of the sum of all servers' $Q$ values.

With the above definition of Q function, we utilize the *actor* part of DDPG [16] to update the gradient:

$$\nabla J = \mathbb{E}\left[\nabla Q^{\mu}\left(s, \hat{u}\right) \nabla \hat{\mu}\right]. \tag{13.23}$$

### 13.4.2.2  Discrete Action Updating

If we get the output of $\hat{u}$, we propose a $Q$ updating procedure with a combination of DQN [17] and the *critic* part of DDPG [16]. We also call it *critic*, but it contains two effects: it acts as a *critic* to evaluate the actions of the *actor*, which is used for updating *actor* as in (13.23); it also acts as a DQN to output the action of maximum $Q$ value, which is the server selection action $m$. Then, we elaborate on the process of finding the optimal policy for the discrete action.

Given the continuous action $\hat{u}$, the optimal Bellman equation can be reformulated as

$$Q^*\left(s, m, \hat{u}\right) = \mathbb{E}\left[r + \gamma \max_{m'} Q^*\left(s', m', \hat{u}'\right) \mid s, m, \hat{u}\right], \tag{13.24}$$

where $s', m', \hat{u}'$ are the next state, the corresponding discrete and continuous action respectively. The basic idea behind it is the optimal policy should choose the action $m'$ that maximizing the expected value of $r + \gamma \max_{m'} Q^*\left(s', m', \hat{u}'\right)$.

Following this idea, the parameterized action-value function $Q\left(s, m, \hat{u}; \phi_Q\right)$ can be iteratively updated to estimate the $Q$ value. Specifically, it can be trained by minimizing the following loss function:

$$L\left(\phi_Q\right) = \mathbb{E}\left[\left(y - Q\left(s, m, \hat{u}\right)\right)^2\right], \tag{13.25}$$

where $y = \mathbb{E}\left[r + \gamma \max_{m' \in \mathcal{M}} Q\left(s', m', \hat{u}'\right) \mid s, m, \hat{u}\right]$ is the target for each iteration.

The gradient for updating the $Q$ function can then be obtained by differentiating the loss function:

$$\nabla L \left( \phi_Q \right) = \mathbb{E} \left[ \left( y - Q \left( s, m, \hat{u} \right) \right) \nabla_{\phi_Q} Q \left( s, m, \hat{u} \right) \right]. \tag{13.26}$$

With the above optimization methods (13.23) and (13.26), two neural networks are used to learn the offloading policy. To stabilize the learning procedure, a target actor network, a target critic network, and a replay memory are used. Besides, an exploration policy $\mu'(s) = \mu(s) + \nu$ for *actor* and an $\varepsilon$-greedy policy for *critic* are used in the training procedure to balance the tradeoff between exploration and exploitation. $\nu$ is a noise process [16]. Algorithm 5 illustrates the training procedure of the proposed hybrid decision controlled DRL based dynamic computation offloading algorithm.

## 13.5  Multi-Device Hybrid Decision Based DRL for Dynamic Computation Offloading

This section considers dynamic computation offloading with multiple devices where the issues of non-stationarity of the environment and coordination among devices are addressed. Directly applying Q-learning to multi-agent settings will result the environment non-stationary from the view of any device [38, 39], which causes the difficulty of stabling the learning procedure and also the use of past experience replay. Policy gradient methods, on the other hand, exacerbate high variance gradient estimates in multi-agent environments. As actions of each device may affect the reward of other devices, the reward conditioned only on its own actions exhibits much more variability, therefore, increasing the variance of its gradients in the policy optimization. So, Hybrid-AC proposed in Sect. 13.4.2 cannot be directly applied in the scenario of multiple devices.

For multi-agent DRL, centralized training with decentralized execution is a standard paradigm [40, 41]. Furthermore, it is reasonable for each device to make its own decision of $(\alpha, f)$, and based on the decisions of all devices to jointly schedule the server selection $m$ for each device. Based on this observation, we utilize multi-agent deep deterministic policy gradient (MADDPG) to introduce an operation mode of centralized training with half-decentralized execution. Built on this operation mode and the hybrid decision based DRL method in Sect. 13.4.2, we propose a multi-device hybrid decision based DRL algorithm.

The proposed method is an extension of the *actor-critic* architecture. Each device has its own *actor* with local information and shares a common *critic* which is augmented with information of all devices. The centralized training with half-decentralized execution works in the following way. At training time, *actor*, which generates continuous actions, can use extra information to ease training, so long as this information is not used at execution time. On the other side, the $Q$ function

---

**Algorithm 5 Hybrid** Decision Controlled **A**ctor-**C**ritic Learning Based Dynamic Computing Offloading Algorithm (Hybrid-AC)

---

1: Initialize actor and critic network $\mu\left(s|\phi_\mu\right)$, $Q\left(s, \hat{u}, m|\phi_Q\right)$.

2: Initialize target actor network $\mu\left(s|\phi'_\mu\right)$ and target critic network $Q\left(s, \hat{u}, m|\phi'_Q\right)$ with weights $\phi'_\mu \leftarrow \phi_\mu, \phi'_Q \leftarrow \phi_Q$.

3: Initialize replay memory $\mathfrak{D}$.

4: **for** *episodes*=1, $\cdots$ **do**

5:     Initialize a random process $v$ for continuous action exploration.

6:     Initialize observation state $s^1$.

7:     **for** $t$=1, $\cdots$, $T$ **do**

8:         Receive current states $s^t$.

9:         Obtain continuous action for all servers $\hat{u}^t = \mu\left(s|\phi_\mu\right) + v^t$.

10:        With probability $\varepsilon$ select a random discrete action $m^t$, otherwise select

$$m^t = \arg\max Q\left(s^t, m, \hat{u}^t|\phi_Q\right). \tag{13.27}$$

11:        Select continuous action $u^t = \hat{u}^t[m]$.

12:        Execute action $m^t$, $u^t$, obtain reward $r^t$ and next state $s^{t+1}$.

13:        Store transitions $\left(s^t, m^t, \hat{u}^t, r^t, s^{t+1}\right)$ into $\mathfrak{D}$.

14:        Sample $K$ transitions $\left(s^i, m^i, \hat{u}^i, r^i, s^{i+1}\right)$ from $\mathfrak{D}$.

15:        Set $y^i = r^i + \gamma \max_{m'} Q\left(s^{i+1}, m', \mu\left(s^{i+1}|\phi'_\mu\right)|\phi'_Q\right)$.

16:        Update actor network using

$$\nabla\phi_\mu = \frac{1}{K}\sum_i \frac{\partial Q\left(s^i, m^i, \hat{u}|\phi_Q\right)}{\partial\hat{u}} \frac{\partial\mu\left(s^i|\phi_\mu\right)}{\partial\phi_\mu}. \tag{13.28}$$

17:        Update critic network using

$$\nabla\phi_Q = \frac{1}{K}\sum_i \left(y^i - Q\left(s^i, m^i, \hat{u}^i|\phi_Q\right)\right) \frac{\partial Q\left(s^i, m^i, \hat{u}^i|\phi_Q\right)}{\partial\phi_Q}. \tag{13.29}$$

18:        Update the target networks

$$\phi'_\mu \leftarrow \epsilon\phi_\mu + (1 - \epsilon)\phi'_\mu, \tag{13.30a}$$

$$\phi'_Q \leftarrow \epsilon\phi_Q + (1 - \epsilon)\phi_Q. \tag{13.30b}$$

19:    **end for**
20: **end for**

---

cannot contain different information at training and execution time, a centralized *critic* with augmented information is used. At execution time, each device's *actor* outputs the offloading ratio and local computation capacity with its states, acting in a decentralized manner. The centralized *critic* takes the states and actions of all devices to coordinate the server selection among devices.

As different device has the same type of actions and they should have a similar policy for continuous actions when observing the same state information, we use a common neural network $\mu\left(s|\phi_\mu\right)$ to generate the policy of continuous actions for all devices, which can accelerate the learning process. The centralized action-value function $\hat{Q}\left(o, \hat{a}\right) = \hat{Q}\left(o, m_1, \hat{u}_1, \cdots, m_n, \hat{u}_n, \cdots, m_N, \hat{u}_N\right)$ takes the observation $o$ and the actions of all devices as input, where $o = \{L_1^t, b_1^t, e_1^t, Z_1, \cdots, L_N^t, b_N^t, e_N^t, Z_N, g_{1,1}^t, \ldots, g_{N,1}^t, f_{server_1}^t, \cdots, g_{1,M}^t, \cdots, g_{N,M}^t, f_{server_M}^t\}$ consists of the states of all devices and servers. $\hat{Q}\left(o, \hat{a}\right)$ outputs the $Q$ values for all devices, where $\hat{Q}_n\left(o, m_n, \hat{u}_1, \cdots, \hat{u}_N\right), n \in \mathcal{N}$ is the output $Q$ value for device $n$. The final actions of the whole system can be obtained by:

$$m_n = \arg\max \hat{Q}_n\left(o, m, \hat{u}_1, \cdots, \hat{u}_N; \phi_{\hat{Q}}\right), n \in \mathcal{N}, \tag{13.31a}$$

$$(\alpha_n, f_n) = \hat{u}_n\left[m_n\right], n \in \mathcal{N}. \tag{13.31b}$$

For each device, it updates the *actor* part independently as the previous Sect. 13.4.2.1. Similarly, the action-value function for updating continuous actions can be constructed with the maximum $Q$ values of device $n$:

$$\hat{Q}_n^\mu\left(o, \hat{u}_1, \cdots, \hat{u}_N\right) = \max_{m_n \in \mathcal{M}} \hat{Q}_n\left(o, m_n, \hat{u}_1, \cdots, \hat{u}_N\right), \tag{13.32}$$

and the *actor* could be updated using the following gradient:

$$\nabla J_n = \mathbb{E}\left[\nabla \hat{Q}_n^\mu\left(o, \hat{u}_1, \cdots, \hat{u}_N\right) \nabla \hat{\mu}\right]. \tag{13.33}$$

By extending the *critic* updating gradient to multi-device, the centralized *critic* is updated by minimizing:

$$L_n\left(\phi_{\hat{Q}}\right) = \mathbb{E}\left[\left(\hat{y} - \hat{Q}_n\left(o, m_n, \hat{u}_1, \cdots, \hat{u}_N\right)\right)^2\right], \tag{13.34}$$

where $\hat{y} = r + \gamma \max_{m'_n \in \mathcal{M}} \hat{Q}_n\left(o', m'_n, \hat{u}'_1, \cdots, \hat{u}'_N\right)$ is the target for each iteration.

Therefore, the gradient for updating *critic* is:

$$\nabla L_n = \mathbb{E}\left[\left(\hat{y} - \hat{Q}_n\left(o, m_n, \hat{u}_1 \cdots \hat{u}_N\right)\right) \nabla \hat{Q}_n\left(o, m_n, \hat{u}_1 \cdots \hat{u}_N\right)\right]. \tag{13.35}$$

Note that the $Q$ values for each device depend on the actions of other devices. It becomes even more unstable and slower for training compared with the proposed algorithm in Sect. 13.4.2 for a single device. However, as we use one neural network to parameterize *actor* and *critic*, they can both be updated $N$ times in one step, which accelerates the learning process. Besides, a replay memory, along with a target actor network and a critic network are also used to stabilize the learning procedure. The action exploration policies are the same as in the proposed algorithm in Sect. 13.4.2.

Algorithm 6 describes the detailed training procedure of the proposed multi-device hybrid decision controlled DRL based dynamic computation offloading algorithm.

At execution time, the *actor* can be run at each device with its own states, the *critic* can be placed at one server-end to distribute server selection results to each device.[4] With such configuration, devices can prepare the system setting for locally processing in advance, which is helpful for computation offloading. Besides, the proposed learning architecture can not only be applied to the system in this section but also adapted to similar hybrid decision based models.

## 13.6   Performance Evaluations

This section analyzes the optimal offloading policy with a simplified environment. The performance is evaluated in one time slot with sufficient battery energy, which can be viewed as a reference of upper bound to assess the DRL-based offloading methods in the dynamic scenario. We will deduce the optimal offloading policy under different environmental conditions, which would help us analyze the performance of the proposed methods in the experiments. To explain what the optimal policy would be, we first analyze the performance in a single device case, and then discuss how devices would coordinate in a multi-device scenario with relatively small scale, which can be similarly extended to more device scenarios.

**Proposition 13.1** *Consider a dynamic computation offloading system with single device, we set the offloading target as the sum of execution time and consumed energy $U = D + E$. The optimal offloading policy is*

$$m^* = \arg\min \frac{1+p}{v_m} + \frac{X}{f_{server_m}}, m \in \mathcal{M}, \tag{13.40a}$$

$$\alpha^* = \frac{X}{X + \frac{f}{v_{m^*}} + \frac{fX}{f_{server_{m^*}}}}, \tag{13.40b}$$

$$f^* = \min\{\sqrt[3]{1/2\kappa}, f_{max}\}. \tag{13.40c}$$

*when* $\frac{p}{\frac{X}{f} + \kappa X f^2} < v_m < \frac{1+p}{-\frac{X}{f_{server_m}} + \kappa X f^2}, m \in \mathcal{M}.$

---

[4] Noted that the cost of energy to execute on device is more expensive than on server, an alternative implementation plan is to place both the *actor* and *critic* at the server-end, which is a simple and centralized way. This configuration can reduce the interaction between devices and servers, but increase the burden of the servers. We need to choose an appropriate implementation plan according to practical situations.

---

**Algorithm 6** Multi-**Device** **Hybrid** Decision Controlled Actor-Critic Learning Based Dynamic Computing Offloading Algorithm (MD-Hybrid-AC)

---

1: Initialize actor network $\mu\left(s|\phi_\mu\right)$ and critic network $\hat{Q}\left(o, m_1, \cdots, m_N, \hat{u}_1, \cdots, \hat{u}_N|\phi_{\hat{Q}}\right)$.

2: Initialize    target    actor    network    $\mu\left(s|\phi'_\mu\right)$    and    target    critic    network    $\hat{Q}\left(o, m_1, \cdots, m_N, \hat{u}_1, \cdots, \hat{u}_N|\phi'_{\hat{Q}}\right)$ with weights $\phi'_\mu \leftarrow \phi_\mu, \phi'_{\hat{Q}} \leftarrow \phi_{\hat{Q}}$.

3: Initialize replay memory $\mathfrak{D}$.

4: **for** *episodes*=1, $\cdots$ **do**

5:     Initialize a random process $\nu$ for continuous action exploration.

6:     Initialize observation state $s_1^1, \ldots, s_N^1$.

7:     **for** $t$=1, $\cdots$, $T$ **do**

8:         **for** device $n$=1, $\cdots$, $N$ **do**

9:             Receive current states $s_n^t$.

10:            Obtain action $\hat{u}_n^t = \mu\left(s_n^t|\phi_\mu\right) + \nu^t$.

11:        **end for**

12:        **for** device $n$=1,$\cdots$, $N$ **do**

13:            With probability $\varepsilon$ select a random discrete action $m_n^t$, otherwise select

$$m_n^t = \arg\max \hat{Q}_n\left(o^t, m, \hat{u}_1^t, \cdots, \hat{u}_N^t|\phi_{\hat{Q}}\right). \tag{13.36}$$

14:            Select continuous action $u_n^t = \hat{u}_n^t\left[m_n^t\right]$.

15:        **end for**

16:        Execute  action  $m_1^t, u_1^t, \cdots, m_N^t, u_N^t$,  obtain  reward  $r^t$  and  observe  next  state $s_1^t, \cdots, s_N^t$.

17:        Store transitions $\left(s_1^t, m_1^t, \hat{u}_1^t, r^t, s_1^{t+1}, \cdots, s_N^t, m_N^t, \hat{u}_N^t, r^t, s_N^{t+1}\right)$ into replay memory $\mathfrak{D}$.

18:        Sample a batch of $K$ transitions $\left(s_1^i, m_1^i, \hat{u}_1^i, r^i, s_1^{i+1}, \cdots, s_N^i, m_N^i, \hat{u}_N^i, r^i, s_N^{i+1}\right)$ from $\mathfrak{D}$.

19:        **for** device $n$=1, $\cdots$, $N$ **do**

20:            Set $y_n^i = r^i + \gamma\max_{m'} \hat{Q}_n\left(o^{i+1}, m', \mu\left(s_1^{i+1}|\phi'_u\right), \cdots, \mu\left(s_N^{i+1}|\phi'_u\right)|\phi'_{\hat{Q}}\right)$.

21:            Update actor network using

$$\nabla\phi_\mu = \frac{1}{K}\sum_i \frac{\partial\hat{Q}_n\left(o^i, m_n^i, \hat{u}_1^i, \cdots, \hat{u}, \cdots, \hat{u}_N^i\right)}{\partial\hat{u}}\Big|_{\hat{u}=\mu\left(s^i|\phi_\mu\right)} \frac{\partial\mu\left(s^i|\phi_\mu\right)}{\partial\phi_\mu}. \tag{13.37}$$

22:            Update critic network using

$$\nabla\phi_{\hat{Q}} = \frac{1}{K}\sum_i \left(y_n^i - \hat{Q}_n\left(o^i, m_n^i, \hat{u}_1^i \cdots \hat{u}_N^i|\phi_{\hat{Q}}\right)\right) \frac{\partial\hat{Q}_n\left(o^i, m_n^i, \hat{u}_1^i \cdots \hat{u}_N^i|\phi_{\hat{Q}}\right)}{\partial\phi_{\hat{Q}}}. \tag{13.38}$$

23:        **end for**

24:        Update the target networks

$$\phi'_\mu \leftarrow \epsilon\phi_\mu + (1-\epsilon)\phi'_\mu, \tag{13.39a}$$

$$\phi'_{\hat{Q}} \leftarrow \epsilon\phi_{\hat{Q}} + (1-\epsilon)\phi'_{\hat{Q}}. \tag{13.39b}$$

25:    **end for**

26: **end for**

---

**Proof** According to (13.11a), when $\frac{X}{X+\frac{l}{v_m}+\frac{fX}{f_{server_m}}} \leq \alpha \leq 1$, the target function $U$ is

$$U = \frac{\alpha l}{v_m} + \frac{\alpha l X}{f_{server_m}} + \kappa (1 - \alpha) l X f^2 + \frac{\alpha l}{v_m} p, \tag{13.41}$$

and

$$\frac{\partial U}{\partial \alpha} = l \left( \frac{1}{v_m} + \frac{X}{f_{server_m}} - \kappa X f^2 + \frac{p}{v_m} \right). \tag{13.42}$$

Similarly, when $0 \leq \alpha \leq \frac{X}{X+\frac{l}{v_m}+\frac{fX}{f_{server_m}}}$, we have

$$U = \frac{(1 - \alpha) l X}{f} + \kappa (1 - \alpha) l X f^2 + \frac{\alpha l}{v_m} p, \tag{13.43}$$

and

$$\frac{\partial U}{\partial \alpha} = l \left( -\frac{X}{f} - \kappa X f^2 + \frac{p}{v_m} \right). \tag{13.44}$$

When $\frac{p}{\frac{X}{f}+\kappa X f^2} < v_m < \frac{1+p}{-\frac{X}{f_{server_m}}+\kappa X f^2}$, the following inequalities can be obtained

$$-\frac{X}{f} - \kappa X f^2 + \frac{p}{v_m} < 0, \tag{13.45a}$$

$$\frac{1}{v_m} + \frac{X}{f_{server_m}} - \kappa X f^2 + \frac{p}{v_m} > 0. \tag{13.45b}$$

It means that the target function $U$ is decreasing when $\alpha \leq \frac{X}{X+\frac{l}{v_m}+\frac{fX}{f_{server_m}}}$, and increasing when $\alpha \geq \frac{X}{X+\frac{l}{v_m}+\frac{fX}{f_{server_m}}}$. Thus, $U$ has a global minimum at $\alpha^* = \frac{X}{X+\frac{l}{v_m}+\frac{fX}{f_{server_m}}}$. We can also find that $\alpha^*$ is the ratio when locally consumed time equals to the totally consumed time for offloading execution.

Next, we find the optimal value for $f$. Differentiate (13.43) against $f$:

$$\frac{\partial U}{\partial f} = (1 - \alpha) l X f \left( -f^{-3} + 2\kappa \right), \tag{13.46}$$

where $f$ changes in $[0, f_{max}]$.

When $\kappa < \frac{1}{2}$, $\frac{\partial U}{\partial f} < 0$ for any $f \in [0, f_{max}]$, and $U$ gets minimum value at $f^* = f_{max}$. When $\kappa \geq \frac{1}{2}$, $U$ gets the minimum when $\frac{\partial U}{\partial f} = 0$, which results in $f^* = \sqrt[3]{1/2\kappa}$. Note that $\kappa$ is the effective capacitance coefficient, which usually has a relatively small value. Thus, the optimal $f^*$ is usually on the boundary value $f_{max}$.

After fixed $\alpha$ and $f$, we can find optimal server from (13.41). $U$ can be rewritten as

$$U = \alpha l \left( \frac{1+p}{v_m} + \frac{X}{f_{server_m}} \right) + \kappa (1 - \alpha) l X f^2. \tag{13.47}$$

As different server has different transmission rate $v$ and computation capacity $f_{server}$, the optimal server is the one which has minimum value of $U$ when $m \in \mathcal{M}$. So, optimal server $m^*$ has minimum value of $\frac{1+p}{v} + \frac{X}{f_{server}}$.

Thus, we prove (13.40) is the optimal offloading policy when $\frac{p}{\frac{X}{f}+\kappa X f^2} < v_m < \frac{1+p}{-\frac{X}{f_{server_m}}+\kappa X f^2}$. This completes the proof of Proposition 13.1.

*Remark 13.1*  When there exists a server, which has a relatively good transmission channel and computation capacity, the time and energy consumption are comparable with that of the device. In this situation, device should offload partial data to save time. When the execution time of local computing and offloading computing are the same, the device achieves minimum execution time, which saves the most time.

**Proposition 13.2**  *The optimal offloading policy is*

$$\alpha^* = 0, \tag{13.48a}$$

$$f^* = \min\{\sqrt[3]{1/2\kappa}, f_{max}\}. \tag{13.48b}$$

*when* $v_m < \frac{p}{\frac{X}{f}+\kappa X f^2}, m \in \mathcal{M}.$

**Proof**  Similar to Proposition 13.1, we can derive that when $v_m < \frac{p}{\frac{X}{f}+\kappa X f^2}$, $\frac{\partial U}{\partial \alpha} > 0$ for $\alpha \in [0, 1]$. Then, $U$ is an increasing function for $\alpha$, and achieves minimum value at $\alpha = 0$. This completes the proof of Proposition 13.2.

*Remark 13.2*  When the transmission channel is in bad condition, the consumption of time and energy during transmission are expensive, which results in the offloading procedure is uneconomical. In this situation, the device prefers to process the whole computation tasks locally.

**Proposition 13.3** *The optimal offloading policy is*

$$m^* = \arg\min \frac{1+p}{v_m} + \frac{X}{f_{server_m}}, \, m \in \mathcal{M},$$   (13.49a)

$$\alpha^* = 1.$$   (13.49b)

*when* $v_m > \dfrac{1+p}{-\frac{X}{f_{server_m}} + \kappa X f^2}, \, m \in \mathcal{M}.$

**Proof** Similar to Proposition 13.1, we can derive that when $v_m > \dfrac{1+p}{-\frac{X}{f_{server_m}} + \kappa X f^2}$, $\frac{\partial U}{\partial \alpha} < 0$ for $\alpha \in [0, 1]$. Then, $U$ is a decreasing function for $\alpha$, and achieves minimum value at $\alpha = 1$. This completes the proof of Proposition 13.3.

*Remark 13.3* When there exists a server, which has an excellent transmission channel and powerful computation capacity, it will consume less time and energy for server to process per task data than that of local processing. Thus, it is more beneficial for the device to offload more computation tasks to the server for computing. In this situation, device prefers to execute all computation tasks remotely.

In the following part, we will analyze the performance of multi-device scenario. For the ease of analysis, we simplify the scenario into a two-device model. If the states of servers are similar with each other, two devices should choose two different servers for the benefit of good performance. For each device, this will be the similar performance as in the previous single device scenario. However, when the transmission channel and computation capacity vary greatly for different servers, choosing a common server will be better for both devices.

Considering that devices have same transmission channel to each server: $v_m, m \in \mathcal{M}$, and the servers satisfy the conditions in Proposition 13.1: $\frac{p}{\frac{X}{f} + \kappa X f^2} < v_m < \frac{1+p}{-\frac{X}{f_{server_m}} + \kappa X f^2}, m \in \mathcal{M}$, therefore, each device shall offload partial computation tasks to one server for a relatively low cost of time and energy. Servers are sorted as $\frac{1+p}{v_{\tilde{m}_1}} + \frac{X}{f_{server_{\tilde{m}_1}}} < \frac{1+p}{v_{\tilde{m}_2}} + \frac{X}{f_{server_{\tilde{m}_2}}} < \cdots < \frac{1+p}{v_{\tilde{m}_M}} + \frac{X}{f_{server_{\tilde{m}_M}}}, \tilde{m}_1 \cdots \tilde{m}_M \in \mathcal{M}.$ Then, we derive the following Propositions.

**Proposition 13.4** *Both two devices will select optimal server $\tilde{m}_1$, when its computation capacity satisfies* $f_{server_{\tilde{m}_1}} > \dfrac{(\alpha_1 l_1 + 2\alpha_2 l_2)X}{\alpha_2 l_2 \left( \frac{1+p}{v_{\tilde{m}_2}} - \frac{p}{v_{\tilde{m}_1}} + \frac{X}{f_{server_{\tilde{m}_2}}} \right) - \left( 2\max\{\frac{\alpha_1 l_1}{v_{\tilde{m}_1}}, \frac{\alpha_2 l_2}{v_{\tilde{m}_2}}\} - \frac{\alpha_1 l_1}{v_{\tilde{m}_1}} \right)}.$

**Proof** When the devices both select server $\tilde{m}_1$, if $\frac{\alpha_1 l_1}{v_1} \geq \frac{\alpha_2 l_2}{v_2}$, the total time and energy consumption $U_t$ is

$$
\begin{aligned}
U_t =& D_1 + E_1 + D_2 + E_2 \\
=& \frac{\alpha_1 l_1}{v_{\tilde{m}_1}} + \frac{(\alpha_1 l_1 + \alpha_2 l_2) X}{f_{\text{server}_{\tilde{m}_1}}} + \kappa (1 - \alpha_1) l_1 X f_1^2 + \frac{\alpha_1 l_1}{v_{\tilde{m}_1}} p \\
&+ \frac{\alpha_1 l_1}{v_{\tilde{m}_1}} + \frac{(\alpha_1 l_1 + \alpha_2 l_2) X}{f_{\text{server}_{\tilde{m}_1}}} + \kappa (1 - \alpha_2) l_2 X f_2^2 + \frac{\alpha_2 l_2}{v_{\tilde{m}_1}} p.
\end{aligned} \tag{13.50}
$$

When $f_{\text{server}_{\tilde{m}_1}} > \dfrac{(\alpha_1 l_1 + 2\alpha_2 l_2) X}{\alpha_2 l_2 \left( \frac{1+p}{v_{\tilde{m}_2}} - \frac{p}{v_{\tilde{m}_1}} + \frac{X}{f_{\text{server}_{\tilde{m}_2}}} \right) - \frac{\alpha_1 l_1}{v_{\tilde{m}_1}}}$, we have

$$
\begin{aligned}
U_t =& 2 \frac{\alpha_1 l_1}{v_{\tilde{m}_1}} + \frac{\alpha_1 l_1 X}{f_{\text{server}_{\tilde{m}_1}}} + \frac{(\alpha_1 l_1 + 2\alpha_2 l_2) X}{f_{\text{server}_{\tilde{m}_1}}} \\
&+ \kappa (1 - \alpha_1) l_1 X f_1^2 + \frac{\alpha_1 l_1}{v_{\tilde{m}_1}} p + \kappa (1 - \alpha_2) l_2 X f_2^2 + \frac{\alpha_2 l_2}{v_{\tilde{m}_1}} p \\
<& 2 \frac{\alpha_1 l_1}{v_{\tilde{m}_1}} + \frac{\alpha_1 l_1 X}{f_{\text{server}_{\tilde{m}_1}}} + \alpha_2 l_2 \left( \frac{1 + p}{v_{\tilde{m}_2}} - \frac{p}{v_{\tilde{m}_1}} + \frac{X}{f_{\text{server}_{\tilde{m}_2}}} \right) - \frac{\alpha_1 l_1}{v_{\tilde{m}_1}} \\
&+ \kappa (1 - \alpha_1) l_1 X f_1^2 + \frac{\alpha_1 l_1}{v_{\tilde{m}_1}} p + \kappa (1 - \alpha_2) l_2 X f_2^2 + \frac{\alpha_2 l_2}{v_{\tilde{m}_1}} p \quad (13.51) \\
=& \underbrace{\frac{\alpha_1 l_1}{v_{\tilde{m}_1}} + \frac{\alpha_1 l_1 X}{f_{\text{server}_{\tilde{m}_1}}} + \kappa (1 - \alpha_1) l_1 X f_1^2 + \frac{\alpha_1 l_1}{v_{\tilde{m}_1}} p}_{U_1} \\
&+ \underbrace{\frac{\alpha_2 l_2}{v_{\tilde{m}_2}} + \frac{\alpha_2 l_2 X}{f_{\text{server}_{\tilde{m}_2}}} + \kappa (1 - \alpha_2) l_2 X f_2^2 + \frac{\alpha_2 l_2}{v_{\tilde{m}_2}} p}_{U_2}.
\end{aligned}
$$

$U_1$ and $U_2$ are the relative costs of time and energy when the device selects server $\tilde{m}_1$ and $\tilde{m}_2$ for offloading.

When $\frac{\alpha_1 l_1}{v_1} \leq \frac{\alpha_2 l_2}{v_2}$, similar results can be found when

$$
f_{\text{server}_{\tilde{m}_1}} > \frac{(\alpha_1 l_1 + 2\alpha_2 l_2) X}{\alpha_2 l_2 \left( \frac{1+p}{v_{\tilde{m}_2}} - \frac{p}{v_{\tilde{m}_1}} + \frac{X}{f_{\text{server}_{\tilde{m}_2}}} \right) - \left( 2\frac{\alpha_2 l_2}{v_{\tilde{m}_2}} - \frac{\alpha_1 l_1}{v_{\tilde{m}_1}} \right)}. \tag{13.52}
$$

This completes the proof of Proposition 13.4.

*Remark 13.4* When servers have different computation capacities, devices should make a wise decision on which server to offload, even a common server is better when the computation capacities of servers vary widely. For example, assuming that the wireless channels are the same for all servers and devices offload same computation data to one server as $\alpha_1 l_1 = \alpha_2 l_2$, if there exists a server with computation capacities $f_{\text{server}_{\tilde{m}_1}} > 3 f_{\text{server}_{\tilde{m}_2}}$ as obtained from Proposition 13.4, the optimal offloading policy for the two devices is selecting the same server $\tilde{m}_1$.

**Proposition 13.5** *Considering the conditions in Proposition 13.4, the optimal offloading ratios of the two devices satisfy* $\alpha_1 = 1 - \frac{f_1 l_2}{f_2 l_1} (1 - \alpha_2)$. *When* $f_1 = f_2, l_1 = l_2$, *the devices have same offloading ratio with* $\alpha_1 = \alpha_2 = \dfrac{X}{X + \frac{f}{\max\{v_{\tilde{m}_1}, v_{\tilde{m}_2}\}} + \frac{2fX}{f_{\text{server}_{\tilde{m}_1}}}}$.

**Proof** Similar to single device scenario as in the Proposition 13.1, the optimal offloading ratio is achieved when the consumed time of local and remote processing are the same:

$$\frac{(1 - \alpha_1) l_1 X}{f_1} = \max\{\frac{\alpha_1 l_1}{v_{\tilde{m}_1}}, \frac{\alpha_2 l_2}{v_{\tilde{m}_2}}\} + \frac{(\alpha_1 l_1 + \alpha_2 l_2) X}{f_{\text{server}_{\tilde{m}_1}}}, \tag{13.53a}$$

$$\frac{(1 - \alpha_2) l_2 X}{f_2} = \max\{\frac{\alpha_1 l_1}{v_{\tilde{m}_1}}, \frac{\alpha_2 l_2}{v_{\tilde{m}_2}}\} + \frac{(\alpha_1 l_1 + \alpha_2 l_2) X}{f_{\text{server}_{\tilde{m}_1}}}. \tag{13.53b}$$

From (13.53), we can get

$$\frac{(1 - \alpha_1) l_1 X}{f_1} = \frac{(1 - \alpha_2) l_2 X}{f_2} \Rightarrow \alpha_1 = 1 - \frac{f_1 l_2}{f_2 l_1} (1 - \alpha_2). \tag{13.54}$$

When $f_1 = f_2 = f, l_1 = l_2 = l$, (13.53) is simplified as

$$\frac{(1 - \alpha_1) X}{f} = \frac{\alpha_1}{\max\{v_{\tilde{m}_1}, v_{\tilde{m}_2}\}} + \frac{(\alpha_1 + \alpha_2) X}{f_{\text{server}_{\tilde{m}_1}}}, \tag{13.55a}$$

$$\frac{(1 - \alpha_2) X}{f} = \frac{\alpha_1}{\max\{v_{\tilde{m}_1}, v_{\tilde{m}_2}\}} + \frac{(\alpha_1 + \alpha_2) X}{f_{\text{server}_{\tilde{m}_1}}}. \tag{13.55b}$$

From (13.54), we get $\alpha_1 = \alpha_2$. Bring it into (13.55), we finally get $\alpha_1 = \alpha_2 = \dfrac{X}{X + \frac{f}{\max\{v_{\tilde{m}_1}, v_{\tilde{m}_2}\}} + \frac{2fX}{f_{\text{server}_{\tilde{m}_1}}}}$. This completes the proof of Proposition 13.5.

**Remark 13.5** When devices selecting the same server for offloading, the relationship between their optimal offloading ratios is related to local computation capacity $f$ and processed task loads $l$. If they are processing the same amount of task data with same CPU frequency, then, they should offload the same computation tasks for optimal performance.

## 13.7 Simulation Results

### 13.7.1 General Setups

In this part, the performance of the proposed Hybrid-AC and MD-Hybrid-AC algorithms are analyzed through simulation experiments. In the experiment, there are $M = 3$ MEC servers to provide edge computing servers. For MD-Hybrid-AC, we use $N = 2$ devices to interact with servers, which is to present in detail how the offloading policy changes with the environment. The system parameters are set as [19]: $L_b = 1000$ bits, $\eta = 1.5$, $X = 737.5$, $L_{max} = 6000$ bits, $\kappa = 10^{-28}$, $\varphi = 0.02s$, $f_{max} = 1.5$ GHz, $b_{max} = 3.2$mJ, $\Omega = 1$ MHz, $p = 2$W, $f_{server}^{max} = 16$ GHz, $f_{server}^{unit} = 2$ GHz. For simplicity, the normalized wireless channel gain $g_{n,m}^t$ is assumed to take values in [5, 14] uniformly [42]. Unless specified, the task requested probability $\zeta = 0.8$, the maximum harvested energy $e_{max} = 0.001$J, $\lambda = 2$ and the reward weighted parameters $\omega_t = 3$, $\omega_e = 1$.

The actor network and critic network in Hybrid-AC both have two hidden layers, whose size are set as 400 and 300. The critic network is the same as the actor network, except that the second hidden layer is concatenated with the control vector and the output layer gives $M$ $Q$ values. The learning rate for the actor network and critic network are 0.0001 and 0.001. We set the soft target updates $\epsilon = 0.001$, the discount factor $\gamma = 0.99$, the maximum time step $T = 20$ in each episode, and the size of replay memory $\mathfrak{D}$ is 10,000. The exploration noise process $v$ is an Ornstein_Uhlenbeck process [16] with $\theta = 0.15$ and $\sigma = 0.2$, and the $\varepsilon$-greedy strategy is set as $\varepsilon = 0.9$. The MD-Hybrid-AC has the similar neural network architecture as in the Hybrid-AC, except the critic network takes states $o$ and continuous actions of all devices as input, and outputs $M \times N$ $Q$ values. In the following experiments, each result in the figure is averaged over 5000 episodes.

### 13.7.2 Performance of Convergence and Generalizability

In this experiment, we will show the convergence and generalization of the proposed methods. The system configurations are set as in the Sect. 13.7.1. The convergence performances of the Hybrid-AC and MD-Hybrid-AC are shown in Fig. 13.3a, where the performance of MD-Hybrid-AC is the averaged reward of two devices. The theoretical result in Sect. 13.6 with $g = 14$ and $f_{server} = 16$ GHz is also shown as a comparison, which is the reward for the single device. Under such simulation conditions, the optimal offloading policy follows Proposition 13.1 and is obtained when the execution time of local computing and offloading computing are the same. The results reveal that the Hybrid-AC converges faster and more stably than MD-Hybrid-AC and also achieves higher rewards. This corresponds to the difficulty of the dynamic computation offloading problem, as the offloading decision of a single device is much simple than that of multiple devices. Besides, with energy

**Fig. 13.3** Illustrations for the convergence and generalization of the proposed algorithms. (**a**) Reward vs episodes. (**b**) Performance vs weighted parameters $\omega_t$, $\omega_e$. (**c**) Performance vs weighted parameters $\omega_t$, $\omega_e$

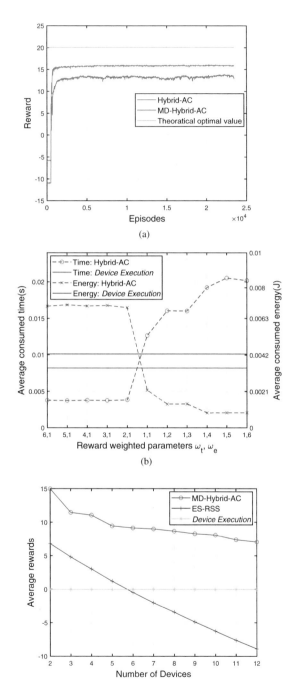

constraints, there is still a gap from the theoretical optimal policy as shown in Proposition 13.1.

The generalizability of Hybrid-AC is demonstrated through the performance with different reward weighted parameters $\omega_t, \omega_e = (6, 1), \ldots, (2, 1), (1, 1), (1, 2), \ldots, (1, 6)$ in Fig. 13.3b. The results reveal the trade-off between time and energy consumption can be easily tuned by assigning different values of $\omega_t, \omega_e$, which means Hybrid-AC adapts to both time-sensitive and energy-sensitive tasks.

The generalizability of MD-Hybrid-AC is demonstrated through the performance with different number of devices in Fig. 13.3c. We set 6 servers with maximum computation capacity $f_{server}^{max} = 32$ GHz. The number of devices $N$ changes from 2 to 12. The performance is compared with *Exhaustive Search with Random Server Selection (ES-RSS)*: The offloading ratio $\alpha$ and computation capacity $f$ are both uniformly discretized into five actions with the consideration of time complexity; Each device randomly selects one edge server for offloading, then, the device enumerate all possible combinations of $\alpha$ and $f$ without considering other devices to find the one with the least time consumption as the solution. The results show that MD-Hybrid-AC deteriorates gradually but can still maintain a relatively good performance when the network size increases. The policy turns to reduce the proportion of executing the task on servers. The reason is that the total computing resources of servers are limited, when more devices share a common server, it may not time and energy efficient compared with locally processing on each device. In comparison, ES-RSS degrades rapidly, which reveals that a wise coordination is important in the multi-device system.

The computational complexity is shown in Table 13.2. The computation time of MD-Hybrid-AC is insensitive to the number of devices, and the actor and critic part cost 1 ms and 1.3 ms on average in our simulations. The time required to output one solution depends on how large the size of the neural network used. For a even larger scale of MEC system, the action dimension will be much larger, and we could use a larger-scale neural network to maintain the performance. In such case, the computation time will increase.

**Table 13.2** Average computation time under different number of devices

| Number of devices | Actor part | Critic part | ES-RSS |
|---|---|---|---|
| 2 | 1.039 ms | 1.292 ms | 0.267 ms |
| 4 | 1.045 ms | 1.305 ms | 0.636 ms |
| 6 | 1.044 ms | 1.309 ms | 0.935 ms |
| 8 | 1.042 ms | 1.307 ms | 1.209 ms |
| 10 | 1.056 ms | 1.314 ms | 1.442 ms |
| 12 | 1.052 ms | 1.303 ms | 1.690 ms |

### 13.7.3  Performance Evaluation of Hybrid-AC with Different System Parameters

To validate the effectiveness of Hybrid-AC, we compare it with four baselines:

- *Device Execution:* The device executes all of their computation tasks locally. The maximum computation capacity is scheduled with maximum possible energy: $f_{D\_exe}^t = \min\{\sqrt{(b_n^t + e_n^t)/\kappa l_n^t X}, f_{\max}\}$.
- *Server Execution:* The device offloads all of their computation tasks to the server for execution. The server is selected according to $\arg\max_{m \in \mathcal{M}} \frac{1+p}{\Omega \log_2\left(1+g_{1,m}^t p\right)} + \frac{X}{f_{server_m}^t}$. If the transmission time exceeds the maximum tolerable delay, the task is dropped.
- *Deep Q-learning based offloading (DQLO):* We implemented a deep Q-learning based offloading method based on the MDP model proposed in Sect. 13.4.1. The action space is uniformly discretized into finite discrete values and we developed two algorithms based on the different number of discretized actions. DQLO(5): The offloading ratio $\alpha$ and computation capacity $f$ are both uniformly discretized into five states, making the action dimension of DQN be 13. DQLO(10): Both $\alpha$ and $f$ are discretized into ten actions. The neural network is the same as the critic network in the proposed method. The learning rate is 0.0001, and the action is chosen by an $\varepsilon$-greedy strategy with $\varepsilon = 0.9$. The discount factor and memory size are set as the same as the proposed method. Besides, we used the same states and reward for a fair comparison.
- *Exhaustive Search:* The offloading ratio and computation capacity are both uniformly discretized into five actions. The server is selected according to $\arg\max_{m \in \mathcal{M}} \frac{1+p}{\Omega \log_2\left(1+g_{1,m}^t p\right)} + \frac{X}{f_{server_m}^t}$. Then, the device enumerate all possible combinations of $\alpha$ and $f$ to find the one with the least time consumption as the solution.

Besides, we build an upper bound to help analyze the performance, which is constructed as: The offloading ratio $\alpha$ and computation capacity $f$ are both discretized into twenty states; We enumerate all possible combinations of actions and select the action corresponding to the largest reward as the current action.

#### 13.7.3.1  Performance vs Different Task Requested Probability $\zeta$

The average computation offloading performance in terms of rewards, consumed time and energy under different computation task requested probability $\zeta \in [0.4, 2]$ is shown in Fig. 13.4. The parameters in the reward function of proposed method and DQLO are set as $\omega_t = 3$, $\omega_e = 1$ with a focus on time-sensitive tasks. For the sake of illustration, the results of the consumed time in Fig. 13.4b are expressed logarithmically. It can be observed that the execution time and consumed energy increase with the amount of computation task. The *Server Execution* method has

**Fig. 13.4** Performance vs
different requested task load
$\zeta$. (**a**) Reward. (**b**) Execution
time. (**c**) Consumed energy

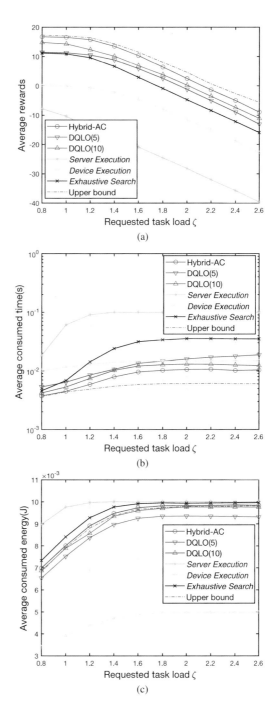

the highest energy consumption and time delay, this can be explained as follows: the device consumes much more power for transmission to offload the computation task than local execution; Besides, the transmission procedure in the offloading operation consumes the primary time consumption compared to the execution time of MEC servers. The *Exhaustive Search* method has a comparable performance with DQLO(5) when the requested task is not heavy, and its performance deteriorates even more with the increase of the task load as it does not consider energy consumption, which leads to more tasks are dropped. Thus, a wise and intelligent decision should be made to reduce the time and energy consumption. The DQLO achieves better performance with the increase of the discretized action dimensions. However, the performance would not be further improved as a finer discretization may make it difficult to converge. Hybrid-AC is closest to the upper bound and outperforms DQLO(10) with 18.75% shorter execution time at the cost of only 1.36% higher energy consumption, which means that Hybrid-AC makes a better decision than the discrete action based DRL methods. As we concentrate more on the execution time, it achieves the lowest averaged execution delay and a relatively low energy consumption under different task loads.

### 13.7.3.2   Performance vs Different Maximum Harvested Energy $e_{max}$

We show the performance under different maximum harvested energy $e_{max} \in \left[6 \times 10^{-4}, 12 \times 10^{-4}\right]$ in Fig. 13.5. The performance of the consumed time in Fig. 13.5b are expressed logarithmically as in the Fig. 13.4b. The results indicate that when the harvested energy is relatively low, the performance of *Server Execution*, DQLO and Hybrid-AC deteriorate at execution delay. This is because the energy consumption of transmission often exceeds the existing energy in the battery, which results in dropping tasks frequently. The energy consumption for local execution is relatively low in all the experiments, so the performance of *Device Execution* usually has a good performance compared with methods using offloading. The *Exhaustive Search* method has similar performance with DQLO(5) when the energy is sufficient, and gets worse rapidly as the harvested energy decreases. It reveals that when making an offloading decision, energy consumption is an important consideration. The learning-based methods also perform better as the harvested energy increases. Specifically, Hybrid-AC has a similar performance of *Device Execution* as soon as $e_{max}$ reaches $7 \times 10^{-4}$J, and even consumes much less time as the harvested energy continues to increase. It gets closer to the upper bound when the harvested energy is sufficient, and outperforms DQLO(10) with 40.06% shorter execution time at the cost of only 1.13% higher energy consumption in an episode for different maximum harvested energy.

**Fig. 13.5** Performance vs
different maximum harvested
energy $e_{max}$. (**a**) Reward. (**b**)
Execution time. (**c**)
Consumed energy

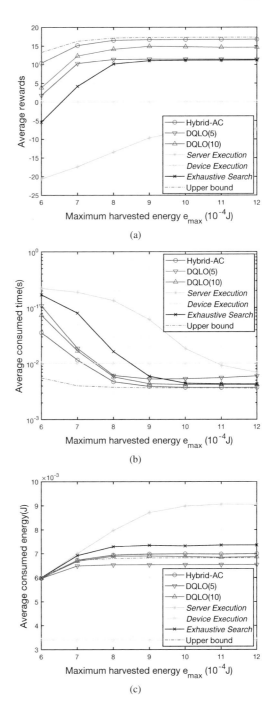

### 13.7.4    Performance Evaluation of MD-Hybrid-AC with Different System Parameters

#### 13.7.4.1    Performance vs Different Server's Occupied Resource Units $\lambda$

The average execution time, energy and offloading ratio are evaluated with different server's occupied units $\lambda = [1, 10]$. The results of two devices with important factor $Z_1 = 1, Z_2 = 2$ are shown in the Fig. 13.6, along with the performance of Hybrid-AC for comparison. Apparently, as the occupied computation resources of servers increase, the device's execution time gets longer. At the meantime, the deteriorated service quality has caused devices to prefer to process locally, which also decreases the energy consumptions. Besides, we can find that the two devices have different performance. Specifically, device 2 with larger important factor has a relatively better performance compared to that of device 1, and behaves more like Hybrid-AC does. This can be explained that the policy pays more attentions on device 2 with higher priority, however, the limited resources constraint the performance to be inferior to that of single device. As we concern more on time with $\omega_t = 3, \omega_e = 1$, device 2 consumes 19.35% shorter time than device 1 and 16.28% longer time than single device. It also consumes 9.38% more energy than device 1 and almost the same energy as single device.

#### 13.7.4.2    Performance vs Differentiated Server Capacities

This part shows the performances when three servers have differentiated available computation resources. We set the occupied resources of servers as $\lambda_1 = 1, \lambda_2 = 10$ and $\lambda_3$, which changes in [1, 10]. Besides, the default setting of important factors is $Z_1 = 1, Z_2 = 2$. We compare its performances with the results of another important factor setting $Z_1 = Z_2 = 1$, and Hybrid-AC. The averaged performances of consumed time, consumed energy, offloading ratio, selected ratio of each server are shown in Figs. 13.7 and 13.8. We will analysis the results from different aspects. For Hybrid-AC with single device, the time and energy consumptions are nearly unchanged, as well as the offloading ratio. We can find evidences from server selection results. When the occupied resource units in server 3 increase, the selected ratio of server 3 reduces consequently. However, the selected ratio of server 1 fills the vacancy. This can also be verified by the selected ratio of server 2, which is almost not changed. Because server 1 has a relatively enough computation resources with $\lambda_1 = 1$, which is the best performance of server 3, the device has a stable performance all the time. For MD-Hybrid-AC with $Z_1 = 1, Z_2 = 1$, the two devices have similar performance. They both increase the execution time as the available computation capacities of server 3 become smaller. Simultaneously, they offload less data for server execution, which also decrease the energy consumptions. However, they behave differently after they decreasing the selected ratio of server 3 as the available computation resources decreasing. Device 1 mainly increases the

**Fig. 13.6** Multi-device:
Performance vs different
occupied server computation
resource unit λ. (**a**) Execution
time. (**b**) Consumed energy.
(**c**) Offloading ratio

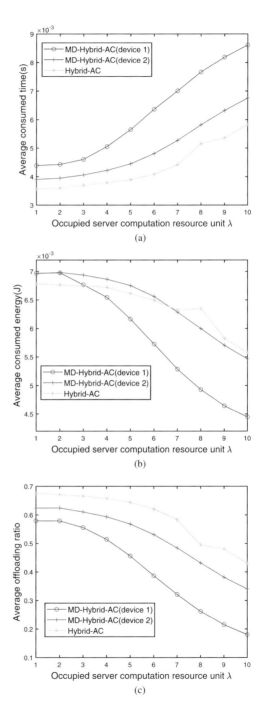

(a)

(b)

(c)

**Fig. 13.7**  Multi-device:
Performance vs different
occupied computation
resource unit $\lambda_3$ in server 3
(Part 1). (**a**) Execution time.
(**b**) Consumed energy. (**c**)
Offloading ratio

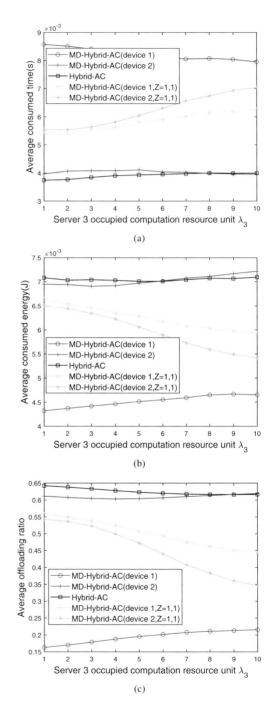

**Fig. 13.8** Multi-device: Performance vs different occupied computation resource unit $\lambda_3$ in server 3 (Part 2). (**a**) Server 1 selection ratio. (**b**) Server 2 selection ratio. (**c**) Server 3 selection ratio

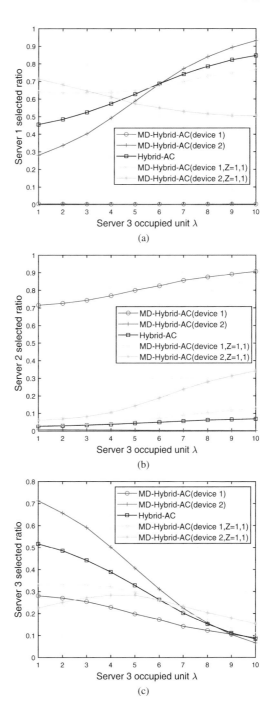

selected ratio of server 1, while device 2 selects server 2. This can be seen as the coordination among devices, which can be explained as follows. As server 1 is more powerful than server 2, when they both choose server 1 for better performance, they may both perform worse. Because server 1 needs to compute the sum of the offloaded data of device 1 and device 2, the two devices may consume more time than that of selecting different servers. For comparison, if Hybrid-AC is utilized in a multi-device scenario, coordination will not exist as different devices will select the same server based on the same states of servers, which results in poor performance. For MD-Hybrid-AC with $Z_1 = 1, Z_2 = 2$, the two devices exhibit different performances and policies. The device 2 with relatively large important factor $Z_2 = 2$ has a better performance than that of device 1, and acts like Hybrid-AC with single device. This is consistent with what we found in Sect. 13.7.4.1. Device 1 behaves very differently from device 2. It mostly chooses server 2 with least computation resources and saves the better servers for device 2. Though, this policy may affect the final performance, it shows the coordination among devices. This policy is intended to put more attention to the performance of device 2. So, it takes a conservative strategy to guarantee the performance of device 2 with the price of sacrificing its own performance. This is also the reason that the performance of multi-device is worse than that of single device.

## 13.8   Conclusion

This part proposed two DRL-based dynamic computation offloading algorithms for MEC systems with EH devices, which addressed the challenges of continuous-discrete hybrid action spaces and coordination among devices. Hybrid-AC is intended to solve the hybrid decision spaces, which is a combination of DQN and DDPG. It utilizes *actor* to output continuous actions: offloading ratio and local computation capacity. At the meantime, it uses *critic* to evaluate the actions of *actor* and output the discrete action: server selection. MD-Hybrid-AC is aimed at providing coordination among devices. Built on the hybrid decision architecture of Hybrid-AC, MD-Hybrid-AC adopts the paradigm of centralized training with decentralized execution to build a centralized *critic*, which considers actions of all devices and achieve coordination on server selection. The simulation results verify the effectiveness and superiority of the proposed algorithms.

# References

1. J. Du, E. Gelenbe, C. Jiang, H. Zhang, and Y. Ren, "Contract design for traffic offloading and resource allocation in heterogeneous ultra-dense networks," *IEEE J. Sel. Areas Commun.*, vol. 35, no. 11, pp. 2457–2467, Nov. 2017.
2. Y. Mao, C. You, J. Zhang, K. Huang, and K. B. Letaief, "A survey on mobile edge computing: The communication perspective," *IEEE Commun. Surveys & Tutorials*, vol. 19, no. 4, pp. 2322–2358, Fourthquarter 2017.
3. M. Satyanarayanan, "The emergence of edge computing," *Computer*, vol. 50, no. 1, pp. 30–39, Jan. 2017.
4. X. Peng, J. Ren, L. She, D. Zhang, J. Li, and Y. Zhang, "Boat: A block-streaming app execution scheme for lightweight iot devices," *IEEE Internet Things J.*, vol. 5, no. 3, pp. 1816–1829, Jun. 2018.
5. J. W. Rittinghouse and J. F. Ransome, *Cloud computing: implementation, management, and security*.   Boca Raton, FL, USA: CRC press, 2017.
6. S. Verma, Y. Kawamoto, Z. M. Fadlullah, H. Nishiyama, and N. Kato, "A survey on network methodologies for real-time analytics of massive iot data and open research issues," *IEEE Commun. Surveys & Tutorials*, vol. 19, no. 3, pp. 1457–1477, Thirdquarter 2017.
7. M. Chiang and Z. Tao, "Fog and iot: An overview of research opportunities," *IEEE Internet Things J.*, vol. 3, no. 6, pp. 854–864, Jun. 2016.
8. X. Chen, "Decentralized computation offloading game for mobile cloud computing," *IEEE Trans. Parallel Distrib. Syst.*, vol. 26, no. 4, pp. 974–983, Apr. 2014.
9. P. Mach and Z. Becvar, "Mobile edge computing: A survey on architecture and computation offloading," *IEEE Commun. Surveys & Tutorials*, vol. 19, no. 3, pp. 1628–1656, Thirdquarter 2017.
10. Z. Ni, R. V. Bhat, and M. Motani, "On dual-path energy-harvesting receivers for iot with batteries having internal resistance," *IEEE Internet Things J.*, vol. 5, no. 4, pp. 2741–2752, Aug. 2018.
11. P. Kamalinejad, C. Mahapatra, Z. Sheng, S. Mirabbasi, V. C. Leung, and Y. L. Guan, "Wireless energy harvesting for the internet of things," *IEEE Commun. Mag.*, vol. 53, no. 6, pp. 102–108, Jun. 2015.
12. L. Yang, X. Chen, J. Zhang, and H. V. Poor, "Cost-effective and privacy-preserving energy management for smart meters," *IEEE Trans. Smart Grid*, vol. 6, no. 1, pp. 486–495, Jan. 2015.
13. S. Ulukus, A. Yener, E. Erkip, O. Simeone, M. Zorzi, P. Grover, and K. Huang, "Energy harvesting wireless communications: A review of recent advances," *IEEE J. Sel. Areas Commun.*, vol. 33, no. 3, pp. 360–381, Mar. 2015.
14. Z. Ding, S. M. Perlaza, I. Esnaola, and H. V. Poor, "Power allocation strategies in energy harvesting wireless cooperative networks," *IEEE Trans. Wireless Commun.*, vol. 13, no. 2, pp. 846–860, Feb. 2014.
15. F. K. Shaikh and S. Zeadally, "Energy harvesting in wireless sensor networks: A comprehensive review," *Renew. Sustain. Energy Rev.*, vol. 55, pp. 1041–1054, Mar. 2016.
16. T. P. Lillicrap, J. J. Hunt, A. Pritzel, N. Heess, T. Erez, Y. Tassa, D. Silver, and D. Wierstra, "Continuous control with deep reinforcement learning," in *Int. Conf. Learn. Represent.*   San Juan, Puerto Rico, 2-4 May 2016, pp. 1–14.
17. V. Mnih, K. Kavukcuoglu, D. Silver, A. A. Rusu, J. Veness, M. G. Bellemare, A. Graves, M. Riedmiller, A. K. Fidjeland, G. Ostrovski *et al.*, "Human-level control through deep reinforcement learning," *Nature*, vol. 518, no. 7540, pp. 529–533, Feb. 2015.
18. C. You, K. Huang, H. Chae, and B.-H. Kim, "Energy-efficient resource allocation for mobile-edge computation offloading," *IEEE Trans. Wireless Commun.*, vol. 16, no. 3, pp. 1397–1411, Mar. 2017.
19. Y. Mao, J. Zhang, and K. B. Letaief, "Dynamic computation offloading for mobile-edge computing with energy harvesting devices," *IEEE J. Sel. Areas Commun.*, vol. 34, no. 12, pp. 3590–3605, Dec. 2016.

20. S. Zhao, Y. Yang, Z. Shao, X. Yang, H. Qian, and C.-X. Wang, "Femos: Fog-enabled multitier operations scheduling in dynamic wireless networks," *IEEE Internet Things J.*, vol. 5, no. 2, pp. 1169–1183, Feb. 2018.

21. R. S. Sutton and A. G. Barto, *Reinforcement learning: An introduction.*   Cambridge, MA, USA: MIT press, Nov. 2018.

22. J. Xiong, Q. Wang, Z. Yang, P. Sun, L. Han, Y. Zheng, H. Fu, T. Zhang, J. Liu, and H. Liu, "Parametrized deep q-networks learning: Reinforcement learning with discrete-continuous hybrid action space," *arXiv preprint arXiv:1810.06394*, 2018.

23. S. Levine, C. Finn, T. Darrell, and P. Abbeel, "End-to-end training of deep visuomotor policies," *J. Mach. Learn. Res.*, vol. 17, no. 39, pp. 1–40, 2016.

24. J. C. Caicedo and S. Lazebnik, "Active object localization with deep reinforcement learning," in *Proc. IEEE Int. Conf. Comput. Vis.*   Santiago, Chile, 7-13 Dec. 2015, pp. 2488–2496.

25. C. Wang, J. Wang, Y. Shen, and X. Zhang, "Autonomous navigation of uavs in large-scale complex environments: A deep reinforcement learning approach," *IEEE Trans. Veh. Technol.*, vol. 68, no. 3, pp. 2124 – 2136, Mar. 2019.

26. C. Wang, J. Wang, J. Wang, and X. Zhang, "Deep reinforcement learning-based autonomous uav navigation with sparse rewards," *IEEE Internet Things J.*, 2020.

27. X. Liu, Z. Qin, and Y. Gao, "Resource allocation for edge computing in iot networks via reinforcement learning," in *IEEE Int. Conf. Commun.*   Shanghai, China, 20-24 May 2019, pp. 1–6.

28. M. Min, L. Xiao, Y. Chen, P. Cheng, D. Wu, and W. Zhuang, "Learning-based computation offloading for iot devices with energy harvesting," *IEEE Trans. Veh. Technol.*, vol. 68, no. 2, pp. 1930–1941, Jan. 2019.

29. X. Chen, H. Zhang, C. Wu, S. Mao, Y. Ji, and M. Bennis, "Optimized computation offloading performance in virtual edge computing systems via deep reinforcement learning," *IEEE Internet Things J.*, vol. 6, no. 3, pp. 4005–4018, Oct. 2018.

30. Z. Wei, B. Zhao, J. Su, and X. Lu, "Dynamic edge computation offloading for internet of things with energy harvesting: A learning method," *IEEE Internet Things J.*, vol. 6, no. 3, pp. 4436–4447, Jun. 2019.

31. C. Qiu, Y. Hu, Y. Chen, and B. Zeng, "Deep deterministic policy gradient (ddpg) based energy harvesting wireless communications," *IEEE Internet Things J.*, vol. 6, no. 5, pp. 8577–8588, Oct. 2019.

32. Y. Wei, F. R. Yu, M. Song, and Z. Han, "Joint optimization of caching, computing, and radio resources for fog-enabled iot using natural actor–critic deep reinforcement learning," *IEEE Internet Things J.*, vol. 6, no. 2, pp. 2061–2073, 2019.

33. L. Huang, S. Bi, and Y. J. Zhang, "Deep reinforcement learning for online computation offloading in wireless powered mobile-edge computing networks," *IEEE Trans. Mobile Comput.*, 2019.

34. G. Qu and G. Qu, "What is the limit of energy saving by dynamic voltage scaling?" in *Proc. IEEE/ACM Int. Conf. Computer-aided design.*   San Jose, CA, USA, 4-8 Nov. 2001, pp. 560–563.

35. T. D. Burd and R. W. Brodersen, "Processor design for portable systems," *J. VLSI Signal Process. Syst.*, vol. 13, no. 2-3, pp. 203–221, Aug. 1996.

36. X. Lyu, H. Tian, C. Sengul, and P. Zhang, "Multiuser joint task offloading and resource optimization in proximate clouds," *IEEE Trans. Veh. Technol.*, vol. 66, no. 4, pp. 3435–3447, Apr. 2017.

37. D. Zhang, Z. Chen, M. K. Awad, N. Zhang, H. Zhou, and X. S. Shen, "Utility-optimal resource management and allocation algorithm for energy harvesting cognitive radio sensor networks," *IEEE J. Sel. Areas Commun.*, vol. 34, no. 12, pp. 3552–3565, Dec. 2016.

38. J. Zhang, J. Du, Y. Shen, and J. Wang, "Dynamic computation offloading with energy harvesting devices: A hybrid-decision-based deep reinforcement learning approach," *IEEE Internet Things J.*, vol. 7, no. 10, pp. 9303–9317, Oct. 2020.

39. J. Zhang, J. Du, C. Jiang, Y. Shen, and J. Wang, "Computation offloading in energy harvesting systems via continuous deep reinforcement learning," in *IEEE Int. Conf. Commun. (ICC'20)*. Dublin, Ireland, 7-11 Jun. 2020.

40. L. Kraemer and B. Banerjee, "Multi-agent reinforcement learning as a rehearsal for decentralized planning," *Neurocomputing*, vol. 190, pp. 82–94, May 2016.

41. J. Zhang, J. Du, J. Wang, and Y. Shen, "Hybrid decision based deep reinforcement learning for energy harvesting enabled mobile edge computing," in *IEEE Int. Wireless Commun. Mobile Comput. Conf. (IWCMC'20)*. Limassol, Cyprus, 15-19 Jun. 2020.

42. J. Du, L. Zhao, X. Chu, F. R. Yu, J. Feng, and I. Chih-Lin, "Enabling low-latency applications in lte-a based mixed fog/cloud computing systems," *IEEE Trans. Veh. Technol.*, vol. 68, no. 2, pp. 1757–1771, Feb. 2019.

# Part V
# Cooperative Resource and Information Sharing Among Users

# Chapter 14
# Introduction of Cooperative Resource and Information Sharing

**Keywords** Information Sharing · User Interaction · Social Networks · Resource Allocation

Recently, the Internet of Things (IoT) enabled 6G networks have penetrated many aspects of the physical world to realize different applications. Resulting from ubiquitous connections of 6G, data traffic from these applications is experiencing unprecedented increases. In addition, these applications generate, exchange, aggregate, and analyze a vast amount of security-critical and privacy-sensitive data, which makes them attractive targets of attacks. The past chapters of this book have investigated a comprehensive study on communication, computing, caching resource allocation of infrastructure. In this chapter, we will focus on the effect of users' interactions and cooperation behaviors on the quality of services.

(1) Cooperative data transaction: The first part considers auction mechanism design and performance analysis for data transactions in mobile social networks. We propose a data transaction mechanism, and summaries the analysis on state transmission, stationary probabilities of the system, and the expected income for data sellers. To make sure that both the data supply and the demand can be satisfied at the same time, a data-demands-driven mobility model is proposed. In addition, the designed mobility model can also improve the efficiency of data transaction. (2) Cooperative trustworthiness evaluation: Users' reporting and sharing of their consumption experience can be utilized to rate the quality of different approaches of online services. How to ensure the authenticity of users' reports and identify malicious ones with cheating reports become important issues to achieve an accurate service rating. The second part proposes a private-prior peer prediction mechanism for a service rating system with a fusion center, which evaluates users' trustworthiness with their reports by applying the strictly proper scoring rule. In addition, to identify malicious users and bad-functioning/unreliable users with high error rate of quality judgement, an unreliability index is proposed to evaluate the uncertainty of reports. By combining the trustworthiness and unreliability, malicious users cannot receive a high trustworthiness and low unreliability at the same time when they report falsified feedbacks. (3) Cooperative privacy protection: User-

J. Du, C. Jiang, *Cooperation and Integration in 6G Heterogeneous Networks*, Wireless Networks, https://doi.org/10.1007/978-981-19-7648-3_14

centric mobile sensing and computing devices, such as smartphones and vehicle sensors, are serving as an emerging category of devices at the edge of networks. The data in a mobile crowdsensing system always contains sensitive private information, and can easily spread over the system via wireless channels and social application platforms, which poses a serious threat to user privacy and causes privacy protection to face serious strain. In response, the third part carries out some preliminary works focusing on the interaction between users based on a community-structured evolutionary game model. In this model, users are players, and their behaviors (i.e., take the secure strategy or not) are the game strategies that will evolve and spread over the system.

# Chapter 15
# Cooperative Data Transaction in Mobile Networks

**Abstract** Mobile data traffic is experiencing unprecedented increases due to the proliferation of highly capable smartphones, laptops and tablets, and mobile data offloading can be used to move traffic from cellular networks to other wireless infrastructures such as small-cell base stations. This work addresses the related issue of data allocation, by proposing a novel infrastructure independent method based on the hotspot function of smartphones. In the proposed scheme, smartphones transfer data allowances among mobile users, so that users with excess data allowances act as accessible Wi-Fi hotspots, selling their data allowance to other users who need extra data allowances. To achieve this objective, we propose to use auctions with single and multiple data sellers. Efficient schemes based on auction models are discussed to sell the data allowances over successive days in a month, and over different time slots during a single day. Overall system performance is considered based on the behavior of mobile users, such as changing demands for the sale or purchase of data allowances. Together with the analytical results presented, our simulation experiments also indicate that knowledge of user behavior can significantly improve the performance of data allowance transactions, leading to highly efficient allocations among users.

**Keywords** Data Transaction · Resource Allocation · Auction · Mobile Networks · Cooperative Particle Swarm Optimization

## 15.1 Introduction

In recent decades, the mobile data traffic is experiencing an enormous growth due to the significant penetration of smartphones, as well as Web 2.0 and a large number of applications with high bandwidth requirements. Researchers have predicted that each person will consume on average as much as 5 GB of data each month by 2020 [1–3]. To meet these increasing and high speed data requirements, many new communication techniques and standards are provided, such as LTE Release 8, which can achieve a high peak data rate of 300 Mbps on the downlink and 75 Mbps on the uplink for a 20 MHz bandwidth [4]. Additionally, ultra-dense heterogeneous

© The Author(s), under exclusive license to Springer Nature Singapore Pte Ltd. 2023
J. Du, C. Jiang, *Cooperation and Integration in 6G Heterogeneous Networks*,
Wireless Networks, https://doi.org/10.1007/978-981-19-7648-3_15

networks, which consists of a large number of small-cell base stations (SBSs) to provide data offloading, have been proposed as an another solution to relief the heavy traffic load brought to the macro base stations [5–7]. Through data offloading, the data traffic from mobile users can be sent over SBSs, such as femto base stations and Wi-Fi hotspots, when these SBSs are available, otherwise traffic is delivered over cellular networks.

However, almost all recent data offloading studies can only be implemented with assistance of external infrastructures, i.e., SBSs. Sometimes these SBSs are operated by the mobile network operators (MNOs), while usually the SBSs are owned by some third parties, which means that MNOs need to rent these SBSs if they want to utilize them for data offloading. Meanwhile, due to the introduction of these external and heterogeneous infrastructures, resource management problems, such as power control and mobile user equipment scheduling, become more complicated and challenging, especially for networks with densely deployed SBSs and a large number of mobile users. Moreover, system stability, protocol compatibility and switching, traffic fairness, network congestion control, etc., will pose great challenge for data offloading. In order to avoid these problems above, in this work, we propose a novel infrastructure-free approach to implement data offloading, i.e., operating the data transaction among the mobile users by turning on the Wi-Fi hotspot function of smartphones.

### 15.1.1  Motivation

Currently, to face the increasing data demands of mobile users, all MNOs, such as AT&T in the US, Giffgaff in the UK and China Mobile, have offered many optional monthly data plans with different amounts of data. While the arbitrary of the data plan regulations made by different MNOs are the same, i.e., if the data in the current data plan is not run out by the end of a month, the remaining data will not be cumulated to the next month data plan.[1] On the other hand, when the monthly data has been used out before the end of a month, mobile users have to purchase some extra data with a higher price than monthly plans, otherwise they will suffer a lower speed of data service. Therefore, the opposite results leading by these regulations come down to the following situations. On the one hand, users buying data plans with a large amount of data might still hold a lot of unconsumed data at the end of a months. On the other hand, users with a small amount of data might use out their data before the end of a month. As a consequence, this contradiction between

---

[1] Currently, most MNOs provide this arbitrary data clear policy, except for those very few operators, such as AT&T and China Mobile, not clearing users data by the end of a month. However, such "data rollover program" provided is limited popularized. Take China Mobile and AT&T for instance, the accumulated data will be cleared by the end of the second month. In addition, the data rollover service of NTT can only be accessed by their users with data plans larger than 5 GB.

the redundancy and demands of data resource make it possible for the two types of mobile users to make an internal data transaction between them, which is operated without any third party infrastructures. According to a certain rule or contract of data transaction, "data owners" can sell their unused data to those "data requesters" at a lower price compared with the market price.

### 15.1.1.1  Feasibility of Data Transaction

The direct data dealing is never allowed among the mobile users, no matter whether they are belonging to the same MNO. Fortunately, the *hotspot* function of current smartphones makes this data transaction between data owners and requesters mentioned above to become a reality. By unlocking the hotspot mode, hotspot phones will allow other mobile phones or wireless devices to access them and phone-to-phone communication via WiFi interface can be realized [8, 9]. The data transaction between mobile users is similar to an offloading process in current heterogeneous networks, which refers to delivering data traffic of the MNO to third-party networks. However, the data transaction is designed to transfer the data of one mobile user to another one to increase the data utilization among users effectively, which cannot reduce the total volume of data traffic mobile networks [10, 11].

### 15.1.1.2  Effective and Efficient Data Transaction

Mobile phones will consume more energy when working as Wi-Fi hotspots. Moreover, such accessible Wi-Fi mode might result in potential threat of personal information. So it is reasonable for data owners to sell their data with a price as high as possible to compensate their costs of energy and privacy. In addition, such incomes brought by unconsumed and to-be-wasted data can encourage data owners to participate in data transaction, if incomes can compensate their costs resulting from information security risks [12]. On the other hand, for data requesters, buying data through this data transaction can obtain high-speed data service by a relatively lower price. So these requesters have the motivation to fuel the transaction and compete for the data resource when there are many data requesters. This competitive relationship can be modeled by auction mechanisms effectively. So in this work, we will introduce the auction models to describe the operation of data transaction.

### 15.1.1.3  Changing Demands of Selling and Buying Data

The demands of selling and buying data always change over time for data owners and requesters, respectively. To be specific, the closer to the end of a month, the more urgent the data owners are to sell their data. Similarly, the number of data requesters will increase when the end of a month is coming, and the willing to buy data through data transaction tends to be much stronger. Then tendency of data selling and buying will further influence the price of data. So how to allocate the amount of data to be sold in every day over a month, and how to schedule data auction in a single day are very important to maximize the utilities of data owners and satisfy the data demands of requesters at the same time. In this work, we will design different data allocation mechanisms to realize high efficient data transaction, based on auction models.

## 15.1.2   Contribution

The main contributions and our main ideas are summarized as follows:

1. We establish a basis auction framework for the data transaction system with only one data seller. This auction model is different from traditional auction models based on game theory [13–15], but established for performance analysis based on stochastic process and queueing theory. Based on this model, the data allocation mechanisms are designed to decide how to sell the extra data in different days to optimize the expected income for the data owner. In this work, we consider that the urgency of data selling and buying are changing over time, when maximizing the income of the data owner. In addition, the transaction efficiency is also considered to achieve a further optimization of the income. Simulation results validate that the designed allocation mechanism can increase the total income of the system, and the data transaction efficiency can be also guaranteed.
2. We propose a networked data transaction system, in which there are multiple data owners operating their own basic data auction, based on a networked auction model. To describe the movement of data bidders among different auctions, a data-demands-driven mobility model is proposed, which can make sure that both the data supply and the demand can be satisfied at the same time. In addition, the designed mobility model can also improve the efficiency of data transaction.
3. Based on the networked data transaction system established, we design three different data allocation mechanisms to decide how to sell data in a single day for every data auctioneer in the system, to maximize the income obtained in each time slot, each auctioneer's income or the entire income of the system. To optimize the system performance, the prediction of the data bidders' movement is considered when designing the allocation mechanisms. Simulation results demonstrate that the prediction based data allocation mechanism can bring more income for data auctioneers than the non-prediction approach.

The reminder of this part is organized as follows. We first review the relevant literature in Sect. 15.2. In Sect. 15.3, the data allocation mechanisms are designed for the basic auction model based data transaction. Data allocation mechanisms based on the networked auction model are proposed in Sect. 15.4. The approximate solution of the optimization problems based on cooperative particle swarm for data allocation is introduced in Sect. 15.5. Simulations are shown in Sects. 15.6, and 15.7 concludes this part.

## 15.2   Related Work

Mobile data offloading by applying SBSs in heterogeneous networks, regarded as a feasible solution to deal with the increasing of mobile data requirement and services, has attracted more and more attention. With assistance of heterogeneous SBSs, the throughput of cellular networks can be greatly improved. However, providing offloading services results in more energy consumption and bandwidth resource occupation for these SBSs, which calls for incentive mechanisms to encourage SBSs to participate in the transmission cooperation. To promote altruistic behaviours among resource providers, many economics theory based mechanisms have been designed to improve resource utilization of communication systems [16]. In [17], a user-centered opportunistic offloading approach was proposed based on a network formation game, in which the users autonomously formed a cooperative network, and promised device-to-device (D2D) sharing with their adjacent users. A coalitional game framework was proposed in [18] to improve the performance of mobile data offloading in wireless mesh networks, by savings the base stations' power consumption and reimbursements for mesh users. A competitive game was established for the mobile computing offloading problem in [19], in which each user pursued to minimize its own energy consumption, and the game formulated was subject to the real-time constraints imposed by the job execution deadlines, user specific channel bit rates, and the competition over the shared communication channel. In [20], the Nash bargaining game combining with the group bargaining theory was analyzed for the mobile traffic offloading in heterogeneous networks (HetNets), in which the social welfare maximization and the fairness of resource sharing were both considered.

Auction has become an important and effective theory in network economics, which can model and analyze the resource supplying and requesting from the aspect of economic. In recent years, auction mechanisms have been introduced to deal with dynamic spectrum optimization, resource sharing, D2D communications and many other issues in wireless networks [13, 21–23]. To deal with the increasing of mobile data traffic, a novel spectrum sharing framework for the cooperation and competition between LTE and Wi-Fi was designed based on an effective auction model in [14]. In [15], a double auction based resource allocation was designed for mobile edge computing in industrial Internet of Things, in order to maximize the system efficiency, while meeting the desired economic properties. To increase the network capacity dynamically and adaptively, a reverse auction model was established to formulate the mobile data offloading problem in [24]. In [25], a hierarchical combinatorial auction was designed for the virtualization issues in 5G cellular networks, based on a truthful and sub-efficient resource allocation framework. In addition, different auction models, such as reverse auction, VCG auction, procurement auction, were introduced to D2D communications for cellular traffic offloading [26–28]. All these auction mechanisms above were modeled based on game theory, where all bidders submit their bids/prices/costs towards a single auctioneer which finally decides only one winner to provide or receive the offloading service. Such game-based models are designed to guarantee the system efficiency and ensure that bidders in the system report bids truthfully. However, these models fail to describe the traditional bidding process and realize an information of bidders' requirements sharing among the system, which results from the incentive capability of game-based auction models. A sequential-price based auction mechanism was designed in [29] to realize data offloading via D2D communications. However, this model still paid less attention to the analysis of auction process.

Moreover, as mentioned above, all these current studies focus on the mobile data offloading depending on applying third party infrastructures. The internal transaction among mobile users to realize data offloading has been hardly investigated. In this user-initiative data offloading, it is necessary to study the influence of human behaviors on data transaction performance. In our previous work, we did some studies focusing on such user-initiative data offloading based on auction in [10, 11] and [30], in which the movement of users was predicted according to the topology of network. However, the changing amount of mobile data can be sold and the auction process in a long time period were not considered. So in this work, the statistics feature of data requests, the rest amount of mobile data owned by data sellers are consider to optimize the efficiency and balancing of data allocation. In addition, for the data transaction system with multiple data sellers, the prediction information of data requester' movement among different data auctions is introduced into the data allocation design to optimize the system performance.

## 15.3  Data Allocation of Single Data Provider

In this section, a classic basic data auction, first established in [31, 32], will be introduced to model a series of successive transaction among a single data owner and multiple potential data buyers as bidders. In each of the successive transaction, the data owner operates an auction to sell a fixed size of its unconsumed and unnecessary data from its data plan. Different from game theory based auction models designed mainly for the truthful biding, this auction model is designed to describe a traditional and practical auction process. In a later section, this basic auction will be extended into a networked auction for the data transaction system with multiple data owners. Before proceeding further, we summarize the main notations used throughout the following sections in Table 15.1.

### 15.3.1  Basic Auction Mechanism

The process of the auction-based data transaction with a single data seller is shown in Fig. 15.1. In this work, we assume that the automatical operation of data auction can be realized through a special application installed in both of the data owners' and bidders' smartphones. This assumption is feasible since

**Table 15.1** List of main notations in the basic auction (B) and networked auction (N) formulation for data transaction

| Parameter | Definition |
|---|---|
| $\lambda$ | Bid arrival rate for data owner (B) |
| $\lambda_i$ | Bid arrival rate for the $i$th data owner (N) |
| $\delta$ | Increment of data bids |
| $h$ | Highest price that data requesters intend to pay |
| $r_c$ | Rate parameter of exponential distributed considering time for the data owner (B) |
| $r_{c,i}$ | Rate parameter of exponential distributed considering time for the $i$th data owner (N) |
| $r_s$ | Rate parameter of exponential distributed service time for the data owner (B) |
| $r_{s,i}$ | Rate parameter of exponential distributed service time for the $i$th data owner (N) |
| $v_0$ | Beginning state of data auction/starting price of data auction |
| $v_j$ | State of that $j$ data bids have arrived/price of the $j$th bid made by data requesters |
| $v_h$ | State that highest bid has been made/highest bid made by data requesters |
| $A_j$ | State that the $j$th bid is accepted |
| $n_i$ | Number of data requesters in auction $i$ (N) |
| $x_i$ | Price has been reached in auction $i$ (N) |
| $f_{i,j}$ | Probability of each bidder in auction $i$ gives a bid (N) |
| $C$ | Total amount of data left (B) |
| $D$ | Total days left to sell data (B) |
| $c_d/c_{id}$ | Amount of data to be sold on day $d$ (B) for data seller $i$ (N) |
| $M$ | Total time slots for a day (N) |
| $N$ | Number of data sellers (N) |
| $z_{im}$ | Data to be sold in time slot $m$ for seller $i$ (N) |

there are already many mature mobile applications which make the auction-based transaction become true, especially on some e-commerce platforms. In addition, user authentication can be implemented through these applications to guarantee the user trustworthiness, which can protect the data transaction from malicious attacks [33]. Moreover, resulting from the requirement of physical proximity to realize the Wi-Fi hot-spots transmission among mobile phones, we consider that the data auction and transaction process is implemented in a limited area, such as offices, halls, classrooms, etc., where data owners and requesters may settle for a while to operate such data transaction. Moreover, there are always a crowd of people gathered in such places, which results in a deterioration of Wi-Fi quality, and mobile users with data requesters may looking for an alternative and effective way to obtain data. Therefore, data owners can be also considered as feasible Wi-Fi resources to meet these data requests. Then we will introduce the elements and operations in the automatic data auction as follows.

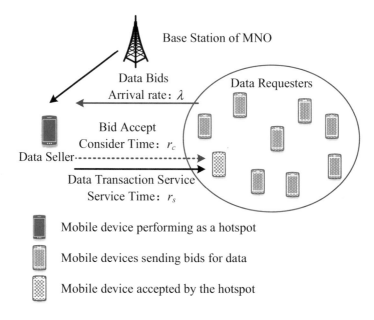

Fig. 15.1  Basic data auction with single auctioneer

1. *Data auctioneer:* The mobile user with unconsumed and needless mobile data. To meet data requirements requested by more mobile users, we assume that, instead of selling all data as a whole, data is cut into "data blocks" with a certain size, and in each round of auction, only one block of data can be sold. Then the data owner will perform as an auctioneer to operate a series of successive data auction.
2. *Data requesters (data bidders, potential data buyers):* The mobile users who have run out their mobile data and have the data requirements. In each round of data auction, all data requesters are potential data buyers and give bids for the data.
3. *Beginning of a basic auction:* The data owner starts an auction by unlocking the hotspot mode of its phone, sets the starting price $v_0$ of one block of data planned to sell, and then waits for bids. In this stage, $v_0$ can be modeled as the monetized compensation to cover the data owner's energy consumption when turning on its hot-spot feature.
4. *Bid arrivals:* The personal mobile business for a single user arrives according to a Poisson process with a certain rate, and therefore the average time between successive arrivals is the reciprocal of the arrival rate [34–36]. During the data auction, every time when the business arrivals to the phone of a data requester, a data bid will be triggered. We assume that the data bid providing is public information which can be observed by both the data owner and other data

requesters.[2] The business arrivals for different mobile users are independent and identically distributed (i.i.d.), so that the bid arrivals for the data owner can be still regarded as a statistic process obeying a Poisson distribution with arrival rate $\lambda$. In a single round of auction, $\lambda$ can be considered constant. If a bid is not accepted by the data owner, then the next bid will increase the value of the offer by fixed $\delta$. In addition, potential buyers will stop increasing the price of bid as long as the highest price $h$, at which data requesters intend to pay, is reached. Consider that data requesters are rational, which means that they will not buy the data from the auctioneer at a price higher than the market price. Then the value of $h$ can be set as a constant lower than the market price of data.

5. *Auctioneer decisions:* After each bid arrives, the data owner waits for a random "considering time" to determine whether to accept the current bid or not. Assume that the consider time has an exponential distribution with average $r_c^{-1}$ and the memoryless property. If the next bid arrives before the end of the considering time, then this considering process is repeated for this new bid. On the contrary, if the considering time expires and still no new bid arrives, the data owner will accept the latest data requester's offer, allow it to access into the hotspot and complete the data transaction with this successful data bidder. Considering the limited transmission capacity of the data owner, this work assumes that only the winning bidder can receive the data transaction service in a single round of auction to guarantee the transmission quality, although a hotspot can be simultaneously accessed by multiple users. Due to the memoryless property of the considering time, the potential data buyers cannot use the ongoing observations of considering time to give bids. Furthermore, the remaining considering time at any time point after an arrived bid has the same distribution as the initial considering time.

6. *Data transaction procedure:* The data transaction will last a "service time" before starting a new round of data auction by the data owner. The service time is modeled as an exponentially distributes time with rate $r_s$, and is i.i.d. and memoryless in different rounds of auctions.

*Remark 15.1* According the "auctioneer decision" step, we notice that if the data owner decides to wait a long time for the next bid, it might have a chance to get a higher-price offer, but the cost is a long time consumption. Conversely, short "considering time" will lead to a frequently repeated auctions, in each of which the data owner tends to get a low-price offer due to its weak patience. So for the data seller, how to select appropriative "considering time" to optimize the income, specifically, the income per unit time that the auctions bring to the data owner? Furthermore, the willing of selling and buying data of the data owner and requesters is always changing over days, as explained in Sect. 15.1.1.3. So how to allocate the amount of rest redundant data to be sold in different days before the end of a

---

[2] Concerning the privacy issues, we consider that only the prices offered by the bid can be shared. Meanwhile, the personal information of data bidders should be preserved or provided anonymously.

month, plays an important role when maximizing the income of the data owner and satisfying demands of data requesters. Next, we will pay attention to the problems above and design the data allocation mechanisms to optimize the performance of the basic auction-base data transaction system.

### 15.3.2  Data Allocation for Single-Auctioneer Transaction

The mathematic model and system performance of the basic auction model have been well established and analyzed in [37], in which many important economical characteristics of the basic auction model are derived and provided with closed-form expressions. In this section, we will first introduce this auction model into the data transaction system, and summarize some of important results obtained in [37] as Lemma 15.1. Then based on these theoretic results, we will design some efficiency and request aware data allocation mechanisms for the single-auctioneer data transaction in the later part of this section.

**Lemma 15.1** *In a data auction system with only one data auctioneer, the starting price of the data is set as $v_0$. The data bids arrive as a sequence of Poisson arrivals with arrival rate $\lambda$, and every bid increases the value of offer by $\delta$. The data requesters stop providing bids when the offer price reaches $h$. The data seller accepts the bid after a considering time, which follows an exponential distribution with average $r_c^{-1}$, and starts a new round of data auction after the service time, which also follows an exponential distribution with average $r_s^{-1}$. Then the average income of the data owner from a single round data auction is*

$$E_I = \sum_{j=1}^{h} (v_0 + j\delta)\, P_a\,(j) = v_0 + \delta \cdot \frac{1 - \rho^h}{1 - \rho}, \tag{15.1}$$

*where $\rho = \frac{\lambda}{\lambda + r_c}$. The total average time that every round of data auction lasts is*

$$T = \lambda^{-1} + r_c^{-1} + r_s^{-1}. \tag{15.2}$$

*The average income per unit time for the data owner is*

$$E_I^0 = \frac{E_I}{T} = \left(\lambda^{-1} + r_c^{-1} + r_s^{-1}\right)^{-1} \cdot \left(v_0 + \delta \cdot \frac{1 - \rho^h}{1 - \rho}\right). \tag{15.3}$$

*To be general, let $v_0 = 0$ and $\delta = 1$, then the average income per unit time can be gotten as*

$$E_I^G = \frac{E_I}{T} = \left(\lambda^{-1} + r_c^{-1} + r_s^{-1}\right)^{-1} \cdot \frac{1 - \rho^h}{1 - \rho}. \tag{15.4}$$

***Proof*** See [37].

Consider that the total amount of data left is $C$, and the data seller plans to sell all of its data in the last $D$ days of the month. Let $c_d$ $(d = 1, 2, \cdots, D)$ denote the amount of data planned to be sold on the $d$th day. As assumed previously, in each round of data auction, only one-unit amount of data can be sold. Then on the $d$th day, $c_d$ rounds of data auction are needed for the data auctioneer to sell the amount of $c_d$ data. We use $\lambda(d)$ to denote the arrival rate of data bid on the $d$th day, and $r_c^{-1}(d)$ to denote the average considering time of the data seller on the $d$th day. As mentioned previously, the urgency of selling and buying data from the data provider and requesters, respectively, change over time. Therefore, we assume that $\lambda(d)$ and $r_c(d)$ satisfy the following settings:

$$\lambda(d_1) \leq \lambda(d_2), \ \forall d_1 < d_2; \tag{15.5a}$$

$$r_c(d_1) \leq r_c(d_2), \ \forall d_1 < d_2, \tag{15.5b}$$

which imply that the closer to the end of a month, the more frequently the data requesters make bids, as well as the data seller accepts the offers.

According to Lemma 15.1, the average income of the data auctioneer from a single round of data auction on the $d$th day is

$$E_I(d) = \frac{1 - \rho^h(d)}{1 - \rho(d)}, \ d = 1, 2, \cdots, D, \tag{15.6}$$

where $\rho(d) = \lambda(d)/[\lambda(d) + r_c(d)]$. Then if all allocated data is sold, the total expected income can be achieved is

$$E(d) = c_d \cdot \frac{1 - \rho^h(d)}{1 - \rho(d)}, \ d = 1, 2, \cdots, D. \tag{15.7}$$

To maximize the total income of $D$ days, we establish the following income maximization problem for the data allocation.

$$\max \ \sum_{d=1}^{D} c_d \cdot \frac{1 - \rho^h(d, c_d)}{1 - \rho(d, c_d)}, \tag{15.8a}$$

$$s.t. \ \sum_{d=1}^{D} c_d \leq C, \tag{15.8b}$$

$$c_d \leq C - \gamma(D - d), \ \forall d = 1, 2, \cdots, D. \tag{15.8c}$$

In (15.8a),

$$\rho(d, c_d) = \frac{\lambda(d)}{\lambda(d) + r_c(d, c_d)}. \tag{15.9}$$

Constraint (15.8c) indicates that the total amount of data allocated to be sold in $D$ days from $d = 1$ to $d = D$ should not exceed the amount of total data $C$ left on the first day ($d = 1$). In constraint (15.8b), $\gamma > 0$ is set as a constant to denote the amount of data consumed by the data owner every day. We can also consider $\gamma$ as the amount of data to be reserved for the seller's own data demands. To guarantee that the data owner will have plenty data to consume in the following $D$ days after allocating its data, the amount of $\gamma (D - d)$ data needs to be reserved on the $d$th day. Therefore, constraint (15.8b) provides the upper limit of the amount of data allocated on the $d$th day.

### 15.3.2.1   Efficiency Aware Data Allocation

According to (15.8), we can notice that if $E_I (d)$, the average income of the data auctioneer from a single round auction, is high, the data allocation mechanism formulated by this income maximization problem will allocate more data to that day to maximize the total expected income the data auctioneer. With a fixed considering time, selling more data means that the data auction will last longer according to (15.2), which reduces the efficiency of data transaction. When the data auctioneer anticipates higher efficiency as well as an optimized income, it is necessary to design a data allocation method which can achieve a tradeoff between the total income and time cost. To realize this tradeoff, we design an efficiency-aware data allocation (EADA), in which the considering time, $r_c$ in (15.9), is modified according to

$$r_c^{EADA} (d, c_d) = r (d) [1 + \varphi_1 (d, c_d)], \tag{15.10}$$

where

$$\varphi_1 (d, c_d) = 1 - e^{-\left[\frac{c_d}{C - \gamma(D-d)}\right]^2}. \tag{15.11}$$

According to the definition in (15.11), $r_c^{ECDA}$ in (15.10) is an increasing function of the allocated amount of data to sell on the $d$th day, $c_d$. In other words, when the amount of allocated data is large, the EADA mechanism will adjust the seller's considering time to improve the data transaction efficiency and increase the average income per unit time of the data seller. On the contrary, when the amount of allocated data is little on a day, the considering time tends to be longer to increase the expected income of a single round of auction. The upper limit of the allocated data, $C - \gamma (D - d)$, performs as a control factor to modify the increasing speed of the bid acceptance rate with increasing $c_d$. In addition, $r (d)$ in (15.11) is the original rate of bid acceptance, which has the property given by (15.5).

### 15.3.2.2  Efficiency and Request Aware Data Allocation

Through (15.6) and (15.9), we can obtain the first order partial derivative of $E_I$, the average income of a round of data auction, with respect to variable $\lambda$: $\partial E_I / \partial \lambda > 0$. This result holds for both mechanisms of original data allocation and EADA. Therefore, a low rate of data bid will reduce the income of the data auctioneer. By the both of the data allocation methods above, little or even no data will be allocated to days with small $\lambda$, especially when the data bid arrival rate is smaller than the rate of bid acceptance. In other words, the requests from data bidders on these days are hardly met if the bid arrival rate is small, which may also result from that there are not too many data requesters. To meet the data requests in days with small $\lambda$, we design an efficiency and request aware data allocation (ERADA) mechanism, in which the considering time is adjusted to fit the data bid arrival rate according to

$$r_c^{ERADA}(d, c_d) = r(d)[1 + \varphi_1(d, c_d)]\varphi_2(\lambda(d), r(d)), \qquad (15.12)$$

where $r(d)$ is defined similar to EADA, $\varphi_1(d, c_d)$ is defined as (15.11), and $\varphi_2(\lambda(d), r(d))$ is obtained by

$$\varphi_2(\lambda(d), r(d)) = e^{\min\left\{\frac{\lambda(d) - r(d)}{r(d)}, 0\right\}}. \qquad (15.13)$$

According to (15.13), when $\lambda(d) < r(d)$, then we can get

$$\varphi_2(\lambda(d), r(d)) = \exp\left\{\frac{\lambda(d) - r(d)}{r(d)}\right\} < 1. \qquad (15.14)$$

Consequently, the considering time to accept the data bid can be expended. Then the expected incomes of the data seller increase potentially, and more data will be allocated to the corresponding days. On the other hand, when $\lambda(d) \geq r(d)$, some data can be allocated to the corresponding days through applying EADA. Therefore, $r_c$ remains the same as EADA, i.e., $\varphi_2(\lambda(d), r(d)) = 1$.

## 15.4  Networked Auction Model for Data Transaction with Multiple Auctioneers

In the previous section, we establish and analyze the data allocation problem for a single data auctioneer. According the proposed mechanisms of EADA and ERADA, the expected incomes of the data seller can be optimized with high efficiency, and the data requests from data buyers can be satisfied as much as possible. By applying EADA and ERADA, the data seller can make decisions that how to allocate its rest data to remaining days before the end of a month. The next problem is that after the amount of data to be sold in a single day has been decided, then how to operate data auctions to achieve further optimization of the daily income?

As analyzed previously, the proposed income maximization problems in (15.8) is based on the fact that the needs of data selling and buying vary over time within a month. However, in a certain day $d$, specifically, during the period of data transaction on this day, the rate of arrival data bids remains relatively stable, i.e., with $\lambda(d)$. In addition, by the designed EADA and ERADA, the average considering time can be optimized according to (15.10) and (15.12), respectively, when $c_d$ has been determined. In a data transaction system with only one data auctioneer, this auctioneer can operate the basis data auction introduced in Sect. 15.3.1, and the expected maximum income can be achieved when applying EADA and ERADA. However, when there are multiple data auctioneers, who share the same community of potential data requesters, then some system status, such as the number of data requesters in a single auction, the data bid arrival rates, etc., may change if the mobility of data requesters among different auctions is allowed. Therefore, it is necessary to analyze the data auction and allocation mechanisms for a networked data transaction system, in which more than one data owners are planning to sell their fixed amounts of data. Fortunately, with assistance of current mobile social networks, some system status in auctions operated by different data owners can be shared among users, i.e., referring to both the data sellers and requesters, through the social platforms. This status information can be very helpful for the further performance improvement. Consequently, how to model the networked data transaction system and how to make advantage of the system status information to design an efficient data auction and allocation mechanism become essential problems to be studied.

In this section, we extend the basis data transaction model into the networked system to discuss the data transaction processes operated by multiple data auctioneers. The networked data transaction system model and the mobility model are shown in Fig. 15.2. A system-status-aware mobility model is designed for data requesters. Then we analyze the stationary probabilities of the networked auction system for the performance estimation. Furthermore, to maximize the income of every data auctioneer, three data allocation mechanisms are proposed in this part.

### 15.4.1 Networked Auction Model

The classic networked auction model has been formulated in [37]. In this part, we first introduce the established mathematical model in [37] as follows.

Consider that there are $N$ data suctions operated by $N$ data sellers in the system at the same time. These sellers are numbered by $i = 1, 2, \cdots, N$. Let $\mathbf{n}(t) = \{n_1(t), n_2(t), \cdots, n_N(t)\}$ denote the numbers of potential data buyers in auction $i$ at time $t$, and $\mathbf{X}(t) = \{x_1(t), x_2(t), \cdots, x_N(t)\}$ denote the price has been reached in auction $i$ at time $t$. Similar to the basic auction, $x_i(t) \in \{v_0, v_1, \cdots, v_{h_i}\}$, and $v_{h_i}$ is the highest price that data requesters intend to pay in auction $i$. Then the

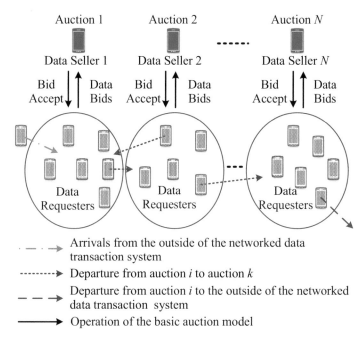

Auction 1            Auction 2                    Auction $N$

Data Seller 1        Data Seller 2                Data Seller $N$

Bid    Data          Bid    Data                  Bid    Data
Accept Bids          Accept Bids                  Accept Bids

Data                 Data                         Data
Requesters           Requesters                   Requesters

— · → Arrivals from the outside of the networked data transaction system

········→ Departure from auction $i$ to auction $k$

— — → Departure from auction $i$ to the outside of the networked data transaction system

——→ Operation of the basic auction model

**Fig. 15.2** Networked data auction system and the mobility model

state of the networked data auction system can be described as the pair of vector $(\mathbf{n}(t), \mathbf{X}(t))$. In each of these $N$ auctions, the auction rule and strategy are similar to the basic auction. We consider that the bid arrival rate in each auction is dependent on the price $x_i(t)$ and the number of bidders $n_i(t)$ in this auction. In addition, for current achieved prices $v_j$ $(j = 1, 2, \cdots, h)$, there are at least one potential data buyer has given a bid and it will not give the next bid. Contrarily, if current price is $v_0$, which means the beginning of a new round of auction, then each of the $n_i(t)$ potential buyers are allowed to give the next bid. As a consequence, we define the bid arrival rate in auction $i$ as

$$\lambda_i(n_i, v_j) = (n_i - 1)\lambda_i f_{i,j}, \tag{15.15a}$$

$$\lambda_i(n_i, 0) = n_i \lambda_i, \tag{15.15b}$$

where $f_{i,j} = P(v_j < v_{h_i})$, and $\lambda_i > 0$ is the rate of that each data bidder in auction $i$ gives a bid. Moreover, similar to the basic auction model, we set $r_{c,i}^{-1}$ to be the average considering time of auctioneer $i$, which is an i.i.d. random variable having an exponential distribution.

## 15.4.2   Mobility Model

In this part, we will design a mobility model for the data requesters based on the mobile bidder model (MBM) established in [37].

In a networked data transaction system operated by $N$ data auctioneers, we consider that the data requesters can enter and leave the whole system, as well as moving from one auction to another in the system. Such movement does not refer to the physical trajectory as analyzed in [12, 38], but is realized through a special social application installed in both of the data owners' and bidders' smartphones. Then how to design a mobility model to describe the moving of these potential data buyers, is an important issue to keep balance of the number of participants in each auction, optimize the efficiency of the system and maximize the expected income of each data sellers. In order to achieve these objectives above, we introduce some prediction-based factors that may affect the user behaiours, and then propose a system-status-aware mobility model for the networked data transaction, which can reflect the mobile users' rationality and further improve the performance of the networked data transaction.

Next, we will formulate the mobility of data requesters. Consider that in auction $i$ at time $t$, the number of potential data buyers is $n_i$ and the current achieved bid is $v_i$ ($v_0 \leq v_i \leq v_{h_i}$). Then the dynamic parameters of the mobile model are defined as follows.

1. **Arrivals from the outside of the system:** Data requesters arrive into auction $i$ from the outside of the networked data transaction system according to a Poisson process with arrival rate $\lambda_i^0$.
2. **Departure from the $i$th auction:** Consider that the bidder providing the current highest price for the data cannot leave auction $i$ until that the next bid arrives or the data seller decides to accept its bid. Moreover, consider the situation that when a new round of the auction is operated by auctioneer $i$, all the data bidders are allowed to depart from this auction. The rate of departure from auction $i$ can be given as follows

$$\mu_i\left(n_i, v_j\right) = (n_i - 1)\,\mu_i, \tag{15.16a}$$

$$\mu_i\left(n_i, 0\right) = n_i\,\mu_i, \tag{15.16b}$$

where $\mu_i > 0$ is the departure rate of each data bidder in auction $i$. The definitions above are similar to the rate of bid arrivals formulated in (15.15).

Next, we consider the two possible actions that a data bidder might take after it departs from a data auction:

• *Departure from auction $i$ to the outside of the system:*

Denote $P_{iD}$ as the probability that the data bidders in auction $i$ leave the entire networked transaction system. In addition, $P_{iD}$ can be estimated by the status changing in each data auction, and then published as a public information for all participants in the networked system.

- *Departure from auction i to auction k:*

In the auction-based networked data transaction system, mobile users with data requests are allowed to shift from one auction to another. We assume that the amount of the rest data to be sold in the day can be updated and observed by other auction participants in the networked system. This assumption is reasonable and feasible because this kind of information can be provided by data owners, and broadcasted to data requested through the relevant mobile applications. Then the data requesters transfer among different auctions according to the amount of remaining data to be sold in these auctions. We represent the transition probability from auction $k$ to auction $i$ with $P_{ki}$

$$P_{ki} = (1 - P_{kD}) \cdot \frac{c_{i,rest}}{\sum_{j=1}^{N} c_{i,rest}}, \quad i, k = 1, 2, \cdots, N, \quad (15.17)$$

where $c_{i,rest}$ denotes the amount of the rest data to be sold for auction $i$. According to (15.17), data requesters more likely tend to participate the data auction operated by the data owners with more remaining data, which can give them more chances and a higher probability to get the data quickly and successfully. In addition, the definition of transition probability in (15.17) essentially guarantees

$$\sum_{k=1}^{N} P_{ik} + P_{iD} = 1. \quad (15.18)$$

### 15.4.3  Expected Income of Networked Systems

To optimize the data auction performance and maximize the incomes of data auctioneers through effective data allocation, it is important to analyze the stationary distribution and the expected income of the networked data transaction system. The stationary probabilities of the number of data requesters in each auction have been analyzed in [37], the main results of which are summarized as Lemma 15.2. Then we will derive some stationary performance such as expected income for the data owners based on results obtained in [37].

**Lemma 15.2** *In a networked data auction system, data bidders arrive and depar-ture from auction $i$ with rate $\lambda_i$ and $\mu_i$, respectively. According to the mobility model of data requesters established in Sect. 15.4.2, the approximate stationary probabilities of data auctioneer $i$ ($i = 1, 2, \cdots, N$) and the stationary probability* *of the networked data transaction system are given by*

$$\pi\left(n_i\right) = \frac{\psi_i^{n_i} e^{-\psi_i}}{\psi_i\left(n_i - 1\right)!}, \tag{15.19a}$$

$$\pi\left(\mathbf{n}\right) \approx \prod_{i=1}^{N} \frac{\psi_i^{n_i} e^{-\psi_i}}{\psi_i\left(n_i - 1\right)!}, \tag{15.19b}$$

*respectively, where* $\psi_i = \varphi_i / \mu_i$, *and* $\varphi_i$ *($i = 1, 2, \cdots, N$) are the solutions of the following linear equations:*

$$\varphi_i = \lambda_i^0 + \sum_{k=1}^{N} \varphi_k P_{ki}. \tag{15.20}$$

*When the bid arrivals and the data transactions are very frequent, which means that for all* $i = 1, 2, \cdots, N$, $\mu_i \ll r_{c,i}$, *then* $\forall n_i > 0$, $k_i > 0$, *the stationary solution* $\pi\left(\mathbf{X} \,|\mathbf{n}\right)$ *is given by*

$$\pi\left(\mathbf{X} \,|\mathbf{n}\right) \approx \prod_{i=1}^{N} \pi_i\left(x_i \,|n_i\right), \tag{15.21}$$

*where*

$$\pi_{j|n_i}^i = \pi_{0|n_i}^i \prod_{l=1}^{j} \frac{\lambda_i\left(n_i - 1\right) f_{i,l-1}}{r_{c,i} + \lambda_i\left(n_i - 1\right) f_{i,l}}, \tag{15.22a}$$

$$\pi_{0|n_i}^i = \left[1 + \sum_{j=1}^{h_i} \prod_{l=1}^{j} \frac{\lambda_i\left(n_i - 1\right) f_{i,l-1}}{r_{c,i} + \lambda_i\left(n_i - 1\right) f_{i,l}}\right]^{-1}. \tag{15.22b}$$

**Proof** See [37].

Based on these results in Lemma 15.2, we further derive the average of the income per unit time for every data owner in the system, and we the results in our Theorem 15.1.

**Theorem 15.1** *In a networked data auction system with $N$ data auctioneers, data bidders arrive and departure from auction $i$ with rate $\lambda_i$ and $\mu_i$, respectively. Consider the situation that the bid arrivals and the data transactions are very frequent. According to the mobility model of data requesters established in Sect. 15.4.2,* *expected income $E_{i,n_i}$ for auction $i$ when there are $n_i$ data requesters in this auction is given by*

$$E_{i,n_i} = \sum_{j=1}^{h_i} j \, P_a^i \, (j \,|n_i \,), \tag{15.23}$$

*where*

$$P_a^i \, (j \,|n_i \,) = \frac{r_{c,i}}{\lambda_i n_i} \prod_{l=1}^{j} \frac{\lambda_i \, (n_i - 1) \, f_{i,l-1}}{r_{c,i} + \lambda_i \, (n_i - 1) \, f_{i,l}}. \tag{15.24}$$

*The average of the income per unit time for the data owner $i$ is given by*

$$E_i^0 = \sum_{n_i=1}^{\infty} \frac{\psi_i^{n_i} e^{-\psi_i}}{\psi_i \, (n_i - 1)!} \frac{r_{c,i} \sum_{j=1}^{h_i} j \prod_{l=1}^{j} \frac{\lambda_i (n_i-1) f_{i,l-1}}{r_{c,i} + \lambda_i (n_i-1) f_{i,l}}}{1 + \sum_{j=1}^{h_i} \prod_{l=1}^{j} \frac{\lambda_i (n_i-1) f_{i,l-1}}{r_{c,i} + \lambda_i (n_i-1) f_{i,l}}}. \tag{15.25}$$

**Proof** The local stationary equations of the networked auction system can be given by

$$\lambda_i \, (n_i, 0) \, \pi_{0|n_i}^i = \lambda_i \, (n_i, 0) \sum_{j=1}^{h_i} \pi_{A_j|n_i}^i = \left( r_{c,i} + \lambda_i \, (n_i, v_1) \right) \pi_{1|n_i}^i, \tag{15.26a}$$

$$\lambda_i \left( n_i, v_{j-1} \right) \pi_{j-1|n_i}^i = \left( r_{c,i} + \lambda_i \left( n_i, v_j \right) \right) \pi_{j|n_i}^i, \; j = 1, 2, \cdots, h_i - 1, \tag{15.26b}$$

$$\lambda_i \, (n_i, v_{h-1}) \, \pi_{h-1|n_i}^i = r_{c,i} \pi_{h|n_i}^i, \tag{15.26c}$$

$$r_{c,i} \pi_{j|n_i}^i = \lambda_i \, (n_i, 0) \, \pi_{A_j|n_i}^i, \; j = 1, 2, \cdots, h_i. \tag{15.26d}$$

In (15.26a), $A_j$ denotes the state that the $j$th bid is accepted.
According to (15.26d) and (15.22b), we get

$$\pi_{A_j|n_i}^i = \frac{r_{c,i} \pi_{j|n_i}^i}{\lambda_i \, (n_i, 0)} = \frac{r_{c,i}}{\lambda_i n_i} \pi_{0|n_i}^i \prod_{l=1}^{j} \frac{\lambda_i \, (n_i - 1) \, f_{i,l-1}}{r_{c,i} + \lambda_i \, (n_i - 1) \, f_{i,l}}, \tag{15.27}$$

and then

$$P_a^i \, (j \,|n_i \,) = \frac{\pi_{A_j|n_i}^i}{\sum_{k=1}^{h_i} \pi_{A_k|n_i}^i} = \frac{r_{c,i}}{\lambda_i n_i} \prod_{l=1}^{j} \frac{\lambda_i \, (n_i - 1) \, f_{i,l-1}}{r_{c,i} + \lambda_i \, (n_i - 1) \, f_{i,l}}. \tag{15.28}$$

When there are $n_i$ potential data buyers in this auction, expected income $E_{i,n_i}$ and the total average time $T_{i,n_i}$ that every round of data auction lasts for auction $i$ is

$$E_{i,n_i} = \sum_{j=1}^{h_i} j P_a^i (j \mid n_i ), \tag{15.29}$$

$$T_{i,n_i} = \frac{1}{n_i \lambda_i} \left( 1 + \sum_{j=1}^{h_i} \prod_{l=1}^{j} \frac{\lambda_i (n_i - 1) f_{i,l-1}}{r_{c,i} + \lambda_i (n_i - 1) f_{i,l}} \right), \tag{15.30}$$

respectively. Consequently, we obtain the average of the income per unit time for the data owner $i$ as:

$$E_{i,n_i}^0 = \frac{E_{i,n_i}}{T_{i,n_i}} = \frac{r_{c,i} \sum_{j=1}^{h_i} j \prod_{l=1}^{j} \frac{\lambda_i (n_i-1) f_{i,l-1}}{r_{c,i}+\lambda_i (n_i-1) f_{i,l}}}{1 + \sum_{j=1}^{h_i} \prod_{l=1}^{j} \frac{\lambda_i (n_i-1) f_{i,l-1}}{r_{c,i}+\lambda_i (n_i-1) f_{i,l}}}. \tag{15.31}$$

Then according to (15.19), the average of the income per unit time for data owner $i$ is given by

$$E_i^0 = \sum_{n_i=1}^{\infty} E_{i,n_i}^0 \pi (n_i)$$

$$= \sum_{n_i=1}^{\infty} \frac{\psi_i^{n_i} e^{-\psi_i}}{\psi_i (n_i - 1)!} \frac{r_{c,i} \sum_{j=1}^{h_i} j \prod_{l=1}^{j} \frac{\lambda_i (n_i-1) f_{i,l-1}}{r_{c,i}+\lambda_i (n_i-1) f_{i,l}}}{1 + \sum_{j=1}^{h_i} \prod_{l=1}^{j} \frac{\lambda_i (n_i-1) f_{i,l-1}}{r_{c,i}+\lambda_i (n_i-1) f_{i,l}}}. \tag{15.32}$$

This completes the proof of Theorem 15.1.

### 15.4.4   Data Allocation for Networked Data Transaction

Based on the obtained analytical results given in the last section, we will design some efficient data allocation mechanisms for the networked data transaction system in the following part.

Consider a networked data transaction system with $N$ data sellers, each of them needs to sell all allocated data in a certain duration $[0, T]$. Data requesters are allowed to enter, departure from any of $N$ data auctions according to the mobility model introduced in Sect. 15.4.2.

For a certain day $d = 1, 2, \cdots, D$, vector $\mathbf{c} = \{c_{1d}, c_{2d}, \cdots, c_{id}, \cdots, c_{Nd}\}$, determined by the EADA or ERADA proposed previously, denotes the amounts of data allocated to be sold for each of the $N$ data sellers. In following work, we apply the ERADA mechanism and the rate of considering time of data auctioneer $i = 1, 2, \cdots, N$ modified by (15.12)–(15.14). Then for data seller $i$, we have

$$r_{c,i}(d, c_{id}) = r_i(d)[1 + \varphi_1(d, c_{id})]\varphi_2(\lambda(d), r_i(d)), \ \forall i = 1, 2, \cdots, N.$$

$$(15.33)$$

When the requests of data is far more than the data can be provided, then $\varphi_2(\lambda(d), r(d)) = 1$, and the rate of considering time for each data auctioneer $i$ is

$$r_{ci}(d, c_{id}) = r_i(d)[1 + \varphi_1(d, c_{id})].$$ $$(15.34)$$

Consider that the duration $[0, T]$ is slotted into $M$ time slots, each of them is indexed by $m = 1, 2, \cdots, M$. We assume that in every time slot, the number of data bidders in every auction is stable. Then let $\mathbf{n}(m) = \{n_1(m), n_2(m), \cdots, n_N(m)\}$ denote the number of data bidders in auction $i$ at time slot $m$, $\forall i$, $m$. For each time slot $m$, we express the allocated amount of data to be sold for every data seller $i$ in the system by $\mathbf{z_m} = \{z_{1m}, z_{2m}, \cdots, z_{Nm}\}$.

### 15.4.4.1 Non-cooperative Distributed Data Allocation (NDDA)

According to Theorem 15.1, when there are $n_i(m)$ data requesters in auction $i$ at time slot $m$, and the amount of data to sold is $z_{im}$, then the expected income of data auctioneer $i$ can be given by

$$z_{im}E_{i,n_i}(m) = z_{im}\sum_{j=1}^{h_i} jP_a^i(j|n_i(m))$$

$$(15.35)$$

$$= z_{im}\sum_{j=1}^{h_i} j\frac{r_{c,i}}{\lambda_i n_i(m)}\prod_{l=1}^{j}\frac{\lambda_i(n_i(m)-1)f_{i,l-1}}{r_{c,i}+\lambda_i(n_i(m)-1)f_{i,l}}.$$

Then at every time slot $m = 1, 2, \cdots, M$, each data requester solves the following optimization problem to maximize the expected income of current time slot:

$$\max f_{z_{im}|\mathbf{z_m}} = z_{im}E_{i,n_i}(m),$$ $$(15.36a)$$

$$s.t. \ z_{im} \geq \min\left\{c_{id} - \sum_{t=1}^{m-1} z_{it}, z_{\min}\right\}, \ \forall i = 1, \cdots, N,$$ $$(15.36b)$$

$$z_{im} \leq \min \left\{ c_{id} - \sum_{t=1}^{m-1} z_{it}, z_{max} \right\}, \forall i = 1, \cdots, N, \qquad (15.36c)$$

$$\sum_{i=1}^{N} z_{im} \leq z_{ch}. \qquad (15.36d)$$

In (15.36b) and (15.36c),

$$z_{min} = \frac{1}{M} \cdot \max \{ c_{1d}, c_{2d}, \cdots, c_{id}, \cdots, c_{Nd} \}, \qquad (15.37a)$$

$$z_{max} = \kappa z_{min}, \quad \kappa > 1. \qquad (15.37b)$$

*Remark 15.2* In constraint (15.36b), lower bound $z_{min}$ shown in (15.37a) ensures that the data owner with the most amount of data to sold on the current day can sold out all of the data before time slot $m = M$ ends. $c_{id} - \sum_{t=1}^{m-1} z_{it}$ in (15.36b) is provided for the case that the remaining data of a auctioneer at time slot $m$ is less than $z_{min}$, then all the remaining data needs to be sold during time slot $m$. The low bound of the allocated amount of data in each time slot can keep the data allocation mechanism efficient. On the other hand, due to the duration of every time slot is limited, the amount of data can be transacted in one time slot is constrained by an upper bound, which is denoted by $z_{max}$ in constraints (15.36c) and (15.37b). In addition, $c_{id} - \sum_{t=1}^{m-1} z_{it}$ in (15.36c) plays a constraining role when the rest data is less than $z_{max}$. Constraint (15.36d) is determined by the channel capacity.

### 15.4.4.2   Prediction-Based Cooperative Distributed Data Allocation (PCDDA)

As mentioned above, the amount of the rest data of different data auctioneers to be sold is accessible information for all data requesters having arrived and planning to enter into the system. According to the mobility model introduced in Sect. 15.4.2, data requesters in auction $i$ will move to auction $k$ with probability $P_{ik}$, which is determined by the amount of remaining data to be sold in auction $k$, i.e., $c_{k,rest}$. In other words, the number of data requester in every auction during the following time slots can be predicted in sense of probability, according to the public information of the rest amount of data. Then by applying the results in Lemma 15.2 and Theorem 15.1, the expected income of the current time slot can be predicted by (15.23) for a fixed number of data requesters. In addition, the potential average income of the next time slot can also be predicted through (15.25) by predicting mobility trend of data requesters. Considering this prediction information, data owners can make better decisions on how much data to sell in the current time slot for maximizing the total income of $M$ time slots.

Consider the previous assumption that the number of data requesters in each auction does not change. Then we assume that the bidder's transition probability from auctions $k$ to $i$ after time slot $m$, $P_{ki}\left(m^+\right)$, is determined by the rest amount of data after finishing the allocation of $m$ time slots:

$$P_{ki}\left(m^+\right) = (1 - P_{kD}) \cdot \frac{c_{id} - \sum_{t=1}^m z_{it}}{\sum_{j=1}^N \left(c_{jd} - \sum_{t=1}^m z_{jt}\right)}. \tag{15.38}$$

If $P_{kD} = P_D, \forall k = 1, 2, \cdots, N$, then

$$P_i\left(m^+\right) \triangleq P_{ki}\left(m^+\right) = (1 - P_D) \cdot \frac{c_{id} - \sum_{t=1}^m z_{it}}{\sum_{j=1}^N \left(c_{jd} - \sum_{t=1}^m z_{jt}\right)}. \tag{15.39}$$

According to (15.20), we can get

$$\boldsymbol{\varphi}(m) = \boldsymbol{\lambda}^0 + \mathbf{P}\left(m^+\right)\boldsymbol{\varphi}(m), \tag{15.40}$$

where

$$\boldsymbol{\varphi}(m) = [\varphi_1(m) \quad \varphi_2(m) \quad \cdots \quad \varphi_N(m)]^{\mathrm{T}}, \tag{15.41a}$$

$$\boldsymbol{\lambda}^0 = [\lambda_1 \quad \lambda_2 \quad \cdots \quad \lambda_N]^{\mathrm{T}}, \tag{15.41b}$$

and

$$\mathbf{P}\left(m^+\right) = \begin{bmatrix} P_{11}\left(m^+\right) & P_{21}\left(m^+\right) & \cdots & P_{N1}\left(m^+\right) \\ P_{12}\left(m^+\right) & P_{22}\left(m^+\right) & \cdots & P_{N2}\left(m^+\right) \\ \vdots & \vdots & \ddots & \vdots \\ P_{1N}\left(m^+\right) & P_{2N}\left(m^+\right) & \cdots & P_{NN}\left(m^+\right) \end{bmatrix}$$

$$\triangleq \begin{bmatrix} P_1\left(m^+\right) \\ P_2\left(m^+\right) \\ \vdots \\ P_N\left(m^+\right) \end{bmatrix} \cdot \underbrace{[1 \ 1 \ \cdots \ 1]}_{N}. \tag{15.42}$$

Then solve (15.40) and we can get the solutions as

$$\varphi_i(m) = \lambda_i + \frac{\sum_{j=1}^N \lambda_j}{P_D} P_i\left(m^+\right), \quad i = 1, 2, \cdots, N. \tag{15.43}$$

Therefore, applying results in Lemma 15.2 and Theorem 15.1, the future expected income of the rest amount of data for data owner $i$ at time slot $m$ can be calculated by

$$E_i\left(m^+\right) \triangleq \left(c_{id} - \sum_{t=1}^{m} z_{it}\right) \sum_{n_i=1}^{\infty} E_{i,n_i}(m)\,\pi\,(n_i), \tag{15.44}$$

where $E_{i,n_i}(m)$ and $\pi(n_i)$ are obtained by (15.23) and (15.19), respectively, and $\psi_i$ in (15.19) is determined by $\varphi_i$ in (15.43).

Based on the analysis above, we consider that at each time slot, every data auctioneer $i$ $(i = 1, 2, \cdots, N)$ is willing to optimize its comprehensive income, which is composed of current income calculated through (15.35) and predictive future income determined by (15.44). Assume that every data auctioneer is selfish and intends to maximize its own total expected income of the $M$ time slots. Then we establish the following income maximization problem for each data auctioneer $i$ at time slot $m$ $(m = 1, 2, \cdots, M)$.

$$\max f_{z_{im}} = z_{im} E_{i,n_i}(m) + \omega^{M-m} E_i\left(m^+\right) \tag{15.45a}$$

$$s.t. \; z_{im} \geq \min\left\{c_{id} - \sum_{t=1}^{m-1} z_{it}, z_{\min}\right\}, \forall i = 1, \cdots, N, \tag{15.45b}$$

$$z_{im} \leq \min\left\{c_{id} - \sum_{t=1}^{m-1} z_{it}, z_{\max}\right\}, \forall i = 1, \cdots, N, \tag{15.45c}$$

$$\sum_{i=1}^{N} z_{im} \leq z_{ch}. \tag{15.45d}$$

In (15.45a), $\omega \in (0, 1]$ denotes the discount rate of the future income considered at the current time slot, which reflects the weight of future income considered in the current comprehensive income.

At the beginning of every time slot, each data owner publishes the rest amount of its data, observes the number of data requesters in its auction, and then solve the optimization problem in (15.45) to determine how much data to be sold in the current time slot.

### 15.4.4.3  Prediction-Based Centralized Data Allocation (PCDA)

According to PCDDA designed in the previous section, if every data auctioneer solves the optimization problem locally to maximize its own expected income instead of a central process, the optimal solution for each data auctioneer cannot ensure that constraint (15.45d) is always satisfied. In other words, the distributed mechanism may cause data auctioneers to fail to get the maximum income they anticipate. Concerning this issue, we design a prediction-based centralized data allocation (PCDA) mechanism to maximize the income of all data auctioneers. To achieve this centralized optimization, we assume that there is a data fusion center which operates the PCDA and determine how much data to be sold for every data seller in each time slot. Wish assistance of current mobile network platform, this assumption is rational and enforceable. The objective function of PCDA is shown as (15.46), and the constraints are the same as those of NDDA and PCDDA, i.e., (15.36b)–(15.36d) and (15.45b)–(15.45d), respectively.

$$\max \sum_{i=1}^{N} \left[ z_{im} E_{i,n_i} (m) + \omega^{M-m} E_i (m^+) \right].  \tag{15.46}$$

*Remark 15.3* According to objective function (15.46) and constrain (15.45d), PCDA tends to allocate more amounts of data to be sold to the data owners with more current expected income in a single round of data auction, which can increase the efficiency of the networked data transaction system.

## 15.5   Operation of Data Allocation for Data Transaction Systems

### 15.5.1   *Approximate Solution of Optimization Problems*

In the previous two sections, we establish the income maximization problems for the data allocation. For a large time scale, the data owner can decide how much data to be sold in the following days before the end of a month, by applying EADA and ERADA. Then for a smaller time scale, data owners can determine how to schedule the data auctions in a single day by considering the peer data auctioneers behaviours, through NDDA, PCDDA or PCDA. However, we can also notice that it is difficult to achieve the closed-form solutions for the optimization problems formulated as (15.8), (15.36), (15.45) and (15.46). To find the optimal solution approximately, an efficient and effective stochastic and cooperation-based optimization technique, called the cooperative particle swarm optimization (CPSO) algorithm [39], will be introduced to our work to solve the income maximization problems. As introduced in Chap. 7, the update process is operated as follows:

$$\begin{aligned} v_{ij} (t+1) = {} & w v_{ij} (t) + c_1 \zeta_{1i} (t) \left[ y_{ij} (t) - x_{ij} (t) \right] \\ & + c_2 \zeta_{2i} (t) \left[ \hat{y}_j (t) - x_{ij} (t) \right], \end{aligned} \tag{15.47}$$

$$\mathbf{x}_i (t+1) = \mathbf{x}_i (t) + \mathbf{v}_i (t+1),  \tag{15.48}$$

---

**Algorithm 7** CPSO algorithm [39].

---

**Initialization:**
 1: Create and initialize $n$ one-dimensional PSOs: $P_j$, $j = 1, 2, \cdots, n$;
 2: Define:
 3: $g(j, z) \equiv \left( P_1 \cdot \hat{\mathbf{y}}, P_2 \cdot \hat{\mathbf{y}}, \cdots, P_{j-1} \cdot \hat{\mathbf{y}}, z, P_{j+1} \cdot \hat{\mathbf{y}}, \cdots, P_n \cdot \hat{\mathbf{y}} \right)$;
 4: Iterations $T$.
 5: **for** $t \leq T$ **do**
 6:     **for** $j = 1, 2, \cdots, n$ **do**
 7:         **for** $i = 1, 2, \cdots, s$ **do**
 8:             **if** $f\left( g\left( j, P_j \cdot \mathbf{x}_i \right) \right) < f\left( g\left( j, P_j \cdot \mathbf{y}_i \right) \right)$ **then**
 9:                 $P_j \cdot \mathbf{y}_i = P_j \cdot \mathbf{x}_i$
10:             **end if**
11:             **if** $f\left( g\left( j, P_j \cdot \mathbf{y}_i \right) \right) < f\left( g\left( j, P_j \cdot \hat{\mathbf{y}} \right) \right)$ **then**
12:                 $P_j \cdot \hat{\mathbf{y}} = P_j \cdot \mathbf{y}_i$
13:             **end if**
14:         **end for**
15:         Update $P_j$ by PSO with (15.47) and (15.48).
16:     **end for**
17: **end for**

---

where $j = 1, 2, \cdots, s$, and $s$ is the swarm size. $\mathbf{x}_i = [x_{i1} \ x_{i2} \ \cdots \ x_{in}]$ is the current position in the search space, $\mathbf{v}_i = [v_{i1} \ v_{i2} \ \cdots \ v_{in}]$ is the current velocity, $\mathbf{y}_i = [y_{i1} \ y_{i2} \ \cdots \ y_{in}]$ is the local best position, $n$ is the number of particles, $c_1$ and $c_2$ are acceleration coefficients, and random sequences $\zeta_1$, $\zeta_{2i} \sim U(0, 1)$ [40].

In PSO, there is only one swarm with $s$ particles, which tries to find the optimal $n$-dimensional vector. While in CPSO, this $n$-dimensional vector is decomposed into $n$ swarms, each of which has $s$ particles. These $n$ swarms cooperatively optimize the one-dimensional vector. The main processes of CPSO are shown in Protocol 7.

## 15.5.2   Data Allocation for Data Transaction

In this part, we will introduce how to operate data allocation in different days and different time slots in one day for the auction-based data transaction system.

As mentioned previously, the elements of the networked data transaction, including data auctioneers, the number of data auctioneers, data requester arrival rates, etc., change over time. Therefore, a certain data owner cannot determine how to allocate its extra data into the rest days through a networked auction mechanism, according to the current data transaction network. In this work, we design a data allocation mechanism based on the basic data auction and networked data auction, for a large time scale (referring to days) and a small time scale (referring to time slots), respectively. Specifically, for every data owner who has extra data and plans to sell the data during the rest days before the month ends, it makes decision on how to allocate the data into different days according to the ERADA mechanism based on the basic data auction model. To achieve a maximum expected total income of

---

**Algorithm 8** Data allocation for data transaction

---

**Initialization:**
 1: Bid arrival rate: $\lambda(d)$;
 2: Original bid acceptance rate: $r(d)$;
 3: Number of days left before the end of the month: $D$.
 4: Number of time slots in a single day to operate data transaction: $M$.
 5:
 6: Each data seller $i$ operates the following processes:
 7: **for** $d = 1, 2, \cdots, D$ **do**
 8:   **if** $\lambda(d) - r_c(d)$ **then**
 9:     $\varphi_2(\lambda(d), r(d)) = \exp\left\{\frac{\lambda(d) - r_c(d)}{r_c(d)}\right\}$;
10:   **else**
11:     $\varphi_2(\lambda(d), r(d)) = 1$.
12:   **end if**
13:   Apply ERADA, solve it by CPSO and obtain optimal $c_{id}$ and $r_{c,i}$.
14:   **for** $m = 1, 2, \cdots, M$ **do**
15:     Recognize and establish the structure of the networked data transaction system;
16:     Submit the amount of the rest data $c_{id} - \sum_{t=1}^{m} z_{it}$;
17:     Predict the stationary probability of the number of data requesters $\mathbf{n}(m)$;
18:     Apply PCDDA/PCDA, solve it by CPSO and obtain optimal $z_{im}$ ($\mathbf{z_m} = \{z_{1m}, z_{2m}, \cdots, z_{Nm}\}$).
19:   **end for**
20: **end for**
**Output:**
21: Amount of allocated data to sell on the day $d$: $c_{id}$; Amounts of allocated data to sell in time slot $m$: $\mathbf{z_m}$; Optimized did acceptance rate: $r_{c,i}$.

---

$D$ days, the data owner optimizes the bid acceptance rate $r_c$ and obtains $c_d$, the amount of data to be sold on a certain day $d$. Then the data owner recognizes and establishes a networked data transaction system with other data owners planning to sell data on this day. Then by applying the networked data allocation mechanism, i.e., DNNA, PCDDA or PCDA designed in Sect. 15.4.4, every data owner can decide how to allocate the amount of $c_{id}$ data into different time slots on day $d$ to maximum the expected income of its own (by DNNA or PCDDA) or the entire networked data transaction system (by PCDA). The operation of the proposed data allocation mechanism is shown in Protocol 8.

## 15.6 Performance Evaluation

In this section, we will evaluate the performance improvement of the designed allocation mechanisms for the data transaction systems with single data auctioneer and multiple data auctioneers.

## 15.6.1   Data Transaction Systems with Single Auctioneer

First of all, we introduce the scenario setup for simulations. We consider a data transaction system with only one data seller, who plans to sell its rest data with an amount of $C = 100$ in following $D = 10$ days. The amount of data to be reserved for the data owner's own consumption every day is set as $\gamma = 5$. The highest price data requesters can accept is set as $h = 10$, and the service time is $r_s = 1$. To reflect the changing demands of buying and selling data from the data requesters and data owner, denoted by (15.5), arrival rates of data bids and original considering time are given by

$$\lambda(d) = \beta_2^\lambda - \left(\beta_2^\lambda - \beta_1^\lambda\right) e^{-(d/\sigma_\lambda)^2}, \tag{15.49a}$$

$$r_c(d) = \beta_2^{r_c} - \left(\beta_2^{r_c} - \beta_2^{r_c}\right) e^{-(d/\sigma_r)^2}, \tag{15.49b}$$

where $\beta_1^\lambda$, $\beta_2^\lambda$, $\beta_1^{r_c}$, $\beta_2^{r_c}$, $\sigma_\lambda$ and $\sigma_r > 0$ are constants, $d = 1, 2, \cdots, 10$. For comparison purposes, we consider four cases with different original bid acceptance rates and bid arrival rates, the parameter settings of which are shown in Table 15.2. These settings for the four cases can reflect different relationships between $\lambda(d)$ and original $r_c(d)$, which will further influence the data allocation policies.

First, we test the convergence of the proposed allocation algorithms and the expected maximum total income for the data owner by applying EADA and ERADA for the data transaction system with a single data seller. For CPSO algorithm, the number of particles is set as 6, and the number of iterations is set as 120. The simulation results of the first 25 iterations for the four cases are shown in Fig. 15.3. As shown in Fig. 15.3, the maximum expected total income for the data seller can be achieved by applying three data allocation mechanisms, after about 15 iterations by applying CPSO algorithm. In addition, for all of the four cases, the data seller can obtain the maximum income when utilizing ERADA (denoted by dotted lines), which is followed by the data allocation method based on the original considering time (denoted by solid lines), and EADA (denoted by dashed lines) performs the worst among the three mechanisms. The weak performance of EADA on data seller's income results from the its efficiency-aware property, which mean that there is a tradeoff between the total income and time cost. On the other hand, when we consider the influence of the data bids arrival rates, and then modify the considering time according to the bid arrivals in ERADA, the total income for the data seller can be also improved, meanwhile the efficiency can be still guaranteed. In addition, we

**Table 15.2** Simulation parameters

| Case | $\beta_1^\lambda$ | $\beta_2^\lambda$ | $\beta_1^{r_c}$ | $\beta_2^{r_c}$ | $\sigma_\lambda$ | $\sigma_r$ |
|------|------|------|------|------|------|------|
| Case 1 | 0.35 | 9 | 0.35 | 9 | 6 | 5 |
| Case 2 | 0.35 | 9 | 0.35 | 9 | 5 | 6 |
| Case 3 | 0.35 | 10 | 1 | 9 | 5 | 5 |
| Case 4 | 0.15 | 10 | 1.25 | 9 | 5 | 5 |

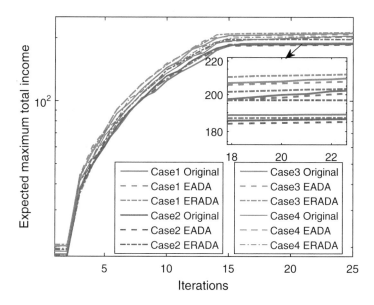

**Fig. 15.3** Achieved maximum total income for the data auctioneer obtained by CPSO, for the four cases when applying three allocation methods

notice that for Case 1, the bid arrival rates are always larger than the bid acceptation rates, then the incomes obtained by the data seller by applying EADA and ERADA are the same, which results from the definition in (15.13).

For a further revelation to see how the proposed data allocation mechanisms optimize the performance of the data transaction system, we analyze the modified considering time and the amount of data allocated for each day. For the four cases illustrated by Table 15.2, we apply the three data allocation methods based on original $r_c$, EADA and ERCDA. Then for the $D = 10$ days, the modified $r_c$ and the amount of data allocated in each day are shown in Fig. 15.4a–d and e–h, respectively.

**For Case 1**, the rates of bid acceptance are always lower than the bid arrival rates in the ten days. Similar to results in Fig. 15.3, the two proposed data allocation mechanisms, i.e., EADA and ERADA, get the same adjusted $r_c$ and the amount of data allocated for every day. Results in Fig. 15.4e also indicate that without modification of $r_c$, the data tends to be allocated to the beginning days when $\lambda(d) > r_c(d)$, $\forall d$, which means that the data requests during the later days cannot be satisfied at all. Through the modification of considering time at the beginning days according to (15.10) and (15.12), the average time of considering time is shortened from $d = 1$ to $d = 5$, as shown in Fig. 15.4a. Then the data are allocated to the ten day with more balance.

**For Case 2**, the opposite situation to Case 1, the results are shown in Fig. 15.4b, f, which indicate that EADA and ERADA will lead to different considering time modification and data allocation when $\lambda(d) < r_c(d)$. Moreover, results in Fig. 15.4b reflect the tradeoff effect of EADA for the income and time cost, i.e.,

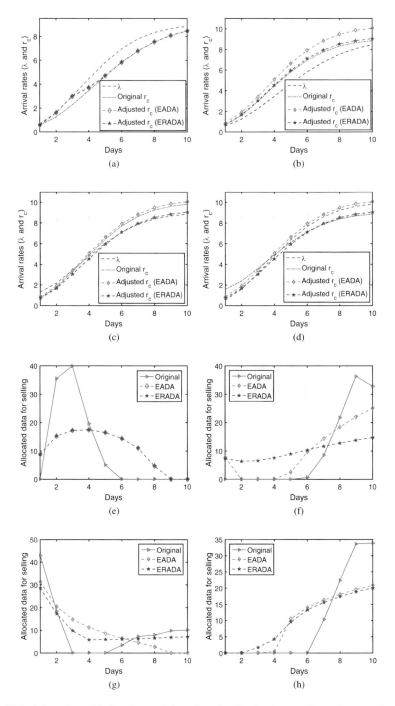

**Fig. 15.4** Adjusted considering time and data allocation for the data auctioneer in every day. (**a**) Case 1. (**b**) Case 2. (**c**) Case 3. (**d**) Case 4. (**e**) Case 1. (**f**) Case 2. (**g**) Case 3. (**h**) Case 4

**Table 15.3** Simulation results of income and efficiency obtained by applying three allocation methods for data transaction system with single auctioneer

|               | Algorithm | Case 1   | Case 2   | Case 3   | Case 4   |
|---------------|-----------|----------|----------|----------|----------|
| Total income  | Original  | 189.2919 | 187.0193 | 210.4264 | 203.9975 |
|               | EADA      | 187.4022 | 185.5160 | 208.0609 | 202.7293 |
|               | ERADA     | 197.4599 | 196.8851 | 212.3980 | 204.2962 |
| Total time    | Original  | 8.5566   | 8.5566   | 6.6608   | 7.3510   |
|               | EADA      | 8.2486   | 7.9895   | 6.3544   | 7.0858   |
|               | ERADA     | 8.2486   | 8.4573   | 6.7222   | 7.5536   |
| Average income| Original  | 22.1223  | 21.8567  | 31.5918  | 27.7509  |
|               | EADA      | 22.7192  | 23.2199  | 32.7428  | 28.6106  |
|               | ERADA     | 23.9386  | 23.2799  | 31.5965  | 27.0462  |

larger $r_c$ means higher frequency of bid acceptance and less time consumption. In addition, ERADA can achieve a further tradeoff between the original $r_c$ and adjusted $r_c^{EADA}$, which can obtain the best data allocation balancing among the three allocation methods and the optimized income for the data seller at the same time.

**For Case 3**, as shown in Fig. 15.4c, $\lambda > r_c$ in the beginning days, and then $\lambda < r_c$ during the rest days, which is opposite to Case 4, as shown in Fig. 15.4d. The optimized balancing effect of ERADA can be also verified by results in Figs. 15.4g, h.

Based on the obtained maximum total income shown in Fig. 15.3, and the adjusted considering time and the amount of data allocated shown in Fig. 15.4, we calculate the total data transaction operation time of the 10 days and the average income per unit time for the data seller by applying the three allocation algorithms. The results are shown in Table 15.3, which indicates that although EADA will bring less total income for the data seller, the average income per unit time is higher than that obtained without considering time modification. Meanwhile, the least time consumption can be achieved by EADA, comparing with another two allocation methods. In addition, ERADA receives the highest total income, which is also shown in Fig. 15.3. This best performance of ERADA results from its adaption to the bid arrival rate, which also leads to larger time cost to operate the transaction than EADA. However, the efficiency can also be improved by ERADA than the original considering time, for most cases. For case 4, as an exception, ERADA does not perform better than the original considering time method, which results from the allocation-balance improvement of ERADA. Specifically, we can notice that for Case 4, without any considering time modification, all data is allocated to the last four days, which means that the data requests of earlier six days cannot be served at all, and in addition, the total time cost tends to be very small comparing to EADA and ERADA, which increases the average income.

## 15.6.2   *Data Transaction Systems with Multi-Auctioneer*

In this section, we will test the performance of the networked data allocation for the data transaction system with multiple auctioneers. In the simulations, we consider that there are $N = 10$ data sellers with different amounts of data to be sold in a single day, i.e., with values from set $\{15, 20, 25, 30, 35, 40, 45, 50, 55, 60\}$, and these sellers are numbered by an ascending sort order according to the amount of data they plan to sell in $M \leq 5$ time slots in a single day. In addition, bids arrival rates $\lambda_i$ $(i = 1, 2, \cdots, 10)$ for different data auctioneers and the bid acceptance rates $r_{c,i}$ are obtained by the single-auctioneer data transaction system in the previous simulations, by applying ERADA. Moreover, the numbers of particle and iterations are set as 20 and 80, respectively, when applying the CPSO algorithm.

By applying the three networked data allocation mechanisms, i.e., NDDA, PCDDA and PCDA designed in Sect. 15.4.4, the allocated amount of data and corresponding income for each of the 10 auctioneer in every time slot are shown in Fig. 15.5. The three mechanisms finish the data transaction in four time slots. Results in Fig. 15.5a–d indicate that NDDA and PCDDA can finish the data transaction faster than PCDA. To be specific, Auction 1 with the least amount of data sells out all of auctioneer's data in the first time slot through NDDA and PCDDA, while two time slots are needed when applying PCDA. In addition, eight auctioneers finish the data transaction in three time slots by NDDA and PCDDA, while three auctioneers with the most data amount still have data to be sold in the fourth time slot by PCDA. Moreover, allocation results also indicate that with some prediction information obtained by the information sharing and cooperation of auctioneers, i.e., the number of data bidders in each auction and their probable movement among different auctions, PCDDA does not operate allocation radically, which means that data owners tend to preserve a certain amount of data and sell it in the later time slots. This behavior can be reflected more obviously in Fig. 15.5d, in which PCDDA has more data allocated to this time slot than NDDA does. With the prediction information and global income optimization for the entire system, this conservative performance is presented much more prominently when applying PCDA.

Then we analyze the economics performance of the three networked data allocation methods. We further process the obtained results in Fig. 15.5e–h, and get the total income of the $N = 10$ data auctioneers and their total income in every time slot. The results are shown in Fig. 15.6. The results in Fig. 15.6a indicate that the higher total income can be obtained for every data seller by PCDDA than by NDDA, which benefits from the prediction information utilization and the income maximization for the entire auction period. Moreover, when applying PCDA, which pursues a global income maximization, the total income of some data auctioneers is sacrificed, meanwhile, the other auctioneers will get more total income.

Without any prediction information, NDDA is operated to maximize the income of the current time slot. As a result, in the beginning time slots, the total income by NDDA is higher than the other two methods, which is achieved by scarifying the income in later time slots. This phenomenon is presents in Fig. 15.6b. Figure 15.6b

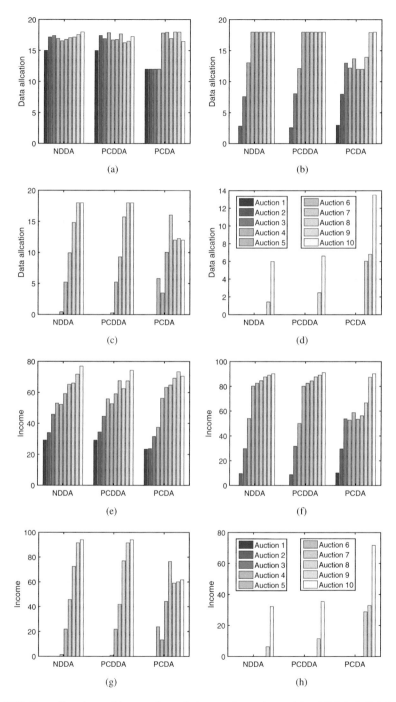

**Fig. 15.5** Data allocation and income for each data auctioneer by applying NDDA, PCDDA and PCDA in networked data auction system. (**a**) Time slot 1. (**b**) Time slot 2. (**c**) Time slot 3. (**d**) Time slot 4. (**e**) Time slot 1. (**f**) Time slot 2. (**g**) Time slot 3. (**h**) Time slot 4

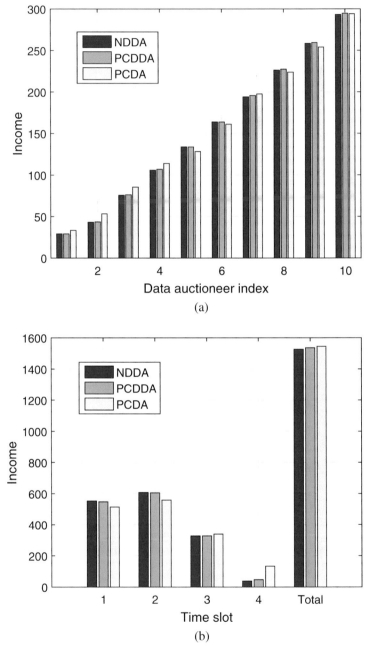

**Fig. 15.6** Income obtained by different auctioneers and in time slots by applying NDDA, PCDDA and PCDA in networked data auction system. (**a**) Income obtained by auctioneers. (**b**) Income obtained in time slots

also indicates the total incomes of the ten data auctioneers in the entire four time slots, which can be considered as the system income. As shown in Fig. 15.6b, higher system income can be achieved by PCDDA than NDDA resulting from the prediction information. In addition, PCDA performs the best on the system income due to its global optimization objective, although the income obtained in some single time slot might be lower than NDDA and PCDDA.

## 15.7 Conclusion

In this part, we have proposed a novel data transaction system for mobile networks based on the basic and networked auction models. In addition, the data allocation mechanisms have been designed to make decisions that how to sell the rest data in different days, and then for each day, how to sell the allocated data in a system with multiple data sellers, to improve the performance of the system, maximize the income of the data sellers and satisfy the demands of data requesters. Simulation results for the system with single data auctioneer indicate that the modification of considering time according to the rest amount of data can improve the efficiency of the data transaction, although the total income for the data seller might decrease due to the tradeoff between the income and time cost. In addition, when the data bid arrival rates are considered, the system efficiency and total income of the data seller can be both guaranteed, and meanwhile, the best data allocation balancing effect can be also achieved. Simulation results for the networked data transaction system demonstrate that the prediction information of data bidders' movement can improve the income for every data auctioneer effectively.

## References

1. R. W. Heath Jr, M. Honig, S. Nagata, S. Parkvall, and A. C. Soong, "LTE-advanced pro: Part 3 [guest editorial]," *IEEE Commun. Mag.*, vol. 54, no. 7, pp. 52–53, Jul. 2016.
2. J. Zhang, G. Fang, C. Peng, M. Guo, S. Wei, and V. Swaminathan, "Profiling energy consumption of DASH video streaming over 4G LTE networks," in *Proc. 8th Int. Workshop Mobile Video*. Klagenfurt, Austria, ACM, 3 May 2016.
3. C. Jiang, H. Zhang, Y. Ren, Z. Han, K.-C. Chen, and L. Han, "Machine learning paradigms for next-generation wireless networks," *IEEE Wireless Commun.*, vol. 24, no. 2, pp. 98–105, Apr. 2008.
4. A. Ghosh, R. Ratasuk, B. Mondal, N. Mangalvedhe, and T. Thomas, "LTE-advanced: next-generation wireless broadband technology," *IEEE Wireless Commun.*, vol. 17, no. 3, pp. 10–22, Jun. 2010.
5. F. Mehmeti and T. Spyropoulos, "Performance analysis of mobile data offloading in heterogeneous networks," *IEEE Trans. Mobile Computing*, vol. 16, no. 2, pp. 482–497, Feb. 2017.
6. K. Poularakis, G. Iosifidis, I. Pefkianakis, L. Tassiulas, and M. May, "Mobile data offloading through caching in residential 802.11 wireless networks," *IEEE Trans. Network Service Manage.*, vol. 13, no. 1, pp. 71–84, Jan. 2016.

7. C. Jiang, H. Zhang, Y. Ren, Z. Han, K.-C. Chen, and L. Hanzo, "Machine learning paradigms for next-generation wireless networks," *IEEE Wireless Commun.*, vol. 24, no. 2, pp. 98–105, Apr. 2017.
8. X. Sun, S. Hu, L. Su, T. Abdelzaher, P. Hui, W. Zheng, H. Liu, and J. Stankovic, "Participatory sensing meets opportunistic sharing: Automatic phone-to-phone communication in vehicles," *IEEE Trans. Mobile Computing*, vol. 15, no. 10, pp. 2550–2563, Oct. 2016.
9. R. Sims and C. Bauer, "Handling a device changing from 3G to Wi-Fi without breaking established connections," in *New Trends Networking, Computing, E-learning, Syst. Sci., Eng.* Springer, 2015, pp. 193–197.
10. J. Du, C. Jiang, Z. Han, H. Zhang, S. Mumtaz, and Y. Ren, "Contract mechanism and performance analysis for data transaction in mobile social networks," *IEEE Trans. Network Sci. Eng.*, vol. 6, no. 2, pp. 103–115, Apr. - Jun. 2019.
11. J. Du, E. Gelenbe, C. Jiang, Z. Han, Y. Ren, and M. Guizani, "Cognitive data allocation for auction-based data transaction in mobile networks," in *14th Int. Wireless Commun. Mobile Computing Conf. (IWCMC)*.  IEEE, Limassol, Cyprus, 25-29 Jun. 2018.
12. I. Rhee, M. Shin, S. Hong, K. Lee, S. J. Kim, and S. Chong, "On the levy-walk nature of human mobility," *IEEE/ACM trans. networking (TON)*, vol. 19, no. 3, pp. 630–643, Jun. 2011.
13. V. S. S. Nadendla, S. K. Brahma, and P. K. Varshney, "Optimal spectrum auction design with 2-D truthful revelations under uncertain spectrum availability," *IEEE/ACM Trans. Networking*, vol. 25, no. 1, pp. 420–433, Feb. 2017.
14. H. Yu, G. Iosifidis, J. Huang, and L. Tassiulas, "Auction-based coopetition between LTE unlicensed and Wi-Fi," *IEEE J. Sel. Areas Commun.*, vol. 35, no. 1, pp. 79–90, Jan. 2017.
15. W. Sun, J. Liu, Y. Yue, and H. Zhang, "Double auction-based resource allocation for mobile edge computing in industrial internet of things," *IEEE Trans. Ind. Informat.*, vol. 14, no. 02, pp. 4692–4701, Oct. 2018.
16. L. Lai and H. El Gamal, "On cooperation in energy efficient wireless networks: the role of altruistic nodes," *IEEE Trans. Wireless Commun.*, vol. 7, no. 5, pp. 1868–1878, May 2008.
17. T. Wang, Y. Sun, L. Song, and Z. Han, "Social data offloading in D2D-enhanced cellular networks by network formation games," *IEEE Trans. Wireless Commun.*, vol. 14, no. 12, pp. 7004–7015, Dec. 2015.
18. A. Apostolaras, G. Iosifidis, K. Chounos, T. Korakis, and L. Tassiulas, "A mechanism for mobile data offloading to wireless mesh networks," *IEEE Trans. Wireless Commun.*, vol. 15, no. 9, pp. 5984–5997, Sept. 2016.
19. E. Meskar, T. D. Todd, D. Zhao, and G. Karakostas, "Energy aware offloading for competing users on a shared communication channel," *IEEE Trans. Mobile Computing*, vol. 16, no. 1, pp. 87–96, Jan. 2017.
20. L. Gao, G. Iosifidis, J. Huang, L. Tassiulas, and D. Li, "Bargaining-based mobile data offloading," *IEEE J. Sel. Areas Commun.*, vol. 32, no. 6, pp. 1114–1125, Jun. 2014.
21. C.-Y. Wang, H.-Y. Wei, and W.-T. Chen, "Resource block allocation with carrier-aggregation: A strategy-proof auction design," *IEEE Trans. Mobile Computing*, vol. 15, no. 12, pp. 3142–3155, Feb. 2016.
22. J. Du, C. Jiang, H. Zhang, Y. Ren, and M. Guizani, "Auction design and analysis for SDN-based traffic offloading in hybrid satellite-terrestrial networks," *IEEE J. Sel. Areas Commun.*, vol. PP, no. 99, pp. 1–1, Sept. 2018.
23. J. Wang, C. Jiang, Z. Bie, T. Q. Quek, and Y. Ren, "Mobile data transactions in device-to-device communication networks: Pricing and auction," *IEEE Wireless Commun. Lett.*, vol. 5, no. 3, pp. 300–303, Jun. 2016.
24. S. Paris, F. Martignon, I. Filippini, and L. Chen, "An efficient auction-based mechanism for mobile data offloading," *IEEE Trans. Mobile Computing*, vol. 14, no. 8, pp. 1573–1586, Aug. 2015.
25. K. Zhu and E. Hossain, "Virtualization of 5G cellular networks as a hierarchical combinatorial auction," *IEEE Trans. Mobile Computing*, vol. 15, no. 10, pp. 2640–2654, Oct. 2016.

26. W. Song and Y. Zhao, "A randomized reverse auction for cost-constrained D2D content distribution," in *IEEE Global Commun. Conf. (GLOBECOM)*. Washington, DC, USA, 4-8 Dec. 2016.

27. M. H. Hajiesmaili, L. Deng, M. Chen, and Z. Li, "Incentivizing device-to-device load balancing for cellular networks: An online auction design," *IEEE J. Sel. Areas Commun.*, vol. 35, no. 2, pp. 265–279, Feb. 2017.

28. Y. Zhu, J. Jiang, B. Li, and B. Li, "Rado: A randomized auction approach for data offloading via D2D communication," in *IEEE 12th Int. Conf. Mobile Ad Hoc Sensor Syst.* Dallas, TX, USA, 19-22 Oct. 2015.

29. S. Huang, C. Yi, and J. Cai, "A sequential posted price mechanism for D2D content sharing communications," in *IEEE Global Commun. Conf. (GLOBECOM)*. Washington, DC, USA, 4-8 Dec. 2016.

30. J. Du, E. Gelenbe, C. Jiang, H. Zhang, Z. Han, and Y. Ren, "Data transaction modeling in mobile networks: Contract mechanism and performance analysis," in *IEEE Global Commun Conf. (GLOBECOM)*. IEEE, Singapore, 4-8 Dec. 2017.

31. E. Gelenbe and K. Velan, "An approximate model for bidders in sequential automated auctions," in *Agent Multi-Agent Syst.: Technol. Applicat. Third KES Int. Symp., KES-AMSTA 2009*. Uppsala, Sweden, 3-5 Jun. 2009.

32. E. Gelenbe and K. Velan, "Analysing bidder performance in randomised and fixed-deadline automated auctions," in *Agent Multi-Agent Syst.: Technol. Applicat., 4th KES Int. Symp., KES-AMSTA 2010*. Gdynia, Poland, 23-25 Jun. 2010.

33. J. Du, C. Jiang, J. Wang, S. Yu, and Y. Ren, "Trustable service rating in social networks: A peer prediction method," in *IEEE Global Conf. Signal Inform. Process. (GlobalSIP)*. Washington, DC, USA, 7-9 Dec. 2016, pp. 415–419.

34. K. Lee, J. Lee, Y. Yi, I. Rhee, and S. Chong, "Mobile data offloading: How much can WiFi deliver?" *IEEE/ACM Trans. Networking*, vol. 21, no. 2, pp. 536–550, Apr. 2013.

35. C. W. Leong, W. Zhuang, Y. Cheng, and L. Wang, "Call admission control for integrated on/off voice and best-effort data services in mobile cellular communications," *IEEE Trans. Commun.*, vol. 52, no. 5, pp. 778–790, May 2004.

36. S. G. Sitharaman, "Modeling queues using Poisson approximation in IEEE 802.11 ad-hoc networks," in *2005 14th IEEE Workshop Local & Metropolitan Area Networks*. Crete, Greece, IEEE, 18-21 Sept. 2005.

37. E. Gelenbe, "Analysis of single and networked auctions," *ACM Trans. Internet Technol.*, vol. 9, no. 2, p. 8, May 2009.

38. Y. Zhou, B. P. L. Lau, C. Yuen, B. Tunçer, and E. Wilhelm, "Understand urban human mobility through crowdsensed data," *IEEE Commun. Mag.*, vol. PP, no. 99, pp. 1–1, 2018.

39. F. Van den Bergh and A. P. Engelbrecht, "A cooperative approach to particle swarm optimization," *IEEE Trans. Evol. Comput.*, vol. 8, no. 3, pp. 225–239, Jun. 2004.

40. R. Eberhart and J. Kennedy, "A new optimizer using particle swarm theory," in *Proc. Sixth IEEE Int. Symp. Micro Machine Human Sci. (MHS'95)*. Nagoya, Japan, 4-6 Oct. 1995.

# Chapter 16
# Cooperative Trustworthiness Evaluation and Trustworthy Service Rating

**Abstract** With the development of online applications based on the social network, many different approaches of service to achieve these applications have emerged. Users' reporting and sharing of their consumption experience or opinion can be utilized to rate the quality of different approaches of online services. How to ensure the authenticity of the users' reports and identify malicious ones with cheating reports become important issues to achieve an accurate service rating. In this chapter, we provide a private-prior peer prediction mechanism based trustworthy service rating system with a data processing center (DPC), which requires users to report to it with their prior and posterior believes that their peer users will report a high quality opinion of the service. The DPC evaluates users' trustworthiness with their reports by applying the strictly proper scoring rule, and removes reports received from users with low trustworthiness from the service rating procedure. This peer prediction method is incentive compatible and able to motivate users to report honestly. In addition, to identify malicious users and bad-functioning/unreliable users with high error rate of quality judgement, an unreliability index is proposed in this chapter to evaluate the uncertainty of reports. Reports with high unreliability values will also be excluded from the service rating system. By combining the trustworthiness and unreliability, malicious users will face a dilemma that they cannot receive a high trustworthiness and low unreliability at the same time when they report falsely. Simulation results indicate that the proposed peer prediction based trustworthy service rating can identify malicious and unreliable behaviours effectively, and motivate users to report truthfully. The relatively high service rating accuracy can be achieved by the proposed system.

**Keywords** Service Rating · Trustworthiness · Reliability · Peer Prediction · Private Prior · Social Networks

## 16.1 Introduction

Information communication and computation technologies have been developing rapidly in recent years. With the growing demands of big data and development of different applications, the emerging fifth generation (5G) mobile communication technology will be a multi-service and multi-technology integrated network, which can enhance the user experience by providing various intelligent and customized services [1]. Moreover, social networks have become important platforms for users to enjoy different kinds of online services. With the rapid development of Internet-based applications, different approaches to achieve these applications have emerged. Take e-commerce for an instance, in which users are allowed to use different online or mobile payment systems, such as PayPal, Google Wallet, Alipay and Apple Pay, to complete payments. In addition, for some file sharing applications, users can use different downloaders to download their favorite music, movies or other media files.

In order to provide accurate and useful suggestions to new users and help them to make choices, the use of service quality ratings for these different services has become an important method [2–4]. Concerning this problem, the feedback and evaluation from users who have experienced a service provide essential reference information for the service rating [5–7]. Meanwhile, social networks provide platforms that collect and share users' feedback, according to which the service rating can be provided through some data fusion mechanism. However, false and dishonest reports from malicious users can destroy the fairness and usefulness of such ratings. Therefore, it is rather necessary to introduce some trust assessment function to such systems and design an incentive mechanism to motivate users to output truthful feedback.

In this part, we will establish a peer prediction based trustworthy service rating system for social networks. With peer prediction based decision, network functions of malicious behavior detection, trustworthiness and unreliability assessment can be achieved. Then the reliable and trustworthy service ratings can be obtained by the feedback from honest and reliable users. In this work, we assume that the service quality is an objective evaluation independent of users' subjective judgements. This assumption is reasonable for many service quality indicators, such as convenience of online payment methods and download speeds [8, 9].

We summarize the major contributions of this part as follows.

1. We introduce private-prior peer prediction in the service rating system of social networks. The user trustworthiness obtained through certain strictly proper scoring rules is formulated to motivate users to report trustfully. We analyze the incentive compatibility of the basic peer prediction mechanism with respect to the false alarm and missed detection probabilities of judgement and report.

2. We propose an unreliability index to eliminate unreliable reports from the service rating system. By applying the unreliability index, malicious users are confronted with a dilemma that they cannot get a high trustworthiness and a low unreliability at the same time when they provide a false report. However, the best choice of honest users is still reporting truthfully even for poorly functioning ones with high error rates of judgement.

3. Based on the proposed user trustworthiness and unreliability index, we design a service rating framework. In this framework, trustworthiness is used to evaluate the possibility of whether the subject user's report is dishonest and the user is a malicious one. On the other hand, the unreliability index is introduced to determine whether the reports are reliable, but does not consider the type of

the users, i.e., honest or malicious. By removing the feedback reports with high unreliability and reports received from users with low trustworthiness, from the final rating procedure, an accurate and trustworthy service rating can be achieved.

The remainder of this part is organized as follows. We review the relevant literature in Sect. 16.2. In Sect. 16.3, the system model is described. The private-prior peer prediction based user trustworthiness evaluation for motivating truthful reports is proposed in Sect. 16.4. Then we analyze the reliability of users' reports and design the service rating system in Sect. 16.5. Simulations are presented in Sect. 16.6, and conclusions are drawn in Sect. 16.7.

## 16.2  Related Works

Service ratings for different application systems have been active research topics over the past decades. Many service evaluation systems have been developed for mobile social networks [10], multiple providers service systems [11] and many other kinds of web services [12, 13]. In [14], researchers designed the objective rating scores of products or services through an iterative rating algorithm. This rating mechanism entirely decoupled the credibility assessment of the evaluations from the ranking itself, which makes it very robust against collusion attacks as well as random and biased raters. A two-phase methodology was proposed in [15] for systematically evaluating the performance and availability of cluster-based Internet services. A service rating scheme that is robust against manipulations by malicious users and services was proposed in [16]. In [16], the service rating made by the target customer was predicted, based on which the system helped this customer to choose a suitable service. The authors of [17] proposed a user-service rating prediction approach for the recommender system by exploring social users' rating behaviors. In [17], the user's social relationships were considered in order to understand social users' rating behavior diffusions.

A social network is a platform that allows its users to obtain services and share their experiences. Based on such feedback gathered, a data processing center (DPC) can provide quality ratings for different services, which can further give suggestions for new users. To ensure the accuracy of service ratings, the trustworthiness and reliability of the feedback from users need to be checked and ensured. Currently, trust and reputation management has become a challenge in many kinds of feedback and decision systems. Many trustworthiness evaluation mechanisms have been proposed for social networks [18, 19], wireless sensor networks [20, 21] and cloud-based service systems [22]. To motivate secondary users (SUs) in a multiple channel cognitive radio network to report truthfully, a Stackelberg game model was designed in [23], according to which trustworthy SUs gain transmission opportunities as rewards. A consumer feedback based service rating system was presented in [24] to evaluate the trustworthiness of a cloud service. In [24], a novel protocol was proposed to improve and ensure the credibility of trust feedback from consumers.

In [25], a dynamic trust evaluation model was proposed to evaluate the user's reputation. The authors of [25] considered both users' preferences for different quality of service attributes and the impact of vicious ratings on trust evaluation. For rating the reputation of the service, different users' ratings were weighted dynamically according to their honesty assessment, and the influence of malicious ratings were thus effectively diluted.

Most of the local and global trustworthiness evaluation methods mentioned above are established by users' own current and/or past behaviors. Further, some researchers have considered relationship and interaction among users of a network for user trustworthiness assessment and prediction [26–30], although the incentive mechanisms for truthful information are not studied much. Originally applied in electronic commerce, common-prior peer prediction with a strictly proper scoring rule [31, 32] was proposed for truthful feedback from users in [33]. To be specific, *Peer Prediction* refers to a scheme using one user's report to update or predict a probability distribution for the report of someone else, whom we refer to as the "peer". The former user is then scored not on a comparison between the likelihood assigned to the peer's possible ratings and the peer's actual rating. Moreover, in the common-prior peer prediction mechanism, the prior probability of the product type or service quality is commonly held, conditional on which, the probability distribution of user's received product type or service quality is also common knowledge. Relaxing the assumption of common-prior, the authors of [34] modified the classical peer prediction method such that only users' subjective and private opinions were needed, and this trustworthiness evaluation mechanism is known as private-prior peer prediction. Both of these two peer-prediction methods estimated the trustworthiness using strictly proper scoring rules, which can provide incentives for truthful reporting. The peer prediction mechanism can be applied efficiently in the scenario where the prior knowledge is subjective and private to each user. For instance, peer prediction has been used in wireless sensor networks [35, 36], cognitive radio networks [37], social and online systems [38, 39] and many other kinds of crowd-sourcing systems [40, 41] to collect truthful reports from users, and has been considered as an effective solution to elicit trustworthy feedback. In this part, we propose a service rating system for social network based services according to honest users with high trustworthiness. Private-prior peer prediction is introduced to evaluate users' trustworthiness and motivate users to provide truthful feedback.

## 16.3  Mathematical Model for Service Rating Based on User Report Fusion

With the boom of online applications based on social networks, different services to support these applications have emerged [42]. As mentioned previously, users are allowed to select different online payment methods to complete their online purchases, or download their favorite music and movies by downloaders those they think are faster and more reliable. To rate the quality of different services, users are required to report and share their consumption experiences or opinions to the social application platform, which can use this valuable feedback for service rating and helping new users to judge whether the applications can provide high quality services. In our work, the quality of the services is considered as an objective evaluation independent of users' subjective judgements. For instance, different users tend to have the same opinion about whether a payment system is convenient or a downloader has a high download speed. Such a social rating system is different from systems such as movie review, in which users' subjective opinions and standards may vary considerably between individuals.

In this case, users' truthful feedback of a service is important for achieving an accurate rating of this service approach's quality and providing helpful suggestions to new users. However, some malicious users in social networks provide untruthful evaluations of the service quality for some purposes. On the one hand, malicious users report to the service rating system that the object approach of service is high in quality when it gives a bad service performance to improve its competitiveness. On the other hand, malicious users report a low service-quality evaluation to lower the rating of the object approach of service, which will encourage new users not to select it. These malicious behaviours undermine the fairness of the service rating and provide unreliable suggestions to new users. Therefore, it is important to make sure that the feedback from users is truthful.

In this work, we design a mechanism to provide incentives for truthful opinions of users. Moreover, we define a trustworthiness management method to identify malicious users, excluding whose untruthful feedback, the service rating with high accuracy can be made.

## 16.3.1    System Model

Consider a population of $N$ users distributed over a social network with a service platform, which can provide different approaches of this service. Quality $Q$ of the service is a binary rating, which is considered as a random variable represented by $\{l, h\}$ referring to the low quality and high quality, respectively. As mentioned previously, this quality is an objective fact. In other words, after experiencing the service, different honest users tend to give the same evaluation or opinion independent of their individual subjective standards. As shown in Fig. 16.1, each user $i$ ($i = 1, 2, \cdots, N$) accepts the service $m$, and then makes a binary opinion of the service quality denoted by $S_i = s_i \in \{l, h\}$. Meanwhile, users are allowed to provide some required QoS reports to the cloud, and these reports will be processed by the DPC. For instance, the opinion report denoted by $x_i \in \{0, 1\}$ is generated by applying a report strategy $r_i : S_i \rightarrow \{0, 1\}$. User $i$ will report $x_i = 1$ when $S_i = h$ (or $x_i = 0$ when $S_i = l$) to the cloud if he/she is honest. We assume that $x_i$ is the semi-public information published to the social service-evaluation platform by the cloud, and can be observed by the DPC and other users over the social network and

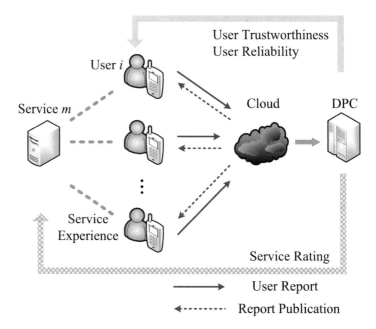

**Fig. 16.1** Peer prediction based service rating and user trustworthiness management system

having the friendship with user $i$. In addition, $S_i$ is the private or local information only known by user $i$, and other users and even the cloud cannot get it.

## 16.3.2  Service Rating Based on User Report Fusion

Define the false alarm of the judgement as that the service is misjudged as a low quality while it is a high quality in fact, and the user $i$'s false alarm probability of judgement is denoted by $P_{fa,i} = P(S_i = l|Q = h)$. In order to simplify the expression, let $P_{fa}$ to denote the false alarm probability of judgement. On the contrary, define the missed detection probability of judgement as $P_{md,i} = P(S_i = h|Q = l)$. Similarly, we use $P_{md}$ to denote the missed detection probability of judgement for a simpler expression. As mentioned previously, the quality is an objective fact, which leads that both honest and malicious users trend to make the similar and accurate judgement for it. So we assume that $P_{fa} < 0.5$ and $P_{md} < 0.5$ hold for all users in the social network.

On the other hand, we consider the false alarm of the report as that a user reports a low quality evaluation to the cloud when the quality of service is high, and the user $i$'s false alarm probability of report is $P_{f,i} = P(x_i = 0|Q = h)$. The missed detection probability of report is $P_{m,i} = P(x_i = 1|Q = l)$. In addition, the simplified expression of the two probabilities of report above are $P_f$ and $P_m$.

We consider that the user type are represented by $\theta_i \in \{0, 1\}$, i.e., $\theta_i = 0$ if $i$ is an honest user, and $\theta_i = 1$ if user $i$ is the malicious otherwise. Assume that if user $i$ is malicious, his/her false alarm cheating rate is $P^c_{f,i} \in [0, 1]$, and the missed detection cheating rate is $P^c_{m,i} \in [0, 1]$. We assume that the honest users always report their real judgement of the service quality, no matter whether his/her judgement is accurate. Then for each user $i$, we have

$$P_{f,i} = \begin{cases} \left(1 - P_{fa,i}\right) P^c_{f,i} + P_{fa,i} \left(1 - P^c_{f,i}\right), & \theta_i = 1; \\ P_{fa,i}, & \theta_i = 0; \end{cases} \quad (16.1)$$

$$P_{m,i} = \begin{cases} \left(1 - P_{md,i}\right) P^c_{m,i} + P_{md,i} \left(1 - P^c_{m,i}\right), & \theta_i = 1; \\ P_{md,i}, & \theta_i = 0. \end{cases} \quad (16.2)$$

As shown in Fig. 16.1, based on the users' reports received by the cloud, the DPC can obtain trustworthiness $T_i$ of each user and make the decision of the service rating by applying the following rule:

$$R = \sum_{i \in T} x_i \begin{cases} < n, & \text{the DPC rates the service } Q = l; \\ \geq n, & \text{the DPC rates the service } Q = h, \end{cases} \quad (16.3)$$

where $n$ is the threshold of service rating. In (16.3), $T = \{i \mid T_i \geq t\}$ is the set of honest users with high trustworthiness $T_i$, which is determined by the threshold of trustworthiness $t$. A simple decision making rule is that the DPC rates the service as high, i.e., $Q = h$, only if more than half of trust users report the service's quality is high, i.e., $n = |T|/2$.

## 16.4   Peer Prediction for User Trustworthiness

In this section, we will introduce the private-prior peer prediction method, which enable to encourage users to provide rating reports truthfully. With some certain strictly proper scoring rules to estimate the users' trustworthiness, the mechanism can identify malicious users those with low trustworthiness. Then users are motivated to report truthfully in order to obtain high trustworthiness and avoid being considered as the malicious.

### 16.4.1   *Private-Prior Peer Prediction Mechanism*

Private-prior peer prediction is an incentive compatible mechanism originally proposed to motivate agents to report their private prior and posterior signal belief on electronic commerce [34]. In the basic private-prior peer prediction mechanism, each agent $i$ coupled with his/her peer agent $j = i + 1$ is required to report his/her prior and posterior signal belief of the state before and after observe the signal, respectively. According to the two reports, the agent $i$'s score can be calculated by a *strictly proper scoring rule*, which will be introduced in the later part of this section.

### 16.4.1.1   Prior Belief Reports to the Cloud

In the system established in this work, any two users accepting the same service can be considered as a pair of peers, which establishes a kind of friendship and topology of all users. For rating the quality of service $m$, we consider that each user $i$ has one peer user $j$ selected randomly from other users who have accepted and will still accept the same service $m$ as $i$. Before experiencing the service, user $i$ is required to report his/her prior belief $y_{ij} \in [0, 1]$, or called *information report*, to the cloud that his/her peer user $j$ will report a high quality signal, i.e., $x_j = 1$. Then $y_{ij}$ can be given by

$$
\begin{aligned}
y_{ij} &= P_i \left( x_j = 1 \right) \\
&= P_i \left( x_j = 1 \,|\, Q = h \right) P_i \left( Q = h \right) + P_i \left( x_j = 1 \,|\, Q = l \right) P_i \left( Q = l \right) \quad (16.4) \\
&\triangleq P \left( x_j = 1 \,|\, S_i = h \right) P \left( S_i = h \right) + P \left( x_j = 1 \,|\, S_i = l \right) P \left( S_i = l \right).
\end{aligned}
$$

In (16.4), $P_i \left( x_j = 1 \,|\, Q = h \right)$ and $P_i \left( x_j = 1 \,|\, Q = l \right)$ can be obtained by the previous report $x_j$ of user $j$ released among the network. $P_i \left( x_j = 1 \,|\, Q = h \right)$ represents the probability that user $j$ gives a report of "high quality" evaluation for the service when user $i$ makes a high quality judgement to the same service, i.e., $S_i = h$. This judgement is a private and local information only known by user $i$, and the prior belief $P_i \left( Q = h \right)$ is user $i$'s subjective prior of the service quality and is identical to $P \left( S_i = h \right)$. Similarly, $P_i \left( Q = l \right)$ is equal to $P \left( S_i = l \right)$. Therefore, we can get the second equivalence relation in (16.4) established by $\triangleq$.

### 16.4.1.2   Posterior Belief Reports to the Cloud

After experiencing the service, user $i$ makes his/her own opinion of the service quality $S_i = s_i$, and then sends the posterior belief, or called *prediction report* to the cloud, denoted by $y'_{ij} (s_i) \in [0, 1]$, that the peer user $j$ will report of a high quality evaluation for the service. Then $y'_{ij}$ can be expressed as

$$
\begin{aligned}
y'_{ij}(s_i) &= P_i\left(x_j = 1 \mid S_i = s_i\right) \\
&= P\left(x_j = 1 \mid Q = h\right) P\left(Q = h \mid S_i = s_i\right) \\
&\quad + P\left(x_j = 1 \mid Q = l\right) P\left(Q = l \mid S_i = s_i\right).
\end{aligned}
\tag{16.5}
$$

Similar to the previous analysis, $y'_{ij}(s_i)$ can be decomposed into two conditions as follows.

$$
y'_{ij}(l) = \frac{\varphi_1\left(1 - P_{f,j}\right) + \varphi_2 P_{m,j}}{\varphi_1 + \varphi_2},
\tag{16.6a}
$$

$$
y'_{ij}(h) = \frac{\varphi_3\left(1 - P_{f,j}\right) + \varphi_4 P_{m,j}}{\varphi_3 + \varphi_4},
\tag{16.6b}
$$

where

$$
\varphi_1 = P_{fa,i} P\left(Q = h\right),
\tag{16.7a}
$$

$$
\varphi_2 = \left(1 - P_{md,i}\right) P\left(Q = l\right),
\tag{16.7b}
$$

$$
\varphi_3 = \left(1 - P_{fa,i}\right) P\left(Q = h\right),
\tag{16.7c}
$$

$$
\varphi_4 = P_{md,i} P\left(Q = l\right).
\tag{16.7d}
$$

As defined previously, $y_{ij}$ is the user $i$'s prior judgement that $x_j = 1$ before user $i$ experiences the service. After user $i$ experiencing the service and sensing that $s_i = h$, it is reasonable for user $i$ to make the judgement that $x_j = 1$ with a larger probability, i.e., $y'_{ij}(h) > y_{ij}$, which means that $i$'s prior belief $x_j = 1$ will be "boosted". On the contrary, $y_{ij} > y'_{ij}(l)$ if user $i$ receives a low-quality service. However, when there are malicious users providing untrustful evaluations of the service, the relation of inequality above cannot always satisfied. Lemma 16.1 provides the sufficient conditions which can ensure $y'_{ij}(h) > y_{ij} > y'_{ij}(l)$.

**Lemma 16.1** *In the private-prior peer prediction mechanism, for each user $i$ with prior and posterior belief reports $y_{ij}$ and $y'_{ij}$ of user $j$, it holds that $y'_{ij}(h) > y_{ij} > y'_{ij}(l)$ if all users satisfy that $P_{fa} + P_{md} < 1$ and $P_f + P_m < 1$.*

***Proof*** In (16.4),

$$
\begin{aligned}
y'_{ij}(h) &= P\left(x_j = 1 \mid S_i = h\right) \\
&= P\left(x_j = 1 \mid Q = h\right) P\left(Q = h \mid S_i = h\right) \\
&\quad + P\left(x_j = 1 \mid Q = l\right) P\left(Q = l \mid S_i = h\right) \\
&= \frac{1}{P\left(S_i = h\right)} \left[\left(1 - P_{f,j}\right)\left(1 - P_{fa,i}\right) P\left(Q = h\right) + P_{m,j} P_{md,i} P\left(Q = l\right)\right].
\end{aligned}
\tag{16.8}
$$

Similarly,

$$
\begin{aligned}
y'_{ij}(l) &= P\left(x_j = 1 \mid S_i = l\right) \\
&= P\left(x_j = 1 \mid Q = h\right) P\left(Q = h \mid S_i = l\right) \\
&\quad + P\left(x_j = 1 \mid Q = l\right) P\left(Q = l \mid S_i = l\right) \\
&= \frac{1}{P\left(S_i = l\right)} \left[\left(1 - P_{f,j}\right) P_{fa,i} P\left(Q = h\right) + P_{m,j}\left(1 - P_{md,i}\right) P\left(Q = l\right)\right].
\end{aligned}
\tag{16.9}
$$

So $y_{ij}$ can be written as

$$
y_{ij} = \left(1 - P_{f,j}\right) P\left(Q = h\right) + P_{m,j} P\left(Q = l\right).
\tag{16.10}
$$

Then,

$$
\begin{aligned}
&y'_{ij}(h) - y_{ij} \\
&= \frac{1}{P\left(S_i = h\right)} \left[\left(1 - P_{f,j}\right)\left(1 - P_{fa,i}\right) P\left(Q = h\right)\right.\\
&\quad \left. + P_{m,j} P_{md,i} P\left(Q = l\right)\right] - \left(1 - P_{f,j}\right) P\left(Q = h\right) - P_{m,j} P\left(Q = l\right) \\
&= \frac{1}{P\left(S_i = h\right)} \left\{\left(1 - P_{f,j}\right)\left(1 - P_{fa,i}\right) P\left(Q = h\right) + P_{m,j} P_{md,i} P\left(Q = l\right)\right. \\
&\quad - \left[\left(1 - P_{f,j}\right) P\left(Q = h\right) + P_{m,j} P\left(Q = l\right)\right] \\
&\quad \left. \cdot \left[\left(1 - P_{fa,i}\right) P\left(Q = h\right) + P_{md,i} P\left(Q = l\right)\right]\right\} \\
&= A_0\left(Q, S_i\right) \left[\left(1 - P_{f,j}\right)\left(1 - P_{fa,i}\right) + P_{m,j} P_{md,i}\right. \\
&\quad \left. - P_{m,j}\left(1 - P_{fa,i}\right) - \left(1 - P_{f,j}\right) P_{md,i}\right] \\
&= A_0\left(Q, S_i\right)\left(1 - P_{fa,i} - P_{md,i}\right)\left(1 - P_{f,j} - P_{m,j}\right),
\end{aligned}
\tag{16.11}
$$

where

$$A_0 (Q, S_i) = \frac{P (Q = h) \, P (Q = l)}{P (S_i = h)}. \tag{16.12}$$

Therefore, when $P_{fa,i} + P_{md,i} < 1$ and $P_{f,j} + P_{m,j} < 1$, inequality $y'_{ij} (h) > y_{ij}$ holds. By symmetry, we have $y'_{ij} (l) < y_{ij}$ under the same conditions. This completes the proof of Lemma 16.1.

*Remark 16.1*  As been assumed that $P_{fa} < 0.5$ and $P_{md} < 0.5$, condition $P_{fa} + P_{md} < 1$ always holds for all users. According to (16.1) and (16.2), for honest user $i$, i.e., $\theta_i = 0$, we have $P_{f,i} + P_{m,i} < 1$. On the other hand, for dishonest user $i$ ($\theta_i = 1$), whether $P_{f,i} + P_{m,i} < 1$ can hold depends on his/her false alarm cheating rate $P^c_{f,i}$ and missed detection cheating rate $P^c_{m,i}$. Notice that outright malicious users with relatively high $P^c_f > 0.5$ and/or $P^c_m$ will have high $P_f > 0.5$ and/or $P_m > 0.5$, respectively. Users with both/either of the two cheating behaviors above can be identified easily according to their former reports with high error report probability. If the rating system removes reports of users having high former $P_f$ and/or $P_m$, these malicious reports will not make sense when the system updates the rating of the service. Consequently, to achieve a continuous trick, malicious users need to manage their $P^c_f$ and $P^c_m$ to disguise themselves as trustful ones sometimes to make sure $P_f < 0.5$ and $P_m < 0.5$. So in our work, we analyze the peer prediction mechanism under the conditions of $P_f < 0.5$ and $P_m < 0.5$. Therefore, the condition of $P_f + P_m < 1$ in Lemma 16.1 is reasonable, and in this case, inequality $y'_{ij} (h) > y_{ij} > y'_{ij} (l)$ can be always satisfied.

### 16.4.1.3   Inferred Opinion Reports

Instead of reporting the private evaluation of the service quality $S_i$ or $x_i$, user $i$ sends his/her prior and posterior probability of belief that peer user $j$ gives report $x_j = 1$. We notice that both report $x_i$ and $x_j$ are not provided directly by the relative user. In basic private-prior peer prediction, user $i$ only sends reports $y_{ij}$ and $y'_{ij} (s_i)$ to the cloud, according to which the DPC infers opinion report $x_i$ and publishes it to the social service-evaluation platform. Inferred opinion report $x_i$ is generated by the following rule:

$$x_i = x \left( y_{ij}, y'_{ij} \right) = \begin{cases} 1, & y'_{ij} > y_{ij}, \\ 0, & y'_{ij} < y_{ij}. \end{cases} \tag{16.13}$$

*Remark 16.2*  According to Lemma 16.1, it holds that $y'_{ij} (h) > y_{ij} > y'_{ij} (l)$ when both user $i$ and $j$ satisfy $P_{fa} + P_{md} < 1$ and $P_f + P_m < 1$. In other words, when user $i$ makes a high-quality judgement of the service after experiencing it ($S_i = h$), inequality $y'_{ij} (h) > y_{ij}$ always holds. Then applying (16.13), the DPC infers the opinion report as $x_i = 1$ because $y'_{ij} > y_{ij}$. So this inferred report $x_i = 1$ is consistent with user $i$'s real judgement $S_i = h$. Symmetrically, when $S_i = l$, (16.13) can also derive the truthful opinion report $x_i = 0$. Therefore, the rule formulated by (16.13) can truthfully reflect the judgement when the user is honest, under the conditions of $P_{fa} + P_{md} < 1$ and $P_f + P_m < 1$.

#### 16.4.1.4   User Trustworthiness

Based on reports $y_{ij}$ and $y'_{ij}$ $(s_i)$, the DPC calculates user $i$'s trustworthiness through a certain scoring rule. Users with low trustworthiness are classified as the malicious, and their reports will be unconsidered in the service rating system. Next, we first introduce the *strictly proper scoring rule*, which can motivate users to provide truthful reports $y_{ij}$ and $y'_{ij}$ $(s_i)$. The strictly proper scoring rule can be defined as Definition 16.1.

**Definition 16.1  Strictly Proper Scoring Rule** [34]: A binary scoring rule is proper if it leads to an agent maximizing his/her score by truthfully providing his/her report $y \in [0, 1]$, and is strictly proper if an agent can maximize his/her score if and only if providing his/her report truthfully.

The binary logarithmic and quadratic scoring rules shown as (16.14) and (16.15), respectively, are strictly proper, which has been proved in [31].

1. The binary logarithmic scoring rule:

$$R_l\,(y, \omega = 1) = \ln y, \tag{16.14a}$$

$$R_l\,(y, \omega = 0) = \ln\,(1 - y)\,. \tag{16.14b}$$

2. The binary quadratic scoring rule:

$$R_q\,(y, \omega = 1) = 2y - y^2, \tag{16.15a}$$

$$R_q\,(y, \omega = 0) = 1 - y^2. \tag{16.15b}$$

In (16.14) and (16.15), $\omega \in \{0, 1\}$ indicates a binary report.

We define the trustworthiness of user $i$ as a function of $y_{ij}$, $y'_{ij}$ and $x_j$:

$$T_i = \alpha R\left(y_{ij}, x_j\right) + (1 - \alpha)\,R\left(y'_{ij}, x_j\right) + \beta, \tag{16.16}$$

where $R\,(y, \omega)$ is a strictly proper scoring rule, $\alpha \in [0, 1]$ is the parameter weighting the importance of the prior and posterior belief. In addition, the trustworthiness will be cumulative as the service and scoring process continues. A negative trustworthiness can be a reflection of either monetary punishment or the limitation of report providing for the corresponding user, and the negative benefits will be transferred as positive benefits to the users as rewards for their honor and accurate reports. Therefore, to keep the budget balanced, $\beta$ is given by

$$\beta = -\frac{1}{N} \sum_{k=1}^{N} \left[ \alpha R \left( y_{kj}, x_j \right) + (1 - \alpha) R \left( y'_{kj}, x_j \right) \right]. \tag{16.17}$$

In (16.16), $y_{ij}$ and $y'_{ij}$ are the reports from user $i$ before and after he/she makes judgement $S_i = s_i$ for the object service approach, respectively, and $x_j$ is the user $j$'s implicit opinion report inferred by the DPC according to user $j$'s reports.

In addition, according to the analysis above, one can notice that the trustworthiness of user $i$ is determined on user $j$'s inferred opinion report $x_j$, user $i$'s prior belief report $y_{ij}$ and posterior belief report $y'_{ij}$. In other words, one user's trustworthiness is irrelevant to reports or inferred reports of the other users in the system. Therefore, the cooperative cheating of malicious users will have little effect on the evaluation of users' trustworthiness, which is defined by (16.16).

## 16.4.2   Incentive Compatibility

As proved in [34], prior belief report $y_{ij}$ and posterior belief report $y'_{ij}$ ($s_i$) given by user $i$ are temporal separated, which results from that they happen before and after making judgement $S_i = s_i$. Therefore, $y_{ij}$ and $y'_{ij}$ ($s_i$) are independent and then we have

$$
\begin{aligned}
E\left[T_i\right] &= E\left[\alpha R\left(y_{ij}, x_j\right)\right] + E\left[(1 - \alpha) R\left(y'_{ij}, x_j\right)\right] + E\left[\beta\right] \\
&= \alpha \left(1 - \frac{1}{N}\right) E\left[R\left(y_{ij}, x_j\right)\right] \\
&\quad + (1 - \alpha)\left(1 - \frac{1}{N}\right) E\left[R\left(y'_{ij}, x_j\right) | S_i = s_i\right] \\
&\quad - \frac{1}{N} \sum_{k=1, k \neq i}^{N} \left[\alpha R\left(y_{kj}, x_j\right) + (1 - \alpha) R\left(y'_{kj}, x_j\right)\right],
\end{aligned} \tag{16.18}
$$

where both $\alpha \left(1 - \frac{1}{N}\right) R\left(y_{ij}, x_j\right)$ and $(1 - \alpha)\left(1 - \frac{1}{N}\right) R\left(y'_{ij}, x_j\right)$ are still strictly proper [33].

### 16.4.2.1   Binary Logarithmic Scoring Rule

We first apply the binary logarithmic scoring rule. Let $p_1 = P\left(x_j = 1\right)$ and $p_2 = P\left(x_j = 1 \,|\, S_i = s_i\right)$, and then we have

$$
\begin{aligned}
E\left[T_i\right] =& \alpha\left(1 - \frac{1}{N}\right)\left[p_1 \ln y_{ij} + (1 - p_1)\ln\left(1 - y_{ij}\right)\right] \\
=& \alpha\left(1 - \frac{1}{N}\right)\left[p_1 \ln y'_{ij} + (1 - p_1)\ln\left(1 - y'_{ij}\right)\right] \\
& - \frac{1}{N}\sum_{k=1,k\neq i}^{N}\left[\alpha R\left(y_{kj}, x_j\right) + (1 - \alpha) R\left(y'_{kj}, x_j\right)\right].
\end{aligned}
\tag{16.19}
$$

Take the partial derivatives with respect to $y_{ij}$ and $y'_{ij}$:

$$
\frac{\partial E\left[T_i\right]}{\partial y_{ij}} = \alpha\left(1 - \frac{1}{N}\right)\frac{p_1 - y_{ij}}{y_{ij}\left(1 - y_{ij}\right)} = 0,
\tag{16.20a}
$$

$$
\frac{\partial E\left[T_i\right]}{\partial y'_{ij}} = \alpha\left(1 - \frac{1}{N}\right)\frac{p_1 - y'_{ij}}{y'_{ij}\left(1 - y'_{ij}\right)} = 0.
\tag{16.20b}
$$

Therefore we get the optimal values as

$$
\hat{y}_{ij} = p_1 = P\left(x_j = 1\right),
\tag{16.21a}
$$

$$
\hat{y}'_{ij} = p_2 = P\left(x_j = 1 \,|\, S_i = s_i\right).
\tag{16.21b}
$$

Then take the second partial derivatives with respect to $y_{ij}$ and $y'_{ij}$, and let $y_{ij} = \hat{y}_{ij}$ and $y'_{ij} = \hat{y}'_{ij}$, then we have

$$
\left.\frac{\partial E^2\left[T_i\right]}{\partial y_{ij}^2}\right|_{y_{ij}=\hat{y}_{ij}} = \alpha\left(1 - \frac{1}{N}\right)\frac{y_{ij}\left(y_{ij} - 1\right)}{y_{ij}^2\left(1 - y_{ij}\right)^2} < 0,
\tag{16.22a}
$$

$$
\left.\frac{\partial E^2\left[T_i\right]}{\partial y'_{ij}{}^2}\right|_{y'_{ij}=\hat{y}'_{ij}} = \alpha\left(1 - \frac{1}{N}\right)\frac{y'_{ij}\left(y'_{ij} - 1\right)}{y'_{ij}{}^2\left(1 - y'_{ij}\right)^2} < 0.
\tag{16.22b}
$$

Therefore, the maximum of $E\left[T_i\right]$ can be achieved when $y_{ij} = p_1$ and $y'_{ij} = p_2$, which means that user $i$ can receive the maximum trustworthiness if and only if he/she reports both $y_{ij}$ and $y'_{ij}$ truthfully.

### 16.4.2.2   Binary Quadratic Scoring Rule

Next, we employ the binary quadratic scoring rule shown as (16.15). Thus we have

$$
\begin{aligned}
E\left[T_i\right] = & \,\alpha\left(1-\frac{1}{N}\right)\left[p_1\left(2y_{ij}-y_{ij}^2\right)+(1-p_1)\left(1-y_{ij}^2\right)\right] \\
& +(1-\alpha)\left(1-\frac{1}{N}\right)\left[p_2\left(2y_{ij}'-y_{ij}'^2\right)+(1-p_2)\left(1-y_{ij}'^2\right)\right] \\
& -\frac{1}{N}\sum_{k=1,k\neq i}^{N}\left[\alpha R\left(y_{kj},x_j\right)+(1-\alpha)R\left(y_{kj}',x_j\right)\right].
\end{aligned}
\tag{16.23}
$$

Take the partial derivatives with respect to $y_{ij}$ and $y'_{ij}$, and set them to zero, we get the same optimal values as (16.21a) and (16.21b). Then take the second partial derivatives, the following inequality

$$
\frac{\partial E^2\left[T_i\right]}{\partial y_{ij}^2}=\frac{\partial E^2\left[T_i\right]}{\partial y_{ij}'^2}=-2\alpha\left(1-\frac{1}{N}\right)<0
\tag{16.24}
$$

can be always satisfied.

*Remark 16.3* Noticing that $\partial^2 E\left[T_i\right]/\partial y_{ij}^2<0$ and $\partial^2 E\left[T_i\right]/\partial y_{ij}'^2<0$ will always be satisfied no matter whether the binary logarithmic or quadratic scoring rule is applied, the maximum of $E\left[T_i\right]$ can be reached when satisfying both (16.21a) and (16.21b). In other words, user $i$ can receive the maximum trustworthiness if and only if he/she provides both $y_{ij}$ and $y'_{ij}$ truthfully, as mentioned previously. Assume that the cooperative cheating exists, which means that malicious users can contact with each other and manage the malicious behaviour. According to Definition 16.1, user $i$ will obtain a lower score by reporting untruthfully than truthfully when his/her peer user $j$ is a malicious one. For example, user $i$ experiences a high-quality service and it means that his/her honest reports satisfy $y'_{ij}>y_{ij}$. However, because of user $j$'s dishonest implicit opinion $x_j=0$, user $i$ will obtain a higher score if he/she gives a lower $y'_{ij}<y_{ij}$ instead of reporting truthfully, according to the binary logarithmic or quadratic scoring rule formulated as (16.14b) and (16.15b), respectively. To make sure that the honest users are predominant even when the cooperative cheating happens in the social network, we assume that the number of malicious users is less than the half of the total. Based on this assumption, the users with accurate information reports and prediction reports will always receive higher trustworthiness in a long term; meanwhile, the malicious users will be punished by a loss of trustworthiness every time they announce dishonest reports resulting in cheating opinion reports.

## 16.5   User Trustworthiness and Unreliability Based Service Rating

### 16.5.1   Unreliability of User Report

In private-prior peer prediction, all users are required to report their prior belief that their peer users will report a high evaluation for the service before experiencing the service $y_{ij} = P_i(x_j = 1)$. This report can be obtained by the past reports $x_j$ inferred by the DPC and published by the cloud, which means that past reports $x_j$ are accessible for $i$'s other friends in the social network, the cloud and DPC. Therefore, it is difficult to fabricate information report $y_{ij}$ for malicious users. To achieve cheating, malicious user $i$ needs to manage his/her information and prediction report according to (16.13), i.e., $y'_{ij} = y_{ij} + \varepsilon$ ($\varepsilon > 0$) with probability $P^c_{m,i}$ when the service quality is low ($Q = l$), and $y'_{ij} = y_{ij} - \varepsilon$ with probability $P^c_{f,i}$ when the service quality is high ($Q = h$). Meanwhile, malicious users have to set $\varepsilon$ as small as possible to avoid being punished by much loss of score and trustworthiness when their peer users are honest ones. In addition, we can notice that the false-alarm report and missed-detection report do not only result from the wrong judgements of honest users, but also due to the dishonest users' cheating behaviours, according to (16.1) and (16.2). Both of the situations above are considered as unreliable behaviours which need to be identified and removed from the final service rating. Therefore, it is necessary to set a threshold to limit the minimum gap between $y_{ij}$ and $y'_{ij}$.

Next, we analyze the influence of false-alarm judgement and missed-detection judgement on the scoring. Taking the derivative of (16.6a) and (16.6b) both with respect to $P_{fa,i}$ and $P_{md,i}$, we can calculate to get

$$\frac{\partial y'_{ij}(l)}{\partial P_{fa,i}} = \Phi_1 \left(1 - P_{md,i}\right)\left(1 - P_{f,j} - P_{m,j}\right), \tag{16.25a}$$

$$\frac{\partial y'_{ij}(l)}{\partial P_{md,i}} = \Phi_1 P_{fa,i}\left(1 - P_{f,j} - P_{m,j}\right), \tag{16.25b}$$

$$\frac{\partial y'_{ij}(h)}{\partial P_{fa,i}} = -\Phi_2 P_{md,i}\left(1 - P_{f,j} - P_{m,j}\right), \tag{16.25c}$$

$$\frac{\partial y'_{ij}(h)}{\partial P_{md,i}} = -\Phi_2 \left(1 - P_{fa,i}\right)\left(1 - P_{f,j} - P_{m,j}\right), \tag{16.25d}$$

where $\Phi_1 = P(Q = h)P(Q = l)/(\varphi_1 + \varphi_2)^2$, $\Phi_2 = P(Q = h)P(Q = l)/(\varphi_3 + \varphi_4)^2$. Based on the previous assumptions of $P_{fa,i} < 0.5$, $P_{md,i} < 0.5$ and $P_{f,j} + P_{m,j} < 1$, we have

$$\frac{\partial y'_{ij}(l)}{\partial P_{fa,i}} > \frac{\partial y'_{ij}(l)}{\partial P_{md,i}} > 0, \quad \frac{\partial y'_{ij}(h)}{\partial P_{md,i}} < \frac{\partial y'_{ij}(h)}{\partial P_{fa,i}} < 0. \tag{16.26}$$

So under both of situations $Q = h$ and $Q = l$, the score of user $i$ goes down with the increasing $P_{fa,i}$ and $P_{md,i}$ when user $j$ reports truthfully, according to (16.14a)/(16.15a) and (16.14b)/(16.15b), respectively. In other words, for fixed $P_f$, $P_j$ and $P(Q = h)$, the honest users with high judgement accuracy will receive higher scores and trustworthiness, compared to those honest users with high judgement error rates and malicious users reporting their prediction inversely and conservatively to give wrong reports and minimize the loss of scores. In the service rating system, neither implicit opinion reports of malicious users nor honest users with low judgement accuracy should be considered. To identify the two kinds of unreliable behaviour, we define an unreliability index to indicate the unreliability of user $i$ by his/her prior belief report $y_{ij}$ and posterior belief report $y'_{ij}$ as follows.

$$
\rho_i = \begin{cases} \dfrac{\left|y'_{ij}-P_{m,j}\right|P(Q=l)}{\left|y'_{ij}-(1-P_{f,j})\right|P(Q=h)}, & y'_{ij} < y_{ij}, \\[3mm] \dfrac{\left|y'_{ij}-(1-P_{f,j})\right|P(Q=h)}{\left|y'_{ij}-P_{m,j}\right|P(Q=l)}, & y'_{ij} > y_{ij}. \end{cases}
\tag{16.27}
$$

*Remark 16.4*  In (16.27), the first situation $y'_{ij} < y_{ij}$ indicates that the more report $y'_{ij}$ is closed to $P\{x_j = 1 | Q = l\}$ when the service quality is low and farther away from $P\{x_j = 1 | Q = h\}$ when the service quality is high, the more reliable $y'_{ij}$ is. Meanwhile, for $y'_{ij} > y_{ij}$, when report $y'_{ij}$ is closed to $P\{x_j = 1 | Q = h\}$ and far away from $P\{x_j = 1 | Q = l\}$, this report can be considered reliable. In addition, according to (16.26), $y'_{ij}(l)$ increases with growing $P_{fa,i}$ and $P_{md,i}$, and is more sensitive to $P_{fa,i}$ than $P_{md,i}$; $y'_{ij}(h)$ decreases with growing $P_{fa,i}$ and $P_{md,i}$, and is more sensitive to $P_{md,i}$ than $P_{fa,i}$. With assumption $P_{fa,i}, P_{md,i} \in [0, 1]$, we can get that $P_{m,j} < y'_{ij}(l)$, $y'_{ij}(h) < 1 - P_{f,j}$, thus the definition of unreliability shown in (16.27) can be rewritten as

$$
\rho_i = \begin{cases} \dfrac{\left[y'_{ij}-P_{m,j}\right]P(Q=l)}{\left[(1-P_{f,j})-y'_{ij}\right]P(Q=h)}, & y'_{ij} < y_{ij}, \\[3mm] \dfrac{\left[(1-P_{f,j})-y'_{ij}\right]P(Q=h)}{\left[y'_{ij}-P_{m,j}\right]P(Q=l)}, & y'_{ij} > y_{ij}. \end{cases}
\tag{16.28}
$$

To calculate the unreliability of users' reports, the DPC needs to observe the report error rates $P_f$ and $P_m$ of each user based on the historical reports and service rating results. In addition, we assume that the service quality, denoted by $P(Q = l)$ and $P(Q = h)$, can also be obtained according to a long time scale and relatively stable historical rating results of services. Such assumptions are feasible and reasonable, considering that most current service-based application systems have the ability to provide such information. By utilizing (16.28), the users with high unreliability $\rho$ are considered to be uncertainty ones who might be honest users with high error judgement rate or malicious users. Reports from these users are not reliable for the DPC to rate the quality of service. Consequently, the DPC needs to set a threshold $\rho_{thr}$, and reports from the users with unreliability exceeding $\rho_{thr}$ will be removed from the service rating procedure. The threshold can be designed by the typical error rates of honest users with relatively high judgement accuracy.

Next, we describe the validity of the user unreliability defined in (16.28). Take situation $Q = h$ for instance, malicious user $i$ has to give the prediction report $y'_{ij} = y_{ij} - \varepsilon < y_{ij}$ to achieve cheating. In order to get a lower unreliability value below the threshold and make his/her cheating make sense in the service rating, user $i$ needs to fabricate report $y'_{ij}$ to make it close to $P_{m,j}$ and away from $1 - P_{f,j}$. With conditions $P_{f,i} < 0.5$ and $P_{m,i} < 0.5$, the smaller $y'_{ij}$ is, the lower unreliability value will be get. On the other hand, the majority honest users trend to report the implicit opinion reports as $x_j = 1$ when $Q = h$. According to (16.14a) and (16.15a), the score of user $i$ decreases with reducing $y'_{ij}$ when his/her peer user $j$ gives an accurate and honest report. Symmetrically, the dilemma still exists when $Q = l$. Therefore, it is difficult for malicious users to get high trustworthiness and low unreliability at the same time, if they report trickly. However, for those "bad functioning" honest users with relatively high error rate of judgement, the best choice is still reporting $y_{ij}$ and $y'_{ij}$ truthfully. It is unnecessary for them to modify their $y'_{ij}$ because their benefit is the score and trustworthiness determined by the information and prediction reports, and this benefit is irrelevant to that whether their reports are accepted by the DPC or not.

## 16.5.2   Peer Prediction Based Service Rating

According to the user's trustworthiness and unreliability analysis above, we design the private-prior peer prediction based service rating method as following procedures.

1. For every user $i$ who accepts the service, choose another non-overlapped user $j$ randomly among his/her friends as $i$'s peer.
2. Ask user $i$ for his/her prior belief report $y_{ij} \in [0, 1]$, i.e. his/her peer $j$ will provide a report to the cloud that $j$ evaluates the service as high-quality.
3. User $i$ experiences the service and then makes his/her judgement $S_i = s_i$ for the quality of the service.
4. Ask user $i$ for his/her posterior belief report $y'_{ij} \in [0, 1]$ to the cloud, with $y'_{ij} \neq y_{ij}$, that his/her peer $j$ will provide a report of receiving a high-quality service.
5. The DPC calculates the unreliability of every user by applying (16.28), and removes reports of users with $\rho_i > \rho_{thr}$ from the service rating system.
6. The DPC infers the implicit opinion report $x_i$ of user $i$ through (16.13), and calculates user $i$'s trustworthiness according to (16.14), (16.15) and (16.16) assisted by user $j$'s inferred opinion report $x_j$. Then remove reports of users with lowest trustworthiness from the service rating system.

7. The DPC makes the rating for the service by implicit opinion reports of users with both high trustworthiness and lower unreliability through (16.3).

## 16.6 Performance Evaluation

In this part, we perform numerical simulation experiments to analyze the properties and performances of the private-prior peer prediction service rating system and its influential factors such as the proportion of the malicious users and $\varepsilon$. First, we analyze the effect of the time accumulation on the trustworthiness and unreliability. In the peer prediction mechanism, we can notice that if the peer user of an honest user is a malicious one who decides to cheat when he/she reports to the cloud, the trustworthiness of the honest user trends to be low because of the strictly proper scoring rule. However, when malicious users are not predominant in the social network, which means that the proportion of the malicious users is less than half of the total, then honest users' accumulative trustworthiness will increase distinctly comparing with malicious ones in a long term.

### 16.6.1 Simulation Settings

The simulation for the service rating system is operated based on the topology of Flickr, a real-world online social network database. The Flickr topology contains 5,899,882 edges connecting 80,513 users, and the edge represents the friendship of the connected two users. In addition, this friendship of users in the Flickr network, also known as the topology, is determined by their favorites. In other words, the connection between any two users is established if the corresponding two users are sharing the common favorites and have followed the same community. Then such two users will be considered as a pair of peers. The topology of the Flickr network are depicted in Fig. 16.2. These users are separated into three types, i.e., reliable honest users with high judgement accuracy rate, malicious users with high judgement accuracy rate and high error report rate, and unreliable honest users with relatively high judgement error rate but always report truthfully. The three types of users exist with some certain percentage. We set that false alarm of judgement $P_{fa}$ and missed detection of judgement $P_{md}$ are uniform distribution variables, and for all reliable and malicious users $P_{fa}$, $P_{ma} \sim U [0.01, 0.02]$, and for unreliable users $P_{fa}$, $P_{ma} \sim U [0.05, 0.06]$. In addition, as analyzed in the Remarks of Lemma 16.1, malicious users need to make sure that their false alarm and missed detection of report, $P_f$ and $P_m$, are both smaller than 0.5 to achieve a continuous trick. Therefore, we set $P_f = P_m = 0.3$ ($< 0.5$) for malicious users in the following experiences. We assume that all honest users always report truthfully, i.e., $P_f = P_m = 0$.

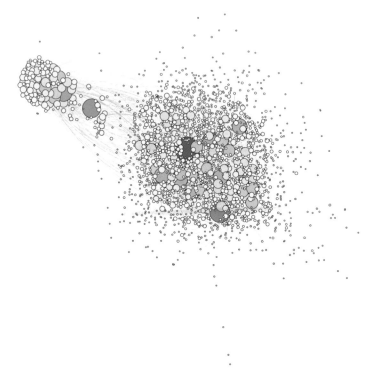

**Fig. 16.2** Graph structures of the Flickr network used for simulation

- *Historical database.* To calculate the unreliability of each user, the DPC needs to obtain their historic error rates of report. So we first establish the report database by allowing each user judge the quality of the service independently and then report to the cloud all according to the type of the user. The process repeats 80 times and in each time, the probability of high service quality is set as $P(Q = H) = 0.6$. In addition, the quality of the service is determined through the majority rule shown as (16.3) by applying reports from all of the users.

### 16.6.2   Accumulative Trustworthiness and Unreliability

Then in the following experiences, the private-prior peer prediction method is introduced, and the peer of each user is updated in every new experience. Then new implicit opinion reports (inferred by $y_{ij}$ and $y'_{ij}$) and service rating results are added into the database and provide the historical data for the DPC. We consider that the trustworthiness and the unreliability of each user can be accumulated with the increasing service times. To calculate the trustworthiness, both of the scoring rules, i.e., binary logarithmic and binary quadratic, are applied. Simulation results of users' accumulative trustworthiness and unreliability in the following 200 times of service are shown in Figs. 16.3 and 16.4, in which the percentages of reliable honest user, malicious user and unreliable honest user are set as 40, 40 and 20%, respectively. In both of the figures, we show the results of some sample users selected from the three types randomly. In Fig. 16.3, the trustworthiness of honest users might be negative at the beginning, when their peer users are the malicious. On the other hand, some malicious ones even obtain larger trustworthiness at the beginning, when their peers are also the malicious. However, resulting from the peer updating after each time of service, as well as the small proportion of the malicious, the predomination of honest users trend to work in a long term. Figure 16.3 indicates that the accumulative trustworthiness of honest users grows with the service rating times or experience time. On the contrary, the accumulative trustworthiness of malicious users drops down and is negative. In addition, we can notice that no matter which scoring rule is applied, the accumulative trustworthiness shows the similar characteristics and tendency.

Similar results of accumulative unreliability are shown in Fig. 16.4, in which the gaps are more obvious among different types of users. Moreover, we can notice that unreliable honest users can be identified through the unreliability index, which cannot be achieved by the trustworthiness. This result demonstrates that the best choice for unreliable honest users is still reporting truthfully, and their unreliability will bring no hazard to their high positive trustworthiness.

### 16.6.3 Influence of ε, Scoring Rules and User Structure

In the basic private-prior peer prediction mechanism, the strictly proper scoring rule leads malicious users to fabricate minimum $\varepsilon$, i.e., $y'_{ij} = y_{ij} + \varepsilon$ ($\varepsilon > 0$) when the quality of the service is low, and $y'_{ij} = y_{ij} - \varepsilon$ when the quality is high. In the trustworthy service racing system, the unreliability index proposed brings the dilemma to malicious users when they set $\varepsilon$ as discussed previously. Next, we test the influence of $\varepsilon$ on the average trustworthiness and unreliability. Considering two cases of user structure, the percentages of reliable honest user, malicious user and unreliable honest user are set as 60, 20 and 20% in one case, respectively, and in another case are set as 40, 40 and 20%. We repeat the service rating experiments for 200 times, and then calculate the average trustworthiness and unreliability of each type of users in these 200 times experiments (not the accumulative trustworthiness or unreliability). Results in Fig. 16.5a, b present the average trustworthiness when applying binary logarithmic and binary quadratic scoring rules, respectively, when $\varepsilon \in [0.1, 0.2]$. In addition, Fig. 16.6 presents how the average unreliability changes when $\varepsilon$ increases. As depicted in Figs. 16.5 and 16.6, both the trustworthiness and unreliability decrease with the increase of $\varepsilon$ for malicious users, which demonstrates the incentive and identification capabilities when combining trustworthiness and

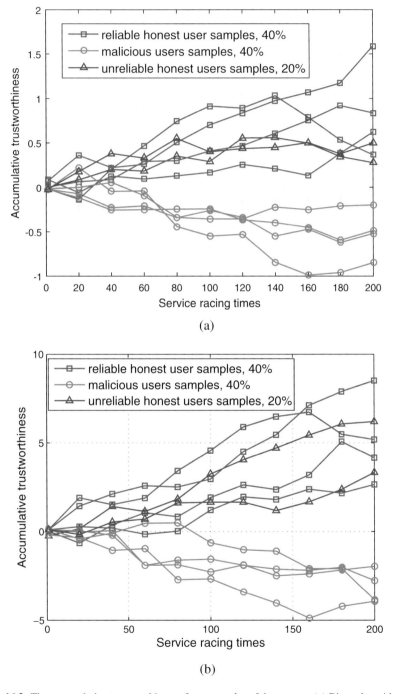

**Fig. 16.3** The accumulative trustworthiness of user samples of three types. (**a**) Binary logarithmic scoring. (**b**) Binary quadratic scoring

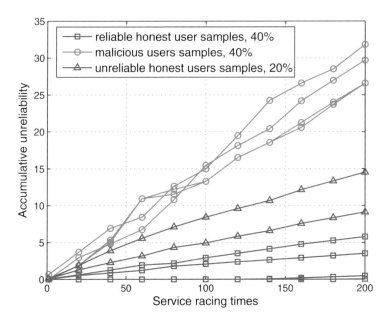

**Fig. 16.4** The accumulative unreliability of user samples of three types

unreliability together to evaluate users reports. On the other hand, the average trustworthiness and unreliability of honest users are not sensitive to changing $\varepsilon$. In addition, we can notice that, when the percentage of malicious users is small, the gaps between the trustworthiness and unreliability malicious and honest user tend to be wide, which will make it much easier to identify the malicious.

Removing unreliable reports and reports from users with low trustworthiness, we rate the service quality by trustful reports to improve the accuracy of rating. In this part, we define the service rating accuracy as the ratio of the number of selected correct reports to the number of all correct reports. In addition, the threshold of unreliability is set as an empirical value obtained from the training of historical database, to be specific, $\rho_{thr} = 5$. Then we test the service rating accuracy over the proportion of malicious users, unreliable honest users' error rates of judgement and $\varepsilon$. Results shown in Fig. 16.7 indicate that the service rating accuracy decreases with the increasing proportion of malicious users. When this proportion is closed to 0.5, the rating accuracy decreases distinctly because of the probable cooperative cheating. In addition, the rating accuracy is higher when unreliable honest users' $P_{fa}$, $P_{ma} \sim U [0.35, 0.45]$ than $P_{fa}$, $P_{ma} \sim U [0.1, 0.2]$, which results from that honest users with higher judgement error rates can be identified more easily by applying the unreliability index. Figure 16.7 also indicates that the lower $\varepsilon$ malicious users set, the harder they can be detected through the trustworthiness.

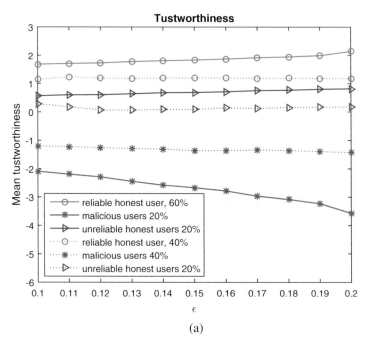

(a)

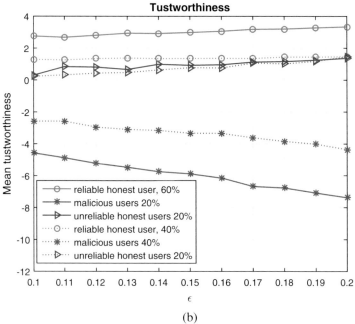

(b)

**Fig. 16.5** The average trustworthiness of different types of users versus the percentage of each user type and $\varepsilon$. (**a**) Binary logarithmic scoring. (**b**) Binary quadratic scoring

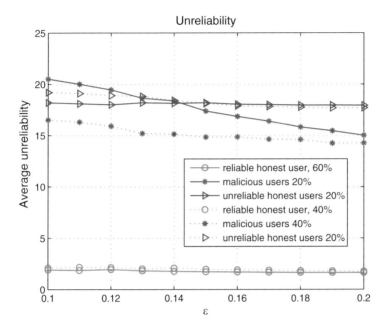

**Fig. 16.6** The average unreliability of different types of users versus the percentage of each user type and $\varepsilon$

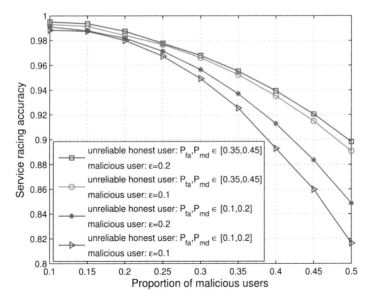

**Fig. 16.7** The service racing accuracy versus the percentage of each user type, error rates of judgement $P_{fa}$, $P_{md}$ and $\varepsilon$

## 16.7   Conclusions

In this part, we proposed a cloud based architecture for the service rating system. To achieve a trustworthy service rating, a private-prior peer prediction based mechanism was designed to identify malicious and dishonest users. Coupled with some certain strictly proper scoring rules, the peer prediction method can evaluate users' trustworthiness and motivate them to report honestly. Moreover, an unreliability index was also designed to ensure the reliability of the users' reports. According to the trustworthiness and unreliability index, untruthful and unreliable reports can be identified and eliminated to improve the accuracy of service rating. Simulation results indicated that the proposed peer prediction based trustworthy service rating system can identify malicious and unreliable behaviours effectively, and achieve relatively high service rating accuracy.

## References

1. M. Chen, Y. Zhang, L. Hu, T. Taleb, and Z. Sheng, "Cloud-based wireless network: Virtualized, reconfigurable, smart wireless network to enable 5G technologies," *Mobile Networks and Applicat.*, vol. 20, no. 6, pp. 704–712, Feb. 2015.
2. G. Zhao, X. Qian, and C. Kang, "Service rating prediction by exploring social mobile users' geographical locations," *IEEE Trans. Big Data*, vol. 3, no. 1, pp. 67–78, Mar. 2017.
3. J. Du, C. Jiang, S. Yu, K. C. Chen, and Y. Ren, "Privacy protection: A community-structured evolutionary game approach," in *IEEE Global Conf. Signal Inform. Process. (GlobalSIP)*. Washington, DC, USA, 7-9 Dec. 2016, pp. 415–419.
4. G. Zhao, X. Qian, and X. Xie, "User-service rating prediction by exploring social users' rating behaviors," *IEEE Trans. Multimedia*, vol. 18, no. 3, pp. 496–506, Mar. 2016.
5. X. Li, H. Ma, F. Zhou, and W. Yao, "T-broker: A trust-aware service brokering scheme for multiple cloud collaborative services," *IEEE Trans. Inf. Forens. Security*, vol. 10, no. 7, pp. 1402–1415, Jul. 2015.
6. Y. Yang, Y. Sun, S. Kay, and Q. Yang, "Securing rating aggregation systems using statistical detectors and trust," *IEEE Trans. Inf. Forens. Security*, vol. 4, no. 4, pp. 883–898, Dec. 2009.
7. L. Xu, C. Jiang, Y. Chen, Y. Ren, and K. J. R. Liu, "Privacy or utility in data collection? a contract theoretic approach," *IEEE J. Sel. Topics Signal Process.*, vol. 9, no. 7, pp. 1256–1269, Oct. 2015.
8. S. Chikkerur, V. Sundaram, M. Reisslein, and L. J. Karam, "Objective video quality assessment methods: A classification, review, and performance comparison," *IEEE Trans. Broadcasting*, vol. 57, no. 2, pp. 165–182, Jun. 2011.
9. K. Seshadrinathan, R. Soundararajan, A. C. Bovik, and L. K. Cormack, "Study of subjective and objective quality assessment of video," *IEEE Trans. Image Processing*, vol. 19, no. 6, p. 1427, 2010.
10. X. Liang, X. Lin, and X. S. Shen, "Enabling trustworthy service evaluation in service-oriented mobile social networks," *IEEE Trans. Parallel Distrib.uted Syst.*, vol. 25, no. 2, pp. 310–320, Feb. 2014.
11. B. Jennings and P. Malone, "Flexible charging for multi-provider composed services using a federated, two-phase rating process," in *2006 10th IEEE/IFIP Network Operations Manage. Symp. (NOMS 2006)*. Vancouver, BC, Apr. 2006, pp. 13–23.

12. S. Sharma and J. Negi, "Customer importance rating of service quality dimensions for automobile service," *Int. J. of Eng. Sci. and Technology*, vol. 6, no. 12, pp. 822–826, Dec. 2014.

13. G. Di Fabbrizio, A. Aker, and R. Gaizauskas, "Summarizing online reviews using aspect rating distributions and language modeling," *IEEE Intell. Syst.*, no. 3, pp. 28–37, May. 2013.

14. M. Allahbakhsh and A. Ignjatovic, "An iterative method for calculating robust rating scores," *IEEE Trans. Parallel Distrib.uted Syst.*, vol. 26, no. 2, pp. 340–350, Feb. 2015.

15. K. Nagaraja, G. Gama, R. Bianchini, R. P. Martin, W. Meira Jr, and T. D. Nguyen, "Quantifying the performability of cluster-based services," *IEEE Trans. Parallel Distrib.uted Syst.*, vol. 16, no. 5, pp. 456–467, May 2005.

16. X. Ye and J. Zheng, "An adaptive rating system for service computing," in *2013 12th IEEE Int. Conf. Trust, Security, Privacy Computing Commun. (TrustCom)*. Melbourne, VIC, Jul. 2013, pp. 1817–1824.

17. G. Zhao, X. Qian, and X. Xie, "User-service rating prediction by exploring social users' rating behaviors," *IEEE Trans. Multimedia*, vol. 18, no. 3, pp. 496–506, Mar. 2016.

18. Y. Kim, E.-W. Jhee, J. Choe, J.-S. Choi, and Y. Shin, "A measurement model for trustworthiness of information on social network services," in *2015 Int. Conf.Inform. Networking (ICOIN)*. Cambodia, Jan. 2015, pp. 437–438.

19. J. Du, C. Jiang, K. C. Chen, Y. Ren, and H. V. Poor, "Community-structured evolutionary game for privacy protection in social networks," *IEEE Trans. Inf. Forens. Security*, vol. 13, no. 3, pp. 574–589, Mar. 2018.

20. J. Jiang, G. Han, F. Wang, L. Shu, and M. Guizani, "An efficient distributed trust model for wireless sensor networks," *IEEE Trans. Parallel Distrib.uted Syst.*, vol. 26, no. 5, pp. 1228–1237, May 2015.

21. D. He, S. Chan, and M. Guizani, "User privacy and data trustworthiness in mobile crowd sensing," *IEEE Wireless Commun.*, vol. 22, no. 1, pp. 28–34, Feb. 2015.

22. B. Kantarci and H. T. Mouftah, "Trustworthy sensing for public safety in cloud-centric internet of things," *IEEE J. Internet of Things*, vol. 1, no. 4, pp. 360–368, Jul. 2014.

23. N. Zhang, N. Cheng, N. Lu, H. Zhou, J. W. Mark, and X. Shen, "Risk-aware cooperative spectrum access for multi-channel cognitive radio networks," *IEEE J. Sel. Areas. Commun.*, vol. 32, no. 3, pp. 516–527, Mar. 2014.

24. T. Noor, Q. Sheng, L. Yao, S. Dustdar, and A. Ngu, "CloudArmor: Supporting reputation-based trust management for cloud services," *IEEE Trans. Parallel Distrib.uted Syst.*, vol. 27, no. 2, pp. 367–380, Mar. 2015.

25. B. Li, L. Liao, H. Leung, and R. Song, "PHAT: A preference and honesty aware trust model for web services," *IEEE Trans. Network and Service Manage.*, vol. 11, no. 3, pp. 363–375, Sept. 2014.

26. H. Dong, C. Wu, Z. Wei, and Y. Guo, "Dropping activation outputs with localized first-layer deep network for enhancing user privacy and data security," *IEEE Trans. Inf. Forens. Security*, vol. 13, no. 3, pp. 662–670, Mar. 2018.

27. L. Xu, C. Jiang, N. He, Z. Han, and A. Benslimane, "Trust-based collaborative privacy management in online social networks," *IEEE Trans. Inf. Forens. Security*, pp. 1–1, May 2018.

28. B. Rashidi, C. Fung, A. Nguyen, T. Vu, and E. Bertino, "Android user privacy preserving through crowdsourcing," *IEEE Trans. Inf. Forens. Security*, vol. 13, no. 3, pp. 773–787, Mar. 2018.

29. L. Xu, C. Jiang, Y. Chen, Y. Ren, and K. J. R. Liu, "User participation in collaborative filtering-based recommendation systems: A game theoretic approach," *IEEE Trans. Cybernetics*, vol. PP, no. 99, pp. 1–14, Feb. 2018.

30. J. Du, C. Jiang, Z. Han, H. Zhang, S. Mumtaz, and Y. Ren, "Contract mechanism and performance analysis for data transaction in mobile social networks," *IEEE Trans. Network Sci. Eng.*, vol. 6, no. 2, pp. 103–115, Apr. –Jun. 2019.

31. R. Selten, "Axiomatic characterization of the quadratic scoring rule," *Experimental Econ.*, vol. 1, no. 1, pp. 43–62, Jun. 1998.

32. T. Gneiting and A. E. Raftery, "Strictly proper scoring rules, prediction, and estimation," *J. of the American Statistical Association*, vol. 102, no. 477, pp. 359–378, Jan. 2012.
33. N. Miller, P. Resnick, and R. Zeckhauser, "Eliciting informative feedback: The peer-prediction method," *Manage. Sci.*, vol. 51, no. 9, pp. 1359–1373, Sept. 2005.
34. J. Witkowski and D. C. Parkes, "Peer prediction without a common prior," in *Proc. 13th ACM Conf. Electron. Commerce*. ACM, Valencia, Spain, Jun. 2012, pp. 964–981.
35. B. Faltings, J. J. Li, and R. Jurca, "Incentive mechanisms for community sensing," *IEEE Trans. Comput.*, vol. 63, no. 1, pp. 115–128, Jul. 2014.
36. G. Radanovic and B. Faltings, "Incentives for truthful information elicitation of continuous signals," in *Twenty-Eighth AAAI Conf. Artificial Intell.*, Jul. 2014, pp. 770–776.
37. Y. Gan, C. Jiang, N. C. Beaulieu, J. Wang, and Y. Ren, "Secure collaborative spectrum sensing: A peer-prediction method," *IEEE Trans. Commun.*, vol. 64, no. 10, pp. 4283–4294, Oct. 2016.
38. X. A. Gao, A. Mao, Y. Chen, and R. P. Adams, "Trick or treat: putting peer prediction to the test," in *Proc. fifteenth ACM conf. Econ. computation*. California, US, Jun. 2014, pp. 507–524.
39. J. Du, C. Jiang, J. Wang, S. Yu, and Y. Ren, "Trustable service rating in social networks: A peer prediction method," in *IEEE Global Conf. Signal Inform. Process. (GlobalSIP)*. Washington, DC, USA, 7-9 Dec. 2016, pp. 415–419.
40. J. Witkowski, Y. Bachrach, P. Key, and D. C. Parkes, "Dwelling on the negative: Incentivizing effort in peer prediction," in *First AAAI Conf. Human Computation Crowdsourcing*, Nov. 2013, pp. 190–197.
41. G. Radanovic and B. Faltings, "A robust bayesian truth serum for non-binary signals," in *Proc. 27th AAAI Conf. Artificial Intell. (AAAI 2013)*, no. EPFL-CONF-197486, Jul. 2013, pp. 833–839.
42. J. Du, E. Gelenbe, C. Jiang, H. Zhang, Z. Han, and Y. Ren, "Data transaction modeling in mobile networks: Contract mechanism and performance analysis," in *IEEE Global Commun. Conf. (GLOBECOM)*. Singapore, 4-8 Dec. 2017.

# Chapter 17
# Cooperative Privacy Protection Among Mobile User

**Abstract** Social networks have attracted billions of users and supported a wide range of interests and practices. Users of social networks can be connected with each other by different communities according to professions, living locations and personal interests. With the development of diverse social network applications, academic researchers and practicing engineers pay increasing attention to the related technology. As each user on the social network platforms typically stores and shares a large amount of personal data, the privacy of such user-related information raises serious concerns. Most research on privacy protection relies on specific information security techniques such as anonymization or access control. However, the protection of privacy depends heavily on the incentive mechanisms of social networks, like users' psychological decisions on security execution and socio-economic considerations. For example, the desire to influence the behaviors of other people may change a user's choice of security setting. In this chapter, a game theoretic framework is established to model users' interactions that influence users' decisions as to whether to undertake privacy protection or not. To model the relationship of user communities, community-structured evolutionary dynamics are introduced, in which interactions of users can only happen among those users who have at least one community in common. Then the dynamics of the users' strategies to take a specific privacy protection or not is analyzed based on the proposed community structured evolutionary game theoretic framework. Experiments show that the proposed framework is effective in modeling the users' relationships and privacy protection behaviors. Moreover, results can also help social network managers to design appropriate security service and payment mechanisms to encourage their users to take the privacy protection, which can promote the spreading of privacy behavior throughout the network.

**Keywords** Community Structure Based Evolutionary Games · Privacy Protection · Behavior Spreading · Social Networks

J. Du, C. Jiang, *Cooperation and Integration in 6G Heterogeneous Networks*,
Wireless Networks, https://doi.org/10.1007/978-981-19-7648-3_17

## 17.1  Introduction

Over the past decade there has been an unprecedented development of social network applications. Online social networks, such as Facebook, Google+, and Twitter are inherently designed to enable people to distribute and share personal and public information [1–5]. In addition, social connections among friends, colleagues, family members, and even strangers with similar interests are established via these online social network platforms. However, as these platforms, as well as other online applications and cloud computing, allow their users to host large amounts of personal data on their platforms, important concerns regarding the security and privacy of user-related information arise [6–10]. How to protect users' personal information, and encourage users to participate the privacy protection to improve the information security of the entire social network, have become one of critical problems for social network managers.

In response, many social networks have provided different privacy protection measures to try to protect their users' personal information. Take "Privacy Setting and Tools" of Facebook for instance, it allows the users to decide who can see their stuff, contact them and look them up to obtain different levels of protection. In addition, the "Privacy and safety" settings of Twitter provide some similar options for its users to determine that who can receive their Tweets, tag them in photos, etc. Furthermore, privacy protection mechanisms have been also studied from many aspects such as information collection [11], information processing [12–14], anonymity [15], access control [16, 17], etc., to improve the security of users' data. However, users' decisions, actions, and preferences regarding personal information security, and social-economic relationships, can critically influence the implementation of privacy protection on online social network platforms. On the one hand, a user's selection of security level can protect her or his own personal information, and help to preserve the privacy of others related to this user. On the other hand, users' behavior to adopt security measures can be affected by the decisions of other users and potentially spread throughout the entire social network, depending of course depends on the network topology. Thus, the privacy protection of users in the network relies on its users to make use of security services to protect their friends' and their own information, and this behavior is conditional. One user has to make a decision on whether or not to undertake privacy protection according to many considerations, such as if and how many friends of his/hers make the same choice. To understand and to model such interactions among users, game theory can be used. Particularly useful is evolutionary game theory which considers that a game is played over and over again by socially conditioned players randomly drawn from large populations. It studies population shift and evolution processes, and pays particular attention to the dynamics and stability of the strategies of the entire population. For the privacy protection issues, evolutionary game theory can be used to model the spreading of users' security behavior over social networks, which heavily depends on the interaction and friendship among the users. Thus, in this part, we establish a community structured evolutionary game theoretic framework to analyze and reveal the interactions between users and the spreading of security behavior throughout a social network.

Most current research on privacy protection and behavior spreading considers a social network of a regular, random, and flattened topology. Based on this assumption, individuals connect with each other, and the influence of users' actions and behaviors is spread over the entire social network accordingly. However, in a real social network, the relationships among users are much more complicated than these simple models. Moreover, the interaction and influence between any two users largely depends on how close the relationship between these two users is. In this research, we model the population of social networks as a community structure in order to characterize the connections of users in a more appropriate and accurate way. By this community structure based model, we may successfully analyze the spreading of the privacy protection behavior over the social network.

The main contributions and our main ideas are summarized as follows:

1. We propose a game theoretic framework to model the interaction and influence when users choose strategies that make use of the privacy protection or not. The framework reveals that the protection of the users' privacy information depends not only on the users' own strategies, but also strategies of other users. In other words, the framework can analyze the information protection through users' interactions and decision making.

2. We establish a community structure based evolutionary game theory to model and analyze the privacy protection over social networks with a community structured population. This framework can characterize the dynamics of the process of the users' behaviors regarding taking the privacy protection or not. In addition, the framework can also predict the final stable behavior spreading state.

3. Based on the proposed community structure based evolutionary game theory framework, we analyze the dynamics of the users' behaviors with regard to taking the privacy protection or not. The critical cost performance is analyzed for both non-triggering game and triggering game scenarios. The critical cost performance is an important parameter, exceeding the value of which the behavior of taking the privacy protection is more frequent than the behavior of not taking the privacy protection in the equilibrium distribution of the deviation-imitation process in the social network.

The remainder of this part is organized as follows. Section 17.2 reviews the existing methods for personal information security. In Sect. 17.3, the community structure based evolutionary game formulation of privacy protection in social networks is described. The privacy protection among users belonging to $K$ communities and evolution of security behavior are analyzed in Sect. 17.4. Then we extend the model to a triggering interaction scenario in Sect. 17.5. Simulations are shown in Sect. 17.6, and conclusions are drawn in Sect. 17.7.

## 17.2  Related Works

Game theoretic models and evolutionary game theoretic models have been introduced in the literature to comprehend and to interpret the interactions among network users regarding personal information security. In [18], the authors organized the presented works on network security and privacy into six main categories: security of the physical and medium access control (MAC) layers, security of self-organizing networks, intrusion detection systems, anonymity and privacy, economics of network security, and cryptography. In each category, they identified security problems, players, and game models, and main results such as equilibrium analysis. In [19], the authors formulated a non-cooperative cyber security information sharing game, the strategies of which are participation and sharing versus non-participation. They analyzed the game from an evolutionary game-theoretic viewpoint, and determined the conditions under which the players' self-enforced evolutionary stability can be achieved. A model of an evolutionary game between social network sites (SNS) and their users was established from the perspective of privacy concerns in [20]. In this work, the SNS tend to decide whether to disclose users' privacy or not for profit, and users tend to decide about privacy disclosure to obtain certain benefits. Authors of [21] proposed an evolutionary game theoretic framework to model the dynamic information diffusion process in social networks, and derived the closed-form expressions of the evolutionary stable network states through analyzing the proposed framework in uniform degree and non-uniform degree networks. For a better understanding of online information exposure, a deception model for online users was proposed in [22] based on a game theoretic approach characterizing a user's willingness to release, withhold or lie about information depending on the behavior of individuals within the user's circle of friends.

As mentioned previously, the influence of users' behaviors also plays an important role on the selection of privacy protection throughout the social network. In other words, behaviors of users can spread over the network according to some kind of natural selection. In the practice of a game, if the utility obtained by one strategy is larger than that by another strategy for a specific player, this strategy will be imitated by other players of high probability, which suggests this strategy is more likely being spread over the entire social network. For the user privacy concerns, it is important to analyze the spreading and influence of user behaviors that make use of the privacy protection or not. Based on such analysis, the benefit-cost mechanism can be designed to promote the use of the privacy protection among the users, and then the information security of the social network can be improved. The spreading of human behaviors has been studied from various aspects. In [23], individuals were separated into interdependent groups, and their different combinations were studied to reveal that an intermediate interdependence optimally facilitates the spreading of cooperative behavior between groups. It has been shown that there is an intermediate fraction of links between groups that is optimal for the evolution of cooperation in the prisoner's dilemma game. Results in [24] suggested that strong ties are instrumental for spreading both online and real-world behavior in human social networks. The authors demonstrated that the messages diffused in the network directly influenced political self-expression, information seeking and real world voting behavior of millions of people. Furthermore, the messages not only influenced the users who received them but also the users' friends, and friends of friends. It was suggested in [25] that if the goal of policy is to adequately protect privacy, then we need policies that protect individuals with minimal requirements of informed and rational decision making that include a baseline framework of protection. In [26], a model for small-world networks regarding information epidemics was proposed to analyze the mixed behaviors of delocalized infection and ripple-based propagation for hybrid malware in generalized social networks consisting of personal and spatial social relations. A number of other works have analyzed the spread of user behavior based on the epidemic spreading theory or social contagion [27–29].

## 17.3   Community Structure Based Evolutionary Game Formulation

### 17.3.1   Basic Concept of Evolutionary Game

Consider an evolutionary game with $r$ strategies $\chi = \{1, 2, \cdots, r\}$ and a payoff matrix $\mathbf{U}$, which is an $r \times r$ matrix with entry $u_{mn}$ denoting the payoff for strategy $m$ versus strategy $n$. The system state of the game can be denoted as $\mathbf{p} = [p_1, p_2, \cdots, p_r]^{\mathrm{T}}$. In this case, the average mean payoff within a population in state $\mathbf{q} = [q_1, q_2, \cdots, q_r]'$ against a population in state $\mathbf{p}$ is $\mathbf{q}'\mathbf{U}\mathbf{p}$.

**Definition 17.1   Evolutionary Stable State, ESS**: A state $p^*$ is an ESS, if and only if $\mathbf{p}^*$ satisfies following conditions for all different states $\mathbf{q} \neq \mathbf{p}$ [30]:

$$\mathbf{q}'\mathbf{U}\mathbf{p}^* \leq \mathbf{p}^{*'}\mathbf{U}\mathbf{p}^*, \tag{17.1a}$$

$$\text{if } \mathbf{q}'\mathbf{U}\mathbf{p}^* = \mathbf{p}^{*'}\mathbf{U}\mathbf{p}^*, \ \ \mathbf{p}^{*'}\mathbf{U}\mathbf{q} > \mathbf{q}'\mathbf{U}\mathbf{q}. \tag{17.1b}$$

In Definition 17.1, first condition (17.1a) is equivalent to the Nash equilibrium condition, and ensures that the average payoff of the population in ESS $\mathbf{p}^*$ is not smaller than the average payoff of the population in a different strategy $\mathbf{q}'$ against $\mathbf{p}^*$. The second condition (17.1b) further guarantees the stability of ESS $\mathbf{p}^*$ in case of equality in the equilibrium condition. Solving the ESS is an important problem in an evolutionary game [31, 32]. An approach to this problem is to find the stable point

$$\mathbf{p}^* = \arg_{\mathbf{p}} \left( \frac{d\mathbf{p}}{dt} = 0 \right) \tag{17.2}$$

of the network mean dynamics, which specifies that the rate of change in the use of each strategy equals to zero [33, 34]. In this work, we analyze the frequency of users taking different strategies over several times of updates in Sect. 17.4. The network evolves and updates according to the following process, which is similar to the Wright-Fisher process [35–37]. The users with evolutionary behaviors and community memberships are considered as discrete and non-overlapping updated generations, and the number of users is constant. All users update at the same time. Users reproduce their own decisions in the new update proportional to their fitness [38], which means that if the user has a higher fitness, he/she tends to maintain his/her current strategy and community memberships in the following update with a high probability. Consider that when an offspring user adopts the imitated user's strategy and community memberships, he/she might select the opposite strategy or different communities, which is similar to the conception of "mutation" in genetic theory [39]. Denote $u$ as the probability with which an offspring adopts a random security strategy, i.e., selecting the security service or not. Then an offspring will adopt the imitated user's strategy with probability $1 - u$. Similarly, denote $v$ as the probability with which a user adopts a random community membership, which includes that of the imitated user. Then a user adopts the imitated user's configuration with probability $1 - v$. Notice that the probability that any possible configuration of community membership is selected is $v/\binom{M}{K}$.

## 17.3.2  Community Structured Evolutionary Game Formulation

Assume that a social network can provide a higher grade of security, i.e., privacy protection, for users' privacy besides the basic services. This additional security service for user privacy protection means applying more advanced encryption and anonymization, secure database management and dissemination, and personal privacy protection techniques to data processing on the user privacy information. When users take this security service, they need to accept terms ruled by the network, such as that users have to provide more personal information, complete real-name authentication or pay for the service. Then they can get more privacy protection when other legitimate or malicious persons and organizations access or use their personal and privacy information. The more personal information provided by one user to the social network managers, the more secure authentication will be required when stealing his/her information, and as a result, the better privacy protection can be achieved for this user. In most real social networks, these kinds of additional services are not mandatory, and users are therefore free to accept such services or not, according to users' own judgements. For instance, users of many forum websites are usually required to provide a mobile phone number or e-mail address to get a higher level of the user information security service, although this service is not mandatory.[1]

---

[1] Current social networks have provide some privacy protection measures for their users. For instance, there are many optional settings in "Privacy Settings and Tools" of Facebook. However, such protection tends to be weak when the users' privacy information is suffering professional or specialized attacks of hackers. The privacy protection service provided by network managers mentioned in our work refers to the high-level technical protection for the user's privacy, not just simple options by users without any pay.

In current social networks, users are allowed to join multiple but a limited number of communities according to their professions, living or touring locations, expertise, or personal interests, etc. For instance, the Google Circles and Facebook Groups can be considered as establishing different communities or some kinds of relationships for their users. We consider that these users are classified by their categories of groups, which can be termed communities of the population structure, and the friendships are established among users in the same community. Therefore, each user holds multiple friendship relations with the users in the same community. The degree of closeness in the relationship between two users can be measured by the number of communities they share. Take users in Google Groups for instance, if User A and User B are both in the Group "Arts and Entertainment" and Group "Schools and Universities" at the same time, then we can consider that the relationship between A and B is stronger than that between B and C, who only belong to Group "Arts and Entertainment" in common. In addition, the information shared between A and B will typically be more than that between B and C. Assume that users are allowed to change communities, which can be influenced by their own or other users' actions. Consider that user interactions can only happen between individuals belonging to the same community, i.e., having some kinds of friend relationships in a social network. Interaction among users in this work refers to the influence of their friends and their own strategies, i.e., the payoff obtained through the game. In addition, some of users' information, such as personal information and status, is accessible only to their friends.

Based on these premises above, we assume that the information of both of a user and his/her friends can be protected by the network to some extent if the user makes use of the privacy protection, even if his/her friends do not take the same action. We assume that, the user taking the privacy protection can obtain this service by paying a price as a deal with the network manager. Meanwhile, the privacy information of his/her friends can also be protected by the network no matter whether these friends take the privacy protection service or not. Taking WeChat for instance, the Moments (similar to the Timeline of Facebook) of a user can only be seen by his/her friends, and the Group Chat can be organized by one user among his/her friends, not matter whether these friends are also friends with each other or not. In addition, Facebook provides an optional setting in "Privacy Settings and Tools": Who can see your friends list, which can illustrate that user's security behavior makes sense on the information protection of his/her friends. The closer the relationship is, the more personal information of the user can be accessed by his/her friends. If the user's friend makes use of the privacy protection or some other information security services, the accessible information of this user can be also protected at some level. The more accessible information for his/her friend, the more information can be protected, even when this user does not take the privacy protection. In this work, we assume that the privacy protection service not only protects the information of the users who select this service, but also the information these users can access, i.e., the information of their friends. Therefore, if the user does not select the service, the personal information of his/her friends will also be threatened by this user's unsafe strategies.

Privacy protection over a social network shares fundamental similarities with the strategy updating in the community-structured evolutionary game theory (EGT). We consider users in a social network as the players in the evolutionary game. Each of these users has two possible strategies, i.e., to take or not take the privacy protection provided by the network:

$$
\begin{cases}
\mathbf{S}_p, & \text{take the privacy protection,} \\
\mathbf{S}_n, & \text{do not take the privacy protection.}
\end{cases}
\tag{17.3}
$$

The strategy taking the privacy protection can be considered as the secure behavior, and otherwise, insecure behavior. Meanwhile, the users' payoff matrix can be defined as

$$
\begin{array}{cc}
 & \mathbf{S}_p \quad \mathbf{S}_n \\
\begin{array}{c} \mathbf{S}_p \\ \mathbf{S}_n \end{array} &
\begin{pmatrix} \beta b - c & b - c \\ b & 0 \end{pmatrix},
\end{array}
\tag{17.4}
$$

where $b > 0$ is the baseline security benefit received by the user resulting from that this user or this user's friends take the privacy protection (security behavior). For existing social networks, $b$ can be set as traditional measurements of privacy, such as disclosure risk and information loss, when applying current encryption, anonymization, secure database management and dissemination techniques. $c > 0$ denotes the cost that users taking the privacy protection need to pay for the protection service, which could be more personal information providing, real-name authentication and payment as required by current social networks. On the other hand, if the privacy protection service is provided through an application (APP) update, which is a common approach adopted by WeChat, Twetter and other existing social networks, the cost for user to take the service can be then measured by the increasing memory occupancy of the latest APP version. In addition, when both of the interacted users take strategy $\mathbf{S}_p$, two of them will obtain higher level privacy safety benefit $\beta b$ as the first entry of the payoff matrix shown in (17.4), where $\beta > 1$. The payoff will be zero when both of the interacted friends are defectors, i.e., neither of them selects the privacy protection service, then no pay or gain for them.

Based on the definitions of the strategies and payoff above, ratio $b/c$ or $\beta b/c$, which can be defined as the *cost performance*, is a crucial parameter. It can help the social network managers to make appropriate security service level and payment mechanism to encourage their users to accept the security service, and then promote the spreading of this secure behavior. In a community structured population well-mixed, any two individuals belonging to the same community interact with equal likelihood. Then as reflected in (17.4), users taking the privacy protection would be out-competed by those users doing not. Therefore, the interaction between users with security behavior and with insecurity behavior needs to be investigated, and the question that whether dynamics on a community structured population allows the evolution of security behavior needs to be figured out.

Consider a social network with $N$ users. The number of communities operated by the social network is $M$. However, users of current social networks are only allowed to join a limited number of these communities. Then we set that each user belongs to exactly $K$ communities, where $K \leq M$. In addition, each user has a strategy index $s_i \in \{0, 1\}$, which is defined as that $s_i = 1$ when user $i$ take the privacy protection strategy $\mathbf{S}_p$, or $s_i = 0$, otherwise. Then the state of the social network can be given by a strategy vector $\mathbf{s} = [s_1, s_2, \cdots, s_N]$ and a matrix $\Theta$. $\Theta$ is an $N \times M$ matrix, whose entry $\theta_{im}$ ($i = 1, 2, \cdots, N, m = 1, 2, \cdots, M$) is 1 if user $i$ belongs to community $m$, and $\theta_{im} = 0$, otherwise. Matrix $\Theta$ can be represented as $\Theta = [\theta_1, \theta_2, \cdots, \theta_N]^T$, where $\theta_i$ is the vector giving the community membership of user $i$. Then the number of communities that user $i$ and user $j$ having in common can be expressed by the dot product of their community membership vector, as $\theta_i \cdot \theta_j$. In addition, based on the definition of $K$, we have $\theta_i \cdot \theta_i = K, \forall i$. The state of the social network can be given as $S = (\mathbf{s}, \Theta)$.

We assume the influence of user $j$ on $i$ ($i \neq j$) is related to the number of communities that they share in common. Specifically, user $i$'s fitness obtained by $j$ is proportional to the total utility according to (17.4), and the proportional coefficient is the number of communities that $i$ and $j$ share in common. In addition, user $i$ interacts with user $j$ only when they share at least one community in common, i.e., $\theta_i \cdot \theta_j \neq 0$. Then the total fitness of user $i$ of the community-structured social network can be written as

$$
\begin{aligned}
\pi_i &= 1 + \alpha \sum_{j \neq i} (\theta_i \cdot \theta_j) \left[ (\beta b - c) s_i s_j + (b - c) s_i (1 - s_j) + b (1 - s_i) s_j \right] \\
&= 1 + \alpha \sum_{j \neq i} (\theta_i \cdot \theta_j) \left[ (\beta - 2) b s_i s_j + (b - c) s_i + b s_j \right],
\end{aligned}
\tag{17.5}
$$

where $\alpha$ represents the selection intensity, i.e., the relative contribution of the game to fitness. The case $\alpha = 1$ denotes the strong selection, which means that the payoff obtained through (17.4), i.e., the game among users with strategies $\mathbf{S}_p$ and $\mathbf{S}_n$, plays an dominant contribution to the total fitness of every user, and then the user with high payoff will be chosen and imitated with high probability. On the contrary, $\alpha \to 0$ denotes the weak selection [40]. Under the weak selection, the payoff obtained through (17.4) has limited contribution to the total fitness of each user. In this work, we only analyze the weak selection case as the results derived from weak selection are often valid approximations for stronger selection [41]. In addition, the weak selection scenario can be more helpful to reveal the user behavior spreading over social networks [21].

As an example shown in Fig. 17.1, there are $N = 5$ users, denoted by $U_1$ - $U_5$, over $M = 4$ communities as ellipses $A$, $B$, $C$ and $D$. Each user belongs to $K = 2$ communities. The community memberships determine how users interact each other, and the broken lines indicate the weighted interaction. The structure changes as users updating in discrete time slot. In this example, $U_1$, $U_3$ and $U_5$ take the same security strategy, and the other users take the opposite strategy at the first

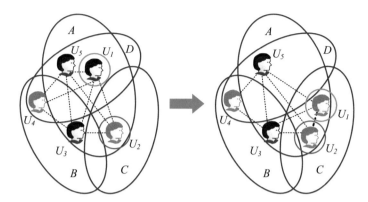

**Fig. 17.1** An example of security strategy and associations evolution over a social network with a community structured population

**Table 17.1** Correspondence between community structured evolutionary theory and social network

| Community-structured EGT | Social network |
| --- | --- |
| Community-structured population | Social network with friendships |
| | Classified by communities |
| Players | Users in the social network |
| Strategies | $S_p$: take the privacy protection |
| | $S_n$: do not take the privacy protection |
| Fitness | Utility from taking the privacy protection or not |
| ESS | Stable security behavior state over users |

time slot. During the update process, imitator $U_1$ picks another user $U_2$, and adopts $U_2$'s security strategy and community associations.

Table 17.1 indicates the correspondence between the elements in the community structured evolutionary game theory and those in the social network, whose users hold relationships according to their interested communities. Based on the definitions above, we can derive the expression for the critical cost performance, which is an important parameter that determines the stable security behavior state of the users among the network. In the following section, we will analyze the critical cost performance for the social network where games exist among all users in the same community. In other words, the security strategy of a user can only influence the payoff of his/her friends who share at least one common community with this user. Moreover, the critical cost performance for the situation named "L-triggering game" will be further analyzed in the later part of this work.

## 17.4 Privacy Protection Among Users Belonging to $K$ Communities

In this section, we study the evolution of users' behaviors that take the privacy protection or not over social networks using the evolutionary game theory based on the community structured population. In the uniform scenario, a social network with $N$ users, each of whom belongs to exactly $K$ communities, is considered in this section. We define the network user state as $(p, 1 - p)$, where $p$ is the frequency of the users those select to take the privacy protection (choose strategy $\mathbf{S}_p$), and $1 - p$ are the others (choose strategy $\mathbf{S}_n$). Our ultimate goal is to derive the evolutionary stable network state $(p^*, 1 - p^*)$ that ensures the evolution of security behavior, i.e., users select strategy $\mathbf{S}_p$ more frequently than $\mathbf{S}_n$.

### 17.4.1 Evolution of Security Behavior on Communities

First, we summarize the key conceptions in the spreading of the privacy protection behavior when applying evolutionary game into social networks.

*Update* Assume that users can change their community memberships and security behaviors to get better user security experience. This change can be considered as the update.

*Imitation* In social networks, the imitation of other users' behaviors plays an important role in behavior spreading. By times of updates, the behavior with higher fitness can spread among the users in the network. Specifically, a user's community membership and security behavior can bring high fitness for the user, which means that the security behaviors of this user and his/her friends obtain a relative high level of privacy protection. Then for the security concern, other users will tend to imitate this user's community membership and security behavior. Similarly, the user will maintain his/her own community membership and security behavior if the current fitness is rather high, which means that the user will imitate hiself/herself.

*Deviation* Deviation means that the user does not imitate the community membership or community membership of the users being imitated, who can be himself/herself and other users with high fitness. The community and strategy deviation can also make sense on the behavior spreading. On the one hand, when a user imitates another one for a better security experience, he/she only imitates the community membership. This user might not change his/her previous security behavior, i.e., he/she still not take the privacy protection to just get security benefit brought by friends, or still take the privacy protection to get more benefit brought by new communities. On the other hand, a user might only imitate another user's security behavior but not change some or all of his/her previous communities because of interests. These two situation bring security behavior (strategy) deviation and community deviation, respectively, as mentioned in Sect. 17.3.1. We still use $u$ and $v$ to denote the rates of strategy deviation and community deviation, respectively.

We consider that user $i$ is an imitated user with the probability proportional to its fitness, which can be given by its payoff relative to the total payoff, i.e., $\pi_i / \sum_j \pi_j$. Assume that both the imitation and deviation are implemented independently $N$ times in each update step. Denote the average number of imitators of user $i$ as $\omega_i$. After one update step, we have

$$\omega_i = \frac{N\pi_i}{\sum_j \pi_j}. \tag{17.6}$$

According to Eq. (17.5), the total payoff can be written as

$$\sum_j \pi_j = \sum_j \left[ 1 + \alpha \sum_{l \neq j} (\theta_j \cdot \theta_l) f(s_j, s_l) \right]$$

$$= N + \alpha \sum_j \left[ \sum_l (\theta_j \cdot \theta_l) f(s_j, s_l) - (\theta_j \cdot \theta_j) f(s_j, s_j) \right]$$

$$= N + \alpha \sum_j \sum_l (\theta_j \cdot \theta_l) f(s_j, s_l) \tag{17.7}$$

$$- \alpha \sum_j K \left[ (\beta - 2) b s_j s_j + (2b - c) s_j \right]$$

$$= N + \alpha \sum_j \sum_l (\theta_j \cdot \theta_l) f(s_j, s_l) - \alpha K (\beta b - c) \sum_j s_j,$$

where

$$f(s_i, s_j) = (\beta - 2) b s_j s_l + (b - c) s_j + b s_l, \tag{17.8}$$

and the last term in Eq. (17.7) is obtained considering $s_j s_j = s_j, \forall j$. We consider the weak selection situation, i.e., $\alpha \to 0$, because of that results derived from the weak selection often remain as valid approximations for large selection strength [42]. In addition, the weak selection assumption helps to achieve a close-form analysis of spreading process and reveal how the behavior spreads over the social network [21]. Then we can rewrite $\omega_i$ in Eq. (17.6) as

$$\omega_i = 1 + \alpha \left[ (\beta - 2) b \sum_j (\theta_i \cdot \theta_j) s_i s_j + (b - c) \sum_j (\theta_i \cdot \theta_j) s_i \right.$$

$$+ b \sum_j (\theta_i \cdot \theta_j) s_j - K (\beta b - c) s_i - \frac{(\beta - 2) b}{N} \sum_j \sum_l (\theta_j \cdot \theta_l) s_j s_l \tag{17.9}$$

$$\left. - \frac{2b - c}{N} \sum_j \sum_l (\theta_j \cdot \theta_l) s_j + \frac{K (\beta b - c)}{N} \sum_j s_j \right] + o(\alpha^2),$$

where the third equality is according to Taylor's Theorem and weak selection assumption with $\alpha \to 0$.

**_Proof_** Similar to the derivation of Eq. (17.7), we get:

$$\pi_i = 1 + \alpha \sum_j (\theta_i \cdot \theta_j) f (s_i, s_j) - \alpha K (\beta b - c) s_i. \qquad (17.10)$$

For $\alpha = 0$, the Taylor expansion of $\omega_i$ can be given by:

$$\omega_i = \frac{N \pi_i}{\sum_j \pi_j} = \omega_i (0) + \alpha \omega_i^{(1)} (0) + o \left( \alpha^2 \right), \qquad (17.11)$$

where $\omega_i^{(1)} (0) = \partial \omega_i (\alpha) / \partial a$.

According to Eq. (17.5) and (17.7), we have

$$\sum_j \pi_j |_{\alpha=0} = N, \quad \pi_i |_{\alpha=0} = 1, \qquad (17.12)$$

$$\frac{\partial \pi_i}{\partial \alpha} |_{\alpha=0} = \sum_j (\theta_i \cdot \theta_j) f (s_i, s_j) - K (\beta b - c) s_i, \qquad (17.13)$$

$$\frac{\partial \sum_j \pi_j}{\partial \alpha} |_{\alpha=0} = \sum_j \sum_l (\theta_j \cdot \theta_l) f (s_j, s_l) - K (b\beta - c) \sum_j s_j. \qquad (17.14)$$

Then $\omega_i^{(1)} (0)$ can be calculated as

$$\omega_i^{(1)} (0) = \frac{N^2 \left[ \sum_j (\theta_i \cdot \theta_j) f (s_i, s_j) - K (\beta b - c) s_i \right]}{N^2}$$

$$- \frac{N \left[ \sum_j \sum_l (\theta_j \cdot \theta_l) f (s_j, s_l) - K (b\beta - c) \sum_j s_j \right]}{N^2}$$

$$= \sum_j (\theta_i \cdot \theta_j) \left[ (\beta - 2) b s_i s_j + (b - c) s_i + b s_j \right] - K (\beta b - c) s_i$$

$$- \frac{1}{N} \left[ \sum_j \sum_l (\theta_j \cdot \theta_l) \left[ (\beta - 2) b s_j s_l + (b - c) s_j + b s_l \right] \right.$$

$$\left. - K (b\beta - c) \sum_j s_j \right]$$

$$= (\beta - 2) b \sum_j (\theta_i \cdot \theta_j) s_i s_j + (b - c) \sum_j (\theta_i \cdot \theta_j) s_i$$

$$+ b \sum_j (\theta_i \cdot \theta_j) s_j - K (\beta b - c) s_i$$

$$- \frac{(\beta - 2) b}{N} \sum_j \sum_l (\theta_j \cdot \theta_l) s_j s_l - \frac{b - c}{N} \sum_j \sum_l (\theta_j \cdot \theta_l) s_j$$

$$- \frac{1}{N} \sum_j \sum_l (\theta_j \cdot \theta_l) s_l + \frac{K (b\beta - c)}{N} \sum_j s_j.$$

Then we can rewrite $\omega_i$ in Eq. (17.11) as

$$
\begin{aligned}
\omega_i = 1 + \alpha \Bigg[ & (\beta - 2)\, b \sum_j (\theta_i \cdot \theta_j)\, s_i s_j + (b - c) \sum_j (\theta_i \cdot \theta_j)\, s_i \\
& + b \sum_j (\theta_i \cdot \theta_j)\, s_j - K\,(\beta b - c)\, s_i - \frac{(\beta - 2)\, b}{N} \sum_j \sum_l (\theta_j \cdot \theta_l)\, s_j s_l \\
& - \frac{2b - c}{N} \sum_j \sum_l (\theta_j \cdot \theta_l)\, s_j + \frac{K\,(\beta b - c)}{N} \sum_j s_j \Bigg] + o\left(\alpha^2\right).
\end{aligned}
$$

This completes the proof of Eq. (17.9).

To find out the ESS of the system state dynamic, we let $p$ denote the frequency of the users those select to take the privacy protection. As assumed previously, there exist two situations in the update process of the social network state, one is the imitation of another user's community membership and security decision or the maintenance of his/her own, and the other is the deviation. So we need to analyze the effect of imitation and deviation on the average change in $p$. Because of that the average value of $p$ is constant, the two effects must cancel [39]. Then we can get

$$
\langle \hat{p} \rangle_{\text{imi}} + \langle \hat{p} \rangle_{\text{dev}} = 0, \tag{17.15}
$$

where $\langle \hat{p} \rangle_{\text{imi}}$ and $\langle \hat{p} \rangle_{\text{dev}}$ denote the effect of imitation and deviation, respectively, and they are both the continuous functions of $\alpha$.

Next, we consider the weak selection situation that $\alpha = 0$, and Taylor expansion of $\langle \hat{p} \rangle_{\text{imi}}$ can be written as

$$
\langle \hat{p} \rangle_{\text{imi}} = 0 + \alpha \langle \hat{p} \rangle_{\text{imi}}^{(1)} + o\left(\alpha^2\right), \tag{17.16}
$$

where $\langle \hat{p} \rangle_{\text{imi}}^{(1)}$ is the first derivative of $\langle \hat{p} \rangle_{\text{imi}}$ with $\alpha = 0$, and $o\left(\alpha^2\right)$ is according to Taylor's Theorem. We notice that when $\langle \hat{p} \rangle_{\text{imi}}^{(1)} > 0$, the amount of users who take the privacy protection due to the imitation increases, which means that the user's decision tends to the security behavior. On the contrary, if $\langle \hat{p} \rangle_{\text{imi}}^{(1)} < 0$, the user's decision tends to not taking the privacy protection.

### 17.4.2 Finding the Critical Ratio

In order to obtain the critical parameter value of cost performance, we must have $\langle \hat{p} \rangle_{\text{imi}}^{(1)} = 0$. The Lemma 17.1 provides the critical cost performance $b/c$ in the limit of weak selection.

**Lemma 17.1** *In a social network with $N$ users, every user belongs to exactly $K$ communities. There are two strategies $\mathbf{S}_p$ and $\mathbf{S}_n$ for users. The state of the social* *network is given as $S = (\mathbf{s}, \Theta)$. Interactions are only allowed among users sharing communities in common. For each user, the payoff matrix is given by (17.4). The critical cost performance that keeps the neutral stationary state, i.e., the frequencies of users selecting strategies $\mathbf{S}_p$ and $\mathbf{S}_n$ approaches stable state, is given by*

$$\left(\frac{b}{c}\right)^* = \frac{Num}{Den}, \tag{17.17}$$

*In Eq. (17.17),*

$$Num = -K \langle f_1 \rangle_0 + \frac{K}{N} \langle f_2 \rangle_0 + \langle f_3 \rangle_0 - \frac{1}{N} \langle f_5 \rangle_0, \tag{17.18}$$

$$Den = -\beta K \langle f_1 \rangle_0 + \frac{\beta K}{N} \langle f_2 \rangle_0 + \langle f_3 \rangle_0$$
$$+ (\beta - 1) \langle f_4 \rangle_0 - \frac{2}{N} \langle f_5 \rangle_0 - \frac{\beta - 2}{N} \langle f_6 \rangle_0, \tag{17.19}$$

*where*

$$f_1 = \sum_i s_i, \qquad f_2 = \sum_{i,j} s_i s_j, \tag{17.20a}$$

$$f_3 = \sum_{i,j} (\theta_i \cdot \theta_j) s_i, \qquad f_4 = \sum_{i,j} (\theta_i \cdot \theta_j) s_i s_j, \tag{17.20b}$$

$$f_5 = \sum_{i,j,l} (\theta_j \cdot \theta_l) s_i s_j, \qquad f_6 = \sum_{i,j,l} (\theta_j \cdot \theta_l) s_i s_j s_l, \tag{17.20c}$$

*In Eqs. (17.18) and (17.19), the angular bracket with a subscript zero represents the average value among all possible states $S$. Take $f_3$ for instance,*

$$\left\langle \sum_{i,j} (\theta_i \cdot \theta_j) s_i \right\rangle_0 = \sum_S \left( \sum_{i,j} (\theta_i \cdot \theta_j) s_i |_{\alpha=0} \right) \cdot q_S^{(0)}, \tag{17.21}$$

*where $q_S$ denotes the probability that the network is in state $S$ [39].*

**Proof** We pursuit the stable state by calculating $\langle \hat{p} \rangle_{imi}$, which can be given by

$$\langle \hat{p} \rangle_{imi} = \sum_S \langle \hat{p} \rangle_S q_S, \tag{17.22}$$

where $\langle \hat{p} \rangle_S$ denotes the average number of users take strategy $\mathbf{S}_p$ (take the privacy protection) in a given state $S$. Then rewrite Eq. (17.22) by introducing the Taylor expression for both $\langle \hat{p} \rangle_S$ and $q_S$ at $\alpha = 0$, we can obtain

$$\langle \hat{p} \rangle_{imi} = \frac{\alpha}{N} \sum_S \left( \sum_i s_i \frac{d\omega_i}{d\alpha} |_{\alpha=0} \right) \cdot q_S^{(0)} + o\left(\alpha^2\right). \tag{17.23}$$

Notice that the sum part of Eq. (17.23) denotes the average of $\sum_i s_i \frac{d\omega_i}{d\alpha} \mid_{\alpha=0}$ among all possible states $S$. Similar to the definition in Eq. (17.21), we express this average value by an angular bracket with a subscript zero referring the weak selection situation ($\alpha = 0$):

$$\sum_S \left( \sum_i s_i \frac{d\omega_i}{d\alpha} \mid_{\alpha=0} \right) \cdot q_S^{(0)} = \left\langle \sum_i s_i \frac{d\omega_i}{d\alpha} \right\rangle_0. \tag{17.24}$$

We plug Eq. (17.9) into $\sum_i s_i \frac{d\omega_i}{d\alpha}$ and get

$$\sum_i s_i \frac{d\omega_i}{d\alpha} = - K (\beta b - c) f_1 - \frac{K (\beta b - c)}{N} f_2 + (b - c) f_3$$
$$+ (\beta - 1) b f_4 - \frac{2b - c}{N} f_5 - \frac{(\beta - 2) b}{N} f_6, \tag{17.25}$$

where $f_i$ ($i = 1, 2, \cdots, 6$) are defined by (17.20). Then plug Eq. (17.25) into (17.23) and with the definition in (17.24), we can calculate and gain the first derivative of $\langle \hat{p} \rangle_{\text{imi}}$ as follows

$$\langle p \rangle_{\text{imi}}^{(1)} = \frac{1}{N} \left[ -K (\beta b - c) \langle f_1 \rangle_0 - \frac{K (\beta b - c)}{N} \langle f_2 \rangle_0 + (b - c) \langle f_3 \rangle_0 \right.$$
$$\left. + (\beta - 1) b \langle f_4 \rangle_0 - \frac{2b - c}{N} \langle f_5 \rangle_0 - \frac{(\beta - 2) b}{N} \langle f_6 \rangle_0 \right]. \tag{17.26}$$

As the derivation process above, we can obtain the critical cost performance $b/c$ when Eq. (17.26) equals zero. The obtained $(b/c)^*$ can be given by

$$\left( \frac{b}{c} \right)^* = \frac{-K \langle f_1 \rangle_0 + \frac{K}{N} \langle f_2 \rangle_0 + \langle f_3 \rangle_0 - \frac{1}{N} \langle f_5 \rangle_0}{f (f_1, f_2, f_3, f_4, f_5, f_6)}, \tag{17.27}$$

where $f (f_1, f_2, f_3, f_4, f_1, f_6) = -\beta K \langle f_1 \rangle_0 + \frac{\beta K}{N} \langle f_2 \rangle_0 + \langle f_3 \rangle_0 + (\beta - 1) \langle f_4 \rangle_0 - \frac{2}{N} \langle f_5 \rangle_0 - \frac{\beta - 2}{N} \langle f_6 \rangle_0$. This completes the proof of Lemma 17.1.

*Remark 17.1* In a social network whose current state $S$, i.e., the users' community memberships and security behaviors, can change with every update, Eq. (17.17) shows the threshold of the security protection cost performance. This parameter can be controlled by the social network manager, either by adjusting the price of the security service that is related to the parameter $c$, or by providing sufficient security services benefit that is related to the parameter $b$. Note that the expression of cost performance provided in Lemma 17.1 hold the weak selection situation i.e., $\alpha \to 0$. Compared with [39], in which a simplified Prisoner's Dilemma game was analyzed, more situations are considered in our game model.

Next, we will analyze the neutral stationary state and get the more general expression of cost performance. Theorem 17.1 states the desired term of cost performance $\beta b/c$.

**Theorem 17.1** *In a social network with $N$ users, every user belongs to exactly $K$ communities. There are two strategies $\mathbf{S}_p$ and $\mathbf{S}_n$ for users. Interactions are only allowed among users sharing communities in common. For each user, the payoff matrix is given by (17.4). The deviate rates of community membership imitation and strategy imitation are given by $v$ and $u$, respectively. The critical cost performance that keeps the neutral stationary state is given by*

$$\left(\frac{\beta b}{c}\right)^* = 1 + \frac{\mu + v + 3}{\mu + v + 1} \cdot \frac{Kv(\mu + v + 2) + M(\mu + 1)}{Kv(\mu + v + 2) + M(\mu + 2v + 3)}, \quad (17.28)$$

*where $v = 2Nv$ and $\mu = 2Nu$.*

**Proof** To proof Theorem 17.1, each term of Eq. (17.20) needs to be analyzed. First, we consider that $\langle f_1 \rangle_0$ is the average number of the users taking the privacy protection and can be given by

$$\langle f_1 \rangle_0 = \frac{N}{2}. \quad (17.29)$$

For $\langle f_2 \rangle_0$, we notice that $\langle f_2 \rangle_0 = N^2 \Pr\left(s_i = s_j = 1\right)$. In a neutral stationary state, the probabilities of both of user $i$ and $j$ select to take the privacy or not are equal, i.e., $\Pr\left(s_i = s_j = 1\right) = \Pr\left(s_i = s_j = 0\right) = \Pr\left(s_i = s_j\right)/2$. User $i$ and $j$ are selected randomly to be analyzed, and the replacement is allowed. So we can get

$$\langle f_2 \rangle_0 = \frac{N^2}{2} \Pr\left(s_i = s_j\right). \quad (17.30)$$

Similar to the analysis above, we can get

$$\langle f_3 \rangle_0 = N^2 \langle \theta_i \cdot \theta_j \mathbf{1}\,(s_i = 1)\rangle_0 = \frac{N^2}{2} \langle \theta_i \cdot \theta_j \rangle_0, \quad (17.31)$$

where $\mathbf{1}\,(\cdot)$ is the indicator function, the value of which is 1 if the argument is true, and 0, otherwise. This indicator function introduces a non-zero contribution. So $\langle \theta_i \cdot \theta_j \mathbf{1}\,(s_i = 1)\rangle_0$ indicates the average number of communities that user $i$ and $j$ belong in common under the situation that the first user $i$ takes the privacy protection. $\langle \theta_i \cdot \theta_j \rangle_0$ represents the average number of communities that the two users belong in common. With the same analysis for Eqs. (17.29)–(17.31), we can get other terms of (17.20) as follows.

$$\langle f_4 \rangle_0 = \left( N^2 / 2 \right) \langle \theta_i \cdot \theta_j \mathbf{1} \left( s_i = s_j \right) \rangle_0, \tag{17.32a}$$

$$\langle f_5 \rangle_0 = \left( N^3 / 2 \right) \langle \theta_j \cdot \theta_l \mathbf{1} \left( s_i = s_j \right) \rangle_0, \tag{17.32b}$$

$$\langle f_6 \rangle_0 = \left( N^3 / 2 \right) \langle \theta_j \cdot \theta_l \mathbf{1} \left( s_i = s_j = s_l \right) \rangle_0. \tag{17.32c}$$

Equation (17.32a) provides the average number of communities that the two random users have in common, and the case that the two users select the same security behavior (both or neither of the users take the privacy protection) give the non-zero contribution to the average. In both of Eqs. (17.32b) and (17.32c), three random users are considered. So the sum has $N^3$ terms. Equation (17.32b) provides the average number of communities that latter two users $j$ and $l$ have in common, and the non-zero contribution to the average is given by the case that first two users $i$ and $j$ select the same security behavior. Equation (17.32c) is the average number of communities that latter two users $j$ and $l$ have in common, and the non-zero contribution to the average is given by the case that all these three users take the same security behavior. The three users are selected randomly and with replacement.

Next, we need to calculate the terms obtained in Eqs. (17.29)–(17.32c) in the case that three users are selected to be analyzed without replacement, i.e., $i \neq j$ and $i \neq j \neq l$. For convenience, we give some notations as follows.

$$\varphi = \Pr \left( s_i = s_j \,|\, i \neq j \right), \tag{17.33a}$$

$$\psi = \langle \theta_i \cdot \theta_j \,|\, i \neq j \rangle_0, \tag{17.33b}$$

$$\gamma = \langle \theta_i \cdot \theta_j \mathbf{1} \left( s_i = s_j \right) |\, i \neq j \rangle_0, \tag{17.33c}$$

$$\xi = \langle \theta_j \cdot \theta_l \mathbf{1} \left( s_i = s_j \right) |\, i \neq j \neq l \rangle_0, \tag{17.33d}$$

$$\eta = \langle \theta_j \cdot \theta_l \mathbf{1} \left( s_i = s_j = s_l \right) |\, i \neq j \neq l \rangle_0. \tag{17.33e}$$

In (17.33), $\psi$ is the average number of communities two different randomly picked users have in common. $\gamma$ is the average number of communities the two users have in common given that only users with the same security behavior. For $\xi$ and $\eta$, there are three different users considered. $\xi$ is the average number of communities the latter two users belonging in common given that only the first two users have the same security behavior. $\eta$ is the average number of communities the latter two users having in common given that there is a non-zero contribution to the average only when all the three users take the same security behavior.

Given two users, the probability that the same user is chosen again in the second selection experience is $1/N$. Then we get

$$\Pr\left(s_i = s_j\right) = \frac{1}{N} + \frac{N-1}{N}\varphi, \tag{17.34a}$$

$$\langle\theta_i \cdot \theta_j\rangle_0 = \frac{K}{N} + \frac{N-1}{N}\psi, \tag{17.34b}$$

$$\langle\theta_i \cdot \theta_j \mathbf{1}\left(s_i = s_j\right)\rangle_0 = \frac{K}{N} + \frac{N-1}{N}\gamma. \tag{17.34c}$$

Then for the situation that three users $i$, $j$ and $l$ are given, the probability that both of the last two users are same as the first selection is $N_1 = 1/N^2$. The probability that none of the users chosen in the second and third selection is same as the one in the first selection is $N_2 = (N-1)(N-2)/N^2$. The probability that the user chosen in the second selection is same as the one in the first selection, and the third selection chooses the different user is $N_3 = (N-1)/N^2$. Then we get

$$\langle\theta_j \cdot \theta_l \mathbf{1}\left(s_i = s_j\right)\rangle_0 = N_1 K + N_2 \xi + N_3\left(\psi + \gamma + K\varphi\right), \tag{17.35a}$$

$$\langle\theta_j \cdot \theta_l \mathbf{1}\left(s_i = s_j = s_l\right)\rangle_0 = N_1 K + N_2 \eta + N_3\left(2\gamma + K\varphi\right). \tag{17.35b}$$

According to (17.34) and (17.35), terms in (17.27) can be calculated as follows.

$$\langle f_2\rangle_0 = \frac{N^2}{2}\left(\frac{1}{N} + \frac{N-1}{N}\varphi\right), \tag{17.36a}$$

$$\langle f_3\rangle_0 = \frac{N^2}{2}\left(\frac{K}{N} + \frac{N-1}{N}\psi\right), \tag{17.36b}$$

$$\langle f_4\rangle_0 = \frac{N^2}{2}\left(\frac{K}{N} + \frac{N-1}{N}\gamma\right), \tag{17.36c}$$

$$\langle f_5\rangle_0 = \frac{N^3}{2}\left[N_1 K + N_2 \xi + N_3\left(\psi + \gamma + K\varphi\right)\right], \tag{17.36d}$$

$$\langle f_6\rangle_0 = \frac{N^3}{2}\left[N_1 K + N_2 \eta + N_3\left(2\gamma + K\varphi\right)\right]. \tag{17.36e}$$

By calculating, $\varphi$ is eliminated, and the critical ratio $b/c$ expressed by $\psi, \gamma, \xi$ and $\eta$ is given as

$$\left(\frac{b}{c}\right)^* = \frac{\psi - \xi + \frac{\psi-\gamma}{N-2}}{\psi - 2\xi - (\beta - 2)\eta + (\beta - 1)\gamma + \frac{\gamma-\psi}{N-2}}. \tag{17.37}$$

When the population of the social network is large, i.e., $N \to \infty$, we have

$$\left(\frac{b}{c}\right)^*_{N \to \infty} = \frac{\psi - \xi}{\psi - 2\xi - (\beta - 2)\eta + (\beta - 1)\gamma}. \tag{17.38}$$

Next, we will calculate each quantity of $\psi$, $\gamma$ $\xi$ and $\eta$. According to the physical interpretations of these parameters, we notice that all of them cannot be written as independent products of the average number of common communities times the probability of taking the same security decision. In response, we introduce a time instant that users' most recent common user being imitated (MRCI). Then if we fix the time to the MRCI, the community deviations and strategy deviations are independent. Take $\gamma$ for instance, if the time to the MRCI of users $i$ and $j$ is $T = t$, then we get

$$\begin{aligned} &\left\langle \theta_i \cdot \theta_j \mathbf{1} \left(s_i = s_j\right) | i \neq j, T = t \right\rangle_0 \\ =&\left\langle \theta_i \cdot \theta_j | i \neq j, T = t \right\rangle_0 \cdot \Pr\left(s_i = s_j | i \neq j, T = t\right). \end{aligned} \tag{17.39}$$

So if the time to users' MRCI is given, we can calculate $\psi$, $\gamma$, $\xi$ and $\eta$. Given some randomly selected users, Lemmas 17.2, 17.3 and 17.4 present the probability of users' MRCI, the probability that users have the same security behavior at the time from their MRCI and the average number of communities two random users have in common, respectively. These results are summarized from [39], in which detailed explanation can be found.

**Lemma 17.2** *Consider a social network with $N$ users. Given two random users, the probability that their MRCI is at time $T = t$ is*

$$\Pr\left(T = t\right) = \left(1 - \frac{1}{N}\right)^{t-1} \frac{1}{N}. \tag{17.40}$$

*Given three random users, the probability that the first merging by imitating the same user's communities and strategy happens at time $t_1 \geq 1$ and the second takes $t_2 \geq 1$ more time steps is*

$$\Pr\left(t_1, t_2\right) = \frac{3}{N^2}\left[\left(1 - \frac{1}{N}\right)\left(1 - \frac{2}{N}\right)\right]^{t_1-1}\left(1 - \frac{1}{N}\right)^{t_2}. \tag{17.41}$$

*When $N \to \infty$, let $\tau = t/N$, $\tau_1 = t_1/N$ and $\tau_2 = t_2/N$, the distributions of $\Pr(T = t)$ and $\Pr(t_1, t_2)$ are given by*

$$p(\tau) = e^{-\tau}, \tag{17.42a}$$

$$p(\tau_1, \tau_2) = 3e^{-(3\tau_1 + \tau_2)}. \tag{17.42b}$$

*Remark 17.2* Lemma 17.2 indicates that the MRCI for random two and three users situations both have exponential distributions. The physical meaning of MRCI is the most current common user affected and imitated by another two users. Note that the introduction of MRCI is for the independence between the community deviations and strategy deviations, which makes the calculation of $\psi$, $\gamma$ $\xi$ and $\eta$ defined in Eq. (17.33) feasible, and time indexes $\tau$, $\tau_1$ and $\tau_2$ will be removed by the integral. Equation (17.42a) and (17.42b) hold for the limit of $N \to \infty$, which is rational for social networks with large number of users.

**Lemma 17.3** *In a social network with $N$ users, every user belongs to exactly $K$ communities, where $K \leq M$. The deviate rate of strategy imitation is given by $u$. The probability that two random users have the same strategy at time $t$ from their MRCI is given by*

$$\varphi(t) = \Pr\left(s_i = s_j \,|\, T = t\right) = \frac{1}{2}\left[1 + (1-u)^{2t}\right].\tag{17.43}$$

*When $N \to \infty$, let $\tau = t/N$, $\tau_1 = t_1/N$ and $\tau_2 = t_2/N$, the distributions is*

$$\varphi(\tau) = \frac{1}{2}\left(1 + e^{-\mu\tau}\right),\tag{17.44}$$

*where $\mu = 2Nu$. Given that the first merging by imitating the same user happens at time $t_1 \geq 1$ and the second takes $t_2 \geq 1$ extra time steps, the distribution of the probability that three random users have the same strategy is given by*

$$\varphi(\tau_1, \tau_2) = \frac{1}{8}\left[(1 - e_1)^2(1 - e_2) + (1 + e_1)^2(1 + e_2)\right],\tag{17.45}$$

*where $e_1 = \exp\left\{-\frac{\mu}{2}\tau_1\right\}$, $e_2 = \exp\left\{-\frac{\mu}{2}(\tau_1 + \tau_2)\right\}$, $\mu = 2Nu$.*

*Remark 17.3* Equation (17.44) in Lemma 17.3 indicates that at $\tau < \infty$ after the time when two users imitated the same other user, these two users have the same security behavior with the probability more than 0.5. The shorter $\tau$ is, the larger the probability is, and the probability is an exponential distribution. Equation (17.45) has similar properties. Both of the two equations hold for the $N \to \infty$ and $u \to 0$ limits. $N \to \infty$ is reasonable for most social networks. $u \to 0$ indicates that if a user imitates another user, he/she selects this user's security behavior with high probability. This means that the security behaviors with high fitness can spread over the social network, which is a favorable state for the social network manager.

**Lemma 17.4** *Consider a social network with $N$ users distributed over $M$ communities. Each user belongs to exactly $K$ communities, where $K \leq M$. The deviate rate of community membership imitation is given by $v$. Then the average number of* *communities that two random users have in common is*

$$\psi(\tau) = Ae^{-\upsilon\tau} + B, \tag{17.46}$$

*where* $A = K - \frac{K^2}{M}$, $B = \frac{K^2}{M}$ *and* $\upsilon = 2N\upsilon$.

*Remark 17.4* Lemma 17.4 holds for the $N \to \infty$ and $\upsilon \to 0$ limits. $\upsilon \to 0$ indicates that the users' community memberships are not stable, and users participate in the imitated user's community memberships with high probability. This corresponds to the scenarios in real social networks, where some communities providing more comfortable service, such as security and information service, can attract more and more users due to the interactions and information sharing among users.

According to Lemmas 17.2 and 17.4, we can calculate that

$$\psi = \langle\theta_i \cdot \theta_j | i \neq j\rangle_0 = \int_0^\infty \psi(\tau) p(\tau) d\tau = \frac{A}{\upsilon + 1} + B. \tag{17.47}$$

Next, we analyze and solve $\gamma$ defined as (17.33c). Let

$$\gamma(\tau) = \langle\theta_i \cdot \theta_j \mathbf{1}\left(s_i = s_j\right) | i \neq j, T = \tau\rangle_0. \tag{17.48}$$

As discussed above, the deviations of community membership and security behavior are independent when the time to the MRCI of users is fixed, i.e., $\gamma(\tau) = \varphi(\tau)\psi(\tau)$. Then plug Eqs. (17.44) and (17.46) in and we can get

$$\begin{aligned} \gamma &= \int_0^\infty \varphi(\tau)\psi(\tau) p(\tau) d\tau \\ &= \frac{1}{2}\left(\frac{A}{\upsilon + 1} + \frac{A}{\mu + \upsilon + 1} + \frac{B}{\mu + 1} + B\right). \end{aligned} \tag{17.49}$$

Then we need to calculate $\xi$, for which three users $i$, $j$ and $l$ are considered. As defined in Eq. (17.33d), $\xi$ indicates the amount of communities the latter two users having in common given that the first two users have the same security behavior, for three distinct random users. For any three random users, they must have an MRCI. Let $T(i, j)$ be the time up to the MRCI of $i$ and $j$, and $T(j, l)$ be the time up to the MRCI of $j$ and $l$. We define that

$$\xi(\tau_1, \tau_2) = \langle\theta_j \cdot \theta_l \mathbf{1}\left(s_i = s_j\right) | i \neq j \neq l, T_1 = \tau_1, T_2 = \tau_2\rangle_0, \tag{17.50}$$

where $T_1$ and $T_2$ denote the time of the first and second merging by imitating other users happen, respectively. As mentioned before, the community deviations and the strategy deviations are independent if the time to the MRCI is fixed. Therefore, $\xi(\tau_1, \tau_2)$ can be expressed as a product. By looking back the time into the past,

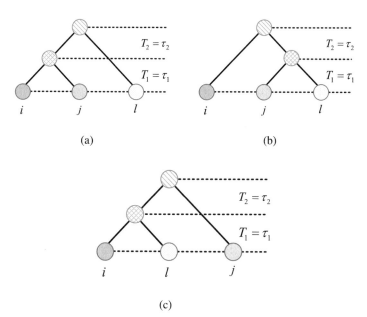

**Fig. 17.2** Three cases of imitated users for three distinct random users $i$, $j$ and $l$ by looking back the time into the past. (**a**) $T(i, j) = \tau_1, T(j, l) = \tau_1 + \tau_2$. (**b**) $T(i, j) = \tau_1 + \tau_2$, $T(j, l) = \tau_1$. (**c**) $T(i, j) = \tau_1 + \tau_2$, $T(j, l) = \tau_1 + \tau_2$

there are three cases shown in Fig. 17.2 for the same imitated users of three distinct users $i$, $j$ and $l$ as follows.

1. user $i$ and $j$ have the same imitated user first, and then they have the same imitated user with $l$:

$$\xi(\tau_1, \tau_2) = \varphi(\tau_1) \psi(\tau_1 + \tau_2);$$ \hfill (17.51)

2. user $j$ and $l$ have the same imitated user first, and then they have the same imitated user with $i$:

$$\xi(\tau_1, \tau_2) = \varphi(\tau_1 + \tau_2) \psi(\tau_1);$$ \hfill (17.52)

3. user $i$ and $l$ have the same imitated user first, and then they have the same imitated user with $j$:

$$\xi(\tau_1, \tau_2) = \varphi(\tau_1 + \tau_2) \psi(\tau_1 + \tau_2).$$ \hfill (17.53)

Each of the three cases happens with probability $1/3$, so we can get $\xi$ as

$$
\begin{aligned}
\xi = &\frac{1}{3} \int_0^\infty d\tau_1 \int_0^\infty p\,(\tau_1, \tau_2)\,(\varphi\,(\tau_1)\,\psi\,(\tau_1 + \tau_2) \\
&+ \varphi\,(\tau_1 + \tau_2)\,\psi\,(\tau_1) + \varphi\,(\tau_1 + \tau_2)\,\psi\,(\tau_1 + \tau_2))\,d\tau_2 \\
= &\frac{1}{2} \left[ \frac{A}{\mu + \upsilon + 3} \left( \frac{1}{\upsilon + 1} + \frac{1}{\mu + 1} + \frac{1}{\mu + \upsilon + 1} \right) \right. \\
&\left. + \frac{A}{\upsilon + 1} + \frac{B}{\mu + 1} + B \right].
\end{aligned} \tag{17.54}
$$

With the similar analysis, we can find $\eta$ as

$$
\begin{aligned}
\eta = &\frac{1}{3} \int_0^\infty d\tau_1 \int_0^\infty \varphi\,(\tau_1, \tau_2)\,(\psi\,(\tau_1) + \psi\,(\tau_1 + \tau_2) \\
&+ \psi\,(\tau_1 + \tau_2))\,p\,(\tau_1, \tau_2)\,d\tau_2 \\
= &\frac{1}{4} \left[ \frac{A}{\mu + \upsilon + 3} \left( 1 + \frac{2}{\upsilon + 1} + \frac{4}{\mu + 2} + \frac{8}{\mu + 2\upsilon + 2} \right) \right. \\
&\left. + \frac{A}{\upsilon + 1} + \frac{3B}{\mu + 3} \left( 1 + \frac{4}{\mu + 2} \right) + B \right].
\end{aligned} \tag{17.55}
$$

According to Eqs. (17.47), (17.49), (17.54), (17.55), and (17.38), we can obtain the critical $(b/c)^*$ as

$$
\left( \frac{b}{c} \right)^* = \frac{1}{\beta} \left( 1 + \frac{\mu + \upsilon + 3}{\mu + \upsilon + 1} \frac{K\upsilon\,(\mu + \upsilon + 2) + M\,(\mu + 1)}{K\upsilon\,(\mu + \upsilon + 2) + M\,(\mu + 2\upsilon + 3)} \right),
$$

which equals to

$$
\left( \frac{\beta b}{c} \right)^* = 1 + \frac{\mu + \upsilon + 3}{\mu + \upsilon + 1} \frac{K\upsilon\,(\mu + \upsilon + 2) + M\,(\mu + 1)}{K\upsilon\,(\mu + \upsilon + 2) + M\,(\mu + 2\upsilon + 3)}. \tag{17.56}
$$

For $\mu \to 0$, we have

$$
\left( \frac{\beta b}{c} \right)^* = 1 + \frac{\upsilon + 3}{\upsilon + 1} \frac{K\upsilon\,(\upsilon + 2) + M}{K\upsilon\,(\upsilon + 2) + M\,(2\upsilon + 3)}. \tag{17.57}
$$

This completes the proof of Theorem 17.1.

*Remark 17.5*

**(1) Properties:** Theorem 17.1 gives the critical cost performance $(b/c)^*$ or $(\beta b/c)^*$. In the equilibrium distribution of the imitation-deviation process, if the cost performance exceeds this critical value, the users in the social network will select the strategy of privacy protection more frequently than the other strategy, i.e., not take the privacy protection, which will promote the diffusion of security behaviors among the network. Moreover, consider $(\beta b/c)^*$ provided in Theorem 17.1 as a function of $K/M$, and we take the derivative of $(\beta b/c)^*$ with respect to $K/M$, then get $\frac{\partial(\beta b/c)^*}{\partial(K/M)} > 0$. So $(\beta b/c)^*$ increases with increasing $K/M$. Hence, for a social network with $M$ communities, the best choice for social network managers to set the minimize $(\beta b/c)^*$ is allowing their users to belong to only one community, i.e., $K = 1$.

(2) ***Feasibility and flexibility:*** The obtained critical cost performance gives suggestions on privacy protection quality and "pricing" strategy for the social network managers from the perspective of economics to incentive their users to take the high quality of privacy protection service. These suggestions are feasible and realizable to be introduced into the social networks, according to the definitions of $b$ and $c$ discussed in the previous section. In addition, it is also flexible to apply these suggestions to the existing social networks, such as WeChat and Facebook. Specifically, the high quality of privacy protection service for the users can be more rigorous backstage verification and authorization when some uncertain users manage to access the personal space, information or photo of legitimate users. In addition, to improve the security benefit or reduce the cost of users, some other security related service can also be provided. Take WeChat for instance, a user can know how many of his/her friends have followed a certain official account, which is a necessary and helpful message for users to choose this official account or not. However, this information can only be obtained if this user has updated his/her APP to the latest version, which can guarantee safe enough privacy protection to provide such personal information of users' friends. This service above can only bring benefit and better experience to users who take the service, and to their friends who also update the APP and take the service. Therefore, the brought benefit can be considered as the reduction of cost $c$, but not as the increasing of $b$. Meanwhile, in this case, the increasing cost of users to obtain the service can be measured by the memory occupancy increment to update the APP.

## 17.5   Privacy Protection Among Users with *L*-Triggering Game

In a social network, the interaction between two users sometimes depends on the strength of their connection, which could be measured by the number of communities that they have in common. In other words, some interactions, especially behavior to take security functions, can only happen among users belonging to multiple common communities. Specifically, user $i$ and $j$ are sharing a close relationship, which means that they have many interested communities in common. As a result, most information of user $j$ are accessed for user $i$. In this case, if user $i$ selects the privacy protection, user $j$'s personal information even privacy information can be protected to a great extent. Conversely, if the amount of the two users' common communities is really small, for instance, user $i$ and $j$ coming from different countries just join the same travel community because of their annual leaves, then the relationship between the two users is actually quite weak and there is little personal information can be accessed for each other. In this case, user $j$ cannot benefit from user $i$'s selection of privacy protection.

In response, we generalize the model, in which the users' interaction happens as long as they have at least one communities in common, into an *L-triggering game* situation in this section. In the extended model, users only influence each other if they have at least a minimum number of common communities, $L$. In a social network, if a user taking the privacy protection $i$ meets another user $j$ in $\theta_i \cdot \theta_j$ communities, then $i$ interact $\theta_i \cdot \theta_j$ times if $\theta_i \cdot \theta_j \geq L$, otherwise, the game between them is not triggered. We call this mechanism as *L-triggering game*. We notice that $L = 1$ degenerates to the previous model. The analysis of cost performance at the end of this section indicates that large values of $L$ lead to that users with security behavior are more imitative in choosing with whom to imitate. Next, we will analyze the impact of $L$-triggering game on the critical cost performance.

### 17.5.1   L-Triggering Game

Given $1 \leq L \leq K$. When $L = 1$ the model is same as of Sect. 17.4. Then the fitness of user $i$ formulated as Eq. (17.5) can be rewritten as

$$\pi_i = 1 + \alpha \sum_{j \neq i} \chi_{ij} \left(\theta_i \cdot \theta_j\right) \left[(\beta - 2) bs_i s_j + (b - c) s_i + bs_j\right], \quad (17.58)$$

where $\chi_{ij} = 1$ if $\theta_i \cdot \theta_j \geq L$, and $\chi_{ij} = 0$, otherwise.

We notice that $\varphi(\tau)$, which indicates the distribution of the probability that two random users have the same security behavior at the time $\tau$ from their MRCI, and $\varphi(\tau_1, \tau_2)$, the distribution of the probability that three random users have the same security behavior, are unchanged. However, $\psi(\tau) = \langle\theta_i \cdot \theta_j | i \neq j, T = \tau\rangle_0$ now changes to $\hat{\psi}(\tau) = \langle\chi_{ij}\theta_i \cdot \theta_j | i \neq j, T = \tau\rangle_0$, which denotes the average number of communities that two random users have in common when they have at least $L$ communities in common. Consequently, $\psi$, $\gamma$, $\xi$ and $\eta$ will all change with the same physical interpretation, but under the constrain that related users have at least $L$ common communities, which can be rewritten as

$$\hat{\psi} = \int_0^\infty \hat{\psi}(\tau) p(\tau) d\tau, \quad (17.59a)$$

$$\hat{\gamma} = \int_0^\infty \hat{\psi}(\tau) \varphi(\tau) p(\tau) d\tau, \quad (17.59b)$$

$$\hat{\xi} = \frac{1}{3} \int_0^\infty d\tau_1 \int_0^\infty \Big( \varphi(\tau_1) \, \hat{\psi}(\tau_1 + \tau_2) + \varphi(\tau_1 + \tau_2) \, \hat{\psi}(\tau_1)$$

$$+ \varphi(\tau_1 + \tau_2) \, \hat{\psi}(\tau_1 + \tau_2) \Big) \, p(\tau_1, \tau_2) \, d\tau_2, \tag{17.59c}$$

$$\hat{\eta} = \frac{1}{3} \int_0^\infty d\tau_1 \int_0^\infty \varphi(\tau_1, \tau_2) \Big( \hat{\psi}(\tau_1) + \hat{\psi}(\tau_1 + \tau_2)$$

$$+ \hat{\psi}(\tau_1 + \tau_2) \Big) \, p(\tau_1, \tau_2) \, d\tau_2. \tag{17.59d}$$

Next, we will find $\hat{\psi}(\tau)$. The probability that two users have $i \leq K$ common communities at time $T = \tau$ from their MRCI is

$$\kappa_i(\tau) = \begin{cases} e^{-\upsilon\tau} + (1 - e^{-\upsilon\tau})/\binom{M}{K}, & i = K; \\ (1 - e^{-\upsilon\tau}) \binom{K}{i} \binom{M-K}{K-i} / \binom{M}{K}, & i < K. \end{cases} \tag{17.60}$$

Then we have

$$\hat{\psi}(\tau) = \langle \chi_{ij}\theta_i \cdot \theta_j \, | i \neq j, T = \tau \rangle_0 = \sum_{i=L}^K i \kappa_i(\tau). \tag{17.61}$$

### 17.5.1.1   Case 1: $L = 1$

According to Vandemonde convolution formula, we have

$$\hat{\psi}(\tau) = \sum_{i=1}^K i \kappa_i(\tau)$$

$$= K e^{-\upsilon\tau} + (1 - e^{-\upsilon\tau}) \sum_{i=1}^K i \binom{K}{i} \binom{M-K}{K-i} / \binom{M}{K} \tag{17.62}$$

$$= K e^{-\upsilon\tau} + (1 - e^{-\upsilon\tau}) K \binom{M-1}{K-1} / \binom{M}{K}$$

$$= e^{-\upsilon\tau} K (1 - K/M) + \frac{K^2}{M}.$$

The result is same as the previous model provided in Lemma 17.4.

### 17.5.1.2   Case 2: $1 < L \leq K$

Let $\hat{K} = \frac{M}{K} \sum\limits_{i=L}^{K} i \binom{K}{i} \binom{M-K}{K-i} / \binom{M}{K}$, we get [39]

$$\hat{\psi}(\tau) = e^{-\upsilon\tau} K \left(1 - \frac{\hat{K}}{M}\right) + \frac{K\hat{K}}{M}. \tag{17.63}$$

Then the critical cost performance formulated in Eq. (17.57) turns to

$$\left(\frac{\beta b}{c}\right)^* = 1 + \frac{\upsilon + 3}{\upsilon + 1} \frac{\hat{K}\upsilon(\upsilon + 2) + M}{\hat{K}\upsilon(\upsilon + 2) + M(2\upsilon + 3)}, \tag{17.64}$$

in case that $N \to \infty$ and $\mu \to 0$. Notice that $\hat{K} = K$, if $L = 1$.

*Remark 17.6*  Comparing with Eq. (17.57), the expressions of $(\beta b/c)^*_{\min}$ for non-triggering game and $L$-triggering game are much the same, except that $\hat{K} \leq K$, and the equality hold up if and only if $L = 1$.

## 17.5.2   *Analysis of Cost Performance*

Setting appropriate cost performance can facilitate the security behavior, i.e., the action of taking the privacy protection, among the entire social network. In this part, we will find the minimum cost performance that can make users to choose the privacy protection more frequently than not.

   According to the last two sections, we notice that the result of the cost performance shown in Eq. (17.64) is general, since that the $L$-triggering game becomes the non-triggering game when $L = 1$. So we only analyze the model with the $L$-triggering game. $(\beta b/c)^*$ given by Eq. (17.64) has a minimum value as a function of $\upsilon$. Then let $r(\upsilon) = (\beta b/c)^*$, and we take the derivative of $r(\upsilon)$ with respect to $\upsilon$. Set the result equal to zero and we get

$$\frac{M}{\hat{K}} = \frac{\upsilon^2 (\upsilon^2 + 4\upsilon + 4)}{\upsilon^2 + 6\upsilon + 6}, \tag{17.65}$$

according to which, optimal solution $\upsilon^*$ must satisfies $\sqrt{M/\hat{K}} < \upsilon^* < \sqrt{M/\hat{K}} + 1$. If $M/\hat{K}$ is large, solution $\upsilon^*$ to obtain the minimum cost performance is

$$\upsilon^* = \sqrt{M/\hat{K}}, \tag{17.66}$$

and the minimum cost performance is

$$\left(\frac{\beta b}{c}\right)^*_{\min} = 1 + \frac{\sqrt{M/\hat{K}} + 3}{\left(\sqrt{M/\hat{K}} + 1\right)^2}. \tag{17.67}$$

*Remark 17.7* According to Eq. (17.67), $(\beta b/c)^*_{\min} \sim \sqrt{\hat{K}/M}$, which means that small values of $\hat{K}$ and large values of $M$ can promote the evolution of security behavior among the social network. For non-triggering game situation, i.e., $\hat{K} = K$, we can notice that given number of communities $M$, it is best if users belong to only one community ($K = 1$). The larger $K$ is, it is harder for users who take the privacy protection to avoid the exploitation by users who do not take the privacy protection. For $L$-triggering game situation, $\hat{K} < K$ if $M$ is fixed according to the definition of $\hat{K}$ in Sect. 17.5.1.2, then smaller $(\beta b/c)^*_{\min}$ can be gotten. So large values of $L$ lead to that users with security behavior are more imitative in choosing with whom to imitate.

## 17.6   Performance Evaluation

The critical cost performance is an important parameter that helps the social network managers to make appropriate security service level and payment mechanism to encourage their users to accept the security service, and then promote the spreading of this secure behavior. In this part, we perform numerical simulation experiments to analyze properties and performances of the critical cost performance and its influential factors such as the community deviate rate, population and number of communities of the social network. First, the community deviate rate $v$ reflects the subjective selectivity for community memberships. If users select communities depending on their own interest mostly, but not on those users with high fitness, then $v$ is large. Otherwise, $v$ is small. Then we analyze the effect of the community deviate rate $\upsilon = 2Nv$ for different selections of $K$ and $L$, which denote the number of communities that a user is allowed to belong to and the minimum number of common communities that game can be triggered. The population of the social network is large, i.e., $N = 10^4$ ($N \to \infty$), and the number of communities is set as $M = 20$. We set the strategy deviate rate as $u = 10^{-4}$ ($u \to 0$). We consider the population of the network is constant. Simulation results of non-triggering game and $L$-triggering game are shown in Fig. 17.3a, b, respectively. As shown in the results, the critical cost performance $(\beta b/c)^*$ is a U-shaped function of community deviate rate $\upsilon$. When $\upsilon$ is small, $(\beta b/c)^*$ tend to be large and all users belong to the same community. Conversely, when $\upsilon$ is large, the community affiliations cannot persist for a long time. Moreover, the results of numerical analysis shown in Eq. (17.66) and (17.67) can be demonstrated by the simulations results shown in Fig. 17.3.

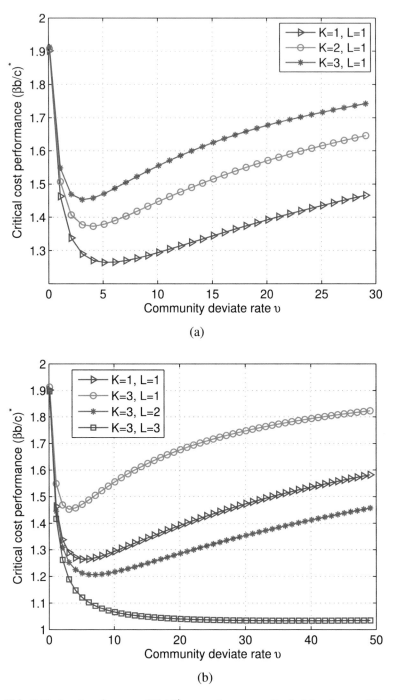

(a)

(b)

**Fig. 17.3** Critical cost performance $(\beta b/c)^*$ versus the community deviate rate $\upsilon = 2N\upsilon$. The population size is large, $N = 10^4$. The strategy deviate rate is $u = 10^{-4}$. The number of communities is $M = 20$. (**a**) Non-triggering game. (**b**) $L$-triggering game

As shown in Fig. 17.3a, we notice that for a fixed number of communities $M$, small values of $K$ can facilitate the evolution of the security behavior, which means that the selection of taking the privacy protection is promoted in the evolution process. This conclusion is consistent with the numerical analysis shown in Eq. (17.67). Consequently, when the number of communities is given, the best choice for users is to belong to $K = 1$ community. With the increasing of $K$, it is hard for users taking the privacy protection to avoid the exploitation by users not taking the privacy protection. But according the results of the $L$-triggering game situation shown in Fig. 17.3b, for $K = 3$, if $L = 2$ or $L = 3$, the critical cost performance is smaller than $K = 1$. These results indicate that belonging to more communities, i.e., $K > 1$, can also facilitate the evolution of the security behavior when the game only happen if users have a certain minimum number of common communities $L$.

We test the effects of the population of the network on the critical cost performance, and the results are shown in Fig. 17.4. We set the strategy deviate rate as $u = 10^{-4}$, and the community deviate rate as $v = 0.01$. Parameter settings of $M$, $K$ and $L$ are shown in the figures. Results illustrate that for both non-triggering game and $L$-triggering game cases, the critical cost performance is a convex function of population $N$. According to the results, we notice that if the population of the network is too small, then the effect of spite tends to be strong, so the critical cost performance $(\beta b/c)^*$ has to be very large. If $N = 2$, it will never pay to users

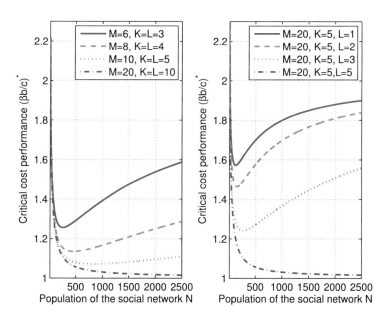

**Fig. 17.4** Critical cost performance $(\beta b/c)^*$ versus the population of the social network $N$ under the non-triggering game and $L$-triggering game, respectively. The strategy deviate rate is $u = 10^{-4}$, and the community deviate rate is $v = 0.01$

with security behavior, which means that users will not take the privacy protection to ensure their information security. When $N$ is large, all the communities that get population by users who take the privacy protection and not take the privacy protection cannot persist for long. In addition, the lower bound of the critical cost performance is 1, which is consistent with the result in Eq. (17.67).

As shown in Fig. 17.5, the cost performance decreases as the number of communities $M$ increasing. These results indicate that more communities is helpful for the spreading of security behavior, which mean that adding community number will help users to take the privacy protection more frequently.

Next, we simulate the evolution process of the strategies that taking the privacy protection or not in the social network. The topology we used in this simulation is based on Flickr, a real-world online social network database. There are 5,899,882 edges connecting 80,513 users in the Flickr graph dataset, and the edge represents the connection between two users. In order to test the performance of the evolutionary game theoretic framework we proposed, the topology of Flickr is modified. The communities are established based on the users with most importance in the network, i.e., with largest values of betweenness or having largest amounts of one-hop and two-hop neighborhoods. In addition, each user is allowed to join limited $K$ communities. If one user belongs to more than $K$ communities, the topology will be modified as the following rules: The connection between user $i$ and community $k$ is established with probability

$$p_{ik} = \frac{M_k}{\sum_{j \in J_i} M_j},\tag{17.68}$$

where $M_k$ is the number of users belonging to community $k$, and $J_i$ is the set of all communities belonged by user $i$. For the network, we use $N = 50,000$ users in Flickr distributing over $M = 15$ or $M = 20$ communities. Each user belongs to $K = 1$ or $K = 2$ communities. The graph structures of the modified Flickr network are depicted in Fig. 17.6. We set the strategy deviate rate and community deviate rate as $u = 10^{-4}$ and $v = 0.01$, responsibility. The parameter settings for the eight cases are shown in Table 17.2. In Table 17.2, $\alpha = 0.05$ and $\alpha = 0.2$ denote different intensities of selection, $p_0 = 0.5$ and $p_0 = 0.4$ indicate the different initialized frequencies of the users who select to take the privacy protection, and $(\beta b/c)^*$ is obtained according to Eq. (17.57). In our simulation, the network updates 100 times. Evolutions of the privacy protection in the network with different cost performance are shown in Fig. 17.7, in which $c = 1$, $\beta = 2$, and $b$ varies to realize different $\beta b/c$. As accepted, the evolutionary stable state of the frequency of users taking the privacy protection is 1 when $\beta b/c > (\beta b/c)^*$, otherwise, 0. These results demonstrate that when the cost performance exceeds the critical cost performance, then users select to take the privacy protection more frequently than not. In addition, the evolutionary stable state of the network cannot be achieved if $\beta b/c = (\beta b/c)^*$, and the frequency is around 0.5.

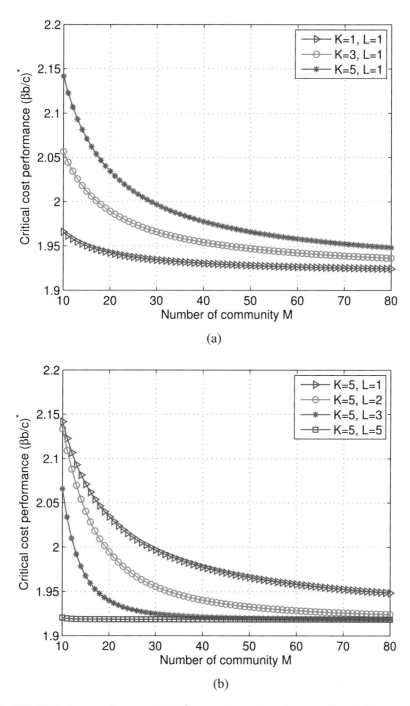

**Fig. 17.5** Critical cost performance $(\beta b/c)^*$ versus the number of communities $M$. The population size of the social network is set as $N = 15$. The strategy deviate rate is $u = 10^{-4}$, and the community deviate rate is $v = 0.01$. (**a**) Non-triggering game. (**b**) $L$-triggering game

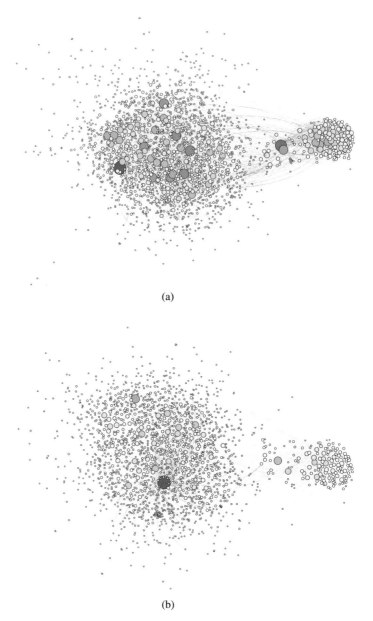

(a)

(b)

**Fig. 17.6** Graph structures of the modified Flickr network used for simulation. (**a**) $N = 50,000$, $M = 15$, $K = 1$. Case 1, 2, 5, 6. (**b**) $N = 50,000$, $M = 20$, $K = 2$. Case 3, 4, 7, 8

**Table 17.2** Parameters setting of the simulation for different cases

| Case | $M$ | $K$ | $L$ | $\alpha$ | $p_0$ | $(\beta b/c)^*$ |
|------|-----|-----|-----|----------|-------|-----------------|
| 1 | 15 | 1 | 1 | 0.05 | 0.5 | 1.8416 |
| 2 | 15 | 1 | 1 | 0.2 | 0.5 | 1.8416 |
| 3 | 20 | 2 | 1 | 0.05 | 0.5 | 1.8850 |
| 4 | 20 | 2 | 1 | 0.2 | 0.5 | 1.8850 |
| 5 | 15 | 1 | 1 | 0.05 | 0.4 | 1.8416 |
| 6 | 15 | 1 | 1 | 0.2 | 0.4 | 1.8416 |
| 7 | 20 | 2 | 1 | 0.05 | 0.4 | 1.8850 |
| 8 | 20 | 2 | 1 | 0.2 | 0.4 | 1.8850 |

**Remark 17.8** For a social network with given number of community and number of community that each user is allowed to belong to, the critical cost performance can be obtained through Theorem 17.1 and Eq. (17.64). Social network managers have to make appropriate security service $b$ and payment mechanism $c$ to ensure that $\beta b/c > (\beta b/c)^*$. Then their users can be encouraged to accept the security service, and the spreading of the secure behavior can be promoted over the social network. Besides, we notice that the convergence speed of evolutionary stable state depends on many factors, such as $M$, the number of communities in the network, and $K$, the number of communities each user belongs to. Given the same cost performance, $L$, $\alpha$ and $p_0$, small values of $M/K$ result in fast convergence. This result is reasonable. On the one hand, if $M$ is fixed, larger values of $K$ increase dimensions of the relationship among users, then each user might have more new friends, and the closeness to his/her old friends might be stronger. These changes can help the spreading of user behaviors, i.e., taking the privacy protection if $\beta b/c > (\beta b/c)^*$, otherwise, not taking the privacy protection. On the other hand, for fixed $K$, smaller $M$ means that there might be more common communities among every two users. Therefore, the closeness between users tends to be stronger, which can help the spreading of user behaviors. After social network managers release a new security service, such as the privacy protection in our work, security service $b$ and cost $c$ for users are determined. Then the speed of revenue for managers and the set up of privacy protection at the network platform depend on how fast that all users take the privacy protection, which is concerned with the convergence speed. It will help the network managers to make network structure and service plan, and the storage and processing capacities of network server can also be planed for the improvement of the user information security.

# 17.7 Conclusions

In this part, we analyze the privacy protection behaviors of social network users by a community structured evolutionary game theoretic framework. The players, strategies, payoff matrix and the topology structure of users are defined in this

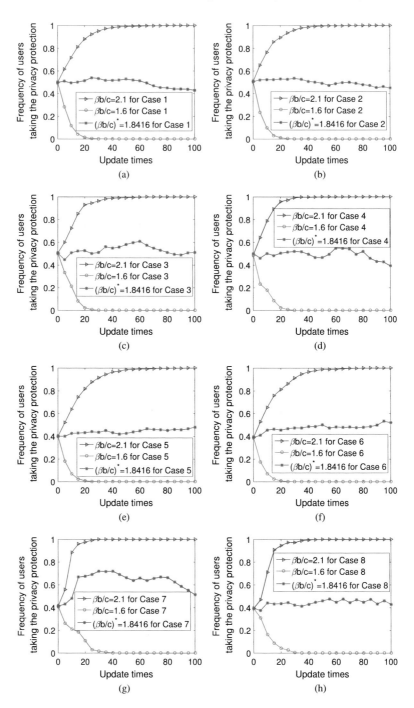

**Fig. 17.7** Evolution of the privacy protection in the network with a population of size $N = 50,000$. The strategy deviate rate is $u = 10^{-4}$, and the community deviate rate is $v = 0.01$. Other parameters values are set as Table 17.2. (**a**) Case 1. (**b**) Case 2. (**c**) Case 3. (**d**) Case 4. (**e**) Case 5. (**f**) Case 6. (**g**) Case 7. (**h**) Case 8

framework. We obtain the critical cost performance, which is an important parameter that can help social networks to design incentive mechanisms to facilitate the privacy protection behavior among their users. Simulation results demonstrate that the proposed theoretic framework is effective in modeling the users' relationship and privacy protection behavior.

# References

1. E. Serrano, C. A. Iglesias, and M. Garijo, "A survey of twitter rumor spreading simulations," in *Computational Collective Intell.* Springer, 2015, pp. 113–122.
2. X. Wang, J. Du, J. Wang, Z. Zhang, C. Jiang, and Y. Ren, "Key issues of security in space-based information network review," in *Int. Conf. Cyberspace Technology (CCT'14).* Beijing, China, 8-10 Nov. 2014.
3. I. Heimbach, B. Schiller, T. Strufe, and O. Hinz, "Content virality on online social networks: Empirical evidence from Twitter, Facebook, and Google+ on German news websites," in *Proc. of the 26th ACM Conf. on Hypertext & Social Media.* Ankara, Turkey, Sept. 2015, pp. 39–47.
4. I. E. Carvajal-Roca, J. Wang, J. Du, and S. Wei, "A semi-centralized dynamic key management framework for in-vehicle networks," *IEEE Trans. Veh. Tech.*, vol. 70, no. 10, pp. 10 864–10 879, Oct. 2021.
5. I. E. C. Roca, J. Wang, J. Du, and S. Wei, "A semi-centralized security framework for in-vehicle networks," in *Int. Wireless Commun. Mobile Computing (IWCMC'20).* 15-19 Jun. 2020.
6. J. Du, C. Jiang, E. Gelenbe, L. Xu, J. Li, and Y. Ren, "Distributed data privacy preservation in IoT applications," *IEEE Wireless Commun.*, vol. 25, no. 6, pp. 68–76, Dec. 2018.
7. I. Krontiris, M. Langheinrich, and K. Shilton, "Trust and privacy in mobile experience sharing: future challenges and avenues for research," *IEEE Commun. Mag.*, vol. 52, no. 8, pp. 50–55, Aug. 2014.
8. C. Bettini and D. Riboni, "Privacy protection in pervasive systems: State of the art and technical challenges," *Pervasive and Mobile Computing*, vol. 17, pp. 159–174, Feb. 2015.
9. Z. Wang, J. Du, Z. Xia, C. Jiang, Z. Fang, and Y. Ren, "Secure routing in underwater acoustic sensor networks based on AFSA-ACOA fusion algorithm," in *IEEE Int. Conf. Commun. (ICC'22).* Seoul, Korea, 16-20 May 2022.
10. Z. Yang, J. Du, Z. Xia, C. Jiang, A. Benslimane, and Y. Ren, "Secure and cooperative target tracking via AUV swarm: A reinforcement learning approach," in *IEEE Global Commun. Conf. (GLOBECOM'21).* Madrid, Spain, 07-11 Dec. 2021.
11. Z. Xia, J. Du, J. Wang, Y. Ren, G. Li, and B. Zhang, "Contract based information collection in underwater acoustic sensor networks," in *IEEE Int. Conf. Commun. (ICC'20).* Dublin, Ireland, 7-11 Jun. 2020.
12. L. Xu, C. Jiang, Y. Chen, J. Wang, and Y. Ren, "A framework for categorizing and applying privacy-preservation techniques in big data mining," *IEEE Computer*, vol. 49, no. 2, pp. 54–62, Feb. 2016.
13. N. Cao, C. Wang, M. Li, K. Ren, and W. Lou, "Privacy-preserving multi-keyword ranked search over encrypted cloud data," *IEEE Trans. Parallel Distrib. Syst.*, vol. 25, no. 1, pp. 222–233, Jan. 2014.
14. L. Xu, C. Jiang, Y. Qian, Y. Zhao, J. Li, and Y. Ren, "Dynamic privacy pricing: A multi-armed bandit approach with time-variant rewards," *IEEE Trans. Inf. Forens. and Security*, vol. 12, no. 2, pp. 271–285, Feb. 2017.
15. R.-H. Hwang, Y.-L. Hsueh, and H.-W. Chung, "A novel time-obfuscated algorithm for trajectory privacy protection," *IEEE Trans. Services Computing*, vol. 7, no. 2, pp. 126–139, Jun. 2014.

16. B. Greschbach, G. Kreitz, and S. Buchegger, "The devil is in the metadatanew privacy challenges in decentralised online social networks," in *2012 IEEE Int. Conf. on Pervasive Computing and Commun. Workshops (PERCOM Workshops)*. Lugano, Switzerland, Mar. 2012, pp. 333–339.

17. H. Hu, G.-J. Ahn, and J. Jorgensen, "Multiparty access control for online social networks: model and mechanisms," *IEEE Trans. Knowl. and Data Eng.*, vol. 25, no. 7, pp. 1614–1627, Jul. 2013.

18. M. H. Manshaei, Q. Zhu, T. Alpcan, T. Bacşar, and J.-P. Hubaux, "Game theory meets network security and privacy," *ACM Computing Surveys (CSUR)*, vol. 45, no. 3, p. 25, Jun. 2013.

19. D. Tosh, S. Sengupta, C. Kamhoua, K. Kwiat, and A. Martin, "An evolutionary game-theoretic framework for cyber-threat information sharing," in *2015 IEEE Int. Conf. on Commun. (ICC)*. London, UK, Jun. 2015, pp. 7341–7346.

20. W. Lian-ren and C. Xia, "Modeling of evolutionary game between SNS and user: From the perspective of privacy concerns," in *2014 Int. Conf. on Manage. Sci. & Eng. (ICMSE),*. Arunachal Pradesh, India, May 2014, pp. 115–119.

21. C. Jiang, Y. Chen, and K. R. Liu, "Graphical evolutionary game for information diffusion over social networks," *IEEE J. Sel. Topics in Signal Process.*, vol. 8, no. 4, pp. 524–536, Aug. 2014.

22. A. C. Squicciarini and C. Griffin, "An informed model of personal information release in social networking sites," in *2012 Int. Conf. on Privacy, Security, Risk and Trust (PASSAT), and 2012 Int. Conf. on Social Computing (SocialCom)*. Amsterdam, Netherlands, Sept. 2012, pp. 636–645.

23. L.-L. Jiang and M. Perc, "Spreading of cooperative behaviour across interdependent groups," *Scientific reports*, vol. 3, Aug. 2013.

24. R. M. Bond, C. J. Fariss, J. J. Jones, A. D. Kramer, C. Marlow, J. E. Settle, and J. H. Fowler, "A 61-million-person experiment in social influence and political mobilization," *Nature*, vol. 489, no. 7415, pp. 295–298, Sept. 2012.

25. A. Acquisti, L. Brandimarte, and G. Loewenstein, "Privacy and human behavior in the age of information," *Science*, vol. 347, no. 6221, pp. 509–514, Jan. 2015.

26. S.-M. Cheng, W. C. Ao, P.-Y. Chen, and K.-C. Chen, "On modeling malware propagation in generalized social networks," *IEEE Commun. Lett.*, vol. 15, no. 1, pp. 25–27, 2011.

27. C. Jiang, Y. Chen, and K. R. Liu, "Evolutionary dynamics of information diffusion over social networks," *IEEE Trans. Signal Process.*, vol. 62, no. 17, pp. 4573–4586, Sept. 2014.

28. C. L. Apicella, F. W. Marlowe, J. H. Fowler, and N. A. Christakis, "Social networks and cooperation in hunter-gatherers," *Nature*, vol. 481, no. 7382, pp. 497–501, 2012.

29. Y. Jiang and J. Jiang, "Understanding social networks from a multiagent perspective," *IEEE Trans. Parallel Distrib. Syst.*, vol. 25, no. 10, pp. 2743–2759, Oct. 2014.

30. C. Jiang, Y. Chen, and K. R. Liu, "Distributed adaptive networks: A graphical evolutionary game-theoretic view," *IEEE Trans. Signal Process.*, vol. 61, no. 22, pp. 5675–5688, Nov. 2013.

31. C. Jiang, Y. Chen, Y. Gao, and K. R. Liu, "Joint spectrum sensing and access evolutionary game in cognitive radio networks," *IEEE trans. wireless commun.*, vol. 12, no. 5, pp. 2470–2483, Mar. 2013.

32. C. Jiang, Y. Chen, and K. R. Liu, "On the equivalence of evolutionary stable strategies," *IEEE Commun. Lett.*, vol. 18, no. 6, pp. 995–998, Jun. 2014.

33. W. H. Sandholm, *Population games and evolutionary dynamics*. MIT press, 2010.

34. R. Cressman, *Evolutionary dynamics and extensive form games*. MIT Press, 2003, vol. 5.

35. W. J. Ewens, *Mathematical Population Genetics 1: Theoretical Introduction*. Springer Science & Business Media, 2012, vol. 27.

36. S. Wright, "Evolution in mendelian populations," *Genetics*, vol. 16, no. 2, p. 97, Jan. 1931.

37. R. A. Fisher, *The genetical theory of natural selection: a complete variorum edition*. Oxford University Press, 1930.

38. J. Hofbauer and K. Sigmund, *Evolutionary games and population dynamics*. Cambridge University Press, 1998.

39. C. E. Tarnita, T. Antal, H. Ohtsuki, and M. A. Nowak, "Evolutionary dynamics in set structured populations," *Proc. of the Nat. Academy of Sci.*, vol. 106, no. 21, pp. 8601–8604, May 2009.

40. M. A. Nowak, A. Sasaki, C. Taylor, and D. Fudenberg, "Emergence of cooperation and evolutionary stability in finite populations," *Nature*, vol. 428, no. 6983, pp. 646–650, Apr. 2004.

41. P. Shakarian, P. Roos, and A. Johnson, "A review of evolutionary graph theory with applications to game theory," *Biosystems*, vol. 107, no. 2, pp. 66–80, Feb. 2012.

42. G. Wild and A. Traulsen, "The different limits of weak selection and the evolutionary dynamics of finite populations," *J. of Theoretical Biology*, vol. 247, no. 2, pp. 382–390, Jul. 2007.

# Part VI
# Conclusion

# Chapter 18
# Conclusion

Resource allocation and networking are classic techniques for the future 6G heterogeneous networks that has gained wide attentions. As a new information network paradigm for heterogeneous networks, the impact of cooperation and integration at different network layers needs to be studied and understood. This book aims to provide a comprehensive cooperation and integration solution for differentiated QoS and Experience of Service (EoS) requirements. Therefore, in this book, we deliver a range of technical issues in cooperative resource allocation and information sharing for the future 6G heterogenous networks, from the terrestrial ultra-dense networks and space-based networks to the integrated satellite-terrestrial networks, as well as introducing the effects of cooperative behavior among mobile users on increasing capacity, trustworthiness and privacy. For the cooperative transmission in heterogeneous networks, we commence with the traffic offloading problems in terrestrial ultra-dense networks, and the cognitive and cooperative mechanisms in heterogeneous space-based networks, the stability analysis of which is also provided. Moreover, for the cooperative transmission in integrated satellite-terrestrial networks, we present a pair of dynamic and adaptive resource allocation strategies for traffic offloading, cooperative beamforming and traffic prediction based cooperative transmission. Later, we discuss the cooperative computation and caching resource allocation in heterogeneous networks, with the highlight of providing our current studies on the game theory, auction theory and deep reinforcement learning based approaches. Meanwhile, we introduce the cooperative resource and information sharing among users, in which capacity oriented-, trustworthiness oriented-, and privacy oriented cooperative mechanisms are investigated.

This book gives a systematic and comprehensive introduction to the resource allocation and networking mechanisms to achieve the cooperation and integration of the future 6G heterogeneous networks, which can provide reliable research ideas and directions for relevant personnel. In the future, we will continue to explore the

© The Author(s), under exclusive license to Springer Nature Singapore Pte Ltd. 2023    459
J. Du, C. Jiang, *Cooperation and Integration in 6G Heterogeneous Networks*,
Wireless Networks, https://doi.org/10.1007/978-981-19-7648-3_18

potential cross-layer and multi-dimensional resource allocation and management for varied and emerging application scenarios requiring high-complexity of computation and huge-amount of communications, and carry out more in-depth research on resource allocation mechanisms combined with cutting-edge technology.

Printed in the United States
by Baker & Taylor Publisher Services